OXFORD MONOGRAPHS ON GEOLOGY AND GEOPHYSICS

OXFORD MONOGRAPHS ON GEOLOGY AND GEOPHYSICS

1. DeVerle P. Harris: Mineral resources appraisal: mineral endowment, resources, and potential supply: concepts, methods, and cases
2. J. J. Veevers (ed.): Phanerozoic earth history of Australia
3. Yang Zunyi, Cheng Yuqi, and Wang Hongzhen (eds.): The geology of China
4. Lin-gun Liu and William A Bassett: Elements, oxides, and silicates: high-pressure phases with implications for the Earth's interior
5. Antoni Hoffman and Matthew H. Nitecki (eds.): Problematic fossil taxa
6. S. Mahmood Naqvi and John J. W. Rogers: Precambrian geology of India
7. Chih-Pei Chang and T. N. Krishnamurti (eds.): Monsoon meteorology
8. Zvi Ben-Avraham (ed.): The evolution of the Pacific Ocean margins
9. Ian McDougall and T. Mark Harrison: Geochronology and thermochronology by the $^{40}Ar/^{39}Ar$ method
10. Walter C. Sweet: The conodonta: morphology, taxonomy, paleoecology, and evolutionary history of a long-extinct animal phylum
11. H. J. Melosh: Impact cratering: a geologic process
12. J. W. Cowie and M. D. Brasier (eds.): The Precambrian-Cambrian boundary
13. C. S. Hutchinson: Geological evolution of south-east Asia
14. Anthony J. Naldrett: Magmatic sulfide deposits
15. D. R. Prothero and R.M. Schoch (eds.): The evolution of perissodactyls
16. M. Menzies (ed.): Continental mantle
17. R. J. Tingey (ed.): Geology of the Antarctic
18. Thomas J. Crowley and Gerald R. North: Paleoclimatology
19. Gregory J. Retallack: Miocene paleosols and ape habitats in Pakistan and Kenya
20. Kuo-Nan Liou: Radiation and cloud processes in the atmosphere: theory, observation, and modeling
21. Brian Bayley: Chemical change in deforming materials
22. A. K. Gibbs and C. N. Barron: The geology of the Guiana Shield
23. Peter J. Ortoleva: Geochemical self organization
24. Robert G. Coleman: Geologic evolution of the Red Sea
25. Richard W. Spinrad, Kendall L. Carder, and Mary Jane Perry: Ocean optics
26. Clinton M. Case: Physical principles of flow in unsaturated porous media
27. Eric B. Kraus and Joost A. Businger: Atmosphere-ocean interaction, second edition
28. M Solomon & D. I. Groves: The geology and origins of Australia's mineral deposits
29. R. L. Stanton: Ore elements in arc lavas
30. P. Wignall: Black shales
31. Orson L. Anderson: Equations of state of solids for geophysics and ceramic science

EQUATIONS OF STATE OF SOLIDS FOR GEOPHYSICS AND CERAMIC SCIENCE

ORSON L. ANDERSON

Institute of Geophysics and Planetary Physics
University of California at Los Angeles

New York Oxford
OXFORD UNIVERSITY PRESS
1995

Oxford University Press

Oxford New York Toronto
Delhi Bombay Calcutta Madras Karachi
Kuala Lumpur Singapore Hong Kong Tokyo
Nairobi Dar es Salaam Cape Town
Melbourne Auckland Madrid

and associated companies in
Berlin Ibadan

Published by Oxford University Press, Inc.,
200 Madison Avenue, New York, New York 10016

Library of Congress Cataloging-in-Publication Data
Anderson, O. L. (Orson L.)
Equations of state of solids for geophysics and ceramic science.
Orson L. Anderson
p. cm. — (Oxford monographs on geology and geophysics: 31)
Includes index.
ISBN 0-19-505606-X
1. Geophysics. 2. Equations of state. 3. Thermodynamics.
I. Title. II. Series:
Oxford monographs on geology and geophysics: no. 31
QE501.3.A53 1995 550—dc20 94-2242

9 8 7 6 5 4 3 2 1
Printed in the United States of America
on acid-free paper

Permissions

Laws die, books never.

—Edward Bulwer-Lytton

The quotation on page ix is from *The Poetry of Robert Frost*, edited by Edward Connery Lathem, copyright 1944 by Robert Frost; copyright 1916© 1969 by Henry Holt and Company, Inc. Reprinted by permission of Henry Holt and Company, Inc. Permission for use also granted by the Estate of Robert Frost and Jonathan Cape, Ltd., Publishers.

The quotation on page 157 is from *The Logic of Modern Physics*, by P.W. Bridgman, copyright 1927, 1960 edition, copyright 1955, by P.W. Bridgman. Reprinted by permission of the MacMillan Publishing Company.

The quotation on page 1 by Professor Julian Schwinger has been used by him in many of his lectures and talks. It is very appropriate for Part I, because we show there that the high temperature experimental data on insulators obey the high temperature limit of the quasiharmonic approximation of statistical mechanics much better than expected. Quotation used with permission of Professor Julian Schwinger.

The quotation on page 221 is from *American Mathematical Physicist Par Excellence*, by Raymond J. Seeger, copyright 1978; copyright 1974 by Pergamon Press, Ltd. Professor Seeger died in 1992. Quotation used with permission of Mrs. Vivian Seeger.

The quotation on page 383 is from *Of All Things! Classic Quotations from Hugh Nibley*, compiled and edited by Gary Gillum. Salt Lake City, Deseret Books and F.A.R.M.S.: copyright 1993, p. 226. Oral permission to use granted by Deseret Books and F.A.R.M.S.

Data in Table A–7, Appendix, reprinted from Anderson et al., *The Reviews of Geophysics*, **30**, 58–63, 1992. Figures 1.1, 1.2, 2.1, 2.2, 2.3, 2.4, 2.5 constructed from data in Table A–7. Figure 2.6 redrafted from a figure in *The Reviews of Geophysics*, **30**, 67, 1992. Figure 4.10 redrafted from a figure in *The Reviews of Geophysics*, **30**, 86, 1992. Figure 5.18 redrafted from a figure in Kieffer *Reviews of Geophysics and Space Physics*, **20**: 866, 1982. All in this paragraph copyright by the American Geophysical Union and used with permission of the American Geophysical Union.

Figure 4.1 redrafted from a figure in Anderson et al., *Geophysical Research Letters*, **19**: 1988, 1992, copyright by the American Geophysical Union. Used with permission of the American Geophysical Union.

Figures 9.2 and 9.3 redrafted from figures in Hemley and Gordon, *Journal of Geophysical Research*, **90**: 7808, 1985. Figures 9.4 and 9.5 redrafted from figures in Anderson and Demarest, *Journal of Geophysical Research*,

TO N. BERNEICE ANDERSON,

WHO KNOWS INSTINCTIVELY

WHEN A FANCY WORD

IS A COVERUP,

WITH LOVE AND AFFECTION

PREFACE

Two roads diverged in a wood, and I—
I took the one less traveled by,
And that has made all the difference.

—Robert Frost

This book is a product of my odyssey in certain branches of materials physics and geophysics. A few years ago, while rereading Frost's poetry, I realized that his words above described very well my decisions at crossroads. My first job as a young Ph.D. physicist was in the research department of Bell Telephone Laboratories, where I was to find elastic constants of solids using physical acoustics under the mentorship of Warren Mason. For several months, I had the good fortune to be in the same corridor as many famous physicists, including Art Schawlow, Walter Bond, Brent Matthais, Phil Anderson, Conyers Herring, Gregory Wannier, William Shockley, Walter Brattain, and John Bardeen. Five of these later obtained (in total) six Nobel prizes. During my temporary stay in this corridor, while my new physical acoustics laboratory was being assembled, I felt a strong impression from these colleagues that the cutting edge of research was in semiconductor physics, and it was tacitly assumed that I would apply the skills of physical acoustics to this field. But I remember thinking that doing my kind of research on semiconductors would be boring. I was fascinated with glass physics and the acoustics of low symmetry silicates, so I took the road "less traveled by." Much of my research was published in the *Journal of the American Ceramic Society* and *Physical Review*, and I was associated with great ceramicists like Jack Wachtman. I found that another market for my research results was in geophysics, and I was soon invited to build a second physical acoustics laboratory at the Lamont Geological Observatory of Columbia University by its famous director, Maurice Ewing. Doing physical acoustics in a geophysical institute was certainly an uncommon endeavor. In 1964, Ed Schreiber, Nahiro Soga, and I nailed a sign over our new laboratory there, proclaiming it the first "Mineral Physics Laboratory."

Ed and I showed the community that precision ultrasonic measurements of sound velocity done under pressure (modest indeed by today's standards) could yield valid equations of state in the shock wave pressure regime. Nahiro and I demonstrated that high temperature elasticity obeyed certain fundamental laws in solid state physics first presented by Max Born and E. Grüneisen in the 1920's and 1930's. Soon a number of other geophysical laboratories were in the ultrasonic sound measurement business. At that time the diamond anvil cell (DAC) was discovered by geophysicists and materials scientists, and there was a scramble to find properties of oxides and ceramics at high pressure by the DAC. It was easy to foresee that the high pressure field with the DAC was going to be popular, and so I gave up pressure measurements altogether and opened up the field of acoustic measurements of elastic constants at higher and higher temperatures.

In high temperature acoustics, I was fortunate to be associated with Mineo Kumazawa, who came to Columbia University as a postdoc from Nagoya. Mineo further developed this field with his students upon his return, naming it "rectangular parallelepiped resonance" (RPR). This was in Japan, but Kumazawa's students worked in my UCLA lab as postdocs in their turn. Even so, in the USA, my laboratory associates were travelling alone until about four years ago, when Al Migliori at Los Alamos independently discovered the technique, naming it "resonance ultrasound spectroscopy" (RUS). I was delighted, of course, to see my dreams fulfilled by the establishment of a new branch of physical acoustics that was being pursued at several laboratories.

Now I choose a sparsely traveled path called physical thermodynamics, once considered as a branch of theoretical physics. (This field became somnolescent shortly after World War II, and I am trying to revive it). *Ab initio* calculations of elastic constants at extreme conditions are producing exciting results for oxides and ceramics and are destined to become a popular research field. Travelers on that road will reap many rewards, but I will instead pursue physical thermodynamics.

Although my odyssey changed directions, I nevertheless found therein a central theme for my research: equations of state (EoS) of solids. I therefore taught a course on EoS in alternate years, first at Columbia and then at UCLA. The lecture notes were refined by student participation and by new research results in my laboratory and in other laboratories. These notes are the basis of this book, which, I believe, is appropriate for first year graduate students in geophysics and ceramic science.

In this book, equation of state means a functional relationship between pressure P, volume V, and temperature T for a solid. Solids of interest here are oxides and silicates, especially those with high density.

ACKNOWLEDGEMENTS

And gladly wolde he lerne,
and gladly teche.

—Geoffrey Chaucer

Most of my former students have contributed to the ideas in this book. To all of them, I owe a great deal. I thank my Columbia University students: L. Thomsen, for teaching me Lagrangian theory; Bob Liebermann, for teaching me persistence, and Harry Demarest, for laying the computational foundation for UCLA and Nagoya's special technique of physical acoustics called rectangular parallelepiped resonance (RPR). I thank my UCLA students: Dave Yuen, who taught me the virial theorem; John Baumgardner, who taught me the value of high-speed computation; Don Isaak, who taught me about *ab initio* calculations and solved the iron problem in RPR; Hyunchae Cynn, who showed that mixing Raman experiments with RPR experiments solved the problem of spinel; Dawei Guo, who corrected a number of algebra errors; and Andy Gratz, whose untimely death cut off a source of inspiration. I thank my former, but unofficial, Cal-Tech students: Hartmut Spetzler, for teaching me the highest standards of ultrasonic interferometry; Susan Kieffer, for teaching me to look again at thermodynamics; and Charles Sammis and Tom Jordan, for vindicating our experimental results on ZnO, which at first sight violated Birch's finite strain theory.

Most of my post-doctoral fellows helped push the measurement of elastic constants to higher and higher temperatures by the RPR experiment. Mineo Kumazawa sent his students one by one, enabling my UCLA laboratory to stay on the cutting edge of research. Mineo's former students included Yoshio Sumino, Isao Suzuki, Ichiro Ohno, Takayasu Goto, Hitoshi Oda, Shigeru Yamamoto, and Koji Masuda. I also thank Nahiro Soga for sending his former student, Rikuo Ota, who worked on phase transformations in glass, and Syun-iti Akimoto, for sending his former student, Yosiko Sato-Sorenson, who helped me find Grüneisen constants for several glasses.

I thank my professional colleagues, who provided inspiration and help on parts of this book. Frank Stacey spent much time critiquing an early version of Part I. Russ Hemley's comments helped with the reorganization of Part I. Tom Shankland and Al Duba's papers made me reconsider parts of the book. Jean-Paul Poirier's incisive comments improved Part I and the chapter on melting. Ann Chopelas and Anne Hofmeister's pioneering work on the applications of optical measurements made me a believer. Ron Cohen inspired three UCLA students to become active in *ab initio* research.

In a special category are six colleagues: Reini Boehler—our fruitful association began many years ago at UCLA, where we developed a healthy, friendly interaction as mutual critics that has benefited many ideas in this book; Mineo Kumazawa—our long association has created a path for ideas and people to flow across the Pacific Ocean steadily for over twenty years; Nahiro Soga, whose early work with me on spherical resonance of spherical glass drops recovered from the lunar surface implanted the idea that resonating small samples of various shapes could lead to precise elastic constant determination; Edward Schreiber, who helped me set up the acoustics lab at the Lamont-Doherty Observatory; Warren Mason, who taught me physical acoustics; and Herb McSkimin, who taught me ultrasonic measurements of velocities of sound.

I thank three University of Utah professors, Walter Elsasser, Henry Erying, and Dan McLachlin, who transformed me from a country boy, raised on a mountain ranch, into a dedicated research physicist.

In addition, I thank my family, and especially my wife, Bernie, for encouraging me to finish this book and for stifling complaints about the time I spent away from the hearth.

The technical production of the typescript was totally in the hands of Judy Hohl. She produced the camera-ready copy, including the tables, using TeX. Her standards of excellence, higher than my own, minimized the technical errors. Richard Sadakane produced the figures with his MacIntosh computer. Joyce Berry, Oxford's editor, was very helpful and extraordinarily patient.

CONTENTS

PART I
THERMAL PHYSICS

There is more joy
in heaven in a good
approximation than in
an exact solution.

—Nobelist Julian Schwinger

Many useful approximations are to be found in this examination of thermal physics. We find, for example, that the experimentally determined thermal pressure is linear with T to quite high temperatures, and that several important thermoelastic parameters are independent of T to high temperatures. These and other good approximations are aspects of the grand approximation called the high temperature limit of the quasiharmonic approximation of statistical mechanics, which is treated in Chapter 2.

Thermal physics is a cornerstone of equations of state at high temperature. This treatment of "thermal physics" is somewhat different from that found in texts with the same name, however, because it is focused on the equation of state of solids, and, as a consequence, certain parts of the traditional "thermal physics" are emphasized and expanded, while others are ignored. There has been increased research emphasis on thermal effects in the EoS, especially thermal pressure P_{TH}, the coefficient of thermal expansivity α, and the Grüneisen ratio parameter γ (which relates pressure and energy at constant V), all of which are treated here.

In the first two chapters I give the thermodynamic and statistical mechanical background needed throughout the book. Definitions of the physical properties required for the EoS, derived in terms of the free energy, are also presented. Emphasis is placed on γ in all its variations, the Debye temperature Θ, and the bulk modulus K_T. The variation of P_{TH} with T and V is developed.

In the course of the derivations, several dimensionless thermoelastic parameters arise. These play a role in physical thermodynamics similar to the role of the Rayleigh number and the Nusselt number in hydrodynamics. In the third chapter I develop theory showing how the thermoelastic parameters vary with T and V at high temperature and high compression. This is emphasized by using thermoelastic parameters that transform the thermodynamic identities into differential equations. Such thermodynamic equations are especially useful in changing from P, T variables to V, T variables, for any given physical property.

In the fourth chapter, I emphasize α and its relationship to many properties in the EoS and cast it in terms of thermoelastic dimensionless parameters. The variation of α with T, V, and P at extreme conditions is presented. Special attention is given to the high P-high T behavior of α for periclase, silicate perovskite, and iron.

In the fifth chapter I use some neglected properties of the Debye theory to find the relationship between mechanical properties (such as sound velocity, bulk modulus, and the pressure derivative of the bulk modulus) and temperature and composition, all at $P = 0$. From this we identify a subclass of oxides and silicates in which many physical properties can be explicitly calculated from Debye theory. This subclass, called "Debye–like" solids, includes those thought to comprise the deep interior of terrestrial planets. The majority of oxides and silicates are not Debye-like solids, and this explains why the Debye theory falls short of explaining physical properties of so many solids of interest to geophysics, ceramics, and mineralogy.

Though few in number, Debye–like solids, for which Debye theory is a useful application, are nevertheless important. Corundum (Al_2O_3), sometimes called sapphire, and periclase (MgO) are emphasized because they are of especial interest to ceramic sciences. Magnesium silicate perovskite ($MgSiO_3$), olivine ($(Fe, Mg)_2SiO_4$), and periclase are emphasized because they of especial interest to solid earth geophysics. Periclase is a linchpin of both disciplines, and as such it receives the most exhaustive treatment.

The whole treatment of Part I rests squarely on the solid state science done in Germany in the late 1920's and 1930's, especially that of Max Born, Peter Debye, Albert Einstein, E. Grüneisen, and Ludwig Boltzmann. The discovery of quantum mechanics diverted these great scientists from thermal physics, and the field was neglected for decades. Recent advances in experimental research on ceramic and geophysical solids have required for their interpretation a respectful revival of the theories of these scientists.

1

THE FREE ENERGY AND THE GRÜNEISEN PARAMETER

1.1. Introduction

A logical way to have constructed this book would have been to use the first chapter to define all required parameters of equations of state by means of derivations arising from a statistical mechanical approach. This is not done here, because it is important first to define the basic parameters of the equation of state in physical terms. I save the statistical mechanics presentation that is necessary to define high temperature and low temperature properties for Chapter 2. In Chapter 1 I also emphasize the relationships of physical properties to the thermodynamic functions.

Although I think this organization will be helpful to students, there is a certain flaw in the logical flow of ideas. On three occasions I must define a property in terms of derivations yet to be made in Chapter 2. My decision is based on the observation that a number of students in geophysical disciplines and ceramics science who want to learn about equations of state have not been trained in statistical mechanics. For them, Chapter 1 will be helpful in learning the contents of Chapter 2.

1.2. The Helmholtz free energy

The Helmholtz energy $\mathcal{F}$ of a solid is used to find the pressure P. For many solids found in geophysics and ceramics, $\mathcal{F}$ has three contributions:

$$\mathcal{F} = E_{ST} + \mathcal{F}_{VIB} + E_{e\ell}, \tag{1.1}$$

where E_{ST} is the potential of a static lattice at absolute zero; $\mathcal{F}_{VIB}$ is the vibrational energy due to the motion of the atoms as each is constrained to vibrate around a lattice point; and $E_{e\ell}$ is the potential arising from free electrons. In some solids other potentials, such as magnetization and optical effects, can arise as contributions, but these are not important in this treatment, which is largely concerned with insulators. Many minerals are insulators; in such cases $E_{e\ell}$ can be ignored. For the case of iron, for example, $E_{e\ell}$ cannot be ignored.

There are three thermodynamic functions by which P or V is related to other thermodynamic variables:

Helmholtz Energy

$$\mathcal{F} = \mathcal{U} - T\mathcal{S} = E_{ST} + \mathcal{F}_{VIB} + E_{el} \tag{1.2}$$

Gibbs Energy

$$\mathcal{G} = \mathcal{U} - T\mathcal{S} + PV = E_{ST} + \mathcal{F}_{VIB} + E_{el} + PV \tag{1.3}$$

Enthalpy

$$\mathcal{H} = \mathcal{U} + PV, \tag{1.4}$$

where $\mathcal{S}$ is the entropy; $\mathcal{U}$ is the internal energy; and $\mathcal{H}$ is the enthalpy. Sometimes $\mathcal{F}$ and $\mathcal{G}$ are called free energies.

In thermodynamic systems where volume V and temperature T are the independent variables, (1.2) is the most convenient energy function. Consequently, the Helmholtz energy $\mathcal{F}$ will be used for most derivations. In equilibrium studies for determining phase boundaries, however, the convenient independent variables are P and T. In that case, the Gibbs energy $\mathcal{G}$ (1.3) is the recommended free energy function because the equilibrium at fixed P is the one with minimum G.

By EoS we mean that pressure is defined in terms of two variables, V and T, as follows

$$P = P(V,T). \tag{1.5}$$

The equation of state (EoS) will be derived from $\mathcal{F}$.

In specifying (1.5) as the EoS, we are ignoring the contribution of shear deformations to $\mathcal{F}$. In planetary interiors this is justified because the pressure P is large compared with the shear stresses. Similarly the time variation of deformation is not significant in our EoS, although it may be important in tectonic processes near the earth's surface, where the deviatorial stresses are no longer small compared with pressure.

In terms of $\mathcal{F}$, the fundamental thermodynamic equations are (Landau and Lifshitz, 1958):

$$S = -\left(\frac{\partial \mathcal{F}}{\partial T}\right)_V \tag{1.6}$$

$$\mathcal{U} = \mathcal{F} - T\left(\frac{\partial \mathcal{F}}{\partial T}\right)_V = -T^2\left(\frac{\partial}{\partial T}\frac{\mathcal{F}}{T}\right)_V \tag{1.7}$$

$$C_V = \left(\frac{\partial \mathcal{U}}{\partial T}\right)_V \tag{1.8}$$

$$P = -\left(\frac{\partial \mathcal{F}}{\partial V}\right)_T \tag{1.9}$$

$$K_T = V\left(\frac{\partial^2 \mathcal{F}}{\partial V^2}\right)_{TT} = -V\left(\frac{\partial P}{\partial V}\right)_T \tag{1.10}$$

$$\left(\frac{\partial P}{\partial T}\right)_V = \alpha K_T = -\frac{\partial}{\partial V}\left[\left(\frac{\partial \mathcal{F}}{\partial T}\right)_V\right]_T = -\left(\frac{\partial S}{\partial V}\right)_T, \tag{1.11}$$

where C_V is the specific heat at constant volume; K_T is the isothermal bulk modulus; and α is the volume coefficient of thermal expansion.

The pressure is also given by

$$P = \left(\frac{\partial \mathcal{U}}{\partial V}\right)_S. \tag{1.12}$$

We also see that the expression for α, the thermal expansivity from (1.10) and (1.11), is

$$\alpha = -\frac{1}{V}\left(\frac{\partial^2 \mathcal{F}}{\partial T \partial V}\right) \Big/ \left(\frac{\partial^2 \mathcal{F}}{\partial V^2}\right)_T. \tag{1.13}$$

1.3. Pressure: The equation of state

For an insulator, from (1.1), $\mathcal{F} = E_{ST} + E_{VIB}$. In Chapter 2, I will show that for an insulator, $E_{VIB} = E_{TH} + E_{ZV}$ and $E_{T=0} = E_{ST} + E_{ZV}$, where E_{TH} is the thermal energy and E_{ZV} is the zero temperature vibrational energy, a small term arising from quantum mechanics. A more convenient expression for $\mathcal{F}$, in which the $T = 0$ effects and the thermal effects are clearly separated, is

$$\mathcal{F} = E_{T=0} + \mathcal{F}_{TH}, \tag{1.14}$$

where $T = 0$ indicates absolute zero.

Using (1.9) and (1.14),

$$P = -\left(\frac{\partial E_{T=0}}{\partial V}\right)_T - \left(\frac{\partial \mathcal{F}_{TH}}{\partial V}\right)_T. \tag{1.15}$$

Following (1.5) and (1.15), P is a function of V and T except at absolute zero, where $P = P_{T=0}(V)$. In subsequent sections, I use (1.15) as the standard definition of P for dielectric solids without free electrons.

Defining

$$P_{TH} = -\left(\frac{\partial \mathcal{F}_{TH}}{\partial V}\right)_T, \tag{1.16}$$

we have the basic formulation

$$P(V,T) = P_{T=0}(V) + P_{TH}(V,T). \tag{1.17}$$

Let V be replaced by the dimensionless variable η, often called the dilation, $\eta = V/V_0$, which shall be called here compression, where the subscript 0 indicates V at zero pressure. Then an alternate form of (1.17) is

$$P(\eta,T) = P_{T=0}(\eta) + P_{TH}(\eta,T). \tag{1.18}$$

For a metal at high temperature, as in iron in the earth's core, to provide an example, an additional term, $P_{e\ell}(\eta,T)$, must be added to (1.18) because the energy of free electrons ($E_{e\ell}$ in (1.1)) becomes important.

Either (1.17) or (1.18) is referred to in this book as the Equation of State (EoS), and effects from the stress deviator are ignored.

1.4. The Grüneisen parameters

1.4.1. *The Grüneisen ratio*

The Grüneisen ratio γ (Grüneisen, 1912, 1926) is a very important thermodynamic parameter used to help quantify the relationship between the thermal and elastic properties of a solid. Sometimes γ is called the thermal Grüneisen parameter (labeled γ_{th} in more recent literature), but in this book we follow the nomenclature of the originator, Grüneisen (1912). The Grüneisen ratio can be considered as the measure of the change of pressure resulting from the increase of energy density at constant V. It is dimensionless, as ΔP and $\Delta\mathcal{U}/V$ have the same units. A convenient form is

$$\gamma = V\left(\frac{\partial P}{\partial \mathcal{U}}\right)_V. \tag{1.19}$$

If the change in pressure were exactly proportional to the change in energy density, γ would be independent of P and T, and experiments show this to be nearly correct if one considers the case for the variation of γ with T at constant V. In fact, the "proportionality constant" usually lies between 1 and 2.

In order to evaluate (1.19) in parameters representing measurable physical properties, start with

$$\left(\frac{\partial P}{\partial \mathcal{U}}\right)_V = \left(\frac{\partial P}{\partial T}\right)_V \Big/ \left(\frac{\partial \mathcal{U}}{\partial T}\right)_V. \tag{1.20}$$

According to (1.8), $(\partial\mathcal{U}/\partial T)_V = C_V$ at unit mass. From calculus,

$$\left(\frac{\partial P}{\partial T}\right)_V = -\left(\frac{\partial V}{\partial T}\right)_P \Big/ \left(\frac{\partial V}{\partial P}\right)_T. \tag{1.21}$$

Equation (1.21) is the ratio of the compressibility to α. Thus

$$\left(\frac{\partial P}{\partial T}\right)_V = \alpha K_T, \tag{1.22}$$

in agreement with (1.11). Equation (1.20) is therefore

$$\left(\frac{\partial P}{\partial \mathcal{U}}\right)_V = \frac{\alpha K_T}{C_V}. \tag{1.23}$$

Using (1.23), (1.19) becomes the most important definition of γ:

$$\gamma = V\left(\frac{\partial P}{\partial \mathcal{U}}\right)_V = \frac{\alpha K_T V}{C_V}. \tag{1.24}$$

Thus γ given by (1.24) is composed of individual measurable physical properties, each of which varies significantly with temperature, but the ratio of these properties as given by (1.24) does not vary greatly with temperature, and sometimes not at all. For this reason (1.24) has been traditionally called the Grüneisen ratio, a name that will be used in this text. Approximations to (1.24) for γ will not involve the word "ratio" and will receive special names and subscripts.

Although the Grüneisen ratio tends to be nearly independent of T, especially at high T, it generally decreases as the volume decreases. An approximation often used for solids is

$$\rho\gamma = \text{constant}, \tag{1.25}$$

where ρ is the density. As we shall see in Chapter 3, (1.25) does not hold for low temperatures ($T < \Theta$, the Debye temperature), but does hold for temperatures in shock waves.

Note that if (1.25) holds true, then from (1.19)

$$\left(\frac{\partial P}{\partial \mathcal{U}}\right)_V = \text{constant}. \tag{1.26}$$

The extent to which the approximation given by (1.25) is valid and useful will be examined in Chapter 3. It is often not true at low T.

From (1.8) and (1.11), γ is defined in terms of the free energy:

$$\gamma = V\frac{\partial^2 \mathcal{F}}{\partial V \partial T}\Big/\left(\frac{\partial \mathcal{U}}{\partial T}\right)_V. \tag{1.27}$$

The derivative in the denominator of (1.27) could consist of a sum of terms. Further, $\mathcal{F}$ could contain separate terms arising from phonons, defects, and anharmonic effects. The exact equation for γ can therefore be complex. As we shall see, it is simplified in the limits of high and low T.

1.4.2. *Adiabatic decompression measurements for* γ

Start with one of Maxwell's relationships:

$$\left(\frac{\partial T}{\partial V}\right)_S = -\left(\frac{\partial P}{\partial S}\right)_V, \tag{1.28}$$

where the cross derivative is held first at constant V, then at constant T. Expand the right side of (1.28): $(\partial P/\partial S)_V = (\partial P/\partial T)_V\,(\partial T/\partial S)_V$. By using (1.8) and (1.11), this becomes $-T\alpha K_T/C_V$. Using (1.24), this becomes $T\gamma/V$, so that (1.28) can also be written as

$$\gamma = -\left(\frac{\partial\, \ell\text{n}\, T}{\partial\, \ell\text{n}\, V}\right)_S, \tag{1.29}$$

where the cross derivative is held fast at constant V, then at constant T. This equation is useful in theoretical formulations. It is the basic equation by which the adiabatic temperature gradient within a planet interior is determined.

Using the definition $K_S = -V\left(\partial P/\partial V\right)_S$, where K_S is the adiabatic bulk modulus (see Section 1.5.1), we obtain from (1.29)

$$\gamma = \frac{K_S\left(\dfrac{\partial T}{\partial P}\right)_e}{T}. \tag{1.30}$$

Equation (1.30) is useful for experimental determination of γ, in which $(\partial T/\partial P)_S$ is a measured adiabatic decompression (Boehler, 1981).

Consider a solid under pressure P at some T. Let an incremental change ΔP be made, and let the system adjust adiabatically, resulting in a measured change of ΔT in the solid. Then γ at that P and T can be measured according to (1.30), provided information on K_S at that particular P and T is available.

1.4.3. *Variation of γ with volume*

Grüneisen (1912) defined a value for the volume derivative of each mode ω_i in the lattice γ_i according to

$$\gamma_i = -\left(\frac{\partial\, \ell n\, \omega_i}{\partial\, \ell n\, V}\right)_T. \tag{1.31}$$

Equation (1.31) is often called a mode gamma. But γ itself is also a simple function of V, at least approximately.

Anderson (1967) suggested that the Grüneisen ratio itself be of a form similar to (1.31) in its relation to volume,

$$\left(\frac{\gamma}{\gamma_0}\right) = \left(\frac{V}{V_0}\right)^q. \tag{1.32}$$

McQueen et al. (1970) specifically took $q = 1$, or (1.25), for their shock wave analysis. This has been the practice in shock compression in many works ever since.

Boehler and Ramakrishnan (1980) and Boehler (1982) measured γ vs. P for a number of solids using the adiabatic decompression method, (1.30), and converted the data to a form like (1.32). A list of their measured values of γ_0 and q is shown in Table 1.1 (Boehler and Ramakrishnan, 1980; Boehler, 1982). This shows that the variation of γ with V/V_0 generally gives $q > 1$ using (1.32). These results hold for relatively low pressure. For quite high pressure and temperature, q is near 1, so that we may expect q to change its value over a large pressure interval (see details in Chapter 3).

Table 1.1. Experimental variation of γ with V

	γ_0	q
Metals		
Al	2.14	1.0
Cu	2.01	1.3
In	2.42	1.8
Fe	1.66	0.8
Pb	2.63	1.2
Na	1.2	0.91
K	1.19	1.7
Bi	1.0	1.4
Alkali Halides		
NaCl	1.62	1.3
KBr	1.43	1.2
KCl	1.15	1.0
Minerals		
MgO	1.39	1.0
Mg_2SiO_4	1.15	1.0
Quartz	0.57	1.0

Source: Boehler and Ramakrishnan, 1980; Boehler, 1982.

1.4.4. *The average of the mode gammas*

The Grüneisen parameter γ is also defined in terms of lattice dynamic theory, where the vibrations of a crystalline lattice are defined in terms of an ensemble of harmonic oscillations known as Einstein functions (see Chapter 2). The general definition of γ then involves a summation and an averaging of the terms describing how each of the lattice frequencies ω_i varies with volume given by (1.31). For each harmonic oscillator (called a mode), there is a mode gamma given by (1.31), there being $3pN$ modes in the crystal (where p is the number of atoms in the basic cell). We shall see this idea developed further in the next chapter, but for purposes of definitions, we anticipate some of those results here.

The most common method of averaging the mode gammas is to assume non-interacting oscillators and require that the energy, the pressure, and the specific heat be thermodynamically consistent (Born and Huang, 1954), and further that ω_i be independent of T (but not of V). This is consistent with the quasiharmonic approximation by which frequencies of the oscillators obey (1.31). The quasiharmonic approximation of $\mathcal{F}$ and its derivatives will be treated in more detail in the next chapter.

It is well known (Barron, 1955) that under this approximation

$$\overline{\gamma} = \frac{\Sigma C_i \gamma_i}{\Sigma C_i}, \tag{1.33}$$

where C_i is the Einstein specific heat capacity defined in terms of ω_i for each mode. The summation is taken over the $3pN$ modes of the lattice in the general case. Here N is Avogadro's number, and p is the number of atoms in the vibrating cell. Equation (1.33) is too general to solve without further approximations.

1.4.5. *The Mie-Grüneisen approximation: γ_{mg} to γ*

In the limiting case where all $(\partial \ln \omega_i / \partial \ln V)$ are equal, γ_i is factored out of the summation in (1.33). Then P_{TH} is proportional to E_{TH} in place of (1.16), and we have the Mie-Grüneisen expression for P_{TH} given by

$$P_{TH} = \frac{\gamma_{mg}}{V} E_{TH} = \rho \gamma_{mg} E_{TH}, \tag{1.34}$$

where the $\overline{\gamma}$ in (1.33) is replaced by γ_{mg}, and where γ_{mg} refers to the Mie-Grüneisen EoS. Spectroscopic measurements of ω_i versus P show that the γ_i are not equal (see Section 1.4.9). So as long as the C_i of (1.33) are not equal, (1.34) is not valid. At high T, where all C_i are equal (see Chapter 2), $\overline{\gamma}$ can be defined as approaching γ_{mg}. But for (1.34) to be useful at high T, even as an approximation, the Grüneisen ratio, (1.24), must have some special properties at high T, which will be discussed in Chapter 3. Altogether there are special limits placed on the reliability of the Mie-Grüneisen EoS.

1.4.6. *The acoustic gamma γ_{ac} approximation to γ*

Barron (1957) showed that in general, for the high temperature approximation, the denominator of (1.33) approaches a limiting value, and (1.33) becomes

$$\gamma = \gamma^{ht} = \frac{1}{3pN} \sum_{j=1}^{3pN} \gamma_j. \tag{1.35}$$

A further approximation is made by ignoring all optic modes of the crystal and also ignoring all acoustic modes except those at low wave number in the summation of (1.35). Thus the summation up to $3pN$ is replaced by a summation up to 3 (for the longitudinal and two shear acoustic modes). Because most of the energy is in the high modes, this approximation appears weak, but (1.35) is an average γ, and the approximation implies that the average of the three acoustic modes is equal to the average of all modes. Barron (1955, 1957) showed that for the acoustic modes (1.31) becomes

$$\gamma_i = \frac{1}{3} - \frac{V}{v_i}\left(\frac{\partial v_i}{\partial V}\right)_T = \frac{1}{3} + \frac{K_T}{v_i}\left(\frac{\partial v_i}{\partial P}\right)_T, \tag{1.36}$$

where $(i = 1, 2, 3)$ and v_i is the sound velocity. An alternative to (1.35) is

$$\gamma_i = -\frac{1}{6} + \frac{1}{2}\frac{K_T}{\mathcal{M}_i}\left(\frac{\partial \mathcal{M}_i}{\partial P}\right)_T, \tag{1.37}$$

where $\mathcal{M}_i$ is the elastic modulus associated with v_i found in (1.36).

The high temperature approximation (1.35) for a single crystal needs to account for the fact that v_i varies with direction. Barron (1957) showed that one must use

$$\gamma^{ht} = \frac{1}{12\pi}\int\left(\sum_{i=1}^{3}\gamma_i\right)d\Omega. \tag{1.38}$$

The reduction of (1.38) to materials isotropic to acoustic waves is obvious because there are only two sound velocities and two values of the mode gamma γ_i. The isotropic γ's are

$$\gamma_s = -\frac{1}{6} + \frac{1}{2}\frac{K_T}{G}\left(\frac{\partial G}{\partial P}\right)_T \tag{1.39}$$

and

$$\gamma_p = -\frac{1}{6} + \frac{1}{2}\frac{K_T}{(\rho v_p^2)}\left(\frac{\partial\left(\rho v_p^2\right)}{\partial P}\right)_T, \tag{1.40}$$

where G is the isotropic shear modulus.

The acoustic approximation to the Grüneisen ratio, called γ_{ac}, is thus the average of the p wave mode γ_p and the two shear wave modes γ_s:

$$\gamma_{ac} = \frac{1}{3}\left(\gamma_p + 2\gamma_s\right). \tag{1.41}$$

Equation (1.41) can be applied immediately to isotropic polycrystalline materials. However, most applications of (1.41) to minerals are made by finding the equivalent isotropic elastic constants, K_S and G, of the single crystal at each pressure and the average associated pressure derivatives of the K_S and G values (see Quareni and Mulargia, 1988).

The comparison of γ_{ac} with γ ((1.24) or (1.30)) for many minerals is shown in Table 1.2. In this table, we see that for dense minerals with high coordination numbers, such as periclase, γ_{ac} is close in value to γ. On the other hand, for minerals with low coordination numbers, such as rutile, γ_{ac} is not close in value to γ. This is explored further in Section 5.9.

Table 1.2. Values of γ and γ_{ac} for minerals, oxides, and three dense rocks at zero pressure

	Experimental Data			γ_0: (1.24)	γ_{ac_0}: (1.42)
Compound	K_0'	γ_s	γ_p	$\dfrac{\alpha K_T V}{C_V}$	$\dfrac{\gamma_p + 2\gamma_s}{3}$
Corundum	3.99	1.20	1.53	1.32	1.31
Periclase	4.20	1.40	1.71	1.54	1.50
NaCl	5.50	0.9	2.20	1.56	1.33
Spinel	4.18	0.49	1.33	1.13	0.79
Garnet	5.45	1.13	1.95	1.21	1.89
Hematite	4.53	0.64	1.53	1.91	0.94
Forsterite	5.37	1.24	1.92	1.15	1.47
Olivine	5.13	1.28	1.88	1.21	1.33
Rutile	6.76	0.55	2.09	1.28	1.07
α-iron	5.29	—	—	1.81	1.81
Eclogite	5.5	0.93	1.90	1.25	1.25
Peridotite	5.3	1.52	2.15	1.72	1.73
Orthopyroxene	9.6	1.48	3.09	1.56	2.02
Trevorite	4.41	0.32	1.44	1.24	0.70
Earth's Lower Mantle	—	—	—	—	1.50

Source: Sumino and Anderson, 1984.

Because both γ_s and γ_p can be expressed in terms of different sets of elastic constant pressure derivatives, alternate forms of (1.39) can be generated. One form uses the pressure derivative of the bulk modulus K and the Poisson ratio ν, so that

$$\gamma_{ac} = \frac{1}{2}\left(\frac{dK}{dP}\right) - \frac{1}{6} - \frac{(4-5\nu)}{(1+\nu)(1-\nu)(1-2\nu)}\left(\frac{K_T}{3}\right)\left(\frac{d\nu}{dP}\right). \quad (1.42)$$

If $d\nu/dP$ is assumed to be zero, (1.42) reduces to the expression of γ known as the Slater gamma, γ_{sl} (Slater, 1939), which is traditionally derived from entirely different premises. We then find that

$$\gamma_{sl} = \left(\frac{1}{2}\right)\left(\frac{dK}{dP}\right) - \frac{1}{6}. \quad (1.43)$$

Both experiments and lattice theory show that ν increases with P, so that (1.43) is only of historical interest.

Data in Table 1.2 indicate that the average of the first three modes often gives a good approximation to the average of all $3pN$ modes of a crystal, especially for the dense minerals. However, the value of γ_{ac} is not close to that of γ for minerals with low coordination, such as hematite.

1.4.7. *Historical account of approximations to the Grüneisen ratio* γ

In the 1950s and 1960s research on γ was characterized by attempts to find the volume dependence of γ without taking possible changes of q, shown in (1.32), into consideration. Simple analytical equations in terms of $K' = (\partial K_T/\partial P)_T$ all arise from severe approximations. Examples will be found in the work of Dugdale and McDonald (1953); Barron (1957); Vashchenko and Zubarev (1963), see next section; and Knopoff and Shapiro (1969). In this era, the volume variation of γ was described by the general function $\gamma(K', P/K_T)$ (See (1.43) and (1.46), for example).

In the 1970s research focused on the parameter q because of an expression derived by Bassett et al. (1968), who presented the idea that q, itself a dimensionless quantity, was a function of other dimensionless quantities, including δ_T (see (1.49a) and (1.63)). Attention thus shifted from describing γ as simply $\gamma(K_T', P/K_T)$ to including q as a variable and to the inclusion of δ_T in q. Most literature of that era was further simplified by the deletion of the volume effect on C_V found in the primary definition of γ, (1.24). The era included contributions by Thomsen (1970), Barsch and Achar (1970), and Anderson (1974). At this time $\gamma(V)$ was found through $q(K', \delta_T)$, where K' and δ_T were taken as independent of V. Neglect of the volume dependence of the C_V term is allowable when $T > \Theta$, but not when $T < \Theta$ (where Θ is the Debye temperature) or when C_V has a large electron component.

In the late 1970s authors using (1.31) or its variants showed that for temperatures below Θ but above room temperature, $q > 1$ for many minerals (a typical value was 1.5) (some q values from low pressure experiments are found in Boehler and Ramakrishnan (1980)). At that time there was insufficient thermodynamic information to find out whether q itself was a variable. However, Palciauskas (1975) investigated the NaCl structure using lattice theory involving an assumed lattice potential. He found that at room temperature q was higher than 1 (~ 1.8) for $1 < \eta < 0.85$ and descended to about 1 for $0.7 < \eta < 0.6$. But until the early 1980s no measurements showing that q itself is a function of V appear to have been made.

Boehler (1983) measured γ to a compression of $\eta = 0.7$ for three alkali metals, and from his plots we find that $q \sim 1$ at high compression where $\eta < 0.8$, and $q > 1$ (about 1.7) at compressions for η from 1 down to 0.9. As a result of Boehler's early work (in the 1980s), the suspicion arose among many researchers that q itself may change with η. Attention has recently shifted to include the possible volume variation and magnitude of q. It turns out that the volume dependence of $\delta_T - K'$ controls the volume dependence of γ.

A more general thermodynamic approach is now possible because a great deal of experimental information on the pressure and high temperature dependence of δ_T (described in detail in Chapter 3) has been acquired recently. This clearly shows that q is a strong function of η, but only at low temperatures ($T < \Theta$). Many earlier approximations have been abandoned.

The "free volume" expression for gamma was derived by at least two different methods (Vashchenko and Zubarev, 1963; Irvine and Stacey, 1975). I refer here to the later derivation by Irvine and Stacey (1975). Following the path suggested by Brillouin (1946), who showed that the thermal pressure may be considered as that required to keep the volume constant as the temperature is raised, they expanded the "mutual forces between the atoms at separation r" in a series expansion around r_0 to quadratic terms and found the pressure necessary to keep ΔV zero at any given T, which is the thermal pressure P_{TH}. Irvine and Stacey (1975) took account of the coupled three-dimensional vibrations, thus implicitly including the effects of high frequency modes. The calculation was for an fcc monatomic solid, but was reported to be valid for all cubic solids.

$$P_{TH} = \left[\frac{\left(\frac{1}{2}\right)K' - \frac{5}{6} + \left(\frac{2}{9}\right)\left(\frac{P}{K_T}\right)}{1 - \left(\frac{4}{3}\right)\left(\frac{P}{K_T}\right)} \right] \rho 3kT. \tag{1.44}$$

They equated the P_{TH} given by (1.44) to the classical high temperature thermal pressure of a solid, $\gamma_{mg}\rho 3kT$, and solved for γ:

$$\gamma_{vz} = \frac{\left(\frac{1}{2}\right)K' - \frac{5}{6} + \left(\frac{2}{9}\right)\left(\frac{P}{K_T}\right)}{1 - \left(\frac{4}{3}\right)\left(\frac{P}{K_T}\right)}, \tag{1.45}$$

where the subscript refers to Vashchenko's and Zubarev's definition.

This was apparently the first derivation of an approximation to γ that included consideration of the high wave number modes and mode interaction. Equation (1.45) has been used for γ extensively in the Soviet literature on EoS (see, for example, Zharkov and Kalinin, 1971).

1.4.8. *The Barton-Stacey approximation* γ_{ba-s}

Stacey was dissatisfied with the P_{TH} in (1.45) because the derivation assumed that the correlation of motions along a bond was the same with respect to dilation as that of those transverse to the bond. Thus, in a later paper (Barton and Stacey, 1985), a correction was made, leading to

$$\gamma_{ba-s} = \frac{\frac{1}{2}K' - \frac{1}{6} - \frac{f}{3}\left[1 - \frac{1}{3}\left(\frac{P}{K_T}\right)\right]}{1 - \left(\frac{4}{3}\right)\left(\frac{P}{K_T}\right)}, \tag{1.46}$$

where $f = 2.35$, not 2, as in (1.45).

At $P = 0$, (1.46) reduces to

$$\gamma_{ba-s} = \frac{1}{2}K_0' - 0.95,$$

which gives values much closer to the measurements of γ than those given by the Slater approximation, (1.43). The central force assumption behind the above equation restricts validity to closely packed materials (see Chapter 5). According to the above equation, the Grüneisen constant is intimately connected with the pressure derivative of the bulk modulus. The thermodynamic foundation for the bulk modulus is examined in Section 1.5. However, we shall see in Section 1.8 that, according to the Debye theory, the effect of $\partial K/\partial P$ upon γ may not be as large as indicated by γ_{ba-s}. The equation for γ_{ba-s} does not show an explicit dependence of γ upon dG/dP.

1.4.9. *On the temperature dependence of* γ

The approximations to γ given by γ_{ht} (1.35), γ_{ac} ((1.41) and (1.42)), and γ_{ba-s} are all members of a class that can be described by an analytical formula using the general function:

$$\gamma = \gamma(\partial K_T/\partial P, K_T/P, \partial\nu/\partial P).$$

In the above equation, we see that there are no temperature effects except those carried by K_T. In the Mie-Grüneisen EoS, where the γ_i are all assumed equal, the resulting γ_{mg} is independent of T at all values of T. Experimentally it is known that at low temperatures ($T < \Theta$), the Grüneisen ratio (1.24) varies with T.

There are many other works in the class described by the above equation for γ. Examples include those of Dugdale and MacDonald (1953), Barron (1957), and Knopoff and Shapiro (1969). In general they are based on the assumption of high T, and thus an explicit temperature dependence of γ cannot be introduced. They are also based on either the quasiharmonic approximation at high T or on the assumption that $E_{TH} = 3pNT$, which means that at high temperature γ must be independent of temperature. They cannot be safely applied to low temperatures (i.e., below the Debye temperature limit, see Section 1.6).

Under the quasiharmonic approximation at high T, the individual mode gammas are independent of temperature. However, the general case of γ is dependent on T because, as shown in (1.35), each γ_j under the summation sign has a different limit in T, passing to the high temperature value.

At low temperatures ($T < \Theta$), γ varies with T, sometimes dramatically, as the individual mode gammas selectively take dominance (Barron et al., 1980). For alkali halides, γ generally diminishes at low T compared with γ at $T = \Theta$. But for oxides, such as Al_2O_3 and MgO, γ is high at T near zero and descends as T increases; the curve smooths out at $T = \Theta$, and stays more or less constant as T further increases (White and Anderson, 1966). The thermodynamic approach shows that γ is an explicit function of two dimensionless parameters: one is $K' = (\partial K_T/\partial T)_P$, and the other, often known as the Anderson-Grüneisen parameter, is

$$\delta_T = -\left(\frac{1}{\alpha K_T}\right)\left(\frac{\partial K_T}{\partial T}\right)_P \tag{1.47}$$

or

$$\delta_S = -\left(\frac{1}{\alpha K_T}\right)\left(\frac{\partial K_T}{\partial T}\right)_P \tag{1.48}$$

(see also (1.63)).

The explicit relationship between γ and T involving δ_T and K' will be presented in Chapter 2 with experimental results. After some foundations in thermodynamic identities have been established in Chapter 3, a thermodynamic theory on the V, T dependence of γ will be presented. Here we emphasize that δ_T and δ_S are the dimensionless representations of the bulk modulus temperature derivative.

1.4.10. *The optic mode approximation to γ*

Recent advances in high pressure spectroscopy (Chopelas, 1990a; Chopelas and Boehler, 1992b) have allowed the determination of the pressure dependence of the optic modes, yielding values for a variant of (1.31) given by

$$\gamma_i = \left(\frac{K_T}{\omega_i}\right)\left(\frac{\partial \omega_i}{\partial P}\right)_T, \tag{1.49}$$

where K_T is the isothermal bulk modulus (see Table 1.3).

A number of optic modes and their pressure derivatives, say L, may be used for finding the average γ, where it is seen that for the evaluation of $\overline{\gamma}$ at a particular T, the evaluation of each Einstein specific heat term for every modal frequency ω_j is required. For L modes,

$$\gamma_{opt} = \frac{\sum_{j=1}^{L} \gamma_j C_j}{\sum_{j=1}^{L} C_j}, \tag{1.50}$$

$$C_V = \sum_{j=1}^{L} C_j \sim 3Nk \int_0^\infty e^x/(e^x - 1)^2 x^2 g(\omega)\, d\omega, \tag{1.51}$$

where $x = \hbar\omega/kT$, and $g(\omega)$ is the density of states. For examples of a density of states $g(\omega)$, see Figs. 5.3–5.8. Lacking a formal solution of $f(\omega)$, one uses an approximate density of states, often called a model of $g(\omega)$. As an example of a model density of states, see Fig. 4 of Chopelas (1990a), who calculated C_V for forsterite.

Table 1.3. Measured infrared frequencies and mode Grüneisen parameters for silicate perovskite. Units, frequencies, cm^{-1} (Chopelas and Boehler, 1992b)

ω_0	$d\omega/dP$	$1/\omega_0(d\omega/dP)$	γ_i
246	0.21	0.875	2.1
251	0.21	0.837	2.1
281	0.137	0.488	1.19
333	0.207	0.621	1.52
379	0.170	0.449	1.10
387	0.210	0.543	1.33
499	0.328	0.657	1.61
$\gamma(IR)^*$			
544	0.25	0.46	1.1
614	0.25	0.41	1.0
683	0.25	0.37	0.91
797	0.30	0.38	0.93

*Williams et al., 1987.

Range of spectroscopic γ's found in other silicates (high energy modes only)

Mineral	γ_i's
Forsterite	0.4–0.6
β-Mg_2SiO_4	0.7–1.1
γ-Mg_2SiO_4	0.9–1.2
Ilmenite	0.75–0.95
Majorite	0.70–0.85

In their theoretical analysis of silicate perovskite, Hemley et al. (1989) found $\gamma_{opt} = 2.0$ at 298 K and 1.63 at 1000 K, approaching 1.59 at very high T. This is also the typical behavior of the more general form of γ: a gradual descent with T below the Debye temperature and independence of T at high T (see Chapter 2).

Table 1.4 shows the modal values of ω_i and γ_i calculated by Hemley et al. (1989) for silicate perovskite. There is a wide spread in the values for the various modal γ_i's. The variation in the modal γ_i's from the spectroscopic data argues against the reliability of the use of γ_{mg} (especially at low T), as γ_{mg} hinges on the assumption that all γ_i are equal. The calculation of $q_i = (\partial \ln \gamma_i / \partial \ln V)_T$ for optic modes has been defeated because the experiments above show strict linearity between ω and P. To obtain a reasonable value of q, there must be some curvature.

Table 1.4. Calculated vibrational frequencies ω_i and mode Grüneisen parameters γ_i at zero pressure for silicate perovskite ($MgSiO_3$). Units, frequency, cm^{-1}.

Mode Identification	Raman ω_i	Raman γ_i	Mode Identification	Infrared ω_i	Infrared γ_i	Mode Identification	Inactive ω_i	Inactive γ_i
A_g	95	6.42	B_{1u}	232	1.03	A_u	242	1.02
B_{1g}	101	5.37	B_{3u}	238	2.06	A_u	294	1.97
B_{3g}	138	6.78	B_{2u}	250	2.03	A_u	399	2.00
A_g	203	4.22	B_{2u}	292	2.29	A_u	502	0.30
B_{1g}	278	3.10	B_{1u}	307	3.09	A_u	648	0.48
B_{2g}	338	2.10	B_{2u}	360	1.18	A_u	889	1.31
B_{1g}	342	2.75	B_{3u}	382	1.28	A_u	908	1.20
A_g	373	2.80	B_{2u}	400	1.16			
B_{3g}	428	0.45	B_{3u}	418	1.60			
B_{2g}	458	1.03	B_{2u}	443	2.25			
A_g	461	0.56	B_{3u}	470	1.87			
B_{1g}	482	1.08	B_{2u}	641	0.40			
B_{3g}	515	1.32	B_{3u}	705	0.45			
B_{2g}	678	1.37	B_{1u}	710	0.50			
B_{3g}	679	1.35	B_{2u}	760	0.33			
A_g	683	1.35	B_{1u}	771	0.61			
B_{2g}	770	0.61	B_{3u}	773	0.46			
A_g	771	0.69	B_{1u}	900	1.25			
B_{1g}	771	0.73	B_{2u}	909	1.28			
B_{1g}	773	1.00	B_{1u}	917	1.26			
B_{1g}	1070	0.84	B_{3u}	920	1.19			
B_{3g}	1113	0.88	B_{2u}	1010	0.47			
B_{2g}	1114	0.97	B_{3u}	1019	0.49			
			B_{2u}	1363	0.61			

After Hemley et al., 1989.

1.5. The bulk modulus K

1.5.1. *The adiabatic bulk modulus K_S*

Equation (1.10) gives K_T in terms of the free energy $\mathcal{F}$. K_T is defined in terms of the pressure and free energy as follows:

$$K_T = -V\left(\frac{\partial P}{\partial V}\right)_T = -\eta\left(\frac{\partial P}{\partial \eta}\right)_T = V\left(\frac{\partial^2 \mathcal{F}}{\partial V^2}\right)_{TT}, \tag{1.52}$$

where $\eta = v/v_0$. The bulk modulus, or incompressibility, is very important because of its direct relationship with seismic or acoustic velocities or as a parameter in the interpretation of compression measurements.

In seismic or acoustic measurements, adiabatic conditions are maintained in the compression-expansion cycle. Heat cannot flow in or out as the wave passes because the wave period is small compared with the thermal relaxation time. Thus, the adiabatic equivalent of (1.48) is the modulus measured by acoustic, seismic, or resonance methods. Using the subscript $\mathcal{S}$ for adiabatic conditions,

$$K_{\mathcal{S}} = -V\left(\frac{\partial P}{\partial V}\right)_{\mathcal{S}} = -\eta\left(\frac{\partial P}{\partial \eta}\right)_{\mathcal{S}} = V\left(\frac{\partial^2 \mathcal{F}}{\partial V^2}\right)_{T\mathcal{S}}. \qquad (1.53)$$

Although $K_{\mathcal{S}}$ is larger than K_T for minerals, the two values generally agree within about one percent at room temperature. This difference is often less than the variation reported by several experimentalists for $K_{\mathcal{S}}$ on the same solid.

Hence the subscripts T and $\mathcal{S}$ can often be dropped. No distinction will be made between $K_{\mathcal{S}}$ and K_T for the majority of EoS cases described in this treatment. However, the adiabatic-isothermal distinction must be retained in some thermodynamic derivations, especially at high temperatures where the difference $(K_{\mathcal{S}} - K_T)$ is no longer small compared with K_T. The thermodynamic relationship used to transform $K_{\mathcal{S}}$ to K_T is shown in Chapter 3.

1.5.2. *The bulk modulus K at $T = 0$*

The isothermal bulk modulus defined by (1.52) is a function of T and V. By taking the volume derivative of $P = P(\eta, 0) + P_{TH}(\eta, T)$ at constant T, we have the general definition of the isothermal bulk modulus

$$K_T(\eta, T) = K_{T=0}(\eta, 0) + K_{TH}(\eta, T). \qquad (1.54)$$

K_{TH} is quite small compared with K_T and is often ignored. Because K_{TH} is a thermal correction and the thermal modes are all quiescent at absolute zero, $(\partial K_{TH}/\partial T) \to 0$ as $T \to 0$. Thus

$$K_T(\eta, 0) = K_{\mathcal{S}}(\eta, 0). \qquad (1.55)$$

The value of $K_T(\eta)$ at room temperature is about one or two percent less than $K_0(\eta)$ at absolute zero for minerals of interest to geophysicists and ceramic scientists, as is shown in Table 1.5.

Table 1.5. Zero-pressure and room temperature values of the adiabatic bulk modulus for minerals relevant to geophysics (in GPa)

Minerals	At room temperature $K_S(1,273)$	At absolute zero (extrapolated) $K_S(1,0)$
MgO	164.0	166.0
Forsterite	128.6	131.0
Corundum	252.1	259.0

Source: Sumino and Anderson, 1984.

The difference between K at room temperature and at absolute zero is about the same as the discrepancy reported by various experimentalists for K at room temperature for the same solid, just as in the comparison of adiabatic and isothermal values. For many oxides and ceramics the room temperature values of K can be used for the zero-degree values of K. Table 1.6 illustrates the range of values reported for NaCl, one of the most commonly measured solids. The data show that the difference between K at room temperature and absolute zero is often within the experimental error of the K measurement at room temperature. Thus for minerals where Θ is much higher than room temperature, one can replace $K\,(1,0)$ by $K\,(1,273)$ in the equation of state. However, this replacement would lead to serious errors were it done for more compressible solids, such as alkali metals, noble gas solids, and some metals. If Θ is much less than 300 K, as, for example, for gold, then $K\,(1,273)$ is not close to $K\,(1,0)$.

Table 1.6. Values of the isothermal bulk modulus at room temperature (GPa) for NaCl from various authors as tabulated by Chhabildas and Ruoff (1976)

Author	K_{S_0}	K_0'	Technique
Chhabildas & Ruoff	23.81	5.64	length change
Ghafelchbashi & Kolievad	23.70	5.37	ultrasonic
Barsch & Chang	23.42	5.39	ultrasonic
Spetzler et al.	23.80	5.34	ultrasonic
Bartels & Schule	23.40	5.25	ultrasonic
Vaidya & Kennedy	23.17	4.92	piston-volume
Fritz et al.	23.73	5.59	shock wave
Haussühl	23.73	—	ultrasonic
Whitfield et al.	23.25	—	Brillouin scattering
Yamamoto et al.	24.02	—	resonance (RPR)

References for authors found in Chhabildas and Ruoff, 1976.

1.5.3. *The pressure derivative of the bulk modulus K′*

The pressure derivative of the bulk modulus K' measures the change of K with P. This derivative is dimensionless. In general

$$K' = \frac{dK}{dP} = -\left(\frac{V}{K}\right)\left(\frac{dK}{dV}\right) = -\left(\frac{\eta}{K}\right)\left(\frac{dK}{d\eta}\right). \tag{1.56}$$

The isothermal definition is

$$K'_T = \left(\frac{\partial K_T}{\partial P}\right)_T = -\left(\frac{V}{K_T}\right)\left(\frac{\partial}{\partial V}\right)\left[V\left(\frac{\partial^2 A}{\partial V^2}\right)_{TT}\right]_T = -\left(1 + V\frac{A'''}{A''}\right). \tag{1.57}$$

The adiabatic definition used in this book is

$$K'_S = \left(\frac{\partial K_S}{\partial P}\right)_T. \tag{1.58}$$

There is another adiabatic definition that corresponds to the conditions in the earth's lower mantle

$$K'_{SS} = \left(\frac{\partial K_S}{\partial P}\right)_S. \tag{1.59}$$

The parameter K' can be defined adiabatically or with mixed adiabatic or isothermal conditions. As Table 1.7 illustrates, the value of K'_{T_0} (K'_T at $P = 0$) is very close to that of K'_{S_0} (K'_S at $P = 0$) for closely packed solids, so this distinction can often be ignored. Unless a particular mathematical derivation requires that the adiabatic and isothermal definitions be maintained separately, the pressure derivative of either K_T or K_S will be given simply by K'. K' at $P = 0$ will often be identified by K'_0. In general the value of K'_0 for minerals and oxides is reported to be between 3.5 and 7.0 (see Table 1.8), although there are a few notable exceptions, such as for bronzite. For dense oxides with high coordination numbers, such as Al_2O_3, K'_0 is near 4, whereas for minerals with low coordination numbers, such as α-quartz, K'_0 is higher.

Table 1.7. Measured room temperature values of K'_{T_0} and K'_{S_0} for several dense minerals, where $K'_{S_0} \equiv (\partial K_S/\partial P)_T$ at $P = 0$

Pressure Derivative $P = 0$	α-quartz	Forsterite	Spinel	Corundum
K'_{S_0}	6.4	5.37	4.18	3.98
K'_{T_0}	6.4	5.39	4.19	3.99

Source: Sumino and Anderson, 1984.

Table 1.8. Representative values of K_0' for silicates and oxides

Solid	K_0'	Solid	K_0'
FeO	4.0	Fayalite, Fe_2SiO_4	6.0
Al_2O_3	3.99	Diopside, $CaMgSi_2O_6$	6.2
Fe_2O_3	4.5	Garnet, ($Py_{22}Al_{16}$ and others)	5.5
MgO	4.2	RutileTiO_2	7.0
CaO	5.3	Spinel, $Mg_2Al_2O_4$	4.2
$Fe_{0.94}O$	4.9	Spinel, $NiFe_2O_4$	4.4
BeO	5.5	Forsterite, Mg_2SiO_4	5.4
ZnO	4.8	Olivine, $Fo_{93}Fa_7$	5.1
$SiO_2(\alpha$-Q)	6.2	Silicate Perovskite ($MgSiO_3$)	4.0
	Average of the above = 5.4		

After Sumino and Anderson, 1984. Perovskite data from Knittle and Jeanloz, 1987. Data for $Fe_{0.94}0$ from Jackson et al., 1990.

1.5.4. *The intrinsic temperature derivative of K_T*

For some solids, such as NaCl, the isochoric temperature derivative of K_T, often called the intrinsic temperature derivative of K_T, is approximately

$$\left(\frac{\partial K_T}{\partial T}\right)_V = 0. \tag{1.60}$$

This can be described also as requiring that $K_T(V = V_0)$ versus T be parallel to the T axis.

To find how K_T varies with T at constant volume in terms of measurable quantities, we use the identity from calculus

$$\left(\frac{\partial K}{\partial T}\right)_V = \left(\frac{\partial K_T}{\partial T}\right)_P + \alpha K_T \left(\frac{\alpha K_T}{\partial P}\right)_T. \tag{1.61}$$

In the case where (1.60) is valid, the above becomes

$$\left(\frac{\partial K_T}{\partial T}\right)_P = -\,\alpha K_T \left(\frac{\partial K_T}{\partial P}\right)_T. \tag{1.62}$$

$(\partial K_T/\partial P)_T$ varies very slowly, if at all, with T, so the temperature dependence of αK_T should be the same as the temperature dependence of $(\partial K_T/\partial T)_P$. I find this to be true for a few solids, which will be shown in Chapter 3.

An important dimensionless parameter is the ratio of $(\partial K_T/\partial T)_P$ to αK_T, $-\delta_T$, given by (1.47). It is a thermoelastic parameter related to the variation of α with density often called the Anderson-Grüneisen parameter (see Barron, 1979), but earlier called the second Grüneisen parameter (see

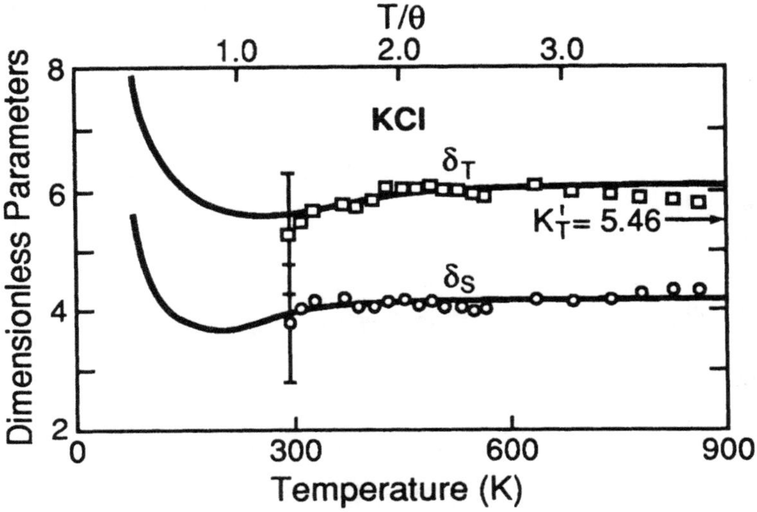

Fig. 1.1. The temperature dependences of the Anderson-Grüneisen parameters, δ_T and δ_S, for KCl. They both approach constant values at high temperatures, but at temperatures below $\Theta/2$ they take on high values (after Anderson and Yamamoto, 1987). (Data in Appendix A-7).

Gilvarry, 1956). The general expressions for this parameter have two values depending on whether one considers the adiabatic or isothermal bulk modulus. Equation (1.47) can also be expressed as

$$\delta_T = -\left(\frac{\partial \ell\text{n}\,\alpha}{\partial \ell\text{n}\,\rho}\right)_T . \tag{1.63}$$

Equation (1.47) is defined by an isobaric condition, whereas (1.63), $x = T$, is defined by an isothermal condition. Data exist for (1.63) for $P = 0$, and it is of some interest to note that δ_T turns out to be virtually independent of temperature for several minerals at high temperatures ($T > \Theta$) (Anderson et al., 1992a). Examples showing the variation of δ_T and δ_S for the cases of KCl and Mg_2SiO_4 are shown in Figs. 1.1 and 1.2. Equation (1.63) and its alternate (1.49) are regarded as the dimensionless forms of $-(\partial K_T/\partial T)_P$. We shall see later that (1.60) is satisfied when $\delta_T = K_0'$. In this case (1.62) is also satisfied. This case is found for NaCl, but is not general.

A related dimensionless parameter arising from the ratio of $(\partial\alpha/\partial T)_P$ to α^2 is

$$\hat{\alpha} = \left(\frac{1}{\alpha^2}\right)\left(\frac{\partial \alpha}{\partial T}\right)_P . \tag{1.64}$$

It arises in derivatives of a function with respect to T when α is involved in the function, as, for example, we shall see later for calculation of γ.

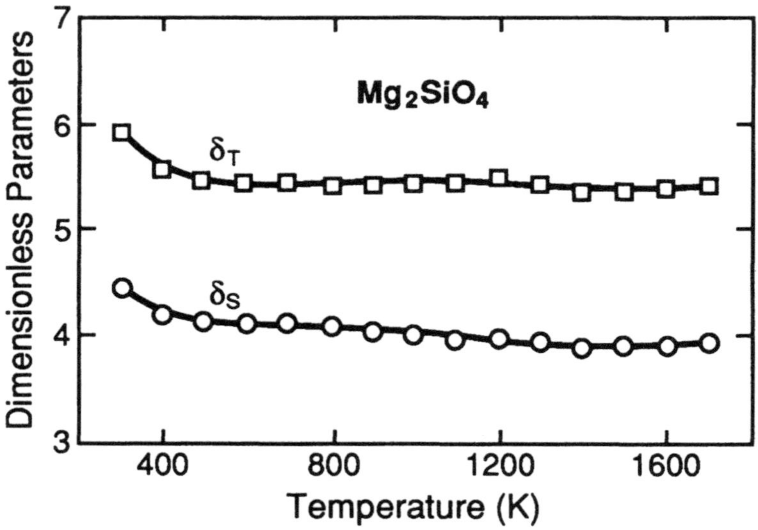

Fig. 1.2. The temperature dependences of δ_T and δ_S for Mg_2SiO_4. Above the Debye temperature, $\Theta = 762$ K, δ_T and δ_S appear to be independent of T (after Isaak et al., 1989a). (Data in Appendix A–7).

1.6. The Debye temperature Θ: A lower bound for high *T*

It is important that the region of high temperature be quantitatively defined and bounded. For this purpose we need a brief discussion of two equations in the Debye theory (a more general discussion of Debye theory will be found in Chapter 2). The Debye temperature is defined in terms of sound velocities by Section 2.8.1

$$\Theta = \frac{\hbar}{k}\left(\frac{9N}{4\pi}\right)^{\frac{1}{3}}\left(\frac{\rho}{\mathbf{M}/p}\right)^{\frac{1}{3}} v_m \;=\; 251.2\left(\frac{\rho}{\mu}\right)^{\frac{1}{3}} v_m. \qquad (1.65)$$

The right side of (1.65) is evaluated where v_m is in km/s and ρ is in g/cc. **M** is the molecular mass; p is the number of atoms in the molecular formula; μ is the mean atomic mass; and k, $\hbar$, and N are the usual atomic constants. v_m is the mean sound velocity given by

$$\frac{3}{v_m^3} = \frac{2}{v_s^3} + \frac{1}{v_p^3}. \qquad (1.66)$$

High temperature is defined here as regions where $T > \Theta$.

We note that seismic data (v_p, v_s, ρ), coupled with an estimate of the mean atomic mass, are sufficient to define Θ for the earth's interior. Values of Θ as computed by (1.65) for a number of solids are shown in Table 1.9. Note that for iron-free silicates, μ is close to 20.5; with a small concentration of iron in silicates, this raises to about 21 (Watt et al., 1975).

Table 1.9. Debye temperature Θ_{ac} estimated from sound velocity values at room T and $P = 0$. Units for Θ_{ac}, °K; for ρ, g/cc

Solid	ρ_0	μ	v_p	v_s	v_m	Θ_{ac}
H_20 (ice)	0.92	6.01	3.76	1.99	2.16	299
NaCl	2.16	29.22	4.63	2.70	2.99	316
KCl	1.99	37.28	3.92	2.26	2.51	228
RbCl	2.82	60.46	3.07	1.65	1.76	168
LiF	2.64	12.97	7.20	4.29	4.48	702
AgI	5.69	11.74	2.29	0.88	0.99	91
BeO	3.01	12.51	12.16	7.41	7.70	1280
MgO	3.58	20.15	9.71	6.05	6.67	942
CaO	3.34	28.04	8.12	4.92	5.44	673
$Fe_{0.92}0$	5.84	35.09	6.54	2.85	3.22	440
ZnO	5.67	15.84	6.00	2.86	3.22	419
NiO	6.79	37.35	6.58	3.65	4.07	579
Al_2O_3	3.99	20.39	10.85	4.40	4.98	1035
Fe_2O_3	5.27	31.94	7.91	4.16	4.65	641
SiO_2 (α-Q)	2.65	20.3	6.05	4.10	4.47	572
SiO_2 (coesite)	2.92	20.3	8.20	4.60	4.87	676
SiO_2 (stishovite)	4.28	20.3	11.00	5.50	6.17	929
TiO_2	4.23	26.63	9.26	5.13	5.71	781
GeO_2 (rutile)	6.28	34.86	8.56	4.90	5.44	740
$CaCO_3$	3.21	20.02	6.57	3.42	3.83	495
Mg_2SiO_4	3.21	20.10	8.58	5.02	5.57	760
$(Mg_{.92}, Fe_{.08})_2SiO_4$	3.31	20.79	8.42	4.89	5.43	739
Fe_2SiO_4	3.98	29.11	6.84	3.40	3.82	511
pyrope garnet	3.31	20.16	9.12	5.02	5.60	788
$MgSiO_3$ (perovskite)	4.21	20.12	10.94	6.69	7.38	1094
Fe (bcc phase)	7.87	55.85	5.90	3.25	3.62	473
W	183.9	183.9	5.20	2.86	3.18	383

Most v_s and v_p data from Sumino and Anderson, 1984. v_s, v_p, and v_m in km/s. Data for perovskite from Yeganeh-Haeri et al., 1989.

From Table 1.9, we see that v_m is about 1.1 v_s, and therefore Θ_{ac} is controlled mostly by v_s. The subscript on Θ is needed because Θ can also be found from the density of states and the specific heat (see Chapter 5).

The value of Θ_{ac} decreases with T as the velocities of sound decrease linearly with T (Anderson et al., 1992a). Hence the change of Θ with T is nearly linear. Values of Θ_{ac} versus T so obtained for four solids are presented in Fig. 1.3. Extrapolation of Θ_{ac} to high T from the 300° value for perovskite gives $\Theta \approx 990$ at 1800 K. This extrapolation is accomplished by making the perovskite line parallel to the corundum line (see Fig. 1.3).

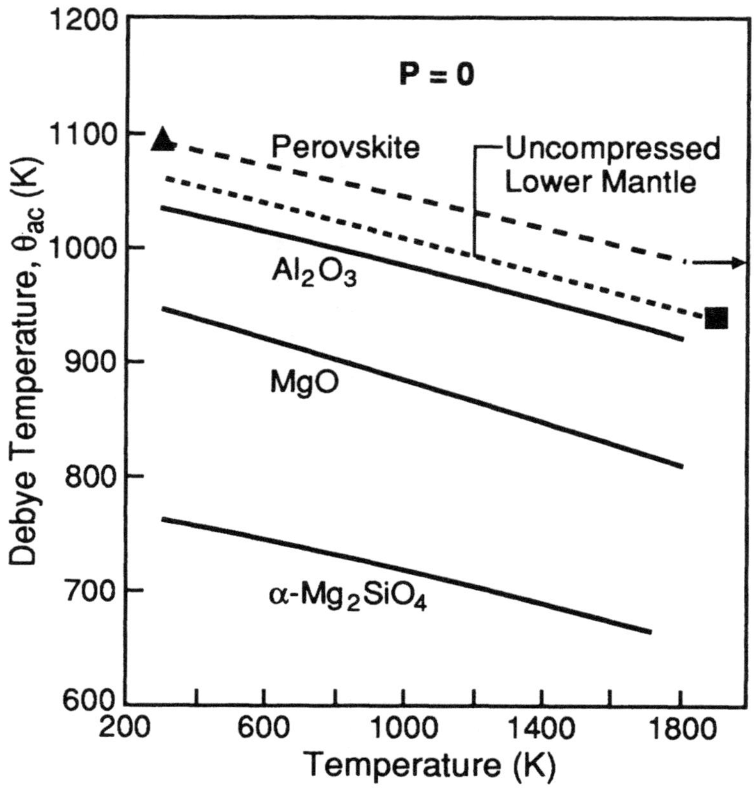

Fig. 1.3. Variation of Θ_{ac} with T for four oxides and the earth's lower mantle. Solid lines represent measurements for three minerals; the triangle on the 300 K axis represents the ambient measurement for silicate perovskite (1094 K) (Table 1.9); the square represents the hot uncompressed lower mantle value at zero depth (940 K) (see Table 1.9). Dashed lines are parallel to the Al_2O_3 curve.

1.7. The Debye temperature of the earth and the moon

The earth's Debye temperature can be computed from its seismic data (1.65) and (1.66), if an assumption is made about the mean atomic weight, μ. The value of μ for minerals important to the earth's mantle does not vary much from $\mu = 21$ (Watt et al., 1975).

Table 1.10 shows Θ for the earth's lower mantle using $\mu = 21$. For the solid inner core, the value of the atomic weight of iron can be used for μ. It is seen from Table 1.10 that Θ steadily increases with depth in much the same way that temperature increases with depth. At all depths in the lower mantle, the value of Θ is slightly more than $1/2T$. This means that the lower mantle is in the high temperature regime at all depths.

Table 1.10. Computation of Θ and T/Θ in the earth's lower mantle

z	v_p	v_s	ρ	T	v_m	Θ	T/Θ	P	P/K_{T_0}
(km)	$\left(\frac{\text{km}}{\text{s}}\right)$	$\left(\frac{\text{km}}{\text{s}}\right)$	$\left(\frac{\text{g}}{\text{cc}}\right)$	(K)	$\left(\frac{\text{km}}{\text{s}}\right)$	(K)		(GPa)	
771	11.096	6.193	4.437	1908	6.887	1031	1.85	23.9	0.19
971	11.415	6.338	4.553	1973	7.054	1066	1.85	37.3	0.29
1171	11.716	6.470	4.668	2034	7.207	1098	1.85	46.5	0.36
1371	12.002	6.590	4.779	2092	7.348	1128	1.85	55.9	0.44
1571	12.271	6.700	4.889	2147	7.468	1155	1.86	65.5	0.51
1771	12.527	6.801	4.996	2199	7.587	1182	1.86	75.3	0.59
1971	12.769	6.893	5.101	2249	7.689	1205	1.87	85.4	0.67
2171	12.998	6.978	5.203	2296	7.794	1231	1.87	95.7	0.75
2371	13.217	7.058	5.303	2341	7.885	1253	1.87	106.3	0.83
2571	13.424	7.133	5.401	2384	7.971	1274	1.87	117.3	0.92
2771	13.622	7.204	5.496	2426	8.048	1294	1.87	128.8	1.01
2871	13.718	7.238	5.543	2449	8.090	1305	1.88	135.4	1.06

Data for depth, z, v_p, v_s, ρ, and P from Dziewonski et al., 1975; data for T from Brown and Shankland, 1981. Θ at $z = 0$ is 940 K (hot) and 1040 K (ambient) (see Fig. 1.3).

The sound velocities of the lunar interior have been measured down to about 900 km (Nakamura et al., 1974). The values recommended by Toksöz (1979) are $v_s = 4.5$ km/s and $v_p = 7.7$ km/s. Assuming $\mu = 22$ and ρ = constant = 3.4 g/cc (Baumgardner and Anderson, 1981), Θ for the lunar mantle was calculated by using (1.65). This calculation results in Θ = constant = 680 K, a constant value from 60 to 400 km, and Θ = constant = 630 K from 400 to 1000 km (see Table 1.11). We see that except for the lunar crust, the interior of the moon is in the high temperature regime, $T > \Theta$.

Table 1.11. P/K_0 and T/Θ for the lunar mantle; $K_0 = 11.2$ GPa

Depth (km)	Pressure (GPa)	Temperature (°C)	Θ	P/K_0	T/Θ
20	0.9	62	680	.008	0.493
70	3.2	153	680	.029	0.645
400	19.2	601	680	.171	1.28
638	28.6	881	630	.255	1.83
838	35.4	1094	630	.316	2.17
1000	40.1	1260	630	.358	2.43

Data from Baumgardner and Anderson, 1981.

1.8. γ_D: the Debye approximation to γ

Evaluating (1.37), where Θ is given by (1.65), we have

$$\gamma_D = \left(\frac{d\,\ell\text{n}\,\Theta}{d\,\ell\text{n}\,\rho}\right) = \frac{1}{3} + \left(\frac{d\,\ell\text{n}\,v_m}{d\,\ell\text{n}\,\rho}\right) \tag{1.67}$$

or

$$\gamma_D = \frac{1}{3} - \frac{1}{3}\left[\frac{d\,\ell\text{n}\left(\frac{1}{v_p^3} + \frac{2}{v_s^3}\right)}{d\,\ell\text{n}\,\rho}\right]. \tag{1.68}$$

Now the relationship between the sound velocities and the bulk modulus K and Poisson's ratio ν and the shear modulus G is given by

$$v_s^2 = \left(\frac{3K}{2\rho}\right)\left(\frac{1-2\nu}{1+\nu}\right) = \left(\frac{G}{\rho}\right) \tag{1.69}$$

$$v_p^2 = \left(\frac{3K}{\rho}\right)\left(\frac{1-\nu}{1+\nu}\right). \tag{1.70}$$

Thus in terms of the elastic moduli, G, K, and ν

$$\gamma_D = -\frac{1}{6} + \frac{1}{2}\left(\frac{dK}{dP}\right) - g(\nu)\left(\frac{dK}{dP}\right) + \left(\frac{2g}{3}\right)(1+\nu)\left(\frac{dG}{dP}\right), \tag{1.71}$$

$$g(\nu) = \left(\frac{1}{(3-3\nu)^{5/2}} + \frac{3.2^{3/2}}{(3-6\nu)^{5/2}}\right)\left(\frac{1}{(3-3\nu)^{3/2}} + \frac{2^{5/2}}{(3-6\nu)^{3/2}}\right)^{-1}, \tag{1.72}$$

where G is the shear modulus. For $\nu = 1/4$, for example, we have from (1.71)

$$\gamma_D = -\frac{1}{6} + 0.024\left(\frac{dK}{dP}\right) + 0.799\left(\frac{dG}{dP}\right). \tag{1.73}$$

As Knopoff and Shapiro (1969) pointed out, the contribution of the dK/dP term is small compared with the contribution of the dG/dP term. This means that the relative contribution of the transverse vibrations to γ must be large. This is seen more explicitly below.

Anderson (1965a) showed that little error is made in the computation of γ_D if v_m is replaced by v_s. This is seen in Table 1.9, which shows that v_m is about 1.1 v_s. He took as an approximation the Cauchy solid

$$v_s = \left(\frac{16}{17}\right) v_m, \tag{1.74}$$

where $\nu = 0.25$. Neglecting the contribution of $0.024 dK/dP$ compared with $0.799\, dG/dP$ in (1.71), we have the approximation

$$\gamma_D \cong \frac{1}{3} + \left(\frac{K_S}{v_s}\right)\left(\frac{dv_s}{d\rho}\right) \tag{1.75}$$

or

$$\gamma_D \cong -\frac{1}{6} + \frac{1}{2}\left(\frac{K_S}{G}\right)\left(\frac{dG}{dP}\right).$$

The ratio K_S/G can be expressed in terms of ν, so that

$$\gamma_D \cong -\frac{1}{6} + \left(\frac{1-\nu}{3(1-2\nu)}\right)\left(\frac{dG}{dP}\right). \tag{1.76}$$

For $\nu = 1/4$, we have $\gamma_D \cong -1/6 + .885\, dG/dP$ to be compared with (1.71). Equation (1.76) is far less cumbersome than (1.71) and gives close to the same result. In other words, a fair approximation to γ_D can be made by ignoring all but the shear vibrational modes. The example of the Debye γ is emphasized here to demonstrate that it may be dangerous to rely on approximations to γ that are functions of dK/dP only, unless dK/dP is proportional to dG/dP and stays so as P and T change.

1.9. Electronic heat capacity contribution to γ for iron

This book emphasizes the properties of insulator-type solids. However, geophysicists are greatly interested in the properties of iron at high temperatures because iron is the dominant solid in the earth's core.

Metals are distinguished from insulators by their large electrical and thermal conductivity. An early idea about the conduction of electrons was to assume that electrons act as a perfect gas. However, band theory clearly shows that electrons cannot be considered as a gas because of electron–electron interaction and the Pauli exclusion principle. This principle has a considerable effect on the thermal behavior of the conduction electrons. In fact, the electronic specific heat C_{V_e} is proportional to T at low T (compared to T^3 for insulators) and also proportional to T (compared to a constant) at high T. The coefficient of T for C_{V_e} is small, so that the effect of C_{V_e} on the total C_V is important only at very low T and very high T for metals.

In the elementary theory (Kittel, 1956), the number of electrons excited by the conduction of heat is limited to the high energy end of the Fermi-Dirac distribution, which is the model of the electronic density of states $g(\mathcal{U}_e)$, where $\mathcal{U}_e$ is the electronic internal energy. Kittel showed that

$$C_{V_e} = \frac{\partial \mathcal{U}_e}{\partial T} \approx \left(\frac{T}{T_f}\right), \tag{1.77}$$

where T_f is the Fermi temperature, which defines the cutoff for $g(\mathcal{U}_e)$. (Since T_f is of the order of 10^4 K, for ordinary temperatures C_{V_e} contributes a small fraction of the total C_V (about 10^{-3} K)). The situation is more complicated for transition metals, however, and (1.77) is not justified.

Bukowinski (1977) calculated the electronic density of states of face-centered cubic (fcc) iron and from first principles found that $g_e(\mathcal{U}_e)$ did not terminate at the boundary of the Fermi-Dirac distribution: there were broad peaks on both sides of $\mathcal{U}_f$, the Fermi energy limit. Bukowinski's calculation of $d\mathcal{U}_e/dT$ led him to conclude that C_{V_e} of iron is

$$C_{V_e} = 3.5 \times 10^{-3} N_e kT, \tag{1.78}$$

e.g., C_{V_e} increases linearly with T. Jamieson et al. (1978) pointed out that at temperatures much larger than 6000 K, C_{V_e}, being linear in T, could exceed the lattice specific heat, and they suggested that C_{V_e}/R is in the range of 1–2 at core temperatures. Boness et al. (1986) found that for hexagonal closely packed (hcp) iron, the density of states had broad peaks on both sides of the Fermi energy limit, just as for fcc iron. They calculated $d\mathcal{U}_e/dT$ and found $C_{V_e} = 1.7 \pm 0.7R$ for core conditions.

To find γ_e, Boness et al. (1986) showed that (1.12) is replaced by

$$\gamma_e = \frac{\partial \ \ell\text{n } g(E_F)}{\partial \ \ell\text{n } \rho}. \tag{1.79}$$

Bonness et al. (1986) found $\gamma_e = 1.27 \pm 0.03R$ for fcc iron and $1.34 \pm 0.02R$ for hcp iron. But, considering the uncertainty of whether the core is hcp or fcc, they reported $\gamma_e = 1.3 \pm 0.7$. Bukowinski's (1977) estimate for fcc iron is $\gamma_e = 1.5 \pm 0.1$ at core conditions. Jeanloz (1979) set the value of $\gamma_e = 1.5 \pm 0.3$ for iron at core conditions. Jamieson et al. (1978) found $1.2 < \gamma_e < 1.8$. The value of the lattice γ for iron is nearly the same. The discussion of how γ_e of iron is related to the γ of the inner core and the measurement of γ of iron for inner core conditions are given in Section 10.8.

1.10. Problems

1. Find γ for an Einstein solid, γ_E (assume that all values of ω are equal, and similarly for $d\omega/dv$, and use a single spring constant).
2. Find γ_E in terms of the potential, $\phi = aw^{-1/3} + bw^{-n/3}$, where $w = \rho/\rho_0$.
3. We shall later find that for the Born-Mie potential, the n in Problem 2 is $3K_0' - 7$. What is γ_E for low pressure and also for very high pressure, in terms of K_0'?
4. Derive (1.36) from (1.31). Hint: Use the condition that at low wave numbers, $\mathbf{k}$, the acoustic frequency, is proportional to k, according to $\omega_i = v_i\mathbf{k} (i = 1, 2, 3)$.

2

STATISTICAL MECHANICS AND THE QUASIHARMONIC THEORY

2.1. Introduction

In a solid the vibration amplitudes of the atoms about their equilibrium position are small, and as a consequence, the quantity kT, representing the thermal energy of the vibrating atom, is small compared with the interaction energy of adjoining atoms, excepting as T approaches the melting point T_m.

If N is the number of molecules (or basic cells) in a solid, and p is the number of atoms per cell, then the total number of degrees of freedom is $3pN$. Now 6 of these are for translation and rotation of the whole body, but as N is a huge number (Avogadro's number for one mole), an infinitesimal error is made by representing the vibrational degrees of freedom by $3pN$. Neglecting the rotational modes is not serious for dense minerals and oxides, but could be serious for less dense minerals like $CaCO_2$.

If we assume the solid to be an insulator, then we do not have to account for the electron degrees of freedom. They are important for metals and for some insulators in which free electron motion necessitates some corrections. But for most discussions we shall assume that we deal only with the lattice part of the thermodynamic quantities of an insulator. However, we shall comment occasionally on calculations for two metals: iron and gold. Iron is important to geophysics because it is the main constituent of the earth's core. Gold is important to high pressure experimental physics because it is often used as a pressure calibrant (Jamieson et al., 1982; Anderson et al., 1989).

In statistical mechanics of solids, it is customary to treat the $3pN$ vibrational degrees of freedom as an ensemble of $3pN$ independent oscillators. That is, each of the $3pN$ oscillators is treated as though it had a separate mode of vibration, where each mode has an energy determined by the modal frequency ω_i. We thus need to have an expression for the energy of vibrational freedom of one oscillator and add these up for all $3pN$ modes.

In this approximation there is no mode interaction, and we need only the total sum of all the vibrational energies. In the classical case the vibrational energy will be equivalent to the vibrational energy of a diatomic gas. The statistical mechanical derivations found in Landau and Lifshitz (1958) are used as models for the derivations found throughout this chapter.

2.2. The vibrational energy and the thermal energy

The derivation of the vibrational energy of a diatomic gas is often found in the literature of physical chemistry. Taking ω as the vibrational frequency, the contribution to the Helmholtz energy due to the diatomic gas vibrational energy for N molecules is (Landau and Lifshitz, 1958)

$$\mathcal{F}_{VIB}^{dg} = NkT\ \ell\text{n}\left(1 - e^{-\hbar\omega/kT}\right),$$

where $\hbar$ is Planck's constant and k is Boltzmann's constant. There will be a term similar to the above equation for each of the $3pN$ modes in our solid, and we must account for the possibility that each of the modes has a different frequency, ω_j. Thus the vibrational energy of the insulating solid (no free electrons) is (Einstein, 1911)

$$\mathcal{F}_{VIB}^{\mathbf{E}} = kT\sum_{j=1}^{3pN} \ell\text{n}\left(1 - e^{-\hbar\omega_j/kT}\right). \tag{2.1}$$

If a quantum mechanical treatment is made, a small term, called the zero point energy, must be added to (2.1).

According to statistical mechanics, a system's quantum mechanical energy levels completely determine the thermodynamic functions through the expression

$$\mathcal{F}_{VIB} = -kT\ \ell\text{n}\ \mathcal{Z}, \tag{2.2}$$

where $\mathcal{Z}$ is the partition function, which is found by taking the sum of all the quantum energy levels. For a collection of independent oscillators, the sum of the energy levels for each mode is

$$\mathcal{Z} = \sum_i e^{-\epsilon_i/kT}, \tag{2.3}$$

where ϵ_i are the eigenvalues of the energy operator, given by

$$\tfrac{1}{2}\hbar\omega_i,\ \tfrac{3}{2}\hbar\omega_i,\ \tfrac{5}{2}\hbar\omega_i,\ \text{etc.}\ldots,$$

so that for each mode, there are many energy levels.

$$\mathcal{Z}_i = e^{-\hbar\omega_i/2kT}\sum_{i=0}^{\infty} e^{-3\hbar\omega_i/kT} = \frac{e^{-\hbar\omega_i/2kT}}{\left(1 - e^{-\hbar\omega_i/kT}\right)}.$$

Defining $\mathcal{F}_{VIB_i}$ in terms of $\mathcal{Z}_i$ by (2.2)

$$\mathcal{F}_{VIB_i} = \hbar\omega_i/2kT\ \ell\text{n}\left(1 - e^{\hbar\omega_i/kT}\right). \tag{2.4}$$

Summing over all modes we find, where each mode is in the form given above,

$$\mathcal{F}_{VIB} = \frac{1}{2}\sum_{j=1}^{3pN} \hbar\omega_j + kT\sum_{j=1}^{3pN} \ell\text{n}\left(1 - e^{-\hbar\omega_j/kT}\right). \tag{2.5}$$

Comparing (2.5) with (2.1), we see that there is an extra term in (2.5), called the zero point energy,

$$E_{ZV} = \frac{1}{2}\sum_{j=1}^{3pN} \hbar\omega_j. \tag{2.6}$$

Note that, unlike the last term in (2.5), there is no T in E_{ZV}. This term arose by considering all the quantum state energy levels of an oscillator.

Replacing $\hbar\omega/kT$ with y, the modal definition is

$$y_j = \hbar\omega_j/kT, \tag{2.7}$$

and the expression for (2.5) becomes

$$\mathcal{F}_{VIB} = E_{ZV} + \mathcal{F}_{TH}, \tag{2.8}$$

where

$$\mathcal{F}_{TH} = kT\sum_{j=1}^{3pN} \ell\text{n}\left(1 - e^{-y_j}\right) = kT\sum_{j=1}^{3pN} \mathcal{F}_{TH_j}. \tag{2.9}$$

$\mathcal{F}_{TH}$ is the energy arising from temperature excitation, called the thermal energy. E_{ZV} is not affected by T, as shown in (2.6). Thus $\mathcal{F}_{TH} \to 0$ as $T \to 0$. However, $\mathcal{F}_{VIB} \to E_{ZV}$ as $T \to 0$, and E_{ZV} is a non-zero number. E_{ZV} is sufficiently small that for most numerical evaluations it could be dropped, but it is useful to keep this term in $\mathcal{F}_{VIB}$ for algebraic manipulations done later on.

The expression for the Helmholtz energy for an insulator is thus

$$\mathcal{F} = E_{ST} + E_{ZV} + \mathcal{F}_{TH}, \tag{2.10}$$

where E_{ST} is the potential of the static lattice at $T = 0$. To divide (2.10) into temperature-dependent and nontemperature-dependent parts, we use

$$\mathcal{F} = E_{T=0} + \mathcal{F}_{TH}, \tag{2.11}$$

where

$$E_{T=0} = E_{ST} + E_{ZV}. \tag{2.12}$$

To divide (2.10) into vibrational and nonvibrational parts, we use

$$\mathcal{F} = E_{ST} + \mathcal{F}_{VIB}, \tag{2.13}$$

where $\mathcal{F}_{VIB}$ is given by (2.8).

2.3. The quasiharmonic approximation

Before we can find thermodynamic properties such as P and C_V from (2.11), we must make decisions regarding the volume and temperature behavior of ω_j. In the quasiharmonic approximation, ω is assumed to be dependent on V but not on T. This makes all the thermodynamic properties directly dependent on V. The temperature behavior of the thermodynamic properties comes from the fact that although ω_i is not dependent on T, the sum $kT \sum \ell\text{n}\ (1 - e^{-y_j})$ depends on T. When ω_j are different, then y_i are also different, and the sum above becomes T dependent, especially at low T.

The internal energy $\mathcal{U}$ is found by applying (1.7) to (2.10), whereupon

$$\mathcal{U} = E_{T=0} + kT \sum_{j=1} \frac{y_i}{(e^{y_i} - 1)} \tag{2.14}$$

$$= E_{T=0} + kT \sum_{j=1} \mathrm{E}_{TH_j}. \tag{2.15}$$

The pressure P is found by applying (1.9) to (2.11), yielding

$$P = -\left(\frac{\partial E_{ST}}{\partial V}\right)_T + \frac{kT}{2V} \sum_j \gamma_j y_j + \frac{kT}{V} \sum_j \gamma_j \frac{y_i}{e^{y_j} - 1}, \tag{2.16}$$

where

$$\gamma_j = \frac{\partial\ \ell\text{n}\ \omega_j}{\partial\ \ell\text{n}\ V}. \tag{2.17}$$

Equation (2.17) defines the mode gammas found in Chapter 1 (1.31). Equation (2.16) is rewritten as

$$P = P_0 + P_{TH}, \tag{2.18}$$

where the thermal energy (2.9) effect on the pressure, called the thermal pressure, becomes

$$P_{TH} = k\frac{T}{V} \sum \gamma_j \frac{y_i}{(e^{y_j} - 1)}. \tag{2.19}$$

Comparing (2.19) with (2.15), we see that

$$P_{TH} = \frac{1}{V} \sum \gamma_j \mathcal{F}_{TH_j}. \tag{2.20}$$

The specific heat is found by applying (1.8) to (2.14), yielding

$$C_V = k \sum_j \frac{y_j^2 e^{y_j}}{(e^{y_j} - 1)^2} = \sum C_{V_j}. \tag{2.21}$$

The expression for entropy is found by applying (1.6) to (2.10), yielding

$$\mathcal{S} = k\sum y_i\left\{\frac{e^{y_j}}{(e^{y_j}-1)}\right\} - k\sum \ell\text{n}\left[1-e^{-y_i}\right]. \tag{2.22}$$

The expression for αK_T is found by applying (1.4) to (2.10), yielding

$$\alpha K_T = \frac{\hbar}{V}\sum y_j^2\left\{\gamma_j\frac{e^{y_j}}{(e^{y_j}-1)^2}\right\}. \tag{2.23}$$

To get α we divide (2.23) by K_T, found by $K_T = -V(\partial P/\partial V)_T$, where P is given by (2.16).

Using (1.23), showing us that $\alpha K_T = \gamma C_V/V$, where C_V is per unit mass, and equating this to (2.23), we find that

$$\overline{\gamma} = \frac{\sum_j \gamma_i C_{V_j}}{C_V} \tag{2.24}$$

(Barron, 1955).

The bar over γ indicates that this result is an approximation to γ resulting from invoking the quasiharmonic approximation to the Helmholtz energy. Equation (2.24) appears in Chapter 1 as (1.33).

2.4. The Mie-Grüneisen equation of state

If we take γ_j out of the sum in (2.20), we must assume that all γ_j are equal to each other and to γ. In this case

$$P_{TH} = \frac{\gamma_{mg}}{V}\mathcal{F}_{TH}, \tag{2.25}$$

which is known as the Mie-Grüneisen expression for thermal pressure, and γ_{mg} stands for the Mie-Grüneisen EoS Grüneisen parameter. The Mie-Grüneisen EoS (Grüneisen, 1912) was written as

$$PV = -VE_0'(V) + \gamma_{mg}E_{TH}, \tag{2.26}$$

where $E_0'(V) = \partial E_{ST}/\partial V$.

Using E_{TH} as defined by (2.15), we see that

$$\gamma_{mg} = \frac{\sum_{j=1}\left\{\gamma_j\left[y_j/(e^{y_j}-1)\right]\right\}}{\sum_{j=1}\left[y_j/(e^{y_j}-1)\right]} = \frac{\sum \gamma_j \mathcal{F}_{TH_j}}{\mathcal{F}_{TH}}, \tag{2.27}$$

which is different from (2.24), unless all γ_j are equal.

Thus, although $\gamma_{mg} \to \overline{\gamma}$ at very high T, at intermediate temperatures they are not the same. Thus γ_{mg} is an approximation to $\overline{\gamma}$ (Barron, 1955).

2.5. The high–temperature limit of the quasiharmonic approximation

2.5.1. *The Helmholtz energy*

At sufficiently high temperatures, the expression for $\mathcal{F}_{TH}$ can be simplified due to the convergence of certain series expansions. The argument in (2.5) becomes small since $\hbar\omega_j << kT$. At high temperatures, we take advantage of the expansion (Zharkov and Kalinin, 1971),

$$\ell n\ \left(1 - e^{-y_j}\right) = \ell n\ y_j - \left(\frac{1}{2}y_j^2 + \ldots\right) = \ell n\ y_j + \ell n\ \left(1 - \frac{1}{2}y_j\right). \quad (2.28)$$

At high T, where y_j is small, replace $\ell n\ (1 - (1/2)y_j)$ with $-y_i/2$. Then

$$\ell n\ \left(1 - e^{-y_j}\right) \simeq \ell n\ y_j - \frac{1}{2}y_j. \quad (2.29)$$

Thus, the high-temperature representation of (2.9) is

$$\mathcal{F}_{TH}^{ht} = kT\sum_{j}^{3pN}\left(\ell n\ y_j - \frac{1}{2}y_j\right), \quad (2.30)$$

and by using (1.2) and (1.12), the high T Helmholtz energy for insulators is

$$\mathcal{F}^{ht} = E_{ST} + E_{ZV} + kT\left[\sum \ell n\ y_j - \frac{1}{2}\sum y_j\right]. \quad (2.31)$$

We see that the last term above just cancels the zero point energy given by (2.6), so that

$$\mathcal{F}^{ht} = E_{ST} + kT\sum_{j=1}^{3pN}(\ \ell n\ \hbar\omega_j\ -\ \ell n\ kT). \quad (2.32)$$

This can be simplified by defining an average frequency $\overline{\omega}$, given by

$$\ell n\,\overline{\omega} = 1/3pN\sum \ell n\ \omega_j, \quad (2.33)$$

in which case (2.32) becomes

$$\mathcal{F}_{TH} = 3pNkT\,[\ \ell n\ \hbar\overline{\omega} - \ \ell n\ kT]. \quad (2.34)$$

2.5.2. *The specific heat*

Thus at high T, $\mathcal{F}$ includes only the static potential and the thermal energy in the quasiharmonic approximation. Applying (1.7) to (2.32), we find the high temperature internal energy to be

$$\mathcal{U}^{ht} = E_{ST} + 3p\mathbf{R}T, \quad (2.35)$$

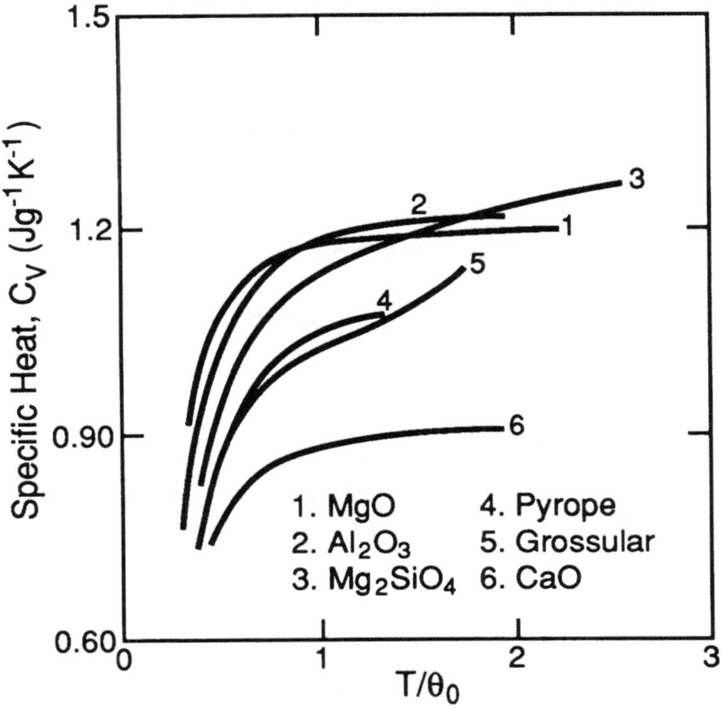

Fig. 2.1. C_V versus T/Θ_0 for 6 oxides and silicates (after Anderson et al., 1991). Note that for Mg_2SiO_4 and two garnets, C_V rises with T/Θ_0 at high T, indicating the onset of anharmonicity effects. (Data in Table A-7).

where **R** is the gas constant, kN. Applying (1.8) to (2.35),

$$C_V^{ht} = 3p\mathbf{R}. \tag{2.36}$$

Equation (2.36), known as the classical Dulong and Petit limit, is the common test for anharmonicity in a solid.

In Fig. 2.1, we plot the data for specific heat versus T for several oxides and silicates. We see that $\partial C_V/\partial T = 0$ at high T for MgO and CaO, where C_V is determined from the measured C_P by $C_P = C_V(1+\alpha\gamma T)$. From Fig. 2.1, we see that olivine and the two garnets apparently violate (2.36) and apparently reflect anharmonicity in $\mathcal{F}$. However, C_V can be determined from Raman and infrared spectra, along with $g(\omega)$, (1.51). Calculating C_V this way led Chopelas (1990b) to deduce that forsterite was not anharmonic. Similarly, Hofmeister and Chopelas (1991) found pyrope and grossular not to be anharmonic. Thus the evidence for anharmonicity of these three solids is marginal. If the calculated C_V from C_P is incorrect, the error probably arises from an inadequate determination of α.

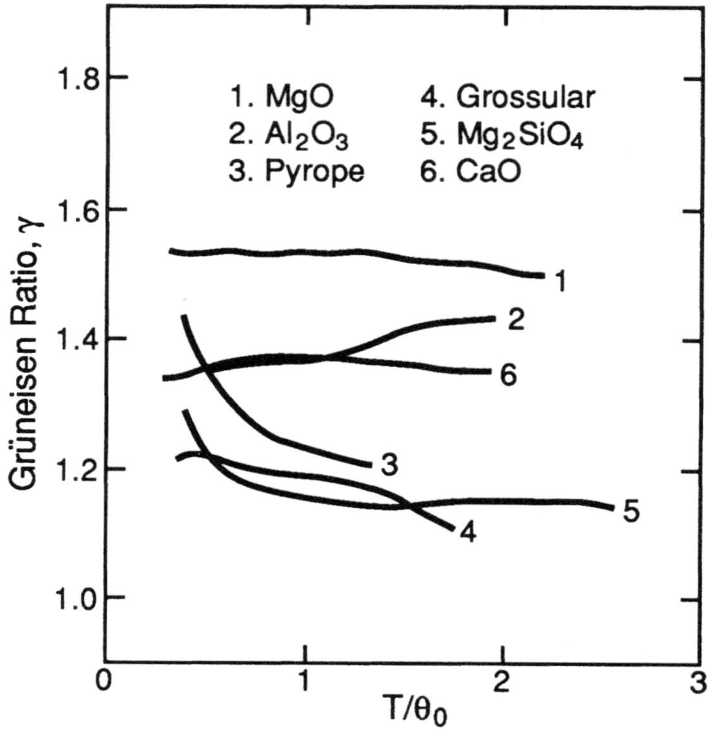

Fig. 2.2. γ versus T/Θ_0 for six minerals. Note that for the same minerals in which γ decreases slightly at high T (grossular, pyrope), C_V increases with T at high T (after Anderson et al., 1991), as shown in Fig. 2.1. (Data in Table A-7).

2.5.3. *The Grüneisen constant* $\overline{\gamma}$

The high temperature Grüneisen parameter is found from (2.25), and considering that as $T \to \infty$, $C_i \to k$,

$$\gamma^{ht} = \overline{\gamma}_{(T\to\infty)} = \frac{1}{3pN}\sum_{j}^{3pN} \gamma_j, \tag{2.37}$$

which, according to theory, is independent of T, as pointed out in Chapter 1 (1.35).

A plot of γ versus T/Θ_0 for several minerals is shown in Fig. 2.2. There is an approximate trend for γ to be independent of T at high T. Where $d\gamma/dT$ is not zero, we see the effect of C_V on γ according to (1.24); that is, departures from quasiharmonic theory in C_V create a departure in the quasiharmonic theory of γ, although by a smaller amount. However, these trends in γ for forsterite, pyrope, and grossular may result from false indications of anharmonicity in C_V (see Section 2.5.1).

2.5.4. *The high temperature thermal pressure*

Applying (1.27) to (2.31), we find the high T pressure, which by using (2.36) is

$$P^{ht}(T,V) = P(0,V) + \frac{3p\mathbf{R}}{V}\gamma^{ht}\left(T - \frac{3}{8}\Theta\right), \tag{2.38}$$

where the $(3/8)\Theta$ term accounts for the approach of P_{TH} to zero at absolute zero. Thus at high T, the thermal pressure is close to the value

$$P_{TH} = \frac{3p\mathbf{R}}{V}\gamma^{ht}T. \tag{2.39}$$

V is a function of T, and it is also equal to (M/p) where M is the molecular weight, so that the above is

$$P_{TH} = \frac{3\mathbf{R}\gamma^{ht}}{(M/p)}\rho_0(1-\alpha T), T \tag{2.40}$$

neglecting the terms independent of T.

We see that the quasiharmonic theory allows for P_{TH} to be virtually linear in T: at the highest measured T, $\alpha T \simeq 0.05$. But γ^{ht} is not, as we shall see, exactly independent of T, although close to it. Thus there are counter effects in T, and the $1-\alpha T$ term is often obscured. Thus to a very good approximation P_{TH} is linear in T under the quasiharmonic approximation.

The thermal pressure relative to $T=\Theta$ for $T>\Theta$ is given by

$$\Delta P_{TH} = P_{TH}(T,V_0) - P_{TH}(\Theta,V_0), \tag{2.41}$$

where the term $(3/8)\Theta$ found in (2.38) is added out.

Fig. 2.3 shows that ΔP_{TH} is linear with T (for $T>\Theta$). Thus for ΔP_{TH} the quasiharmonic approximation appears adequate up to $T/\Theta \approx 2.0$. This range covers values of T/Θ for the earth's mantle. The linearity of P_{TH} with T is not unique to silicates and alkali halides; it is also found for gold (Anderson et al., 1989), in spite of the electronic specific heat in gold.

The slope of the $\Delta P_{TH} - T$ curve from (2.41) is empirically a constant above Θ. Because $P_{TH} = \int \alpha K_T\, dT$, we can define an average value of αK_T, $\overline{\alpha K_T}$, good for T above Θ, such that

$$\left(\frac{\partial \Delta P_{TH}}{\partial T}\right)_P = \overline{\alpha K_T} = \frac{3\mathbf{R}\gamma^{ht}}{\mu}\rho_0(1-\alpha T) = \text{const.}, \tag{2.42}$$

where $\mu = M/p$ is the mean atomic weight (per atom). For non-iron bearing silicate minerals, μ varies little from mineral to mineral and is near the value 21 (Watt et al., 1975). Also γ^{ht} does not vary greatly from mineral to mineral, but it is often slightly temperature dependent. Using (2.42) for

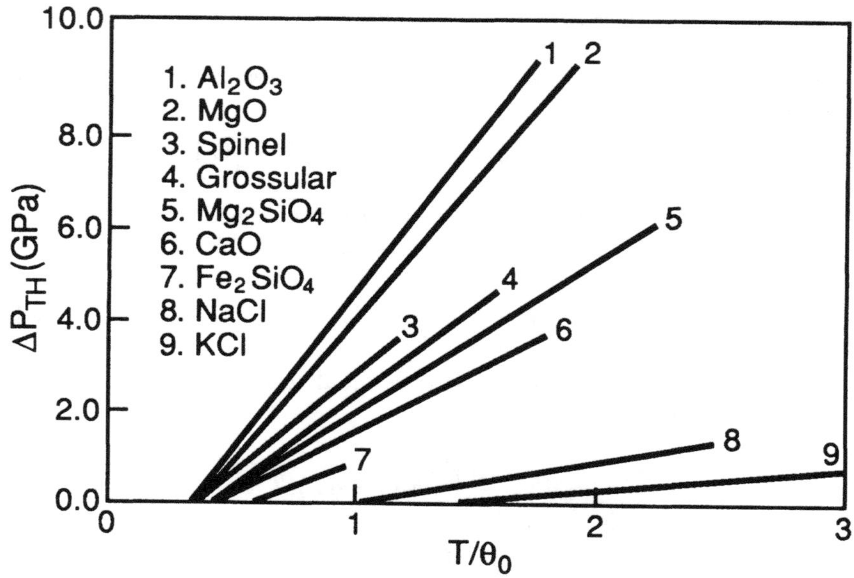

Fig. 2.3. $\Delta P_{TH} = P_{TH}(T) - P_{TH}(300)$ versus T/Θ_0 (subscript 0 stands for ambient temperature) for nine solids. For $T/\Theta_0 > 1$, ΔP_{TH} is strictly linear. There is no evidence of a quadratic law, which indicates that anharmonicity is not affecting P_{TH} at $T > \Theta_0$ (Anderson et al., 1991). The slopes are proportional to ρ_0/μ as given by (2.42). Excluding solids with heavy cations, Fe, Ca, and K, the parameter ρ_0/μ scales as ρ_0. The intercepts of the curves with the T/Θ_0 axis depend on the value of Θ_0 (from Anderson et al., 1992a). (Data in Table A-7).

μ = constant, the slope increases from mineral to mineral as the ambient density increases. Because to a good approximation, for silicates of constant M/p, $\Theta = \text{const}\, \rho^{4/3}$ (Anderson, 1965b), we see that the value $\partial P_{TH}/\partial T$ should increase as $\Theta^{3/4}$: In general, the mineral with the highest value of $\rho_0/(M/p)$ will have the highest slope, according to (2.42), and that is borne out by the experimental results shown in Fig. 2.3.

2.5.5. *The bulk modulus at high temperature*

Using the operator $-(\partial/\partial V)_T$ on (2.38), we find that the bulk modulus K_T along an isochore is

$$K_T^{ht}(T,V) = K_T(0,V) - \frac{3\mathbf{R}\rho}{\mu}\gamma^{ht}\left(1 - q^{ht}\right)T, \tag{2.43}$$

where $q^{ht} = \left(\partial\, \ell\text{n}\, \gamma^{ht}/\partial\, \ell\text{n}\, V\right)$. The last term in (2.43) is often small.

Thus at high T, the isochoric bulk modulus is decreasing with T according to the slope

$$\left(\frac{\partial K_T^{ht}}{\partial T}\right)_V = -\frac{3\mathbf{R}\rho}{\mu}\gamma^{ht}(1 - q^{ht}). \tag{2.44}$$

For $q^{ht} = 1$, $\left(\partial K_T^{ht}/\partial T\right)_V$ vanishes. To obtain the isobaric derivative, $(\partial K_T/\partial T)_P$, we use the calculus expression:

$$\left(\frac{\partial K_T}{\partial T}\right)_P = \left(\frac{\partial K_T}{\partial T}\right)_V - \alpha K_T \left(\frac{\partial K_T}{\partial P}\right)_T . \tag{2.45}$$

Thus at high T, along an isobar,

$$\left(\frac{\partial K_T^{ht}}{\partial T}\right)_P = -\frac{3\mathbf{R}\rho}{(M/p)}\gamma^{ht}(1 - q^{ht}) - \alpha K_T \left(\frac{\partial K_T}{\partial P}\right)_T . \tag{2.46}$$

The term on the far right of (2.43) is the thermal bulk modulus, K_{TH}. Using (2.42) in (2.46), we have

$$\left(\frac{\partial K_T}{\partial T}\right)_P = \overline{\alpha K_T}\left(q^{ht} - 1\right) - \alpha K_T \left(\frac{\partial K_T}{\partial P}\right)_T .$$

Replacing αK_T with $\overline{\alpha K_T}$ in the above,

$$\left(\frac{\partial K_T}{\partial T}\right)_P = -\overline{\alpha K_T}\left[K' - q^{ht} + 1\right] . \tag{2.47}$$

As an example we evaluate (2.47) for the case of MgO from the data plotted in Fig. 2.3 (Anderson et al., 1992a). The average value of $\overline{\alpha K_T}$ is found from the values of ΔP_{TH} between 1000 K and 1800 K. This shows $\overline{\alpha K_T}$ to be 0.0064 GPa/K, which is close to actual measured values of αK_T at high T reported by Anderson et al. (1992a). The high T value of $K' = 4.5$ (Anderson and Isaak, 1993) and $q^{ht} = 1.4$ (Anderson et al., 1993). Thus, we find from (2.47), a predicted value of $(\partial K_T/\partial T)_P = 0.031$ GPa/K. This result agrees well with the measurement showing $(\partial K_S/\partial T)_P$ to be 0.03 GPa/K in the high T range (Anderson et al., 1992a). Isobaric data for K_T versus T are shown in Fig. 2.4.

Two important questions for geophysical interpretation of the earth's mantle are: do the temperature derivatives of K_T^{ht} change with T, and for solid solutions, do they change as a consequence of cation substitution in a solid solution?

Isaak (1993) found that the value of $\left(\partial K_T^{ht}/\partial T\right)_P$ was not affected by substituting Fe for Mg (forsterite to olivine) provided one uses small concentrations of iron (less than 1%) for the substitution. He also found that the slope $\left(\partial K_T^{ht}/\partial T\right)_P$ was more sensitive to the calcium content (pyrope

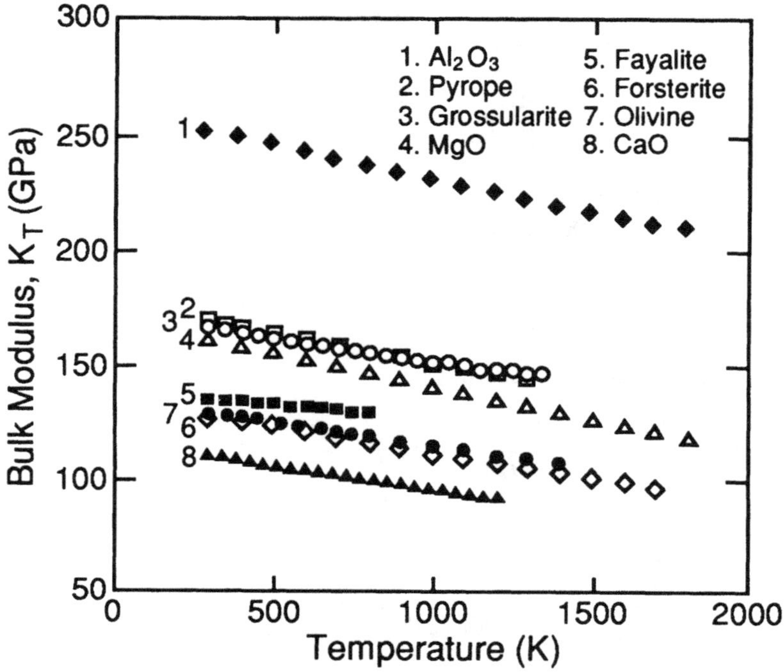

Fig. 2.4. A plot of K_T versus T for eight minerals at $P = 0$. Here we see that the curves are linear (Anderson et al., 1992a). (Data in Table A–7).

to grossular garnet). A quick look at Fig. 2.4. indicates that temperature variations other than linear are small for the solids and oxides shown. I believe this to be a general result, because it has been known for some time that highly accurate ultrasonic data over narrower temperature ranges are generally insufficient to detect higher order T dependence of elasticity. Linearity begins at a T slightly less than Θ, so curvature of K_S versus T at low T is out of the range of the high T limit of the quasiharmonic approximation.

2.5.6. *Entropy and* αK_T

The entropy is found by applying the operator $(\partial/\partial T)_V$ to (2.34), obtaining

$$\mathcal{S}^{ht} = -3p\mathbf{R}\,[\,\ell\text{n}\; kT - \;\ell\text{n}\; \hbar\overline{\omega} + 1]\,. \tag{2.48}$$

The entropy increases as $\ell\text{n}\, T$ in the quasiharmonic approximation at high T. In contrast, C_V is independent of T in this approximation at high T, as shown by (2.36). This also follows from the definition $(\partial S/\partial T)_V = C_V/T$.

We note that if P is linear in T, then $(\partial P/\partial T)_V = \alpha K_T$ is independent of T at high T. This is true (or approximately so) only at T above Θ (see

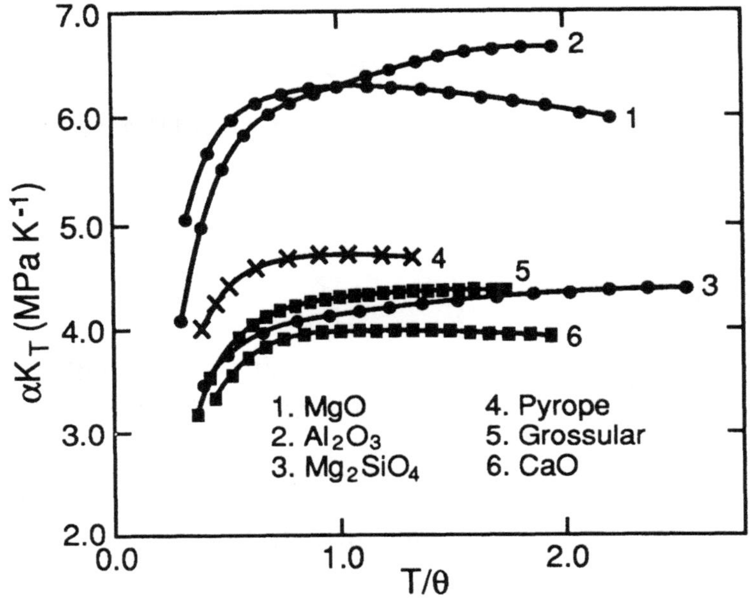

Fig. 2.5. A plot of αK_T versus T/Θ for six minerals. αK_T tends to be independent of T at high T (after Anderson et al., 1992a). (Data in Table A–7).

Fig. 2.5). The functional form of αK_T resembles a C_V curve over the full T range (Fig. 2.6).

Applying the operator $(\partial/\partial T)_V$ to the pressure, given by (2.38), we have for isochores,

$$\left(\frac{\partial P}{\partial T}\right)_V = (\alpha K_T)^{ht} = \frac{3p\mathbf{R}\gamma^{ht}}{V} = \frac{3\mathbf{R}\gamma^{ht}}{(M/p)}\rho_0(1 - \alpha T). \tag{2.49}$$

This is to be compared with (2.42). The actual measured αK_T will have more structure than $\overline{\alpha K_T}$ determined from P_{TH}. The demonstration of equality for the quasiharmonic approximation in the high T limit shows that P is indeed linear with T. Further, this is confirmed experimentally.

To a fair approximation, αK_T is parallel to the T axis for $T > \Theta$, as shown in Fig. 2.5. However, we see from (2.49) that $(\alpha K_T)^{ht}$ is proportional to $\rho_0/(M/p)$. Thus the slope of the ΔP_{TH} versus T curve at $P = 0$ will be proportional to $\rho_0/(M/p)$. In Fig. 2.3, Al_2O_3, with the largest value of $\rho_0/(M/p)$, has the largest slope, and KCl, with the smallest value of $\rho_0/(M/p)$, has the smallest slope. For solids with values of M/p between 20.5 and 21.5, the slopes are ordered according to the value of ρ_0.

The value of T/Θ_0 for the earth's lower mantle is between 1.85 and 1.87 (Table 1.10). Fig. 2.3 shows that much of the data for ΔP_{TH} come close to or exceed $T/\Theta = 1.8$. We conclude that anharmonicity effects

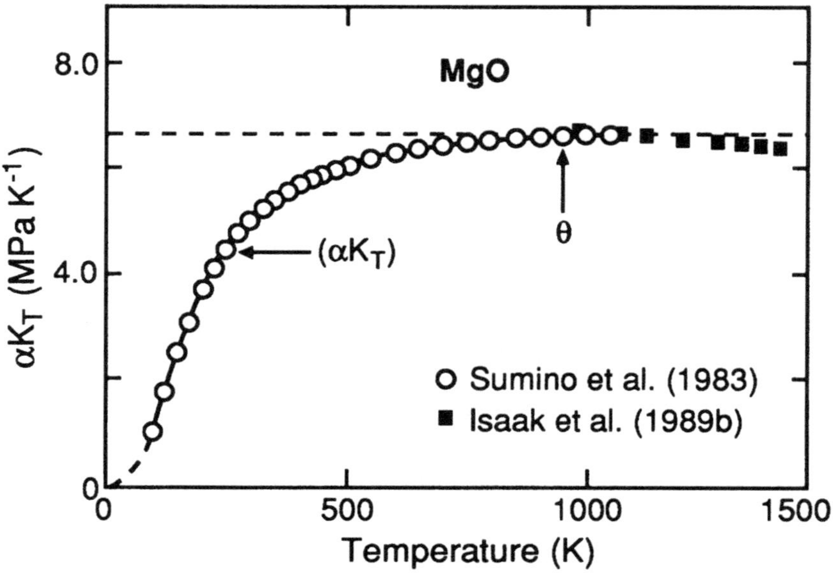

Fig. 2.6. A plot of αK_T at $P = 0$ versus T from $T = 0$ to high T for MgO illustrating the resemblance of the αK_T curve to the C_V curve in Fig. 2.1. (after Anderson et al., 1992a) (Data in Table A–7).

beyond those incorporated into the quasiharmonic approximation need not be considered for P for mantle minerals at mantle conditions.

2.5.7. *α versus T at high temperature along isobars*

Using the expression for K_T^{ht} versus T given by (2.43) and the expression for αK_T given by (2.42), we have

$$\alpha = \frac{\overline{\alpha K_T}}{K_{T_0} + \left(\frac{\partial K_T^{ht}}{\partial T}\right)_P T},$$

which is close to, and for some solids exactly equal to,

$$\alpha = \frac{\text{const}}{K_{T_0} - aT}, \tag{2.50}$$

for the $P = 0$ isobar where $a = \left(\partial K_T^{ht}/\partial T\right)_P$, given by (2.47). We note that (2.51) is the quasiharmonic approximation in the high T limit. The values of the constants in the denominator of (2.51) are such that α is approximately linear with T above $T = \Theta$. This linear increase of α with T at high T does not necessarily require the assumption of additional anharmonic terms in $\mathcal{F}$ beyond those in the quasiharmonic high T approximation.

Using (3.4) and (3.14) and assuming

$$\left(\frac{\partial \alpha}{\partial T}\right)_V = 0, \tag{2.51}$$

we have $(\partial \alpha/\partial T)_P = \delta_T \alpha^2$. Integrating, expanding, and truncating,

$$\alpha = \alpha_{T^*}\left[1 - \alpha_{T^*}\delta_{T_0}(T - T^*)\right], \tag{2.52}$$

where T^* is the beginning T in the integration of $(\partial \alpha/\partial T)_P$, and $T^* > \Theta$.

Thus we conclude that linearity in $\alpha(T)$ for $T > \Theta$ requires that (2.51) holds. In Fig. 2.7, α versus T over a wide temperature range for a number of minerals is shown, demonstrating this linearity.

2.5.8. δ_T *and* q *from the quasiharmonic theory at high* T

From the definition of δ_T, (1.48), and using (2.47), we find

$$\delta_T^{ht} = (q^{ht} - 1) + K'.$$

The above equation tells us that δ_T should be virtually independent of T at high T. This is demonstrated in Fig. 2.8, which shows that roughly speaking, δ_T is parallel to the T axis. We also note that if $(\partial K_T/\partial P)_T = K_0'$ is virtually independent of T, and the experimental evidence is in favor of this, δ_T at high T should be slightly higher than K_0', providing $q > 1$.

From the quasiharmonic theory in the high T limit, we have derived a very significant relationship between three important dimensionless thermoelastic constants at high T. The above equation can be expressed as

$$q^{ht} = \delta_T^{ht} - K' + 1. \tag{2.53}$$

Equation (2.52) is appropriate for high T only. It can also be derived as derivative. For the low temperature equation corresponding to (2.52), we take the logarithmic derivative of γ, as defined by (1.24), yielding

$$q = \left(\frac{\partial \ln \gamma}{\partial \ln V}\right)_T = \delta_T - K' + 1 - \left(\frac{\partial \ln C_V}{\partial \ln V}\right)_T.$$

At high T the above equation reduces to (2.53) because C_V is independent of V above the Debye temperature.

2.6. Anharmonic corrections to the Helmholtz energy

The definition of harmonic and anharmonic varies from one treatment to another. In this book I use the word harmonic to mean that the modal

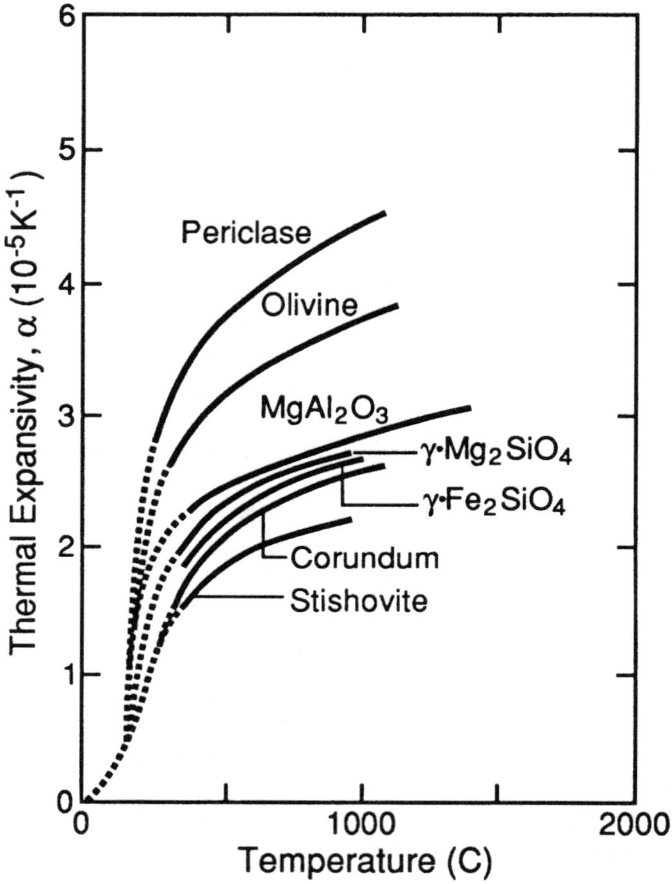

Fig. 2.7. Measured thermal expansivity α versus T for seven minerals at high T and $P = 0$ (modified from Anderson and Sumino, 1980). The knee of the curve is near Θ. Also see Fig. 4.6 for extrapolations. (Data in Table A–7).

frequencies are independent of both V and T. The properties of a three-dimensional harmonic lattice have been determined by Barron (1955) and Stacey (1993). Barron and Stacey find α to be negative and $\gamma_0 = -0.74$ and -0.45, respectively, for a closely packed harmonic lattice. In addition, Stacey finds $K_0' = 1$, $q = 1.23$, and he points out that bond oscillations are sinusoidal in a harmonic lattice. Further, I define the quasiharmonic approximation by the assumption that the modal frequencies are independent of T but depend on V; i.e., $(\partial \ln \omega / \partial \ln T)_V$ is zero. It can be asserted that there is some anharmonicity in the quasiharmonic approximation because the bond oscillations are no longer strictly sinusoidal. But here I define anharmonicity as contributions in T terms to the free energy beyond

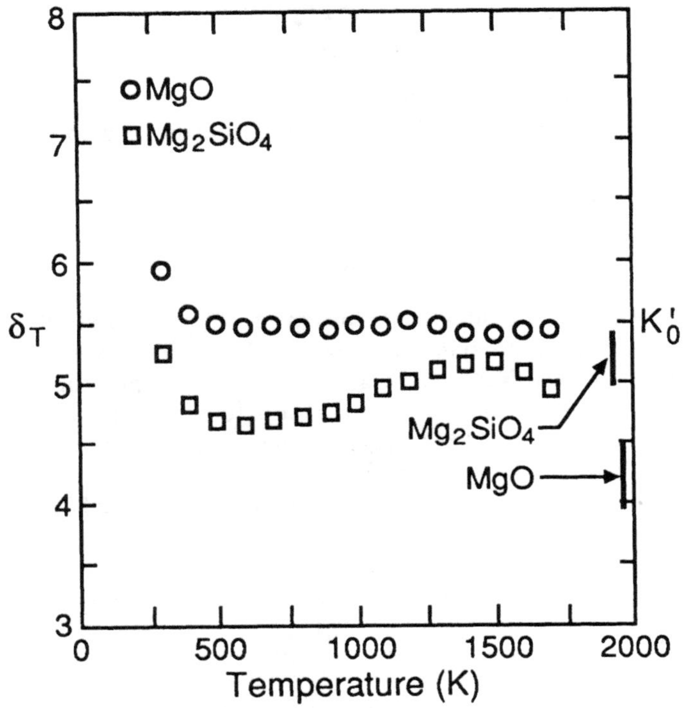

Fig. 2.8. δ_T versus T at $P = 0$ showing two cases: for MgO, δ_T is larger than K_0'; for Mg_2SiO_4, δ_T is approximately K_0' (after Anderson et al., 1993).

those arising from the quasiharmonic thermal energy. As a consequence, I consider a harmonic lattice, a quasiharmonic lattice, and an anharmonic lattice as three distinct behavior patterns. Under my definition, anharmonic properties are dormant except at quite high T, arising as a correction to properties of a quasiharmonic lattice if and only if T is sufficiently large.

We have found that the quasiharmonic approximation is quite adequate for thermal pressure and therefore for K_T, the equation of state, and for α. The quasiharmonic approximation is adequate for C_V of some solids, but is inadequate for other solids. The quasiharmonic approximation may lead to slight modification of the quasiharmonic prediction for those properties where C_V enters into the equations: for example, for γ, $\partial K_T/\partial T$ and αK_T.

I now correct the Helmholtz energy to account for the departures of C_V from the quasiharmonic theory at very high T, and so I invoke some additional anharmonic terms in T. It is known that strong vibrations of atoms at very high T begin to affect the thermodynamic properties of a solid in ways that are beyond those calculated by the quasiharmonic theory. The first three corrections to the Helmholtz energy arising from these strong, temperature–induced motions are given by (Wallace, 1972)

$$\mathcal{F}_{anh} = A_1(V)T^2 + A_0(V) + A_{-2}(V)T^{-2} \tag{2.54}$$

(see also Landau and Lifshitz, 1958).

Using (1.7), we find that the anharmonic contribution to the internal energy is

$$\mathcal{U}_{anh} = -A_1(V)T^2 + A_0(V) + 3A_{-2}(V)T^{-2}.$$

This gives a contribution to the specific heat of

$$C_{V_{anh}} = -2A_1(V)T - 6A_{-2}(V)T^{-3}. \tag{2.55}$$

In Fig. 2.1, we see a possible contribution of $C_{V_{anh}}$ to forsterite and the two garnets. There is no evidence, however, for a significant anharmonic contribution to Al_2O_3, MgO, and CaO, at least for temperatures less than $T = 2\Theta$, so this conclusion holds over a wide T/Θ range. For forsterite and garnets the two terms in (2.55) may combine to give nonlinear variations in $C_{V_{anh}}$ for T slightly larger than Θ. Because the anharmonic corrections in C_V lead to a positive $C_{V_{anh}}$, and as the first term in (2.55) dominates over the second, the sign of $A_1(V)$ must be negative.

The anharmonic contribution to the thermal pressure is found from using (1.9) on (2.54), giving possible additional terms to P_{TH}:

$$P_{TH_{anh}} = -\left[\frac{\partial A_1}{\partial V}T^2 + \frac{\partial A_0}{\partial V} + \frac{\partial A_{-2}}{\partial V}T^{-2}\right]. \tag{2.56}$$

If the coefficients above are non-zero, then extra terms are added to P_{TH} (2.36) at high T, making it of the form

$$P_{TH}^{ht} = a - \frac{\partial A_0}{\partial V} + bT - \frac{\partial A_1}{\partial V}T^2 - \frac{\partial A_{-2}}{\partial V}T^{-2}, \tag{2.57}$$

where the $a - (\partial A_0/\partial V)$ and bT terms emphasize linearity at high T (note that (2.57) is not to be used for $T < \Theta$). At high T the last term in (2.57) would be negligible compared with bT, reducing (2.57) to a quadratic equation. This quadratic dependence of P_{TH} on T is shown by formula (19.49) in Wallace (1972).

Figures 2.3 and 2.4 clearly show that P_{TH} is linear in T over a wide range in T/Θ for $T/\Theta > 1$. The range of linearity of P_{TH} vs. T is sufficiently large that the cT^2 and hT^{-2} terms in (2.57) cannot be significant compared with the bT term. We therefore conclude that either dA_{anh}/dV is negligible, or the terms in (2.56) add up to zero in a unique way. Thus anharmonicity is not significant for the Equation of State for our selected rocks and silicates, at least for pressures and temperatures corresponding to those of the earth's deep mantle. This means also that subsequent volume derivatives of P, such as the elastic constants, would also be independent of anharmonicity effects in the T-P ranges of the earth's mantle. However, in temperature derivatives of the free energy where a prior volume derivative has not been made, such as used in finding the entropy and specific heat, the anharmonic contribution needs to be considered, as shown in Fig. 2.1.

2.7. The free energy and its physical properties at very low temperature

2.7.1. *The thermal contribution to $\mathcal{F}$*

In classical mechanics, all atoms are motionless at absolute zero, and their interactive potential energy, E_{ST}, must be a minimum for equilibrium conditions. (Quantum mechanics allows extra terms in $\mathcal{F}$. There is the exceptional case of liquid helium, which remains liquid at absolute zero, and $P = 0$ arising solely from a quantum mechanics term). For insulators we can examine the consequence to $\mathcal{F}$ when $\omega = kT/\mathbf{h}$ is small, that is, at a low value of the wave number $\mathbf{k}$. This means that the vibrational spectra of the solid are restricted to sound waves. If v is the velocity of sound, and λ is the wavelength of the sound wave, we have $v = \lambda\omega$ and thus $kT = \mathbf{h}v/\lambda$. Because the wavelength of sound is large compared with the lattice constant a, we have

$$\mathbf{k} << \frac{\mathbf{h}v}{a}.$$

In this region of the vibrational spectrum, the frequency is related to the wave vector $\mathbf{k}$ of three sound waves by the linear relation $\omega_i = v_i\mathbf{k}$, $(i = 1, 2, 3)$. For these long wavelengths, the phonon distribution is exactly analogous to the photon distribution of black body radiation, except that there are three separate waves in the phonon distribution.

The number of allowed overtones in $d\omega$ for one direction of vibration is

$$dN = \frac{4\pi\omega^2 d\omega}{v^3}V, \tag{2.58}$$

where V is the specific volume.

In a solid there are two transverse wave vibrations of velocity for each longitudinal wave vibration, so that we have three separate summations to make using (2.58).

Assuming an isotropic body with two equal shear velocities, the frequency distribution is given by

$$g(\omega)d\omega = 4\pi V\left[\frac{1}{v_p^3} + \frac{2}{v_s^2}\right]\omega^2 d\omega,$$

or simply

$$g(\omega)d\omega = \frac{4\pi V}{v_m^3}\omega^2 d\omega, \tag{2.59}$$

where v_m, the mean sound velocity, is given by

$$\frac{1}{v_m^3} = \frac{1}{v_p^3} + \frac{2}{v_s^3}. \tag{2.60}$$

Another version of (2.60) is

$$v_m = v_s \left[\frac{3}{2 + (v_s/v_p)^3} \right]^{1/3},$$

which emphasizes the point that v_m is dominated by the value of v_s as $(v_s/v_p)^3$ is small compared with $\mathcal{Z}$. Equation (2.60) is applicable to $\mathcal{F}_{TH}$ insofar as Debye theory is applicable.

The Planck distribution of photons over different energy levels is well known. Using this for phonons, we change from a summation notation given by (2.9) to an integral,

$$\mathcal{F}_{TH} = \frac{kT3V}{2\pi^2 v_m^3} \int_0^\infty \log\left(1 - e^{-y}\right) \omega^2 d\omega. \tag{2.61}$$

The upper limit of ∞ is allowed because of the rapid convergence of the sum in (2.9) due to the small values of y_i (kT is small compared with $\hbar\omega$).

Although (2.61) is strictly true only for an isotropic body, it can be used to calculate $\mathcal{F}_{TH}$ for low symmetry crystals, provided v_s and v_p in (1.70) are considered as space averages of the single-crystal velocities. The factor 3/2 arises from the fact that sound oscillations have three possible directions of oscillation.

Integrating by parts, (2.61) becomes

$$\mathcal{F}_{TH} = -\frac{V(kT)^4}{2\pi^2 \hbar^3 v_m^3} \int_0^\infty \frac{x^3\, dx}{e^x - 1}. \tag{2.62}$$

The definite integral has the value $\pi^4/15$, so the low temperature value of $\mathcal{F}$ becomes

$$\mathcal{F} = E_{ST} + E_{ZV} - \frac{\pi^2 V(kT)^4}{30(v_m \hbar)^3}, \tag{2.63}$$

where

$$\mathcal{F}_{TH} = -\frac{\pi^2 V(kT)^4}{30(v_m \hbar)^3}. \tag{2.64}$$

Equation (2.64) is applicable to insulators. Consideration of free electrons adds an additional term.

2.7.2. *Physical properties at low temperature*

From (1.7) we see that $\mathcal{U}$ is proportional to T^4, and from (2.64) that $\mathcal{A}_{TH}$ is proportional to V and v_m^{-3}. From (1.6) and (1.8) we see that both $\mathcal{S}$ and C_V are proportional to T^3, and so both must vanish as $T \to 0$ (Barron et al., 1980), as E_{ST} and E_{ZV} are independent of T.

Now, assuming that v_m changes slowly with T as $T \to 0$, we have a T^4 term in P_{TH} (Zharkov and Kalinin, 1971),

$$P = P_0 + P_{ZV} + \frac{\gamma \pi^2 (kT)^4}{10(v_m \hbar)^3}. \tag{2.65}$$

Thus the thermal pressure will vanish according to a T^4 law, but the pressure itself will have a finite value at absolute zero.

Similarly, the T^4 term will occur in the thermal correction to K_T, and as T approaches zero, K_T approaches a constant non-zero value found from the derivative of P_0 and P_{ZV}.

The value of the parameter γ at $T = 0$ is found by applying (1.31) to a single value of ω, ω_D, so that in terms of Θ,

$$\gamma_{T=0} = -\frac{\partial\, \ell\mathrm{n}\left(v_m / v^{1/3}\right)}{\partial\, \ell\mathrm{n}\, V} = \frac{\partial\, \ell\mathrm{n}\, \Theta}{\partial\, \ell\mathrm{n}\, V}. \tag{2.66}$$

Thus we see that at low T, γ is independent of T and does not vanish at absolute zero, but approaches a limiting value, $\gamma_{T=0}$ (Barron, 1955). For alkali halides, $\gamma_{T=0}$ is smaller than γ^{ht}, but for oxides and silicates, $\gamma_{T=0}$ is often larger then γ^{ht}.

By taking the temperature derivative of P (2.65) at constant V, we find that αK_T is proportional to T^3, just as is C_V. Therefore at low T the curve of αK_T will resemble a specific heat curve. In fact, it resembles a C_V curve both at low and high T, as shown in Figs. 2.6. and 2.7.

If we take the temperature derivative of K_T, we find that $(\partial K_T/\partial T)_P$ is proportional to T^3 and must vanish as $T \to 0$. Similarly the second derivative of K_T with respect to T at constant P must also vanish at absolute zero. The plot of $(\partial K_T/\partial T)_P$ must be sigmoidal in shape, going from zero at $T = 0$ but approaching a constant negative value at $T > \Theta$. The plot of the experimental results of $(\partial K_S/\partial T)_P$ for Mg_2SiO_4, shown in Fig. 2.9, illustrates sigmoidal behavior.

At low temperatures there is little difference between $(\partial K_S/\partial T)_P$ and $(\partial K_T/\partial T)_P$. We see that K_T, V, and γ approach positive nonzero values as $T \to 0$, but that C_V, αK_T, and $(\partial K_T/\partial T)_P$ all approach zero as $T \to 0$.

From (2.50) we see that the expression for α as $T \to 0$ has a constant non-zero value in the denominator, but the term $(\partial P/\partial T)_V$ is in the numerator. Taking the temperature derivative of (2.65) eliminates the terms independent of T, leaving a T^3 term controlling α at low T. Thus the temperature dependence behavior of α resembles closely the temperature dependence behavior of C_V and αK_T near absolute zero: all three vanish at absolute zero, and there is a low temperature T^3 region. A plot of α versus T shown in Fig. 2.7 illustrates the point that α resembles a C_V curve at lower temperatures. But at $T > \Theta$, α increases steadily with T at constant P, unlike either αK_T or C_V. However, for constant V, $(\partial \alpha/\partial T)_V$ is often zero at high T, parallel to the case where $(\partial C_V/\partial T)_V$ is zero.

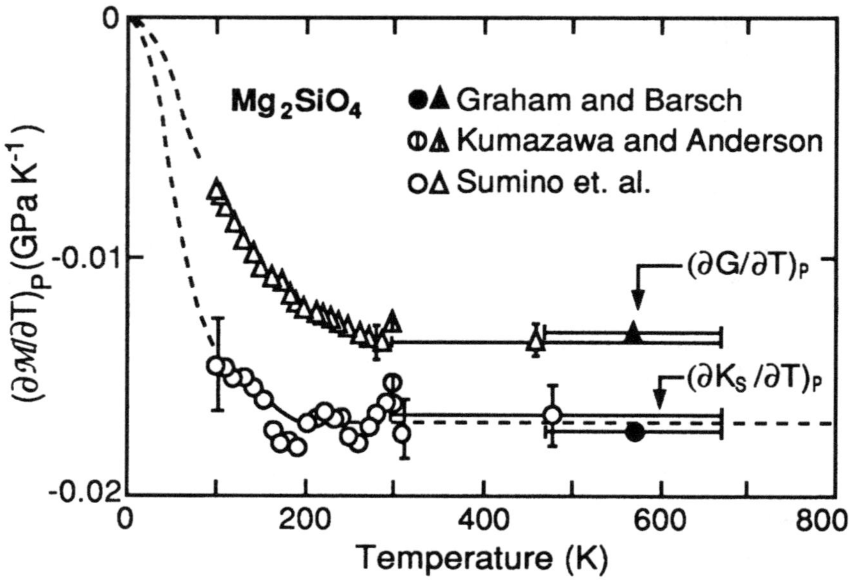

Fig. 2.9. A plot of $(\partial G/\partial T)_P$ versus T and $(\partial K_S/\partial T)_P$ for forsterite, where G is the shear modulus, showing that these parameters are sigmoidal in shape approaching zero at absolute zero (modified from Sumino et al., 1977).

2.8. The Debye theory interpolation

2.8.1. *The Debye temperature*

We have seen that at both low and high temperature within the realm of the quasiharmonic approximation, a calculation of the thermodynamic properties of a solid can be made with a fairly good degree of accuracy, using (2.10) and (2.9) and adding an anharmonic term, (2.53), where required.

For intermediate temperatures the expression for $\mathcal{F}_{TH}$ cannot be exactly evaluated because y_j must be known for all modes. Further approximations are required, and they typically take the form of constructing an approximate frequency spectrum. We saw that at low T, a frequency spectrum was constructed by replacing the sum by an integral in (2.61). At high temperature, where all vibrations are excited, the equations reduce to simple forms.

In the Debye approximation to the quasiharmonic theory, the expression for A_{TH} given by (2.61) is virtually exact at low T. We propose to extend A_{TH} to higher temperatures, but for these the upper limit ∞ in (2.61) has to be replaced by a finite frequency ω_D because y_i is no longer a small number at higher T.

There being a large but finite number of modes, $3pN$, the distribution function, when integrated over all frequencies, must equal this number.

Thus using (2.58)

$$\frac{4\pi V}{v_m^3}\int_0^{\omega_D}\omega^2\,d\omega = \frac{4\pi V}{v_m^3}\frac{\omega_D^3}{3} = 3pN, \tag{2.67}$$

or

$$\omega_D = \left(\frac{3pN}{4\pi V}\right)^{1/3} v_m, \tag{2.68}$$

where v_m is given by (1.66).

The Debye temperature is the temperature corresponding to ω_D, and by using $k\Theta = \hbar\omega_D$, we find that

$$\Theta = \frac{\hbar}{k}\left(\frac{9N}{4\pi}\right)^{1/3}\left(\frac{\rho}{M/p}\right)^{1/3} v_m \quad K = 251.2\left(\frac{\rho}{\mu}\right)^{1/3} v_m \quad K, \tag{2.69}$$

where $M/p = \mu$ is the mean atomic mass, ρ is in g/cc, and v_m is in km/s. When Θ is defined by the acoustic velocities, as in (1.65), as we shall designate it as Θ_{ac}.

2.8.2. *The Debye function*

The expression for $\mathcal{F}_{TH}$ in (2.9) through (2.54) becomes

$$\mathcal{F}_{TH} = \frac{4\pi VkT}{v_m^3}\int_0^{\omega_D}\ell n\left(1 - e^{-\hbar\omega/kT}\right)\omega^2 d\omega. \tag{2.70}$$

Changing to the variable y, given by (2.7) rewritten in terms of Θ and using (1.65),

$$\mathcal{F}_{TH} = 9p\mathbf{R}T\left(\frac{T}{\Theta}\right)^3\int_0^{\Theta/T} y^2\,\ell n\,\left(1 - e^{-y}\right)\,dy, \tag{2.71}$$

where $y = \Theta/T$ and $\Theta = (\hbar/k)\omega_D$. Integrating by parts and introducing the Debye function $D(\Theta/T)$,

$$D\left(\frac{\Theta}{T}\right) = 3\left(\frac{T}{\Theta}\right)^2\int_2^{\Theta/T}\frac{z^3\,dz}{e^{z-1}}, \tag{2.72}$$

we have

$$\mathcal{F}_{TH} = p\mathbf{R}T\left[3\,\ell n\,\left(1 - e^{-\Theta/T}\right) - D\left(\frac{\Theta}{T}\right)\right]. \tag{2.73}$$

Equation (2.72) cannot be represented by an analytical expression, but $D(x)$ is tabled in many standard handbooks for ease in computation. The zero vibrational energy, given by (2.6), becomes

$$E_{ZV} = \frac{9}{8}p\mathbf{R}\Theta. \tag{2.74}$$

The Helmholtz energy for the Debye approximation is consequently

$$\mathcal{F} = E_{ST} + p\mathbf{R}T\left[\frac{9}{8}\left(\frac{\Theta}{T}\right) + 3\,\ell\text{n}\,\left(1 - e^{-\Theta/T}\right) - D\left(\frac{\Theta}{T}\right)\right]. \quad (2.75)$$

Determining the internal energy by using (1.7) on (2.75), we have

$$\mathcal{U} = E_{ST} + p\mathbf{R}\left[\frac{9}{8}\Theta + 3TD\left(\frac{\Theta}{T}\right)\right]. \quad (2.76)$$

The recursion function of $D(\Theta/T)$ is (Zharkov and Kalinin, 1971)

$$\frac{\partial D(\Theta/T)}{\partial(\Theta/T)} = \frac{3}{e^{\Theta/T}-1} - \frac{3}{(\Theta/T)}D\left(\frac{\Theta}{T}\right), \quad (2.77)$$

which is useful for determining P, C_V, and αK_T from (2.75).

2.8.3. *The Debye expressions for thermodynamic functions*

Under the quasiharmonic approximation, ω is a function of V, and therefore Θ is a function of V. We use (1.9) to find P (Zharkov and Kalinin, 1971)

$$P = P(V,0) + \frac{p\mathbf{R}}{V}\gamma_D\left[\frac{9}{8}\Theta + 3TD\left(\frac{\Theta}{T}\right)\right], \quad (2.78)$$

where γ_D is the same as (2.66) except that Θ is Θ_D. For high T limits, (2.72) becomes

$$D\left(\frac{\Theta}{T}\right) = 3T - \frac{9}{8}\Theta + \frac{3T}{2\Theta}\left(\frac{\Theta}{T}\right)^2. \quad (2.79)$$

Operating on (2.75), we have the thermodynamic functions in terms of $D(\Theta/T)$:

$$\mathcal{S} = p\mathbf{R}\left[4D\left(\frac{\Theta}{T}\right) - 3\,\ell\text{n}\,\left(1 - e^{-\Theta/T}\right)\right] \quad (2.80)$$

$$C_V = 3p\mathbf{R}\left[4D\left(\frac{\Theta}{T}\right) - \frac{3(\Theta/T)}{e^{\Theta/T}-1}\right] \quad (2.81)$$

$$K_T = K_T(V,0) + p\mathbf{R}\frac{\gamma_D}{V}\left\{\left[\frac{9}{8}\Theta + 3TD\left(\frac{\Theta}{T}\right)\right]\right\}$$
$$\div\left[(1+\gamma_D - q) - 12\gamma_D TD\left(\frac{\Theta}{T}\right) + \frac{9\Theta\gamma}{e^{\Theta/T}-1}\right] \quad (2.82)$$

$$\alpha K_T = \frac{3p\mathbf{R}\gamma_D}{V}\left[4D\left(\frac{\Theta}{T}\right) - \frac{3(\Theta/T)}{e^{\Theta/T}-1}\right]. \quad (2.83)$$

The Debye expression for α is given by $\alpha K_T/K_T$, which is (2.83) divided by K_T, (2.82). K_T divided into αK_T makes α decrease as V decreases.

2.9. Thermodynamic functions from the partition function

In Section 1.2, we found that the partition function $\mathcal{Z}$ is related to the accessible quantum states by (2.3). In a more general treatment, (2.3) is replaced by

$$\mathcal{Z} = \Sigma \mathbf{g}_i e^{-\epsilon_i/kT}. \tag{2.84}$$

We have placed $\mathbf{g}_i$ in (2.84) to account for possible degeneracy of the ith state (the population may unevenly occupy the states). Here $\mathbf{g}_i^{-\epsilon_i/kT}$ indicates the occupation n_i of a state, where

$$\frac{n_i}{N} = \frac{e^{-\epsilon_i/kT}}{\mathcal{Z}}. \tag{2.85}$$

Whereas quantum mechanics deals with the detailed arrangement of atoms and molecules, thermodynamics deals with their average behavior. The average of the total energy of the system is the internal energy $\mathcal{U}$, so that

$$\mathcal{U} = \frac{E_{\text{total}}}{N} = \frac{\Sigma n_i \epsilon_i}{N}. \tag{2.86}$$

A mathematical property of the function in (2.84) is useful. We find that

$$\frac{\partial \mathcal{Z}}{\partial T} = \Sigma k T^2 \epsilon_i \mathbf{g}_i e^{-\epsilon_i/kT},$$

so that

$$\frac{1}{\mathcal{Z}} \frac{d\mathcal{Z}}{dT} = kT^2. \tag{2.87}$$

Placing (2.85) and (2.87) in (2.86), we find

$$\mathcal{U} = kT^2 \left(\frac{\partial\, \ell\text{n}\, \mathcal{Z}}{\partial T} \right). \tag{2.88}$$

The equation for C_V yields

$$C_V = \left(\frac{\partial \mathcal{U}}{\partial T} \right)_V = \frac{\partial}{\partial T} \left[kT^2 \frac{\partial\, \ell\text{n}\, \mathcal{Z}}{\partial T} \right]_T. \tag{2.89}$$

The entropy is defined by

$$\mathcal{S}(T) - \mathcal{S}(T=0) = \int_0^T \frac{C_V}{T}\, dT. \tag{2.90}$$

Placing (2.89) in (2.90) and recognizing that $\mathcal{S}(T=0) = \left| k\, \ell\text{n}\, \mathcal{Z} \right|_{T=0}$,

$$\mathcal{S}(T) = kT \left(\frac{\partial\, \ell\text{n}\, \mathcal{Z}}{\partial T} \right) + k\, \ell\text{n}\, \mathcal{Z}. \tag{2.91}$$

The Helmholtz energy $\mathcal{F}$, which is really $\mathcal{F}_{VIB}$, is found by applying (1.2) to (2.88) and (2.91), yielding the same expansion as (2.2). The pressure given applying (1.9) to (2.2) is

$$P_{VIB} = kT \left(\frac{\partial \,\ell\text{n}\, \mathcal{Z}}{\partial V} \right)_T . \tag{2.92}$$

2.10. Problems

1. Derive (2.71).
2. Derive $\mathcal{U}_{anh}$ using all the terms in $\mathcal{F}_{anh}$ shown in (2.54).
3. Prove (2.77).
4. Prove that at high T, from (2.34)

$$\mathcal{F}^{ht} = E_{ST} + kT \sum_{j}^{3pN} \left[\ell\text{n} \left(\frac{\hbar\omega_j}{kT} \right) + \frac{1}{24} \left(\frac{\hbar\omega_j}{kT} \right)^2 - \frac{1}{2880} \left(\frac{\hbar\omega_j}{kT} \right)^z \right].$$

5. Prove at high T, from (2.48)

$$\mathcal{S}^{ht} = k \sum_{j=1}^{3pN} \left[\ell\text{n} \left(\frac{kT}{\hbar\omega_i} \right) + 1 + \frac{1}{24} \left(\frac{\hbar\omega_j}{kT} \right)^2 - \frac{1}{960} \left(\frac{\hbar\omega_j}{kT} \right)^4 + \cdots \right].$$

6. Show that at low T

$$S = V \frac{2\pi^2 k (kT)^3}{15\hbar^3 v_m^3}.$$

7. Prove at high T

$$C_V^{ht} = k \sum_{j=1}^{3pN} \left[1 - \frac{1}{12} \left(\frac{\hbar\omega_j}{kT} \right)^2 + \frac{1}{240} \left(\frac{\hbar\omega_j}{kT} \right)^2 + \cdots \right],$$

where $k \sum_j^{3pN} 1 = 3p\mathbf{R}$.

8. Show that at low T

$$\mathcal{U} = E_{ST} + V \frac{\pi^2 (kT)^4}{10\hbar^3 v_m^3}.$$

9. Show that at low T

$$C_V = \frac{2\pi^2 kV}{5\hbar^3 v_m^3} (kT)^3.$$

10. Show that at $T >> \Theta$, from (2.78)

$$C_V = 3pR \left[1 - \frac{1}{20} \left(\frac{\Theta}{T} \right)^2 \right].$$

11. Show that at $T >> \Theta$ from (2.78)

$$P_{TH} = P - P(V,0) = 3pR \frac{\gamma^{ht}}{V} T \left[1 + \frac{1}{20} \left(\frac{\Theta}{T} \right)^2 \right].$$

3

THERMOELASTIC PARAMETERS AT HIGH COMPRESSION

3.1. Thermodynamic identities

I begin by listing the thermodynamic identities used in this book, many of which are well known (see Stacey, 1977a; Brennan and Stacey, 1979; Zemansky, 1943; and Bassett et al., 1968). Relationships of classical thermodynamics are very helpful in determining physical properties at extreme conditions because many of the needed experimental data cannot be obtained. The chief experimental information on the interior of the earth is taken from seismic data, so the thermodynamic relationships appropriate to geophysical problems are often recast emphasizing elastic properties. Compensating for the limited kinds of measurements available in high P–high T materials sciences is a set of relationships and approximations, set forth below, not customarily emphasized in thermodynamics as found in physics and chemistry textbooks. With these identities a temperature measurement, done isobarically, can be converted to pressure information. Conversely, an isothermal pressure measurement can be converted to temperature information. As we shall see, some thermodynamic identities can easily be transformed into differential equations to help construct a theory aimed at showing how a physical property can be defined at extreme P and T.

3.1.1. *Basic identities*

$$\left(\frac{1}{T}\right)\left(\frac{\partial C_V}{\partial V}\right)_T \equiv \left[\frac{\partial(\alpha K_T)}{\partial T}\right]_V \tag{3.1}$$

$$\left(\frac{\partial K_T}{\partial T}\right)_V \equiv -V\left[\frac{\partial(\alpha K_T)}{\partial V}\right]_T \tag{3.2}$$

$$\alpha K_T \equiv \left(\frac{\partial P}{\partial T}\right)_V \tag{1.11}$$

$$K_T^2\left(\frac{\partial \alpha}{\partial P}\right)_T \equiv \left(\frac{\partial K_T}{\partial T}\right)_P \tag{3.3}$$

$$K_T\left(\frac{\partial \alpha}{\partial T}\right)_V \equiv \left[\frac{\partial(\alpha K_T)}{\partial T}\right]_P. \tag{3.4}$$

3.1.2. *Identities using definitions of δ_T, K', $\hat{\alpha}$, and η*

$$\frac{\partial^2 K_T}{\partial P \partial T} \equiv \delta_T\left(\frac{\eta}{K_T}\right)\left[\frac{\partial(\alpha K_T)}{\partial \eta}\right]_T + \alpha\eta\left(\frac{\partial \delta_T}{\partial \eta}\right)_T \tag{3.5a}$$

$$\equiv \alpha\delta_T(\delta_T - K' + \kappa) \tag{3.5b}$$

$$\kappa = \left(\frac{\partial \ \ell\text{n}\ \delta_T}{\partial \ \ell\text{n}\ \eta}\right)_T \tag{3.6}$$

$$\left[\frac{\partial \ \ell\text{n}\ (\alpha K_T)}{\partial \ \ell\text{n}\ V}\right]_T \equiv \delta_T - K' \tag{3.7}$$

$$\left[\frac{\partial \ \ell\text{n}\ (\alpha K_T)}{\partial P}\right]_T \equiv (K' - \delta_T)/K_T \tag{3.8}$$

$$\left[\frac{\partial(\alpha K_T)}{\partial T}\right]_P \equiv \alpha^2 K(\hat{\alpha} - \delta_T), \tag{3.9}$$

where $\hat{\alpha}$ is given by (1.64)

$$\left(\frac{\partial C_V}{\partial V}\right)_T \equiv \alpha\,(\alpha K_T)\,T(\hat{\alpha} + K' - 2\delta_T) \tag{3.10}$$

$$\left(\frac{\partial C_V}{\partial P}\right)_T \equiv -\alpha^2 TV(\hat{\alpha} + K' - 2\delta_T) \tag{3.11}$$

$$\left(\frac{\partial K_T}{\partial T}\right)_V \equiv \alpha K_T(K' - \delta_T) \tag{3.12}$$

$$\left[\frac{\partial(\alpha K_T)}{\partial T}\right]_V \equiv \alpha^2 K_T(\hat{\alpha} + K' - 2\delta_T) \tag{3.13}$$

$$\left(\frac{\partial \ \ell\text{n}\ \alpha}{\partial \ \ell\text{n}\ \eta}\right) \equiv \delta_T. \tag{1.63}$$

3.1.3. *Arising from calculus*

$$\left(\frac{\partial X}{\partial T}\right)_V = \left(\frac{\partial X}{\partial T}\right)_P + \alpha K_T\left(\frac{\partial X}{\partial P}\right)_T \tag{3.14}$$

$$\left(\frac{\partial X}{\partial P}\right)_S = \left(\frac{\partial X}{\partial P}\right)_T + \frac{\gamma T}{K_S}\left(\frac{\partial X}{\partial T}\right)_P \tag{3.15}$$

$$\left(\frac{\partial \eta}{\partial T}\right)_P = \eta(\alpha - \alpha_0) \tag{3.16}$$

$$\left(\frac{\partial \eta^{\delta_T}}{\partial T}\right)_P = \delta_T \eta^{\delta_T - 1}\left(\frac{\partial \eta}{\partial T}\right)_P + \ \ell\text{n}\ \eta^{\delta_T}\left(\frac{\partial \delta_T}{\partial T}\right)_P. \tag{3.17}$$

3.1.4. *Approximations to identities in terms of η (high order terms ignored, assuming $\delta_T > 3$ and δ_T independent of η, i.e., η near 1)*

$$\alpha = \alpha_0 \eta^{\delta_T} \tag{3.18}$$

$$\left(\frac{\partial \alpha}{\partial T}\right)_P = \alpha_0 \eta^{\delta_T} \left[\left(\frac{\partial \ln \alpha_0}{\partial T}\right)_P + \delta_T \eta (\alpha - \alpha_0) + \ln \eta \left(\frac{\partial \delta_T}{\partial T}\right)_P\right] \tag{3.19}$$

$$\left(\frac{\partial C_V}{\partial V}\right)_P = \alpha_0 \eta^{\delta_T} \left[\left(\frac{\partial \ln \alpha_0}{\partial T}\right)_P + \delta_T \eta (\alpha - \alpha_0) + \ln \eta \left(\frac{\partial \delta_T}{\partial T}\right)_P\right] \tag{3.20}$$

$$\left(\frac{\partial \ln C_V}{\partial \ln V}\right)_T = \gamma T \left[\alpha_0 (\hat{\alpha} - \delta_T) + \ln \eta \left(\frac{\partial \delta_T}{\partial T}\right)\right] \tag{3.21}$$

$$\left(\frac{\partial C_V}{\partial P}\right)_T = -VT\alpha \left[\alpha_0 (\hat{\alpha} - \delta_T) + \ln \eta \left(\frac{\partial \delta_T}{\partial T}\right)\right]. \tag{3.22}$$

3.1.5. *Identities involving the temperature derivative of K_T*

The relationship between K_T and K_S is given by

$$K_S = K_T (1 + \alpha \gamma T). \tag{3.23}$$

Taking the derivative of (3.23) with respect to Y at constant Z, we find

$$\left(\frac{\partial K_S}{\partial Y}\right)_Z = (1 + \alpha\gamma T)\left(\frac{\partial K_T}{\partial Y}\right)_Z (\alpha\gamma K_T) \left[\frac{T}{\gamma}\left(\frac{\partial \gamma}{\partial Y}\right)_Z + \frac{T}{\alpha}\left(\frac{\partial \alpha}{\partial Y}\right)_Z + \left(\frac{\partial T}{\partial Y}\right)_Z\right]. \tag{3.24}$$

When $Y = T$ and $Z = P$, we use (1.63) in (3.24) to find

$$\delta_T - \delta_S = \gamma \left[\left(\frac{\partial \ln \alpha}{\partial \ln T}\right)_P + \left(\frac{\partial \ln \gamma}{\partial \ln T}\right)_P + 1\right] - \alpha\gamma T \delta_T. \tag{3.25}$$

$\delta_T > \delta_S$ because $(\partial \ln \alpha / \partial \ln T)_P$ is a positive number; $(\partial \ln \gamma / \partial \ln T)_P$ is close to zero or a small negative number; and $\alpha\gamma T \delta_T$ is small compared with 1. When $Y = P$ and $Z = T$, we use (1.11) and (3.24) to find

$$K'_T = K'_S + \alpha\gamma T (\delta_T + q - K'_T), \tag{3.26}$$

where K'_S is as defined by (1.58) and

$$q = \left(\frac{\partial \ln \gamma}{\partial \ln V}\right)_T. \tag{3.27}$$

Because $\alpha\gamma T > 0$, and if, as is usually found, $\delta_T \cong K'_{T_0}$ and $q > 0$, then $K'_T > K'_S$. The difference is small, often close to 0.1. The identities for δ_S and K_S are

$$\delta_S = K'_S - 1 + q - \gamma - (\partial \ln C_V / \partial \ln V)_S$$

and

$$K'_S = K'_T(1 + \gamma\alpha T) - \gamma\alpha T(\delta_S + \delta_T + q).$$

3.1.6. *Identities involving the volume derivative of γ*

A modification of a general equation derived by Bassett et al. (1968) is (Yamamoto et al., 1987)

$$q = \left(\frac{\partial \ln \gamma}{\partial \ln V}\right)_T = \frac{1 + \gamma + (1 + \alpha\gamma T)\delta_S - K'_S + T\left(\dfrac{\partial \gamma}{\partial T}\right)_P}{1 + \alpha\gamma T}. \tag{3.28}$$

Equation (3.28) is particularly useful because all of its terms are experimentally determinable, as high T measurements are often adiabatic. An equivalent identity for q in terms of isothermally determined quantities is found from (1.24). It is

$$q = \delta_T - K'_T + 1 - \left(\frac{\partial \ln C_V}{\partial \ln V}\right)_T. \tag{3.29}$$

In the high T region, the last term in (3.29) is dropped. Thus at high T,

$$q = \delta_T - K' + 1. \tag{3.30}$$

Further, if δ_T is close to K'_T, q is close to unity.

3.1.7. *Identities involving the thermal pressure using the general equation of state*

Using the general EoS,

$$P = P_0(V) + P_{TH}(V,T), \tag{3.31}$$

where P_{TH} is the thermal pressure, we see that $(\partial P/\partial T)_V = (\partial P_{TH}/\partial T)_V$, and from (3.3)

$$\alpha K_T = \left(\frac{\partial P_{TH}}{\partial T}\right)_V. \tag{3.32}$$

Thus, by integrating at constant volume,

$$P_{TH} = \int_0^T (\alpha K_T)\, dT, \tag{3.33}$$

which defines thermal pressure in terms of measurable parameters.

3.1.8. *Maxwell's relationships*

Useful relationships are easily derived from Maxwell's relations (Slater, 1939):

$$-\left(\frac{\partial P}{\partial S}\right)_V = \left(\frac{\partial T}{\partial V}\right)_S \tag{3.34}$$

$$\left(\frac{\partial V}{\partial S}\right)_P = \left(\frac{\partial T}{\partial P}\right)_S \tag{3.35}$$

$$\left(\frac{\partial P}{\partial T}\right)_V = \left(\frac{\partial S}{\partial V}\right)_T \tag{3.36}$$

$$-\left(\frac{\partial V}{\partial T}\right)_P = \left(\frac{\partial S}{\partial P}\right)_T . \tag{3.37}$$

Rewriting Maxwell's relations so as to incorporate the Grüneisen ratio γ and other measurable parameters, we have

$$\left(\frac{\partial T}{\partial V}\right)_S = -\left(\frac{\partial P}{\partial S}\right)_V = -\frac{\gamma T}{V} \tag{3.38}$$

$$\left(\frac{\partial T}{\partial P}\right)_S = \left(\frac{\partial V}{\partial S}\right)_P = \frac{\gamma T}{K_S} \tag{3.39}$$

$$\left(\frac{\partial V}{\partial T}\right)_P = -\left(\frac{\partial S}{\partial P}\right)_T = \alpha V \tag{3.40}$$

$$\left(\frac{\partial P}{\partial T}\right)_V = \left(\frac{\partial S}{\partial V}\right)_T = \alpha K_T . \tag{1.22}$$

3.2. The mean atomic mass, $\mu = M/p$

The numerical evaluation of V found in thermodynamic equations is awkward because molecular weight is sometimes not well specified, especially in geophysical problems. It is often convenient to replace V with the ratio of mass to density, where the mass is the molecular weight divided by the number of atoms per molecule ($\mu = M/p$). This ratio is particularly useful, as for silicate minerals μ does not vary significantly from mineral to mineral if the iron content of the minerals remains unchanged. The value of μ for a few minerals is given in Table 1.9. For mantle minerals it is assumed, following Watt et al. (1975), that

$$\mu = 21. \tag{3.41}$$

Stacey (1977b) pointed out that it is convenient to identify thermodynamic parameters whose values are proportional to mass and specify the dimensions in terms of mass rather than moles. Thus $\mathcal{S}$, $\mathcal{H}$, $\mathcal{U}$, $\mathcal{F}$, $\mathcal{G}$, and V are proportional to mass, but P and T are not.

3.3. The cases for $(\partial K_T)/\partial T)_V = 0$ and $(\partial(\alpha K_T)/\partial T)_P = 0$

Important simplifications of the equations in the previous sections result if, as in the case for NaCl and Mg_2SiO_4 at zero pressure, we find $\mathbf{w} = 0$ where

$$\mathbf{w} = \left(\frac{\partial K_T}{\partial T}\right)_V . \tag{3.42}$$

The parameter $\mathbf{w}$ can be measured in a pressure experiment (Boehler and Kennedy, 1980; Yagi, 1978), but, using K_T for X, (3.2) and (3.14) show that it can also be found in a temperature experiment where elastic constants and the thermal expansivity versus temperature are measured at one bar (at sufficiently high temperature).

$(\partial K_T/\partial T)_V$ is called the intrinsic temperature effect (D.L. Anderson, 1987). It is smaller than the extrinsic term, $(\partial K_T/\partial T)_P$, and often close to 0 at $P = 0$. When $\mathbf{w} \neq 0$, the bulk modulus arises from an interatomic potential that is a function of T and not interatomic distance alone. In general, as we shall see, $\mathbf{w}$ becomes smaller with compression at constant T. It also becomes larger as T increases along an isochore. When $\mathbf{w} \neq 0$, a special term is added to the entropy at high T (see Chapter 13).

Supposing that $\mathbf{w} = 0$ is valid, we see from (3.7) that

$$\left[\frac{\partial\,(\alpha K_T)}{\partial V}\right]_T = 0. \tag{3.43}$$

We see from (3.6) that if (3.43) is true,

$$K' - \delta_T = 0. \tag{3.44}$$

Conversely, if (3.44) is experimentally shown to be true, then it follows that (3.43) also holds.

From (3.32) it is seen that when (3.43) or (3.44) is true, $(\partial P_{TH}/\partial T)_V$ is not a function of V. In Section 3.5 we shall see that the condition $\delta_T = K'$ or $(\partial K_T/\partial T)_V = 0$ often holds for a trace in the T, η plane, but for many solids (3.44) cannot hold generally throughout the T, η plane. For one solid, NaCl, (3.44) is true over a wide range of T at $P = 0$ (Birch, 1986). But for most solids $\mathbf{w} \neq 0$ at ambient conditions and high T for $P = 0$.

Figure 2.6 shows that αK_T is virtually independent of T at high T (above Θ) for many solids. We note from (3.4) that this is equivalent to requiring that α be independent of T along an isochore. But if $(\partial\alpha/\partial T)_V$ is zero, then by means of (3.14), we find that

$$\left(\frac{\partial\alpha}{\partial T}\right)_P = -\alpha K_T\left(\frac{\partial\alpha}{\partial\rho}\right)_T = \alpha^2\delta_T, \tag{3.45}$$

which should be true for $T > \Theta$. This is further developed in Chapter 4.

It may appear strange that at high T, the slope $(\partial\alpha/\partial T)_P$ increases with the power α^2. Yet this is derived from the quasiharmonic theory, without recourse to added anharmonic terms in the free energy.

In the quasiharmonic approximation $(\partial P/\partial T)_V$ is independent of T, so that using (2.50)

$$\alpha = \frac{\text{const.}}{K_T}.$$

Now K_T is linear with T above the Debye temperature (see Fig. 2.4), so that the above equation becomes

$$\alpha = \text{const.}\left(K_{T_0} - bT\right)^{-1}.$$

The above becomes $\partial\alpha/\partial T = (b/\text{const.})\,\alpha^2$ on differentiation with respect to T. Thus the nonlinear rise in α at high T results from the fact that α is roughly inversely proportional to K_T at high T, or αK_T = constant.

3.4. Theoretical insight into the change of K' with T

In many EoS problems, it is of interest to know how $(\partial K/\partial P)_T$ varies with temperature. This knowledge is often required to successfully extrapolate room temperature data to higher temperatures.

We start with the identity given by (3.5*b*)

$$\left(\frac{\partial^2 K_T}{\partial P \partial T}\right) = \alpha\delta_T(\delta_T - K' + \mathbf{k}). \qquad (3.5b)$$

In order to obtain a qualitative value of $\partial K'/\partial T$, we will examine how each of the quantities on the right changes with η and T in this chapter. We shall find that α, δ_T, $(\delta_T - K')$, and $\mathbf{k}$ from (3.6) each descend as η decreases, so that we can expect the cross derivative to go down as η decreases. Further insight can be found from the data computed from the theory of Isaak et al. (1990) using an *ab initio* method called the PIB (potential induced breathing) model for MgO. From Isaak et al.'s data, values of $K' = (\partial K_T/\partial P)_T$ versus η are found at various isotherms. Five such isotherms are plotted in Fig. 3.1.

We note that $\partial^2 K_T/\partial P\partial T$, $(\partial K'/\partial T)$, is almost always positive. Only when $\delta_T - K'$ becomes negative and is greater in magnitude than $\mathbf{k}$ can $\partial^2 K_T/\partial P\partial T$ be negative. $\delta_T - K'$ can be a small negative number at high P (see Table 3.5), but it is then dominated by the value of $\mathbf{k}$, which is greater than 1. Typical values of the ambient $\partial^2 K/\partial P/\partial T$ are $1.2 \times 10^{-3}\text{K}^{-1}$ for NaCl computed from (3.5*b*) compared with the experimental value $1 \times 10^{-3} K^{-1}$; and for MgO, $0.43 \times 10^{-3} K^{-1}$ from (3.5b) compared with the experimental value $0.3 \times 10^{-3}\text{K}^{-1}$ (Isaak, 1993).

Values of K_T, K', and P for MgO, also computed from the PIB model, are given in Table 3.1. Using the PIB method, the Helmholtz energy is computed over small increments of volume and temperature. From tables of T,

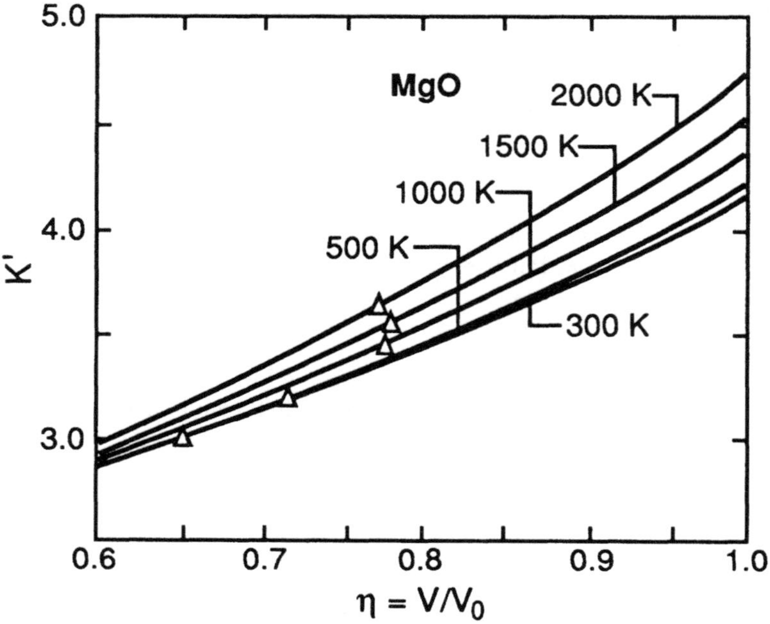

Fig. 3.1. The value of K' versus η along isotherms for MgO (modified from Anderson and Isaak, 1993). Note that the value of $\partial^2 K_T/\partial T \partial P$ diminishes as the value of η decreases. The triangles show the coordinate space (T, η) where $\delta_T = K'$. This constitutes the basis for the $\delta_T - K' = 0$ trace $(w = 0)$ shown in Fig. 3.3.

V, and $\mathcal{F}$, the various properties are found numerically by taking the appropriate derivatives of $\mathcal{F}$, according to the equations given in Sections 1.2 and 1.5.3. Comparing the results of the PIB calculations for K_T and K' with experimental results on MgO, it turns out that the value of K_T is off by 10%, but that of K' agrees within better than 2% (within the experimental error (Isaak et al., 1990)).

The parameter $\partial^2 K/\partial P \partial T$ can be interpreted as the temperature derivative of K', $\partial K'/\partial T$. For MgO, K', which is a function of both T and η, is nearly constant along an adiabat (or geotherm in the earth) up to about 20 GPa. Isaak (1993) has shown that the same occurs for G'. Beyond 20 GPa at high T, K' decreases to a limiting value near 3.4, where change in temperature produces a minimal effect. For oxides and silicates there is little change of $\partial^2 K/\partial P \partial T$ from one mineral to another. An increase of 1000 K at fixed V will increase K' by about 10%. Unless K' is known to better than 20%, there is little point in pursuing the effects of $\partial^2 K/\partial T \partial P$ at high P and high T. Experimentalists seldom agree on K' for a given solid to better than 20% (see Table 1.6).

The value of $\partial^2 K_T/\partial T \partial P$ decreases with increasing P; this is related to the strong dependency of α on η. At low values of η, $\partial^2 K_T/\partial P \partial T$ for MgO has diminished greatly. At P and T corresponding to lower mantle conditions, $\partial^2 K_T/\partial P \partial T$ decreases by about 1/3 from ambient conditions. Thus the mixed derivative may not be a significant correction to the EoS unless K' is known with precision. Uncertainties in K' may mask the effect of the cross derivative on the value of P calculated from the EoS.

Table 3.1. Isotherms of $K'(\eta)$ and $K_T(\eta)$ for MgO. Isotherms of $P(\eta)$—the EoS—for MgO. Units of K_T and P, GPa

η	K' (300)	K' (500)	K' (1000)	K' (1500)	K' (2000)
0.60	2.85	2.86	2.88	2.91	2.96
0.65	2.99	3.01	3.04	3.09	3.16
0.70	3.13	3.15	3.20	3.27	3.35
0.75	3.27	3.30	3.36	3.44	3.54
0.80	3.42	3.45	3.53	3.62	3.73
0.85	3.57	3.61	3.72	3.82	3.94
0.90	3.74	3.79	3.90	4.03	4.18
0.95	3.94	3.98	4.11	4.26	4.44
1.00	4.15	4.21	4.36	4.53	4.74
η	K_T (300)	K_T (500)	K_T (1000)	K_T (1500)	K_T (2000)
0.60	1019	1002	954	901	844
0.65	806	792	753	709	661
0.70	643	631	597	560	520
0.75	515	505	476	444	410
0.80	415	406	381	354	324
0.85	336	328	306	283	257
0.90	273	266	246	226	204
0.95	222	215	199	181	161
1.00	180	175	160	144	128
η	P(300)	P(500)	P(1000)	P(1500)	P(2000)
0.60	260	255	240	224	207
0.65	187	183	172	160	147
0.70	133	130	122	113	103
0.75	94	91	85	79	72
0.80	64	62	58	53	48
0.85	41	40	37	34	30
0.90	24	23	21	19	17
0.95	10.3	10.0	9.2	8.3	7
1.00	0	0	0	0	0

Source: From ***ab-initio*** MgO calculations; Isaak et al., 1990.

3.5. The condition $\delta_T = K'$ in η, T space

Several important physical properties are controlled by the difference of two thermoelastic dimensionless parameters: $\delta_T - K'$. Here we deal with two, αK_T (3.7) and $\partial^2 K_T/\partial P \partial T$ (3.5*b*).

Though both δ_T and K' have been explored at low pressure, it is desirable to know these parameters over a large range of η and T space. At the present time, the experimental approach is not rewarding, because to define the parameters, we need both high pressure and high accuracy.

But by using the theoretical tools of first principles calculations, we can find $\delta_T(\eta, T)$ and $K'(\eta, T)$. The oxide MgO has been explored by these techniques because its band structure is simple, and it is an insulator (Isaak et al., 1990). From the identities given by (3.2) and (3.7) we find

$$\left(\frac{\partial K_T}{\partial T}\right)_\eta \equiv \alpha K_T \left(\delta_T - K'\right). \tag{3.46}$$

From the above equation, $(\partial K_T/\partial T)_V = 0$ results from $\delta_T = K'$, or vice versa. A good theoretical method for finding δ_T at high compression is to find the temperature at which $(\partial K_T/\partial T)_V$ vanishes along an isochore, because in this case $\delta_T = K'$ and K' can be found at the value of V and T from the EoS. If the volume and temperature dependences of the Helmholtz energy are known, then the search for the condition at which $(\partial K_T/\partial T)_V = 0$ is straightforward. The same data set that produced values of K, K', and P for MgO (Isaak et al., 1990) shown in Table 3.1 can be used to find the T at which $(\partial K_T/\partial T)_V$ vanishes along preselected isochores (Anderson and Isaak, 1993), examples of which are given in Fig. 3.2.

Anderson and Isaak (1993) computed $\delta_T(\eta)$ for isotherms using results of *ab initio* calculations for MgO obtained by the PIB approximation, in which δ_T and η were found at a high compression value of η called η_c. The original PIB calculations for MgO are from Isaak et al. (1990).

Because $K' \equiv (\partial K_T/\partial P)_T$ versus η can be found along an isotherm by equations of state, Anderson and Isaak (1993) evaluated δ_T in terms of K'. They found η (called η_c) at which $\delta_T = K'$. It is seen that $\delta_T = K'$ at a value of T where K_T is at a minimum along a selected isochore. Anderson and Isaak (1993) used the PIB calculation to find the value of T where $(\partial K_T/\partial T)_\eta$ is zero for a preselected value of η, which is η_c. In this way pairs of values of η_c, T were found throughout the η, T plane where $\delta_T = K'$, resulting in the plot of the trace when $\delta_T = K'$ (Fig. 3.3).

Birch (1986) found that for NaCl, δ_T is close to K' (Table 3.2) over a large compression range. Thus for NaCl, $(\partial K_T/\partial T)_V = 0$ holds over a wide range in both V and T. For gold (Anderson et al., 1989), in contrast to NaCl, $(\partial K_T/\partial T)_V$ is a large negative number, an order of magnitude greater than found for minerals, no doubt because the conduction electrons in gold yield non-zero values of $(\partial C_V/\partial V)_T$ at high P and T. MgO and KCl are intermediate to the cases of gold and NaCl.

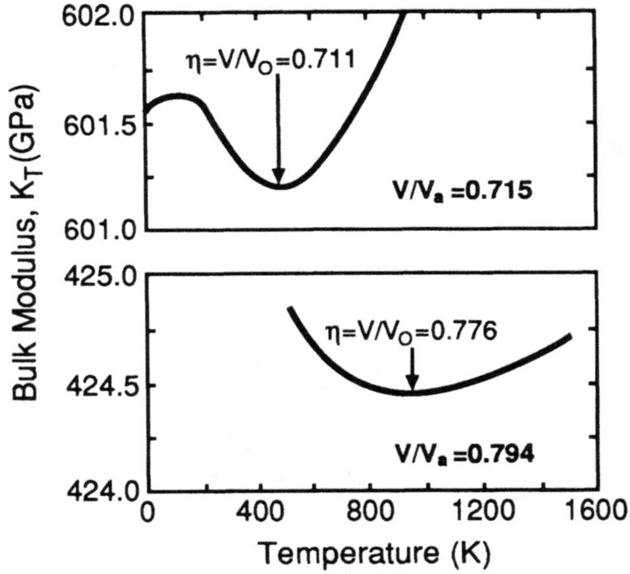

Fig. 3.2. Calculated (PIB) values of bulk modulus K_T at constant volume versus T for MgO. The figure shows the minimum in K_T for two different volumes. V_a is the zero pressure volume at room temperature, and V_0 is the zero pressure volume at the temperature indicated (after Anderson and Isaak, 1993).

Table 3.2. Dimensionless thermoelastic parameters for NaCl versus pressure

P, kbar	η	K'	δ_T	$\left(\frac{1}{\alpha^2}\right)\left(\frac{\partial\alpha}{\partial T}\right)_P$	γ
0	1.000	5.5	5.3	4.8	1.62 ± 0.02
10	0.963	5.0	4.9	4.6	1.55 ± 0.01
20	0.932	4.8	4.7	4.5	1.51 ± 0.01
30	0.907	4.6	4.5	4.5	1.46 ± 0.01
40	0.885	4.4	4.3	4.5	1.43 ± 0.01
50	0.865	4.2	4.2	4.4	1.40 ± 0.01
100	0.791	3.8	3.8	4.3	1.27 ± 0.01
150	0.740	3.4	3.5	3.9	1.19 ± 0.02
200	0.700	3.2	3.1	3.5	1.12 ± 0.02
250	0.669	3.0	2.9	3.1	1.07 ± 0.02
300	0.642	2.8	2.7	2.8	1.03 ± 0.01

Mean values with respect to temperature

Source: After Birch, 1986.

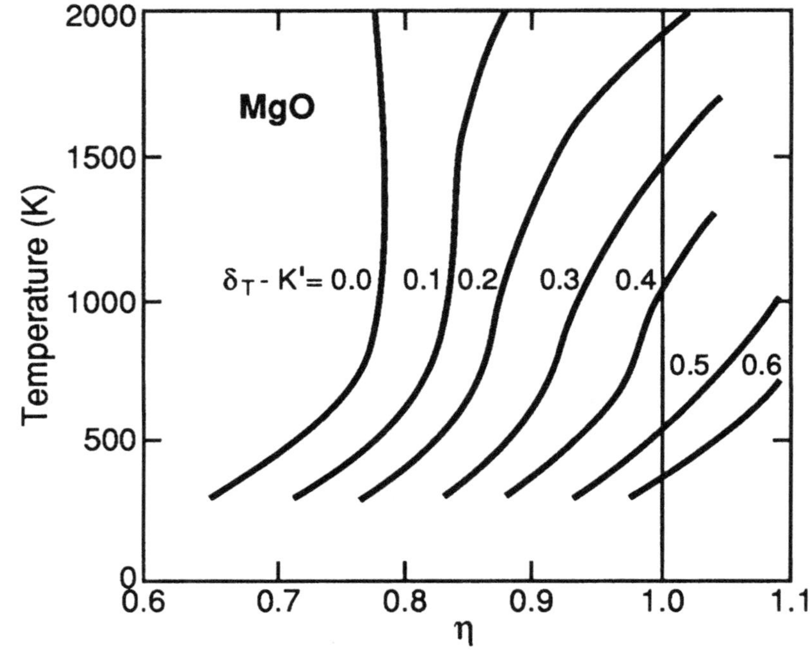

Fig. 3.3. The variation of $\delta_T - K' = -(1/\alpha K_T)\mathbf{w}$ in the η, T space for MgO. $\delta_T - K'$ is the trace in the η, T plane where $\mathbf{w} = 0$, which was found knowing the value of T at which the slope of the K_T versus T isochore vanished $((\partial K_T/\partial T)_V = 0)$, as in Fig. 3.2. Along lines of constant T, $\delta_T = K'$ becomes larger as η becomes larger. At high compression, $\delta_T - K'$ changes very little as T increases along an isobar. Dimensions of $\mathbf{w}$ are $\mathrm{MPa\,K^{-1}}$.

Using the information calculated from a number of cases, such as shown in Fig. 3.2, the set of values of the coordinates (η, T) is found in η, T space for the condition $\delta_T = K'$. The trace of the $\delta_T - K'$ line ($\mathbf{w} = 0$) in the T, η plane for MgO is shown in Fig. 3.3. For $\delta_T > K'$, $\mathbf{w} > 0$, and for $K'_T > \delta_T$, $\mathbf{w} < 0$.

Birch (1986) analyzed the NaCl compression data of Boehler (1981) and the NaCl shock compression data of Fritz et al. (1971), and he found that up to a compression of $\eta = 0.64$, the value of δ_T is close to that of K' (Table 3.2). This means that for this solid, $(\partial K_T/\partial T)_V$ is close to zero throughout a large pressure range at room temperature.

Using the data in Table 3.1 and the curve $\delta_T - K'$ in Fig. 3.3, the value of η at which $\delta_T = K'$ can be placed on the $K'(\eta)$ isotherm as shown by the triangles on solid lines in Fig. 3.1. This point is one coordinate in the δ_T versus η isotherm that δ_T must pass through. The other point on the $\eta = 1$ axis is given by a high $T, P = 0$ experiment, as shown in Fig. 3.4.

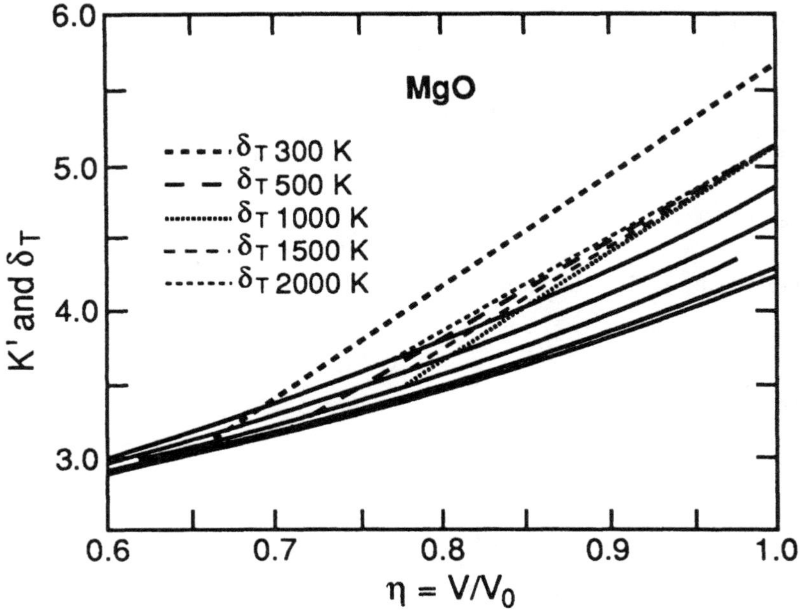

Fig. 3.4. The lower 5 curves for K' (solid lines) were taken from Fig. 3.1. Dashed lines represent $\delta_T(\eta)$ for the various isotherms, all for MgO. Values of η where $\delta_T = K'$ are found at the end of the dashed lines where $\delta_T(\eta)$ intersects the corresponding isotherm of $K'(\eta)$. The positions of these intersections agree with positions of triangles in Fig. 3.1. and with calculations represented by Fig. 3.2 (data from Anderson et al., 1991) (modified from Anderson and Isaak, 1993).

3.6. The isothermal variation of δ_T with compression

3.6.1. *δ_T is linear in η from experiments*

It is clear from several experiments that over a wide range of compression, δ_T decreases approximately linearly as η decreases. However, within experimental error δ_T appears to be independent of η in the range $0.9 < \eta < 1$ (Chopelas, 1990a,b) (see Fig. 4.1). Compressions less than 0.9 correspond to those of the upper mantle and crust. Thus the dependence of δ_T on η becomes important for deep planet interiors and for pressures above about 40 GPa when dealing with oxides and silicates. Below this pressure level, the pressure dependence of δ_T may be ignored.

Some clue to the function of $\delta_T(V)$ might be found in the experiments of Boehler (1983) on alkali metals. His measurements on Li, Na, and K showed that $[\partial \ell n \, (\partial T/\partial P)_S/\partial V]_T$ is linear with η down to a compression of 0.6. If the measured data of NaCl $(\partial T/\partial P)_S$ versus P (Boehler, 1981) are plotted as $\left[\partial \ell n \, (\partial T/\partial P)_S \, /\partial V\right]_T$ versus η, the data for the isotherms are straight lines down to $\eta = 0.86$. $(\partial T/\partial P)_S$ is related to γ using (1.3). Equating the two definitions of γ, we have

$$\gamma = \frac{K_S}{T}\left(\frac{\partial T}{\partial P}\right)_S = \frac{\alpha K_S V}{C_P}.$$

Solving for α from the above equation, we have

$$\alpha = \frac{C_P}{TV}\left[\left(\frac{\partial T}{\partial P}\right)_S\right]. \tag{3.47}$$

δ_T is given by (1.63). Differentiating (3.47) to obtain δ_T,

$$\delta_T = \left(\frac{V}{V_0}\right)m + \left(\frac{\partial \ \ell\mathrm{n}\ C_P}{\partial \ \ell\mathrm{n}\ V}\right) - 1, \tag{3.48}$$

where

$$m = \left[\frac{\partial \ \ell\mathrm{n}\left(\frac{\partial T}{\partial P}\right)_S}{\partial V}\right]_T. \tag{3.49}$$

The dimensionless m has been measured by Boehler and collaborators for many different materials. Boehler reports that m is independent of compression and apparently is virtually material independent. Values of m published by Chopelas and Boehler (1992a) are shown in Table 3.3.

Table 3.3. $m = \left[\partial \ \ell\mathrm{n}\ (\partial T/\partial P)_S / \partial V\right]_T$, a dimensionless parameter

	T (K)	m	Solid	T (K)	m
Pb	298	6.6	In	298	6.9
Al	298	6.3	Cu	298	5.5
Fe	298	6.1	Hg	298, 398	7.3
Li	298	4.8	Na	298	4.8
"	473	6.1	"	473	4.7
"	673	4.5	"	673	5.8
K	298	5.1	Rb	298	6.8
"	473	5.9	Cs	298	6.9
"	673	6.3	KBr	298	6.4
NaCl	298	6.9	RbCl	298	6.5
"	473	7.1	MeOH	298	6.1
"	673	6.9	EtOH	298	6.0
"	873	6.8	pentane	298	6.8
"	1073	6.7	isopentane	298	7.0
MgO	298	7.0	olivine	298	7.8
"	673	6.1	"	673	8.1
"	973	6.1	"	973	6.7

Source: From Chopelas and Boehler, 1992a.

At high T, $\partial C_P/\partial V \to 0$, so that (3.48) becomes

$$\delta_T = m\eta - 1. \tag{3.50}$$

Chopelas (1990b) showed from spectroscopic work that $(\partial \ell n\, C_V/\partial \ell n\, V)_T$ approaches 1 below the Debye temperature, so that at low T

$$\delta_T \to m\eta.$$

There is a temperature dependence in δ_T at high pressure near the Debye temperature, so we use the parameter a in

$$\delta_T = m\eta + a. \tag{3.51}$$

We see that a will be between 1 and zero, depending on the relative value of T with Θ.

This shows that according to the best available experimental data, δ_T is linear in η. However, the rate of change of δ_T with η is small, and for experiments where the limit in η is slightly less than 1, this linear dependence cannot be discerned because of experimental limits. Thus the determination of δ_T from α versus P experiments by Chopelas (1990a), $0.9 < \eta < 1$ indicates that δ_T may be independent of η over the range $0.9 < \eta < 1$, but allows δ_T to descend with further decreasing of η according to (3.51).

3.6.2. $\delta_T(\eta)$ *follows a power law in* η

Birch's (1986) analysis of the NaCl compression data reported by Boehler and Kennedy (1980) shows that δ_T is linear with compression (see Table 3.2). It can be shown that a power law given by (3.52) fits the data as well as a linear curve.

$$\delta_T = \delta_{T_0}\eta^{1.4}. \tag{3.52}$$

A statistical analysis shows that Birch's data for δ_T versus η for NaCl are satisfied with either $\delta_T = -1.696 + 6.879\eta$ or $\delta_T = 5.224\eta^{1.45}$; the correlation coefficient is 0.997 for each solution.

There is very little distinction between the linear law of $\delta_T(\eta)$ represented by (3.47) and the power law given by (3.48). The power law form of $\delta_T(\eta)$, (3.52), is very convenient for use in the mathematical expressions found in thermodynamic identities. We assume, therefore, that the δ_T versus η expression along isotherms for oxides and silicates is of the form

$$\delta_T = \delta_{T_0}\eta^{\kappa}, \tag{3.53}$$

where κ is a positive number near 1.4. In general, κ may change with T. Equation (3.49) is equivalent to the definition of κ given by (3.6). Knowing two points of δ_T for each isotherm in the $\delta_T - \eta$ plane, we can solve for $\delta_T(\eta)$ and evaluate κ in (3.53), as demonstrated below.

3.6.3. *Finding the value of* κ

To use (3.6) in a calculation of δ_T, the value of κ must be known, but this value will depend on the mineral and on temperature. We demonstrate how κ can be determined using MgO as an example.

To find κ we need values of δ_T and η at two widely spaced values of η at each isotherm. One set of δ_T values will be found from the experimentally determined δ_T at $\eta = 1$, setting up the boundary condition (3.53). The other set can be determined by theory at low η. Anderson and Isaak (1993) determined δ_T for MgO at low η by finding the condition where $\mathbf{w}$, (3.46), vanishes, giving the value η at a designated T where $\delta_T = K'$. They used the *ab initio* calculation of free energy (PIB) by which K_T is evaluated from the isothermal variation of energy versus volume. They found $K' = \delta_T$ at a given set of η and T where $\mathbf{w} = 0$. These two sets of δ_T for MgO yield κ in (3.53), which is evaluated for each isotherm. These isotherms are plotted by using (3.54) and are illustrated by the dashed lines in Fig. 3.4.

The two values for $\delta_T\,(\eta)$ obtained for MgO for each isotherm are shown in Table 3.4. For completeness the values of $\delta_T(1)$ for MgO corresponding to the isotherms are listed as follows: 300 K, 5.50; 400 K, 5.2; 500 K, 5.08; 600 K, 5.02; 700 K, 5.0; and for all higher temperatures $\delta_{T_0} = 5.0$.

Three equations for $\delta_T\,(\eta, T)$ are, for example,

$$300\text{ K}: \delta_T\,(\eta, T) = 5.50\,(1,\ 300)\,\eta^{1.41} \tag{3.54}$$

$$1000\text{ K}: \delta_T\,(\eta, T) = 5.00\,(1,\ 1000)\,\eta^{1.48} \tag{3.55}$$

$$2000\text{ K}: \delta_T\,(\eta, T) = 5.00\,(1,\ 2000)\,\eta^{1.23}. \tag{3.56}$$

The plot of $\delta_T\,(\eta)$ for 5 isotherms is shown in Fig. 3.5. Although the value of κ varies from one T to another, such variation has very little consequence in the computation of α. The plot of δ_T versus T at various isochores for MgO is shown in Fig. 3.6.

Table 3.4. Two values of $\delta_T\,(\eta)$ along 18 isotherms for MgO

T	$\delta_T\,(\eta_c)$	$\delta_T\,(1)$	κ	T	$\delta_T\,(\eta_c)$	$\delta_T\,(1)$	κ
300	3.00 (0.650)	5.50	1.41	1200	3.50 (0.781)	5.00	1.44
400	3.10 (0.697)	5.20	1.34	1300	3.50 (0.783)	5.00	1.46
500	3.20 (0.715)	5.08	1.39	1400	3.55 (0.782)	5.00	1.39
600	3.30 (0.741)	5.02	1.40	1500	3.55 (0.782)	5.00	1.39
700	3.35 (0.758)	5.00	1.45	1600	3.55 (0.781)	5.00	1.38
800	3.40 (0.769)	5.00	1.52	1700	3.60 (0.779)	5.00	1.32
900	3.45 (0.775)	5.00	1.50	1800	3.60 (0.778)	5.00	1.31
1000	3.45 (0.778)	5.00	1.48	1900	3.60 (0.776)	5.00	1.29
1100	3.50 (0.780)	5.00	1.44	2000	3.65 (0.774)	5.00	1.23

Source: Anderson and Isaak, 1993.

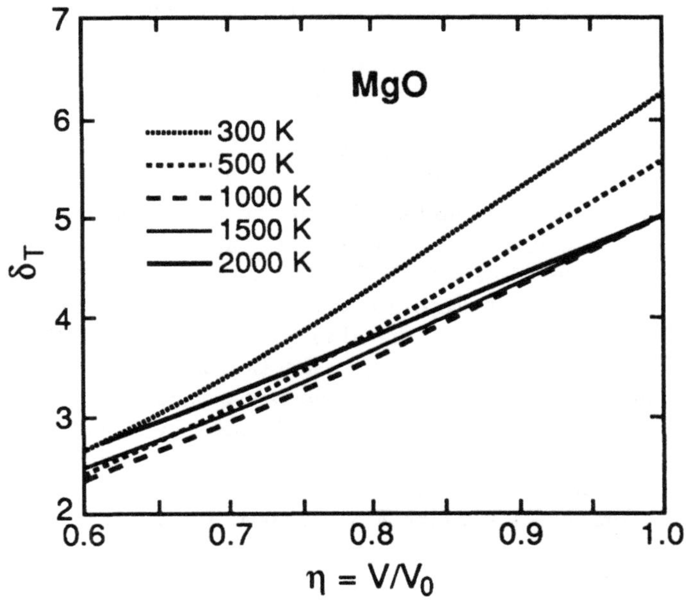

Fig. 3.5. $\delta_T(\eta)$ along various isotherms for MgO, arising from information plotted in Fig. 3.4.

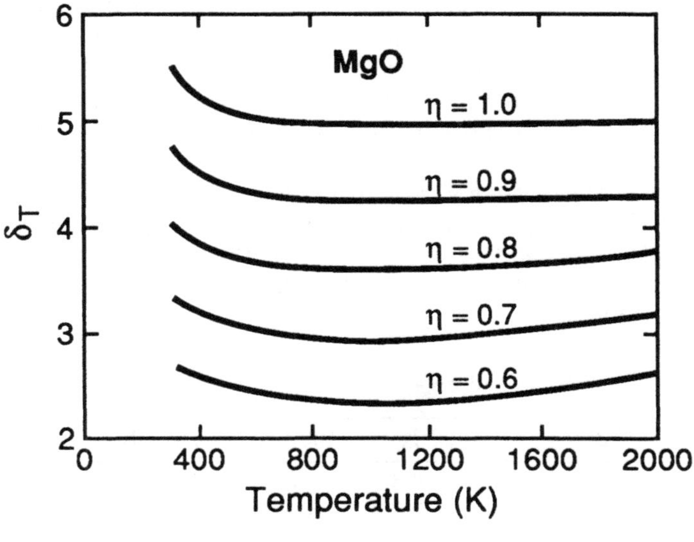

Fig. 3.6. $\delta_T(T)$ along various isochores for MgO, arising from information plotted in Fig. 3.4 (modified from Anderson and Isaak, 1993).

3.7. The isothermal variation of αK_T with compression

Equation (3.7) shows how αK_T must vary with the parameters δ_T and K'. Rewriting (3.7) in an integrable form, we have

$$\frac{\partial \alpha K_T}{\alpha K_T} = (\delta_T - K')\,\frac{\partial \eta}{\eta}. \tag{3.57}$$

Equation (3.57) can be solved for values in the η, T plane for regions where $\delta_T - K'$ is known in the η, T plane. The values of $(\delta_T - K')$ for each set (η, T) are needed to solve (3.57). The value $(\delta_T - K')$ can be found from data represented in Figs. 3.1 and 3.5. The data $\delta_T - K'$ versus η along various isotherms for are listed in Table 3.5. We note that $\delta_T - K'$ decreases with η, and for high compression (low η), $\delta_T - K'$ is negative.

The value of $\delta_T - K'$ affects the important parameter $w = (\partial K_T/\partial T)_V$ given by (3.46). Thus the calculated $\delta_T - K'$ plotted in Fig. 3.3 affects values of w. From this figure, it is noted that at $\eta = 0.7$, the value of $\delta_T - K'$ is close to zero at all values of T above Θ (see Table 3.5). For low compression, or η slightly less than 1 and high T, we see from Fig. 3.4 that $\delta_T > K'$, and $q > 1$. At high compression and high T we see from Fig. 3.4 that $\delta_T < K'$, and therefore $q < 1$.

To find how $\delta_T - K'$ affects αK_T, (3.57) must be integrated using the results shown in Table 3.5, giving αK_T versus η (Fig. 3.7). This shows that αK_T varies very little with η within the lower mantle compression range ($0.74 < \eta < 0.9$) and is consistent with the last column of Table 3.5 where $\delta_T - K'$ for MgO varies slightly around zero. We see that along isotherms at low compression, αK_T descends as η decreases, but after the $\delta_T = K'$ line is crossed, αK_T changes very little as η decreases. Along the 300 K isotherm, $q > 1$, and αK_T is a strong function of η. From Fig. 3.7, it is readily apparent that the results for $T < \Theta$ are substantially different from those for $T > \Theta$.

The constancy of αK_T with η for lower mantle conditions verifies Birch's (1952) assumption that αK_T could be taken independently of the earth's depth. It should be noted, however, that the "mantle conditions" values of αK_T are substantially higher than the ambient value of αK_T, according to Fig. 3.7.

Table 3.5. Values of $\delta_T - K'$ versus $\eta = V/V_0$ for MgO

η	300 K	500 K	1000 K	1500 K	2000 K
0.60	− 0.173	− 0.323	− 0.542	− 0.452	− 0.296
0.70	0.199	− 0.041	− 0.264	− 0.340	− 0.123
0.80	0.604	0.272	0.057	0.044	0.066
0.90	0.995	0.563	0.376	0.293	0.213
1.00	1.350	0.792	0.642	0.466	0.256

Source: From *ab-initio* MgO calculations Isaak et al., 1990.

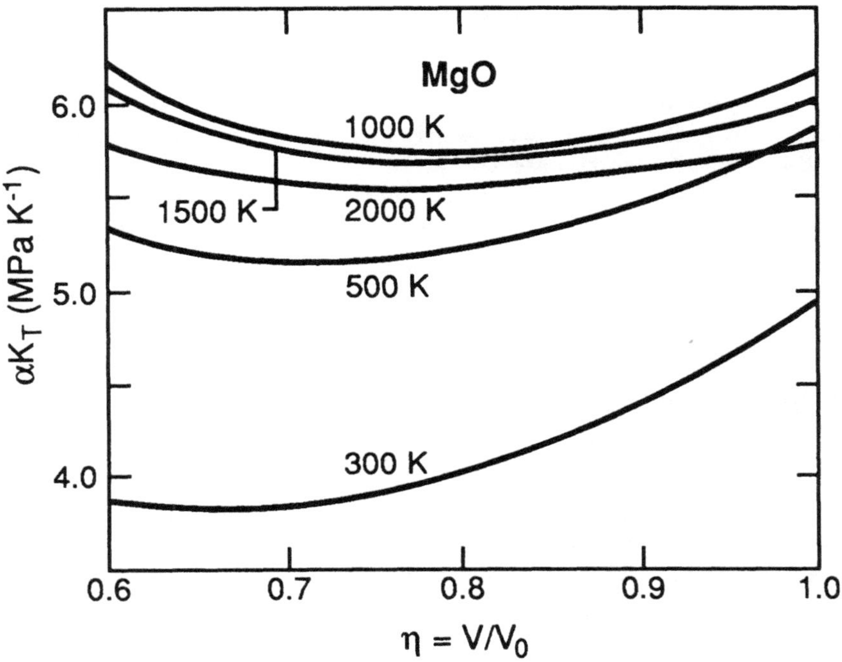

Fig. 3.7. αK_T versus η along isotherms for MgO found by using δ_T plotted in Fig. 3.5 in the solution of (3.57). The path along a low T isotherm shows that αK_T = constant is not valid for $T < \Theta$ (after Anderson et al., 1993).

Most silicates and oxides will be intermediate to the case of NaCl and gold, and there will be regions in which αK_T is not independent of η, whereas in other regions the converse is true. For MgO at P, T conditions throughout the lower mantle, αK_T is roughly independent of η and T. However, any path from low T to high T goes through regions where αK_T varies greatly. Therefore for MgO it is not valid to assume that the value of αK_T found at ambient conditions holds for conditions at high P and high T. Other solids where $\delta_T(\eta, T)$ behaves as for MgO will probably have the αK_T behavior shown in Fig. 3.7. There are two extreme cases.

Birch (1986) showed that for NaCl above room temperature (and therefore above Θ), $\Delta P/\Delta T$ is constant along isochores, and hence αK_T is independent of T, and similarly $\Delta P/\Delta T$ is constant along isochores, and hence αK_T is independent of η. His analysis extended down to $\eta = 0.65$ because he incorporated shock compression data. Birch (1952) assumed that αK_T could, as a first approximation, be treated as independent of density in the deep earth. As an alternate example, gold is a solid where αK_T strongly depends upon η (Anderson et al., 1989). This behavior probably depends on the conduction electrons, not found in insulators.

3.8. The isothermal variation of γ and η with compression

In this section we derive the functional relationship of γ, η, and T. Because $\gamma = V_0 \alpha K_T \eta / C_V$, and as in the previous section we found the dependence of αK_T on η and T, our attention is focused on the dependence of C_V on T and η. At high compression and high T, C_V is invariant with both variables, at least under the quasiharmonic approximation.

As we need to find $(\partial \ln C_V / \partial \ln V)_T$, we start with the identity given by (3.1)

$$\left(\frac{1}{T}\right)\left(\frac{\partial C_V}{\partial V}\right)_T \equiv \left[\frac{\partial(\alpha K_T)}{\partial T}\right]_V. \tag{3.1}$$

Thus we must handle the temperature derivative of αK_T at constant V. After some algebraic manipulations, we arrive at

$$\left(\frac{\partial C_V}{\partial V}\right)_T \equiv \alpha\left[\alpha K_T\, T\,(\hat{\alpha} + K' - 2\delta_T)\right]. \tag{3.10}$$

We see the importance of $\hat{\alpha}$ in (3.10). Because α goes down with temperature, we expect that $(\partial C_V/\partial V)_T$ will be smaller at high T than at lower T. Because $(\partial C_V/\partial V)_T$ vanishes at $T > \Theta$, we observe that

$$\hat{\alpha} + K' - 2\delta_T \rightarrow 0 \quad \text{at } T > \Theta.$$

In Fig. 3.8, it is shown that $\alpha' \rightarrow \delta_T$ at about the Debye temperature. Further, as we have seen from Figs. 2.8 and 3.3, δ_T approaches K' at high pressure. The terms in the brackets in (3.10) add to a very small number at high T, so the above equation is satisfied for $T >> \Theta$. We can express (3.10) in terms of quantities measured at zero pressure, and we find as a good approximation

$$\left(\frac{\partial \ln C_V}{\partial \ln V}\right)_T = \gamma T\left[\alpha_0(\hat{\alpha}_0 - \delta_T) + \ln \eta\left(\frac{\partial \delta_T}{\partial T}\right)\right]. \tag{3.21}$$

From Fig. 3.8 we see that $\hat{\alpha}$ is large near room temperature and plunges rapidly to a value near δ_T at T a little less than Θ. It is the behavior of $\hat{\alpha}$ that influences the volume dependence of C_V at T below Θ.

At high T, $\hat{\alpha} \rightarrow \delta_T$ and $\partial\delta_T/\partial T \rightarrow 0$, which agrees with the usual boundary conditions $(\partial C_V/\partial V)_T \rightarrow 0$ at high T. A plot of $\hat{\alpha}$ and δ_T versus T for MgO is shown in Fig. 3.8. Using data represented by Fig. 3.8 in (3.21), C_V was determined and plotted as isochores in Fig. 3.9.

When the quantities in the brackets of (3.21) do not add to zero, then $(\partial \ln C_V/\partial \ln V)_T$ tends to increase with γT. But with increasing T, the value in the brackets gets smaller. Due to opposing effects, the pressure effect on C_V does not change much with T for a temperature region substantially above Θ. Near and above Θ, the volume effect on Θ is small.

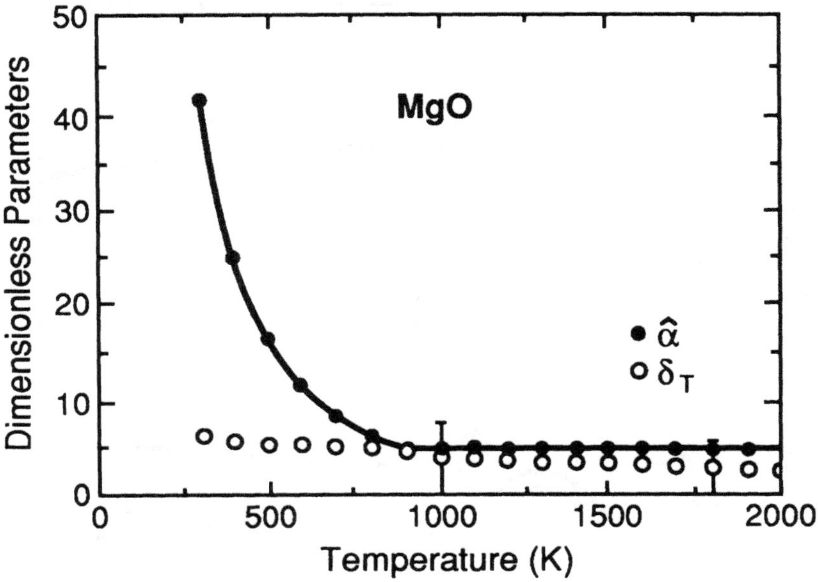

Fig. 3.8. A plot of $\hat{\alpha}$ versus T and δ_T versus T at $\eta = 1$ for MgO. We note that $\hat{\alpha}$ and δ_T converge at high T (after Anderson et al., 1993).

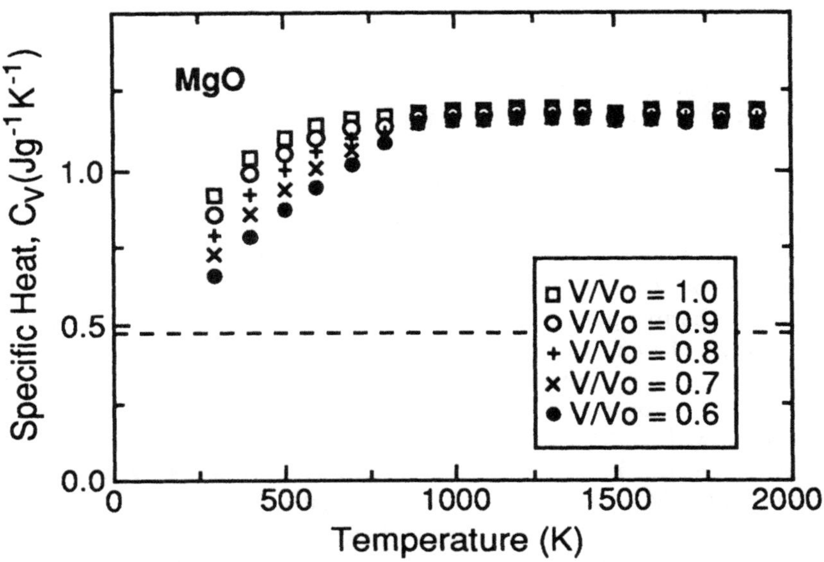

Fig. 3.9. C_V versus T for MgO at various isochores, indicating the shift from the classical C_V curve for $T < \Theta$ (after Anderson et al., 1993).

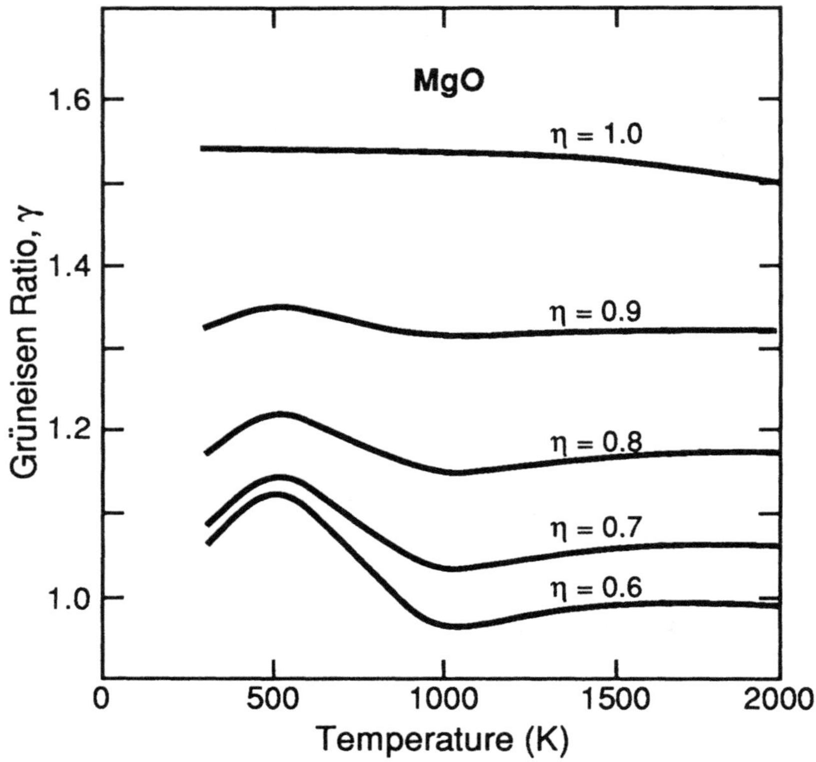

Fig. 3.10. The plot of $\gamma(T)$ along various isochores for MgO (after Anderson et al., 1993).

The data represented in Figs. 3.7 and 3.9 suffice to make the plot of $\gamma(\eta, T)$, according to (3.46), shown in Fig. 3.10. For $T > \Theta$, ($\Theta = 940$ K), γ is independent of T, or nearly so, along isochores. But for $T < \Theta$, γ varies substantially with T at large compressions.

The isothermal plots of γ versus η are shown in Fig. 3.11. In the low T–low P region, the effect of the variation of C_V with η and T is noticeable. There is a significant difference between the results for η for T below and above Θ. The main conclusion is that for $T > \Theta$, T has a small effect on $\gamma(\eta)$. Thus at high T, $(\partial\gamma/\partial T)_V = 0$ is a close approximation: in turn, the Mie-Grüneisen EoS is valid (see Section 1.4.5.) at high T and low η. Conversely, the Mie-Grüneisen EoS is not reliable at high P and low T.

Isaak et al. (1990), using an approximation to first principles methods (PIB), calculated γ (P, T) for MgO. Their values of γ $(\eta, 1000)$ are plotted in Fig. 3.11 as a dashed line that falls quite close to the calculated 1000 K isotherm at high compression, but shows more departure at $\eta = 1$. Departure of the two methods at low compressions arises because laboratory

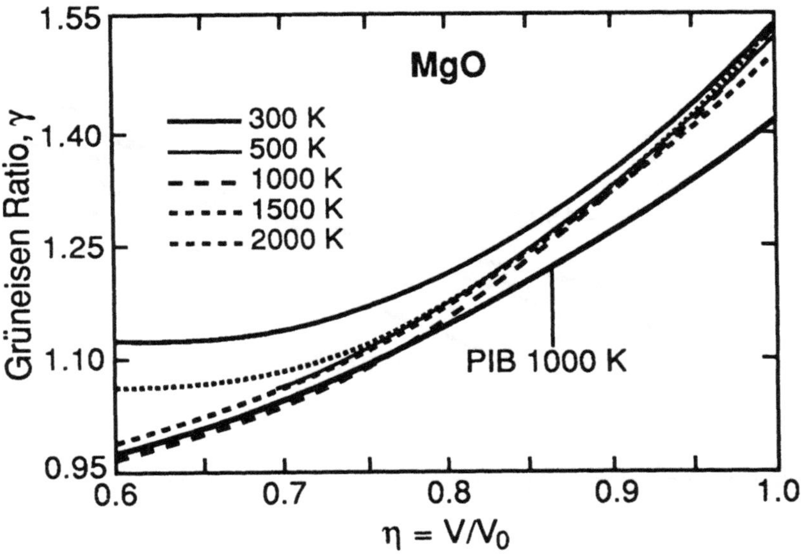

Fig. 3.11. The plot of $\gamma(\eta)$ along various isotherms for MgO arising from (1.24). αK_T and C_V come from data plotted in Figures 3.7 and 3.8, respectively. The computed value of $\gamma(\eta)$ at 1000 K from Isaak et al. (1990) is also shown as the lowest solid line labeled PIB (after Anderson et al., 1993).

data (Anderson et al., 1992a) are used to find γ at $\eta = 1$. Uncertainties in the anharmonic correction to $\mathcal{F}$ make the theory weaker at $\eta = 1$, although it is quite robust at low values of η. At $\eta \sim 0.72$, which corresponds to the compression at the earth's core–mantle boundary pressure, γ is close to unity at all temperatures above Θ, and therefore $\delta_T \approx K'$ or $q \approx 1$.

The data shown in Fig. 3.11 are sufficient to evaluate $q(\eta, T)$ using (3.28). A plot of q versus η for various isotherms for MgO is given in Fig. 3.12. It is clear that near ambient conditions, q is much larger than for the shock wave regime. Conversely, the value of q is smaller than 1 if a large P is exerted at $T = 300$ K. This points out the fallacy of expecting a single value of q to hold for all pressure and temperature regimes, even for the same solid. For the 2000 K isotherm, q is close to 1, though often less than 1, over a wide compression range.

At high T, such as found in shock waves, $q = 1$, and

$$\gamma\rho = \text{constant}. \tag{3.58}$$

At lower values of T, the value of q is sensitive to η; it is often suggested that q obeys a power law in η. $\ell n\ q$ versus $\ell n\ \eta$ is plotted in Fig. 3.13. The plot shows that $\ell n\ q$ versus $\ell n\ \eta$ is not linear in η, but has curvature. Thus a power law in η is inadequate for $q(\eta, T)$.

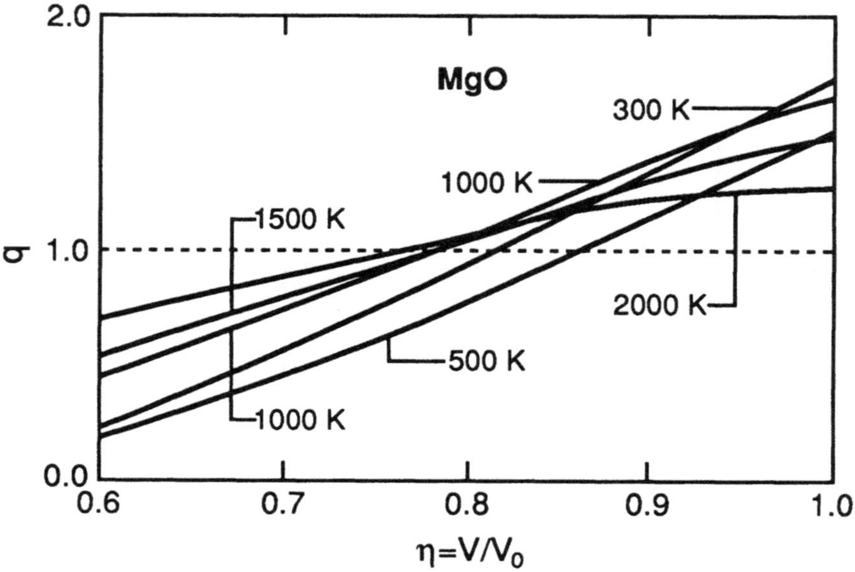

Fig. 3.12. The plot of $q(\eta)$ for various isotherms for MgO. Note that at high temperature, q is close to unity (after Anderson et al., 1993).

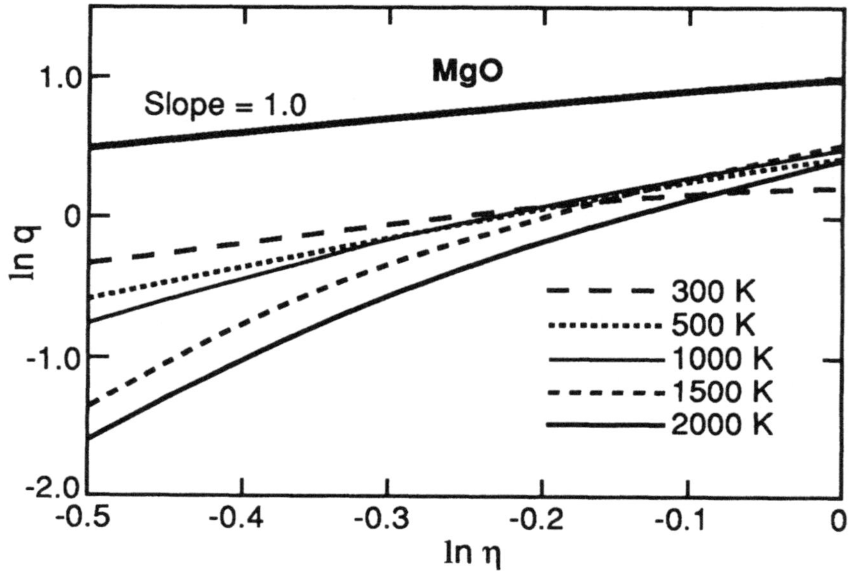

Fig. 3.13. The variation of $\ell n\ q$ with $\ell n\ \eta$ along isotherms, illustrating the difficulty in representing q with a simple power law because of the curvature.

3.9. Experimental insight into the value of $\partial^2 K_T/\partial P\partial T$

In Section 3.4, we discussed the dimensionless thermoelastic parameters

$$\left(\frac{\partial^2 K_T}{\partial P \partial T}\right) = \alpha\delta_T\left(\delta_T - K' + \kappa\right). \tag{3.5b}$$

We noted that theory suggested $\partial^2 K_T/\partial P\partial T$ is almost always positive. This follows from the fact that each parameter in (3.5*b*) is a positive quantity. Thus, only if $\delta_T - K'$ is negative and has magnitude greater than κ can $\partial^2 K_T/\partial P\partial T$ be negative. However, at ambient conditions, δ_T is usually larger than K' (Anderson et al., 1992a). In his Table 1, D.L. Anderson (1988) listed dimensionless parameters for 54 minerals at ambient conditions. We find $K' < \delta_T$ (where $\delta_T \equiv \{K_T\}_P$ in the notation of D.L. Anderson) for all but eight of the 54 minerals. Of these eight, $\delta_T - K' > -0.60$ for five, and $\delta_T - K' < -2.0$ for three (including α-quartz). Since $\kappa \geq 1.0$ for all solids studied to date (see discussion below), we conclude that $\partial^2 K_T/\partial P\partial T$ is positive at ambient conditions (the usual case).

Quantitative assessments of $\partial^2 K_T/\partial P\partial T$ require κ values. Isaak (1993) reviewed the Chopelas and Boehler (1992a) work, derived from the original work of Boehler (1982), showing that $\delta_T(\eta)$ may be expressed as

$$\delta_T = \eta m + \left(\frac{\partial \ln C_P}{\partial \ln V}\right)_T - 1, \tag{3.59}$$

where C_P is the constant pressure heat capacity, and m is defined by

$$m = V_0\left[\frac{\partial \ln\,(\partial T/\partial P)_S}{\partial\eta}\right]_T. \tag{3.60}$$

From (3.59) we have $(\partial\delta_T/\partial\eta)_T = m$ and

$$\kappa = \left.\frac{\partial \ln \delta_T}{\partial \ln \eta}\right|_{\eta=1} = \left.\frac{m}{\delta_T}\right|_{\eta=1}. \tag{3.61}$$

In Table 3.6 are listed κ for several solid compounds calculated from (3.60) and (3.61), all taken from Isaak's (1993) Table 1. α-quartz is anomalous; all the others have positive values of dK'/dT.

Using the $\partial K'/\partial T$ value for MgO from Table 3.6, we find $K'(T)$ by

$$K'(T) = K'(T_0) + \left(\frac{\partial K'}{\partial T}\right)(T - T_0).$$

In the case of MgO, $K'(T_0)$ is 3.8–4.2. Thus at 2000 K, we might expect K' to increase by 0.55 and $K'(2000)$ to be close to 4.4. From Fig. 2.8, for MgO, the discrepancy between K_0 at $T = 300$ K and δ_T at 1800 K is about 0.8. Therefore we conclude that $K'(T)$ increases with T, and, at high T,

$$K'\,(2000) \rightarrow \delta_T\,(2000). \tag{3.62}$$

Table 3.6. Calculated $\partial^2 K_T/\partial P \partial T$ at ambient P, T for several compounds (from Isaak, 1993)

	α (10^{-6}K^{-1})	δ_T	K'	κ	$\partial^2 K_T/\partial P \partial T$ (10^{-3}K^{-1})
NaCl	118	5.67	5.3	1.22	1.1 (0.4)
NaF	98	5.80	5.0	1.64	1.4 (0.3)
KBr	116	5.64	5.1	1.13	1.1 (0.3)
RbCl	119	5.81	5.4	1.10	1.0 (0.3)
MgO	31	5.26	3.8–4.2	1.33	0.39 (0.10)
Olivine	27	5.94	5.2	1.31	0.33 (0.09)
α-Quartz	35	3.28	6.4	2.13	−0.11 (0.07)

From (3.30), equality between δ_T and K' leads to $q^{ht} = 1$, but Fig. 3.3 shows that while K'/δ_T increases with T, it cannot reach 1 ($\mathbf{w} = 0$), even at very large T, when η is close to unity (but it could for small η).

3.10. Comments

I have shown the power of the use of thermodynamic identities combined with experimental data ($P = 0$)at high T to find solutions over the whole η, T field for the thermoelastic parameters, δ_T, γ, q, $\partial K'/\partial T$, and K'.

3.11. Problems

1. Prove (3.47).
2. Prove (3.21) is an approximation to (3.10) at values of η near 1.

4

THERMAL EXPANSIVITY AT HIGH *P* AND *T*

4.1. Introduction

The value of α, the volume thermal expansivity, defined as $(1/V)(\partial V/\partial T)_P$, enters into so many properties and thermoelastic parameters that it must be especially emphasized. Consequently a full chapter is devoted to α. In several thermoelastic parameters, α is a factor in the equation; e.g., $\gamma = (\alpha K_T V/C_V)\,; \delta_T = -(1/\alpha K_T)(\partial K_T/\partial T)_P$. Thus errors in the value of α strongly affect computation of thermodynamic quantities. A thorough understanding of α is necessary in dealing with several problems arising in equations of state.

The importance of knowing the value of the thermal expansivity of candidate materials of the earth's mantle at P, T conditions of the lower mantle is well recognized. Attempts to find the high T, high P values of α for mantle minerals have been made by Anderson (1967), Knittle et al. (1986), Mao et al. (1991), Chopelas and Boehler (1992a), Hemley et al. (1992), Anderson et al. (1992b), and Wang et al. (1994).

The main thermodynamic expression giving the relationship between α and V is a thermodynamic identity, as discussed by Anderson (1967) and Birch (1968), but strongly emphasized by D.L. Anderson (1987):

$$\delta_T \equiv \left(\frac{\partial \,\ell\text{n}\, \alpha}{\partial \,\ell\text{n}\, V}\right)_T . \tag{1.63}$$

This parameter is identical to Anderson-Grüneisen parameter,

$$\delta_T = -\left(\frac{\partial \,\ell\text{n}\, K_T}{\partial \,\ell\text{n}\, V}\right)_P = -\frac{1}{\alpha K_T}\left(\frac{\partial K_T}{\partial T}\right)_P , \tag{1.47}$$

discussed by Barron (1979). The thermodynamic identity (1.63) is derived by combining (1.47) with the better known identity $K_T^2(\partial\alpha/\partial P)_T \equiv (\partial K_T/\partial T)_P$ (Anderson, 1967; Birch, 1968).

4.2. Thermal expansivity at high *T* and constant *η*

We use the equation $\alpha = (\alpha K_T)/K_T$ to evaluate the temperature dependence of α. From (2.45) and (3.33), which define the temperature dependence of α along an isochore, we have

$$\alpha = \frac{1}{V_0}\left(\frac{\gamma C_V}{\eta}\right)\frac{[1 + (\partial \,\ell\text{n}\, C_V/\partial \,\ell\text{n}\, T)_V]}{[K_T(0,\eta) - (\partial K_T/\partial T)_V\, T]} . \tag{4.1}$$

If there is an anharmonic contribution to C_V, it will show up as a nonzero value of $\partial \ln C_V/\partial \ln T$ at high T, which itself is not temperature dependent. A tendency of C_V to increase by anharmonic effects will be exactly compensated for by a corresponding decrease in γ, so the product γC_V will be nearly independent of T. The temperature dependence of α at constant V, if any, will be found in the denominator, and in particular in $(\partial K_T/\partial T)_V T$.

We have seen in Chapter 3 that at high compression and high temperature, $\delta_T \to K'$ and $q \to 1$, which means that $(\partial K_T/\partial T)_V \to 0$. Thus we can expect that α will change very little with T at constant η in this region.

The identity given by (3.4) shows us that if αK_T is independent of T at constant P (true for the quasiharmonic approximation), then

$$\left(\frac{\partial \alpha}{\partial T}\right)_V = 0. \tag{4.2}$$

Thus we can expect that in the high compression regions, where $\delta_T \to K'$, the α versus T curve along isochores will be flat.

4.3. Thermal expansivity versus T at high temperature and constant pressure

In regions where (4.2) holds, it follows from (3.19) that

$$\left(\frac{\partial \alpha}{\partial T}\right)_P = -\alpha K_T \left(\frac{\partial \alpha}{\partial P}\right)_T. \tag{4.3}$$

Using (3.3) to eliminate $(\partial \alpha/\partial P)_T$, we obtain

$$\left(\frac{\partial \alpha}{\partial T}\right)_P = -\frac{\alpha^2}{\alpha K_T}\left(\frac{\partial K_T}{\partial T}\right)_P. \tag{4.4}$$

When placed in (4.4) the definition of δ_T given by (1.47) yields

$$\left(\frac{\partial \alpha}{\partial T}\right)_P = \alpha^2 \delta_T; \tag{4.5}$$

$$\frac{\partial \alpha}{\alpha^2} = \delta_T \partial T. \tag{4.6}$$

Equation (4.6) indicates that at high T along an isobar, where δ_T is independent of T, the slope $(\partial \alpha/\partial T)_P$ increases with T. The fact that δ_T is independent of T above Θ follows from the high temperature limit of the quasiharmonic approximation and is confirmed by experiments, as illustrated in Fig. 2.8. A simple integration of (4.6) yields

$$\frac{1}{\alpha} = \frac{1}{\alpha_{T^*}} - \delta_T (T - T_0) \tag{4.7}$$

or

$$(\alpha/\alpha_{T^*}) = [1 - \alpha_{T^*}\delta_T (T - T_0)]^{-1},$$

where the subscript T^* refers to the beginning point in the high temperature range above Θ. We can approximate the above equation as

$$\alpha = \alpha_{T^*} + \alpha_{T^*}^2 \delta_T(T - T_0), \quad (4.8)$$

so that the high temperature α will be linear in T. Along higher pressure isobars, α_{T^*} will be smaller, which results in a flatter curve of α versus T.

The independence of δ_T from T is a result of the high temperature limit of the quasiharmonic approximation (see Chapter 2). Therefore the linear dependence of α on T is not a result of the assumption of anharmonicity beyond anharmonic effects implied in the quasiharmonic theory.

4.4. Thermal expansivity versus η at constant T

4.4.1. *The high temperature–high compression expression for $\alpha(\eta)$*

As we have seen from Figs. 2.5 and 2.6, at high T αK_T is roughly independent of η and T, a concept proposed by Birch (1952) as a computational device to use when data are lacking. Thus α must decrease with P approximately as K_T increases with P. This can be seen by manipulating the equation for α, (1.24):

$$\alpha = \gamma C_V/(V_0\eta K_T), \quad (4.9)$$

where $V_0 = V$ at $\eta = 1$. Consider now the variables in (4.9) as functions of η. From (1.32) we have

$$\gamma = \gamma_0\eta^q. \quad (4.10)$$

At high P and low η, C_V is independent of T and η, at least in the quasiharmonic approximation. Thus, taking $\alpha_0 = \alpha$ at $\eta = 1$,

$$\alpha = \left[\alpha_0\eta^{q-1}\right]\left[K_T(\eta)/K_T(1)\right]^{-1}. \quad (4.11)$$

In the high pressure region, K' changes but little with η, so keeping K' constant as an approximation, K is linear in P (Anderson, 1967)

$$K_T(\eta)/K_T(1) = \eta^{-K'}. \quad (4.12)$$

Thus (4.11) becomes

$$\alpha/\alpha_0 = \eta^{K'+q-1}. \quad (4.13)$$

From (2.52) at high T and low η, $\delta_T = K' + q - 1$, and thus (4.13) is

$$\alpha/\alpha_0 = \eta^{\delta_T}. \quad (4.14)$$

Figure 3.5 shows that δ_T is in the range of 3 to 5. Thus the effect of η on α is rather strong. From Fig. 3.1 there are many regions in the shock wave regime where $\delta_T = K'$, or $q = 1$. Thus at high T, δ_T can sometimes be replaced by K' in (4.14) (Anderson, 1967; Birch, 1968), so that in this case

$$\alpha/\alpha_0 = \eta^{K'}. \tag{4.15}$$

The product of (4.12) and (4.15) is $\alpha K_T = \alpha_0 K_{T_0}$.

If δ_T in (4.14) is replaced by the average value, $\overline{\delta_T}$, as in the mantle, a useful approximate solution is (see Section 4.12)

$$(\partial \ell\text{n}\ \alpha/\partial \ell\text{n}\ \eta)_T = \overline{\delta_T}. \tag{4.16}$$

To discover how α varies with η over a wide range of η, we must consider (4.15), where δ_T is allowed to be a function of η.

4.4.2. *An equation for $\alpha(\eta)$ along isotherms*

To solve (4.6) for the general case, we need $\delta_T(\eta, T)$. Anderson and Isaak (1993) found that along isotherms,

$$\delta_T = \delta_{T_0}\eta^{\kappa}, \tag{3.53}$$

where δ_{T_0} is the measured value of δ_T at $P = 0$. The justification of (3.49) was presented in Section 3.6.

We eliminate δ_T between (3.49) and (4.6) and integrate to find the following equation for $\alpha(\eta)$ along isotherms:

$$\frac{\alpha}{\alpha_0} = \exp\left[-\frac{\delta_{T_0}}{\kappa}(1-\eta^{\kappa})\right], \tag{4.17}$$

where α_0 is the value of α at $\eta = 1$ for the isotherm considered. According to (4.17) the slope of the ℓn (α/α_0) versus ℓn (ρ/ρ_0) curve is given by

$$\left[\frac{\partial\ \ell\text{n}\ (\alpha/\alpha_0)}{\partial\ \ell\text{n}\ (\rho/\rho_0)}\right]_T = -\delta_{T_0}\left(\frac{\rho}{\rho_0}\right)^{-\kappa}. \tag{4.18}$$

The behavior of the ℓn (α/α_0) versus ℓn (ρ/ρ_0) curve of a typical mineral is shown in Fig. 4.1. Here $\kappa = 1.5$, $\delta_{T_0}\,(T > \Theta) = 5$, and $\delta_{T_0}\,(T < \Theta) > 5$ are assumed.

In regions of η, T space where the measured δ_T appears independent of η $(0.9 < \eta < 1)$ within the error limits, it is correct to take

$$(\alpha/\alpha_0) = (\rho/\rho_0)^{-\delta_T}. \tag{4.19}$$

This formula was derived by Anderson (1967), discussed by Birch (1968), and recently used by Chopelas and Boehler (1989).

For compression values corresponding to the wide pressure range of the lower mantle, δ_{T_0} is not independent of η, even as an approximation, so that (4.14) does not apply, as demonstrated in Fig. 4.1.

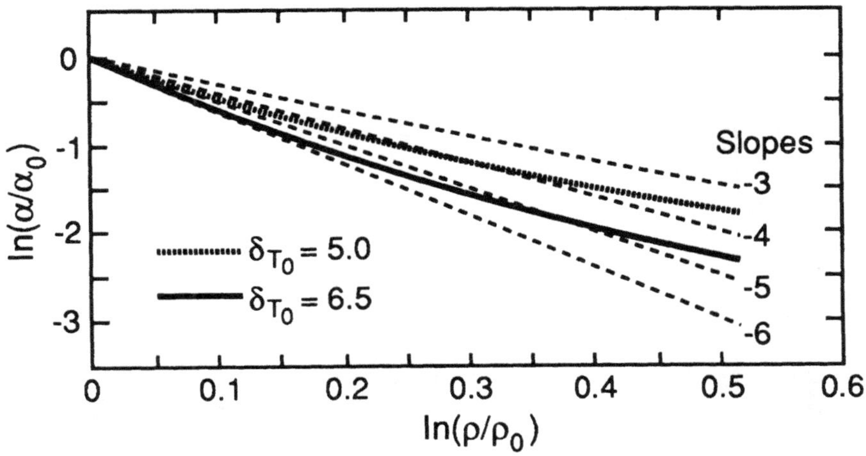

Fig. 4.1. Two cases for $\ell n\ (\alpha/\alpha_0)$ versus $\ell n\ (\rho/\rho_0)$ showing that the slope, δ_T, decreases as the compression increases (after Anderson et al., 1992b). Dashed lines are for constant values of δ_{T_0} where δ_T is unchanged from δ_{T_0}.

4.4.3. *Solution for α at high T and low η for MgO*

Measurements of α are along isobars, usually at $P = 0$. Figure 4.2 shows the measured α for MgO at $P = 0$ and the calculated α at constant η, where the origins of the $\eta = 1$ curve and the $P = 0$ curve coexist at 300 K. Using the observation that αK_T is virtually independent of T at $T > \Theta$ (see Figs. 2.5 and 2.6), we see that (4.2) is approximately valid. Hence the isochoric α (labelled $\eta = 1$) is parallel to the T axis at high T. We use (4.17) to find isochoric α's versus T for lower values of η. These lead to a series of isochores (not shown) parallel to the $\eta = 1$ line.

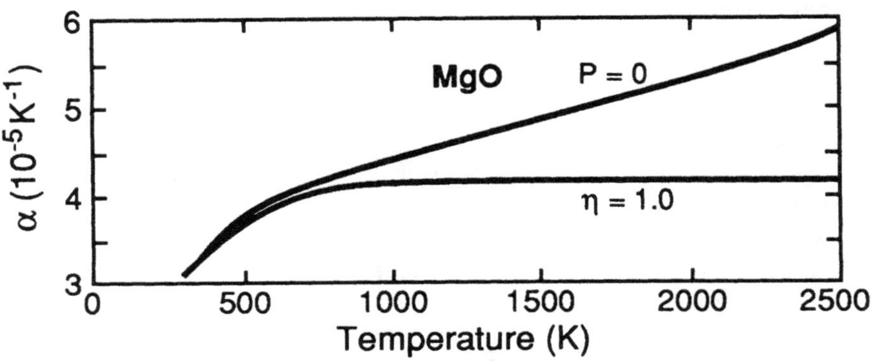

Fig. 4.2. Experimental α versus T for MgO at $P = 0$ (Anderson et al., 1992a). The calculated isochore labelled $\eta = 1.0$ is tied to the $P = 0$ curve at 300 K.

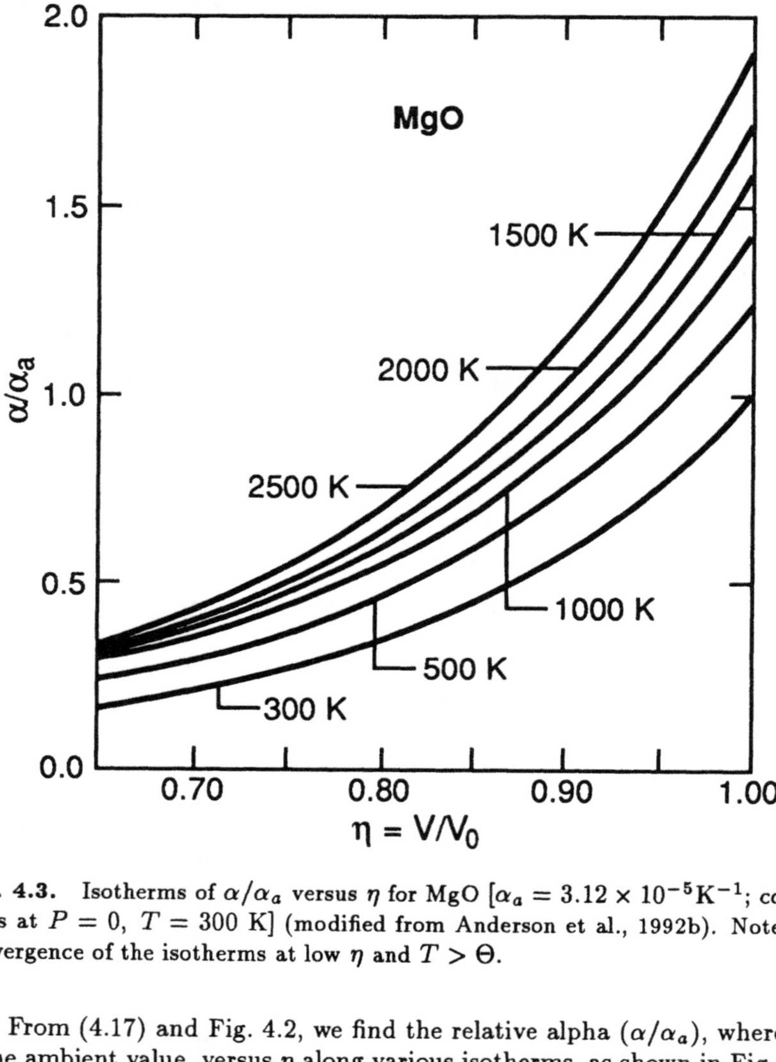

Fig. 4.3. Isotherms of α/α_a versus η for MgO [$\alpha_a = 3.12 \times 10^{-5}\mathrm{K}^{-1}$; conditions at $P = 0$, $T = 300$ K] (modified from Anderson et al., 1992b). Note the convergence of the isotherms at low η and $T > \Theta$.

From (4.17) and Fig. 4.2, we find the relative alpha (α/α_a), where α_a is the ambient value, versus η along various isotherms, as shown in Fig. 4.3. The convergence of the isotherms at high T and low η suggests that the 3000 isotherm will be easily estimated. To get the value of P corresponding to η and T, we use the data from the PIB calculations for MgO (Isaak et al., 1990) that directly define a value of $P(\eta)$ along an isotherm. That the 4th degree EoS method and the PIB model produce the same values of K' and K'' for MgO was demonstrated by Isaak et al. (1992).

To find estimates of α/α_a at high pressures, we must convert from η to P, requiring an EoS. This EoS must be of sufficient degree that K' may be defined as a function of T. Table 4.1 lists corresponding values of η, P and α/α_a shown in Figure 4.3, using EoS data from Table 3.1.

Table 4.1. η, P (GPa) and α/α_a for MgO ($\alpha_a = 3.12 \times 10^{-5}K^{-1}$)

T(K)	η	P	α/α_a	T(K)	η	P	α/α_a	T(K)	η	P	α/α_a
300	1.00	0	1.00	500	1.00	0	1.23	1000	1.00	0	1.42
	0.95	10	0.76		0.95	10	0.96		0.95	9	1.11
	0.90	24	0.58		0.90	23	0.76		0.90	21	0.87
	0.80	64	0.35		0.80	62	0.46		0.80	58	0.55
	0.70	133	0.21		0.70	130	0.29		0.70	122	0.36
	0.65	187	0.16		0.65	183	0.24		0.60	240	0.24
1500	1.00	0	1.56	2000	1.00	0	1.70	2500	1.00	0	1.89
	0.95	8	1.22		0.95	7	1.32		0.95	6	1.47
	0.90	19	0.95		0.90	17	1.04		0.90	15	1.15
	0.80	53	0.60		0.80	48	0.64		0.80	42	0.70
	0.75	78	0.48		0.75	72	0.51		0.75	64	0.55
	0.70	113	0.38		0.70	103	0.40		0.70	93	0.43
	0.65	160	0.31		0.65	147	0.32		0.65	133	0.34
	0.60	224	0.25		0.60	207	0.26		0.60	188	0.27

4.4.4. *Geophysical implications*

We observe from Fig. 4.3 that at high compression, α is virtually independent of T at $T > \Theta$. In the same figure, we see that α decreases rapidly with increasing compression, and for isotherms where T is larger than Θ (920 K) and for $\eta < 0.8$, 80 GPa or so, there is little temperature effect on α.

It would appear that at the pressures corresponding to the deepest part of the lower mantle (100–120 GPa), $\eta \sim 0.68$, the expansivity ratio α/α_a is near 0.33 independent of the temperature (as long as $T > \Theta$). This gives $\alpha = 1 \times 10^{-5}K^{-1}$ at the P, T conditions at the base of the mantle, which agrees well with the findings of Chopelas and Boehler (1992a).

The adiabatically uncompressed (hot) lower mantle is thought to be at a temperature near 1800–2000 K. For $P = 0$ at 2000 K, $\alpha/\alpha_a = 1.70$. The pressure at the top of the lower mantle (670 km) is 24 GPa or about $\eta = 0.88$ at 2000 K from Table 4.1. For these conditions, we see from Table 4.1. that the value of α/α_a is close to 1. This means that at the conditions at the top of the lower mantle, 667 km, the pressure and temperature effects on α have nearly cancelled, and α (670 km) must be close to α_a. Although V is nearly quadratic with T at $P = 0$ (because α is increasing with T), V is approximately linear with T at pressures and temperatures near the top of the lower mantle (because here α is independent of T).

For pressures greater than 80 GPa or so, there is no significant distinction between α along the 1000, 2000, and 3000 K isotherms. Thus, α is independent of T at high T. Generally α spans the range $0.33\alpha_a$–$1.4\alpha_a$ throughout upper and lower mantle P, T conditions.

4.5. Measurements of *V* versus *T* for silicate perovskite

The effect of temperature on volume $P = 0$ for Fe-bearing silicate perovskite was first reported by Knittle et al. (1986), as shown by the squares in Fig. 4.4, where V/V_{298} versus T at constant η is plotted. The value of α is found by the appropriate slope of a curve passing through these data. Because of the extent of the error bars, various interpretations have led to quite different values of α, all of which satisfy the Knittle et al. data. A curve passing through the lower parts of the error bars at low T and the higher parts of the error bars at high T has relatively high curvature and leads to a high value of α. Such a curve (not shown) was chosen by Knittle et al., leading to their announced value of $\alpha = 4 \times 10^{-5}\text{K}^{-1}/\text{deg}$ for α_a.

Other experimenters have satisfied the Knittle et al. measured data with lower values of α. Ross and Hazen (1989) found $\alpha_a = 1.8 \times 10^{-5}\text{K}^{-1}$. Navrotsky (1989), using the pressure dependence of entropy, found V/V_{298} versus T to be $\alpha_a = 1.8 \times 10^{-5}\text{K}^{-1}$ (see Fig. 4.4).

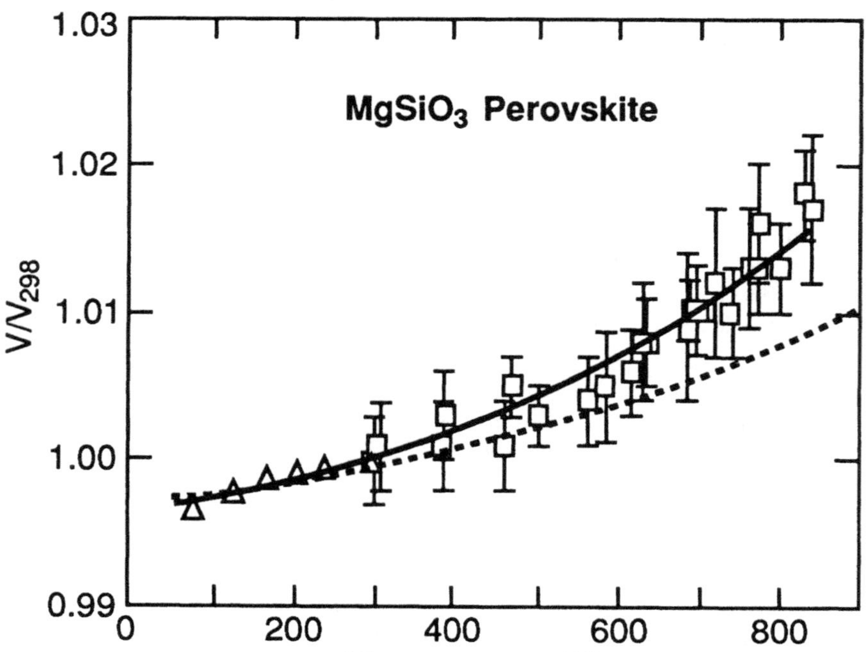

Fig. 4.4. Relative volume, V/V_{298}, of $MgSiO_3$ perovskite at various temperatures. The squares represent data of Knittle et al. (1986), the triangles those of Ross and Hazen (1989). The solid curve is the high T extrapolation by Ross and Hazen (1989); the dashed curve is calculated from $(dS/dP)_T$, Navrotsky (1989). The figure is modified from Navrotsky (1989).

Such low values of α have been reported by others. Parise et al. (1990) reported $\alpha_a = 1.9 \times 10^{-5}\mathrm{K}^{-1}$ at $P = 0$, and Mao et al. (1991) reported $\alpha = 1.3 \times 10^{-4}\mathrm{K}^{-1}$ at 10 GPa. Funamori and Yagi (1993) reported measuring V versus T isobarically from 300 K to 900 K. This resulted in their report of $\alpha \geq 1.7 \times 10^{-5}\mathrm{K}^{-1}$ at 36 GPa for $MgSiO_3$ perovskite.

Wang et al. (1994) measured the volume expansion of $(Mg, Fe)SiO_3$-perovskite from 400 K to 1200 K along several isobars (see Fig. 4.5). They found $\alpha = 2.7\times 10^{-5}\mathrm{K}^{-1}$ for Fe/(Mg + Fe) = 0.12 at 800 K at $P = 0$. Wang et al. measured $(\partial K_T/\partial T)_P = -0.023$ GPa/degree, and, as $K_{T_0} = 246$ GPa at $P = 0$ (Yeganeh-Haeri et al., 1989) or $K_T = 276$ GPa at 8 GPa, δ_T is 4.2 by (1.47). This value compares well with those for solids at modest pressure (see Fig. 3.5). Wang et al. (1994) reported that $\gamma_0 = 1.3$ was consistent with their data.

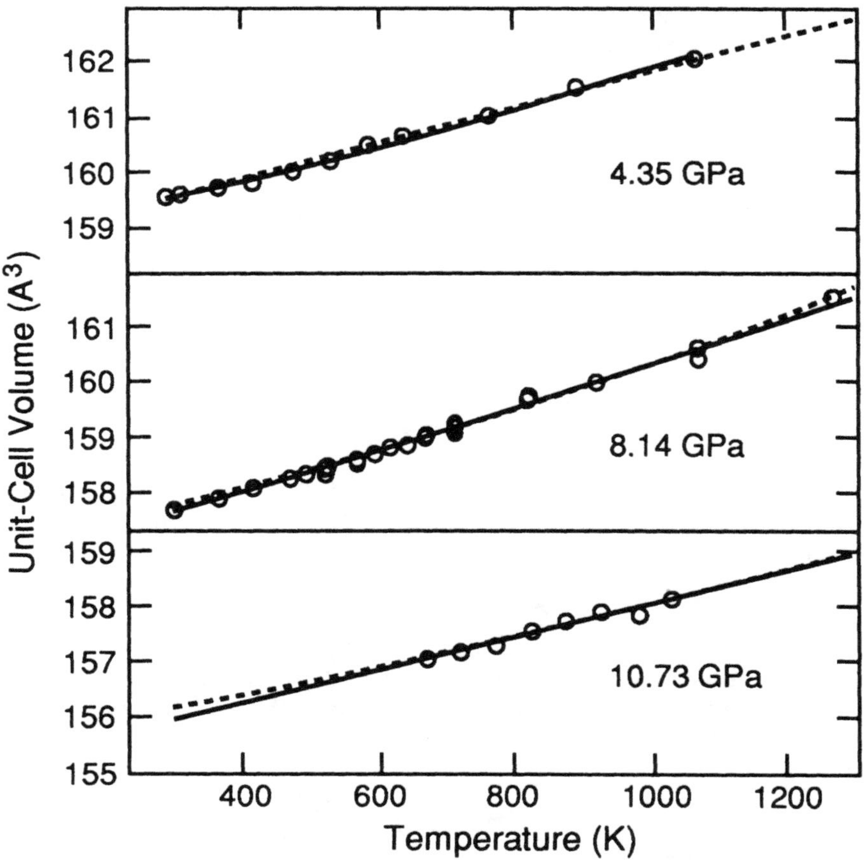

Fig. 4.5. V versus T at 3 isobars for orthorhombic perovskite $(Mg,Fe)SiO_3$ (modified from Wang et al., 1994). The dashed lines represent a Suzuki equation fit (see Section 4.7).

4.6. Grüneisen's theory of thermal expansivity ($P = 0$)

Many experimentalists use a variant of Grüneisen's (1912) theory to represent their measured data on $\alpha(T)$ at $P = 0$. The primary data consist of changes in length with regard to some standard length, L_0.

There are three directions of measurement in a crystal

$$Y_i = \frac{L_i - L_0}{L_0}, \tag{4.20}$$

and the change in volume is given by $Y = Y_1 + Y_2 + Y_3$.

Grüneisen's theory (1912) involved using $Y = \Delta/V_0$ as a primary variable, where

$$\Delta = V - V_0.$$

He assumed the Mie-Grüneisen EoS, with its simplified expression for thermal energy (see Section 1.34),

$$P = P_0(V) + \frac{\gamma_{mg} E_{TH}}{V}. \tag{4.21}$$

Another way of representing (4.21) is

$$VP = VP_0(V) + \gamma_{mg} E_{TH} = G(V). \tag{4.22}$$

Assuming infinitesimal elasticity, Grüneisen got rid of P_0 by taking

$$K_{T_0}\Delta = P_0 V.$$

He then expanded $G(V)$ as a MacLaurin's series to the second order in Δ, which leads to a quadratic equation in Δ, of the form

$$\Delta\left(1 - \frac{\mathrm{k}}{V_0}\Delta\right) = \frac{\gamma_{mg} E_{TH}}{K_{T_0}}, \tag{4.23}$$

where

$$\mathrm{k} = \left(\frac{1}{2}\right)(K' - 1). \tag{4.24}$$

He reduced (4.23) to a linear equation in Δ by reapplying the infinitesimal formula for Δ to the second term in the parentheses. This was justified because the thermally induced dilation is small. Therefore (4.23) is replaced by

$$\Delta\left(1 - \mathrm{k}\gamma_{mg}\frac{E_{TH}}{K_{T_0}}\right) = \frac{\gamma_{mg} E_{TH}}{K_{T_0}}. \tag{4.25}$$

Solving for Δ versus T,

$$Y = \frac{\Delta}{V_0} = \frac{E_{TH}(T,\Theta)}{Q - \mathrm{k}E_{TH}(T,0)}, \tag{4.26}$$

where

$$Q = \frac{K_{T_0} V_0}{\gamma_{mg}}. \tag{4.27}$$

The thermal expansivity is found by differentiating (4.26) with respect to T and with the conditions that $\partial \mathbf{k}/\partial T = 0$ and $\partial K'/\partial T = 0$. These are not exactly zero, but nevertheless are small compared with $\partial E_{TH}/\partial T$. We have

$$\alpha = \left(\frac{\partial Y}{\partial T}\right)_P = \frac{\left(\dfrac{\partial E_{TH}}{\partial T}\right)_P}{Q\left(1 - \dfrac{\mathrm{k} E_{TH}}{Q}\right)^2}. \tag{4.28}$$

Equation (4.26) or (4.28) is used semi-empirically to fit the experimental data of Y versus T, and the parameters Q, E_{TH}, and k are varied until the fit is maximized. Once the parameters are known, (4.28) is used to extrapolate or interpolate α. It is customary to use a Debye function for E_{TH}, so that

$$E_{TH} = D\left(\frac{\Theta}{T}\right), \tag{4.29}$$

where Θ is the Debye temperature. A procedure fitting measured V-T data to the Grüneisen equation will produce empirical values of Θ, K_0', and γ_0. These values should then be checked against other determinations of Θ, K_0', and γ_0 for the same mineral. If the values are reasonably close, the Grüneisen equation may be used with confidence to extrapolate for values of α at a temperature above the measurement range. If they do not agree, the resulting Grüneisen equation is unreliable for extrapolation purposes.

The empirical values of the variables found in (4.28) corresponding to the graph for forsterite, Fig. 2.7, found by Suzuki (1973) are: $Q = 5.14 \times 10^6 J$, k $= 2.67$, and $\Theta = 705°$. From k, $K_0' = 4.34$, which is a little lower than the acoustically determined value, 5.4 (Table 1.8). Some difference is to be expected in view of the approximations and experimental errors in K'. The value of the Debye temperature is considerably lower than the specific heat value, $\Theta = 900°$, but close to the acoustically determined value, $\Theta = 767°$. The value of Q can be used to estimate the ratio of K_{T_0}/Y. Using the measured value, $K_{T_0} = 126.2$ GPa, γ_{mg} is found to be 1.1 compared with the measured value for forsterite, $\gamma = 1.15$ (Table 1.2.). The values of K_0', Θ, and γ_{mg} thus determined from the fit to the Suzuki equation for fosterite are nevertheless reasonable, so (4.28) is justifiably used for representing α with T at high T.

After cross checking the parameters with independent measurements, the Suzuki equation can be used to extrapolate to T beyond the limits of measurements. Alternatively, (2.52) can be used for the extrapolation. There is little difference between the results of the two methods. When extrapolated, data from Fig. 2.7 lead to the dashed lines in Fig. 4.6.

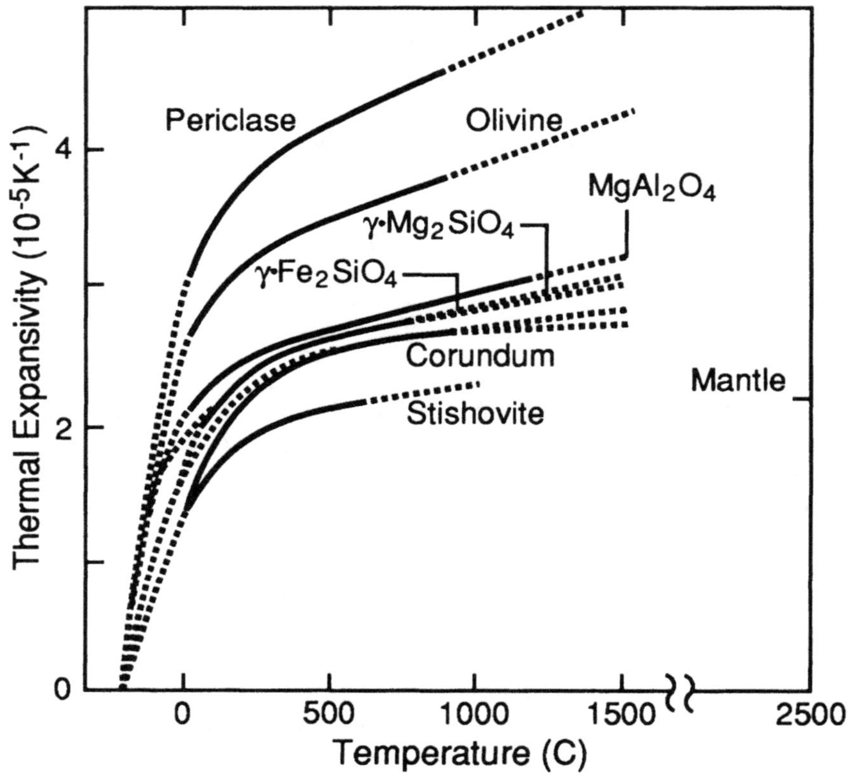

Fig. 4.6. Extrapolation of α for the solids in Fig. 2.7, as shown by the dashed lines using (2.52). At higher temperatures (above Θ) the expansivity is essentially linear with a positive slope (modified from Suzuki et al., 1979).

4.7. Suzuki's theory of thermal expansivity

The Grüneisen theory of thermal expansion has drawbacks that prompted Suzuki et al. (1979) to reconsider (4.23) without the drastic assumption induced by going from (4.23) to (4.25). Equation (4.23) is a quadratic equation in Δ. Solving for the real root of (4.23) determines Δ, which is referenced to the volume at $T = 0$. The expansivity, based on the volume at an arbitrary reference temperature, is

$$Y = \frac{\left[1 + 2\text{k} - \left(1 - \dfrac{4\text{k}E_{TH}}{Q}\right)^{1/2}\right]}{2\text{k}\, a_V} - 1, \tag{4.30}$$

where a_V is the ratio of volume at the reference T, T_r, and at $T = 0$ K; $a_V = V(T_r)/V(0)$. Because $\eta = V/V_0 = 1 + Y$, the solution in terms of η is

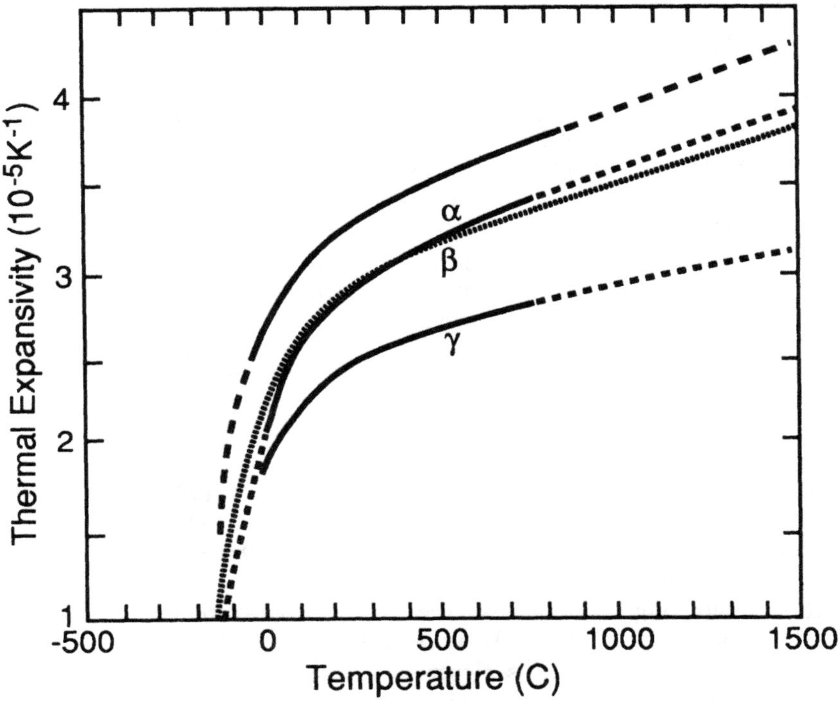

Fig. 4.7. Comparison of thermal expansivities of three polymorphs of Mg_2SiO_4. α : α-phase; olivine of Suzuki (1975); although this specimen includes some fraction of Fe component, the data are regarded as the most reliable and representative ones for α-Mg_2SiO_4. β : β-phase; Solution II, shown by a dotted line, is for a lower value of Θ (830 K). The solid line shows Mg_2SiO_4 for the preferred Θ (925 K). γ : γ-phase (modified from Suzuki et al., 1980).

$$\eta = \frac{\left[1 + 2k - \left(1 - \frac{4kE_{TH}}{Q}\right)^{1/2}\right]}{2k\, a_V} - 1. \tag{4.31}$$

If we wish to refer to the volume at room temperature, $V(T_0)$, where $V/V(T_0) = (V/V_0)[V_0/V(T_0)]$, then (4.31) is replaced by

$$\frac{V}{V(T_0)} = \frac{\left[1 + 2k - \left(1 - \frac{4kE_{TH}}{Q}\right)^{1/2}\right]\left(\frac{V_0}{V(T_0)}\right)}{2k\, a_V}. \tag{4.32}$$

The analytical expression for α found by differentiation of (4.32) is quite complicated. Numerical differentiation of (4.32) is recommended for $\alpha(T)$. Equations 4.30, 4.31, and 4.32 are known as the Suzuki equations.

Table 4.2. Values of V/V_0 versus T and P for NaCl

P(GPa)	T(C) 25	100	200	300	400	500
0	1	1.00928	1.02253	1.03682	1.05226	1.06908
5	.98020	.98833	.99989	1.01247	1.02575	1.04014
10	.96268	.96999	.98024	.99124	1.02290	1.01538
15	.94690	.95356	.96277	.97248	.98287	.99382
20	.93248	.93857	.94699	.95571	.96509	.97478
25	.91913	.92471	.93251	.94056	.94913	.95776
30	.90667	.91172	.91907	.92678	.93468	.94241
35	.89492	.89943	.90647	.91416	.92150	.92845

Data reported by Boehler and Kennedy, 1980.

The thermal expansivities of three polymorphs of Mg_2SiO_4 have been measured and examined by Suzuki et al. (1979). The thermal expansion was analyzed by (4.32), and the resulting thermal expansion coefficients were plotted in Fig. 4.7.

It is interesting to note that the high temperature value of α for forsterite is near $3.8 \times 10^{-5}K^{-1}$, whereas that for γ-Mg_2SiO_4—the high pressure phase of forsterite—is closer to $2.6 \times 10^{-5}K^{-1}$. We might expect, therefore, that a phase change going to the denser phase ordinarily decreases α if a transition to a denser phase increases K_{T_0}. This further suggests that across a phase boundary, αK_T may change little.

4.8. High temperature expansivity of NaCl

Boehler and Kennedy (1980) measured the compression of NaCl over a pressure range of 3.5 GPa and a temperature range of 25 – 500°C. These data are shown in Table 4.2, where η is listed for a given P and T. All the data are for temperatures above the Debye temperature, where the approximation $E_{TH} = 6RT$ is good.

From these data we can test for the limits of applicability of the Suzuki equation (4.32). Using (4.24) and $K'_{T_0} = 5.35$ at 300 K (Spetzler et al., 1972), we find k = 2.18. Q is found by requiring that (4.31) be unity at $T_0 = 298$ K, which reveals that $Q = 21.57E_{TH_0}$, where $E_{TH_0} = 6RT_0$, and where $V(T = 0)/V(298) = 0.95$.

From (4.30) we find an imaginary root when $E_{TH} > (1/4k)Q$, or when $T/T_0 > 2.47$. Thus, if we fix the parameters of (4.30) at room temperature conditions, we see that we cannot account for the volume expansion at the highest measurements ($T = 2.63T_0$). What this means is that the expansion $G(V)$ into only two terms in $(V - V_0)$ in (4.23) leading to (4.25) is inadequate for very high temperature applications.

We conclude that (4.26) and (4.28) are useful at intermediate temperatures and at temperatures up to about 2Θ, but not much higher, if reasonably accurate values of the material parameters Θ, K'_{T_0}, and γ from α versus T measurements are required.

On the other hand, k and Q can be considered as arbitrary parameters to make (4.26) or (4.32) fit measured expansivity data up to rather high temperatures. This may be considered as a useful empirical representation for interpolation of the data. But in that case, the derived values of K'_{T_0}, Θ, and γ from k and Q will probably not have physical meaning. For example, γ_0 may be too high, or Θ too low. The use of such values in other thermodynamic formulas may be dangerous.

4.9. The uncompressed value of α in the lower mantle

The uncompressed thermal expansivity is intimately connected with the choice of γ_0, the uncompressed Grüneisen ratio. Let ρ_0 be the uncompressed (hot) value of ρ at some T, and ρ_{0_c} be the uncompressed cold value of the density. Then at $P = 0$,

$$Y = \frac{\rho_{0_c} - \rho_0}{\rho_{0_c}} \tag{4.33}$$

defines the difference between the hot uncompressed and cold uncompressed densities, where ρ_{0_c} represents the cold value of ρ_0.

From seismological studies, $K'_{T_0} = 4.2$ (Bukowinski and Wolf, 1990) in the lower mantle, so by using (4.24) we find k $= 1.60$. This low value of k will insure the use of (4.31) or (4.32) to relatively high temperatures, avoiding the problem of imaginary roots described above. E_{TH} is the Debye function in (4.29), specified at various temperatures by the value of Θ. Extrapolating the results of Θ of the mantle as listed in Table 1.10 back to zero pressure or zero depth, we find $\Theta = 940$ (hot) and 1040 (cold). Because $Q = K_{T_0} V_0/\gamma_0 = K_{T_0}\mu/(\rho_{0_c}\gamma_0)$, a choice is needed for ρ_{0_c} and μ for the lower mantle. Values of $\rho_{0_c} = 4.22$ and $\mu = 21$ are used. We will take the value of K_{T_0} to be 210 GPa (hot) (Bukowinski and Wolf, 1990) and 255 GPa (cold).

Once γ_0 has been chosen, Q_0 is determined for the mantle. The uncompressed volume ratio can thus be computed by differentiating (4.30) at temperatures throughout the lower mantle, given an assumed value of γ_0. The calculated $V/(V(T_0))$ is plotted in Fig. 4.8, where $V(T_0)$ is the volume at 300 K for various reasonable values of γ_0. The uncompressed expansivity coefficient, $\alpha = (\partial Y/\partial T)_P$, is obtained by temperature differentiation of the data shown in Fig. 4.8. The result for α versus T for uncompressed lower mantle conditions is plotted in Fig. 4.9. Note from Fig. 4.9 that α for 1000°K at $P = 0$ is about $2.3 \times 10^{-5}\mathrm{K}^{-1}$, which is above the value found by Wang et al. (1994), but below that recommended by Knittle et al. (1986).

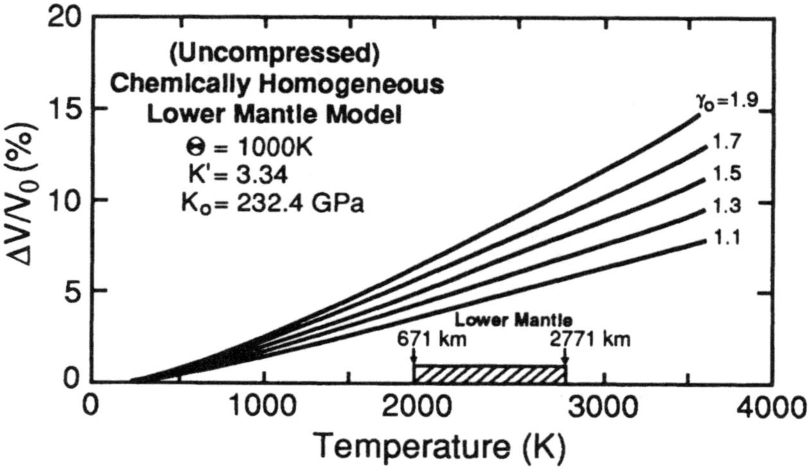

Fig. 4.8. Expansion of the uncompressed lower mantle vs. T: $(V - V_0)/V_0$, according to (4.30) for various assumed values of γ_0. For $\gamma_0 = 1.5$, $\Delta V/V_0 = 5\%$ at 1900 K, the top of the lower mantle (modified from Anderson and Sumino, 1980).

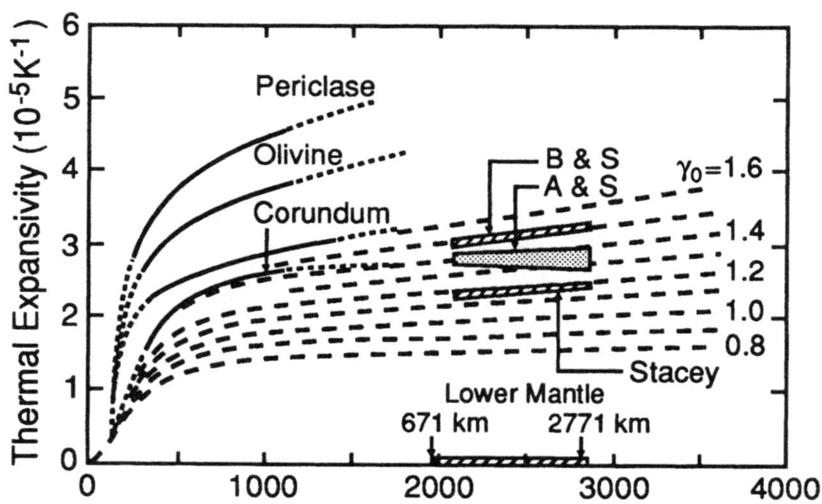

Fig. 4.9. The thermal expansivity of the mantle calculated at zero pressure for the thermodynamic parameters of the mantle used in (4.37) and from data plotted in Fig. 4.6. Calculated α depends on an assumed value of the uncompressed Grüneisen ratio, γ_0. Values of γ_0 from the thermal models of Brown and Shankland (1981), Anderson and Sumino (1980), and Stacey (1977b), are indicated. α of a few minerals is shown (modified from Anderson and Sumino, 1980).

Considering that the temperature of the upper-lower mantle boundary is about 1900 K, the $\Delta V/V_0$ of the uncompressed mantle from 300 to 1900 K is about 4.5% according to Fig. 4.8, depending on the assumed value of γ_0. The Brown and Shankland (1981) study and the Anderson and Sumino (1980) study gave $\gamma_0 = 1.5$ and $\gamma_0 = 1.4$, respectively (Fig. 4.9). A value of γ much less than $\gamma_0 = 1.3$ will seriously limit the uncompressed expansion of the lower mantle material.

The value of γ decreases with pressure. From (4.17) we observe that if α decreases from, say, $3 \times 10^{-5}\text{K}^{-1}$ at the surface to about $1 \times 10^{-5}\text{K}^{-1}$ at the base of the mantle, γ must also decrease from 1.5 to something less than unity, say 0.9, at the base of the mantle.

4.10. Obtaining α from γ using data from seismic models

If γ is known as a function of depth in the mantle, as obtained from the seismic model solutions, α can be solved from (4.32). In the mantle, ρ_0 and K_S are provided by seismic models (however, there is no information on C_P). From the equation of γ, we obtain

$$\alpha = \frac{C_V \gamma}{V K_T}. \tag{4.34}$$

Using as an approximation $K_T = K_S(1 - \alpha\gamma T)$, and using $V = \mu/\rho$ where $\mu = M/p$, the mean atomic weight, (4.34) becomes

$$\alpha = \frac{\dfrac{\gamma \rho C_V}{\mu K_S}}{1 - \dfrac{\gamma^2 \rho T C_V}{\mu K_S}}. \tag{4.35}$$

Thus α is determined at every depth if γ, T, μ, and C_V are known. Watt et al. (1975) showed that throughout the lower mantle, μ is constant and close to $\mu = 21$. Neglecting specific anharmonic contributions and taking $T > 1.87\Theta$ (see Fig. 1.7), C_V should be close to its Dulong and Petit limit (24.44 J/K-g atom) if the Debye model is assumed. The value taken for C_V means that we are assuming there is no anharmonicity on C_V in the mantle. Even if there were anharmonic effects at $P = 0$, they would be suppressed by pressure (Hardy, 1980).

The difficulty with applying (4.35) is that γ is not known accurately with pressure (or depth). However, as shown in Chapter 1, if acoustic velocity data are available, several useful approximations to the Grüneisen ratio can be employed. Two of them are γ_{ac}, given by (1.52), and the "free volume" γ_{vz}, given by (1.61).

Another method uses the entropy formula in the Debye theory (Brown and Shankland, 1981). The lower mantle is approximately adiabatic, and assuming that Debye's thermal energy equation satisfies the lower mantle, the temperature can be calculated so as to insure that the entropy S is constant throughout the mantle. In the Debye model, Θ depends upon v_s and v_p, and thus Θ is determined at every depth. It turns out that Θ/T is virtually a constant with z (see Table 1.10). Brown and Shankland (1981) chose the Brillouin (1946) version of the Debye theory in which there are two Θ's, Θ_s and Θ_p, for the corresponding v_s and v_p. The entropy depends now on Θ_s and Θ_p, both of which are found by the seismic model to increase with depth. The change in T required to keep S_D invariant is computed, and the temperature distribution is found throughout the lower mantle, providing T is specified at the reference depth (Table 4.3) (See also Shankland and Brown, 1985). Brown and Shankland took as the reference T at 771 km the results of the experiments of Akaogi and Akimoto (1979).

Table 4.3. Calculation of T, α, and γ_{D_S} by the Debye model in the lower mantle of the earth

Depth km	Temperature K	Thermal Expansivity 10^5/K α	Grüneisen Parameter γ_{D_S}
771	1908	2.19	1.29
871	1941	2.09	1.27
971	1973	1.98	1.25
1071	2004	1.91	1.24
1171	2034	1.83	1.23
1271	2064	1.74	1.20
1371	2092	1.67	1.19
1471	2120	1.60	1.17
1571	2147	1.55	1.17
1671	2174	1.47	1.14
1771	2199	1.43	1.13
1871	2225	1.35	1.10
1971	2249	1.31	1.09
2071	2273	1.25	1.06
2171	2296	1.21	1.06
2271	2319	1.17	1.04
2371	2341	1.12	1.02
2471	2363	1.08	1.00
2571	2384	1.04	.99
2671	2405	1.00	.97
2771	2426	.96	.94
2871	2446	.92	.93
2885	2449	.95	.95

After Brown and Shankland, 1981.

Choosing $T(670) = 1600$ C from the results of Akaogi and Akimoto (1979), T is thus specified at all depths in the lower mantle (see Table 4.3). Once $T(z)$ is known, then using $P(z)$ from the seismic model, the data are numerically differentiated to obtain dT/dP. To complete the calculation, an expression for γ along an adiabat is needed.

Using the form of γ given by (1.30), we find the value of γ_{D_S} to be

$$\gamma_{D_S} = K_S \left(\frac{d\,\ell\mathrm{n}\,T}{\partial P}\right)_S, \tag{4.36}$$

where the subscript D_S stands for the Debye function at constant entropy. If the values of γ_{D_S} and T are placed in (4.35), α is determined at all depths. Calculated values of α, T, and γ for the lower mantle are listed in Table 4.3. PREM data is the source for data on ρ, v_s, $dv_s/d\rho$, and K_S.

Stacey (1977b) identified T and γ independently and then calculated α by (4.35). He assumed T at the inner core–outer core boundary (4168 K), this temperature arising from the estimated solid–liquid equilibrium temperature of the iron–sulfur eutectic. This was found by extrapolating to this high pressure the melting point measurements of Usselman (1975). Stacey computed the temperature drop through the outer core assuming adiabatic conditions, using from the Grüneisen constant the value computed by γ_{vz}, which requires only seismic data. At the core–mantle boundary he found $T = 3157$ K. He then calculated an adiabatic state in the lower mantle, arriving at $T = 2290$ K at $z = 671$ km. The α distribution for the lower mantle is given in Table 4.4. The values of γ found for the lower mantle by Stacey are slightly lower than those found by Brown and Shankland. At a depth of 2771 km, he found $\gamma = 0.91$, which rises steadily to $\gamma = 1.02$ at 771 km. The value of γ is really γ_{vz}, given by (1.46), and is thus wholly determined by the seismic model, PREM.

Table 4.4. Thermal expansivity for the lower mantle

z km	α 10^6/K	z km	α 10^6/K
2771	9.7	1771	12.0
2671	9.7	1671	12.4
2571	9.9	1571	12.7
2471	10.1	1471	13.2
2371	10.4	1371	13.6
2271	10.6	1271	14.0
2171	10.8	1171	14.5
2071	11.1	1071	15.0
1971	11.4	971	15.5
1871	11.7	871	16.1
		771	16.7

Calculated by Stacey, 1977b.

4.11. Finding α from the assumption αK_T = constant at high P

In Section 3.7, we found that at high T and high compression the value of αK_T varies little with pressure. In 1952 Birch suggested that as an approximation, αK_T could be treated as independent of density in the deep earth, so the idea that αK_T is independent of P has historical justification. Anderson (1982) assumed that $\alpha K_T = b$ is invariant with volume in a thermal model where the thermal pressure at various depths is obtained by the approximation

$$P_{TH} = a + bT. \tag{4.37}$$

When K_S from the seismic model is known, α can be obtained as a function of depth. Anderson (1982) computed the properties of an earth model using (4.37) and a set of differential equations describing density, pressure, and temperature, from which the value of αK_T and thus α was determined at each region of the earth. The elastic constant data for the core were arbitrarily adjusted so that the moment of inertia and the mass vanish at the earth's center. This enabled the best value of αK_T to be computed for each zone of the depth. The other physical properties were all consistent with α, as they were constrained by interlocking differential equations. The model allowed the values of ρ, T and K_S to be computed. The consistency of the computed ρ_0 and the computed K_S with seismic data on these constants helped to verify the model.

Table 4.5. Calculated parameters for the earth, using the assumption that $P_{TH} = P_0 + (a + bT)$ and (4.45) for α

Depth km	ρ Density	Temperature K	P GPa	K_S GPa	α 10^6/K
671	4.391	1707	23.89	303.4	21.43
771	4.450	1757	28.32	316.6	20.53
971	4.564	1855	37.32	343.2	18.94
1171	4.675	1949	46.52	370.1	17.56
1371	4.784	2040	55.92	397.3	16.36
1571	4.890	2128	65.53	424.8	15.30
1771	4.995	2214	75.36	452.7	14.36
1971	5.098	2298	85.42	481.0	13.51
2171	5.199	2381	95.74	509.8	12.75
2371	5.300	2461	106.36	539.3	12.05
2571	5.401	2541	117.31	569.4	11.42
2771	5.502	2619	128.67	600.5	10.82
2886	5.561	2664	135.42	618.9	10.50

After Anderson, 1982.

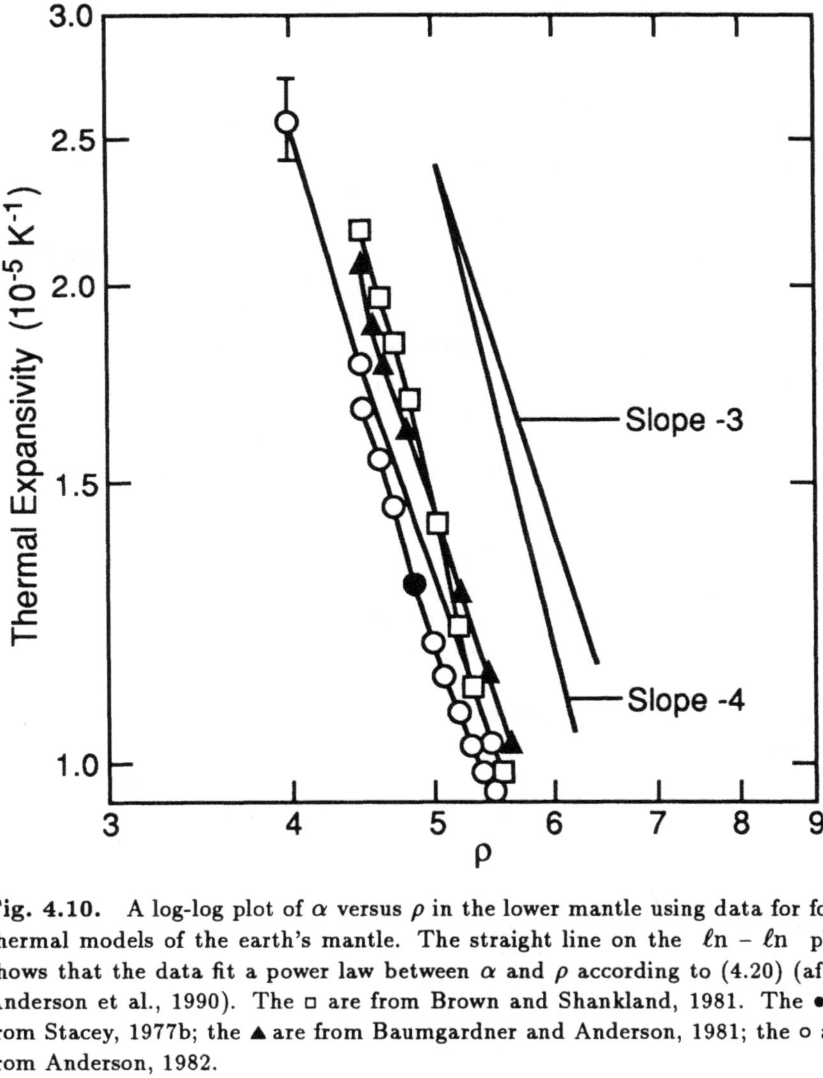

Fig. 4.10. A log-log plot of α versus ρ in the lower mantle using data for four thermal models of the earth's mantle. The straight line on the ℓn – ℓn plot shows that the data fit a power law between α and ρ according to (4.20) (after Anderson et al., 1990). The □ are from Brown and Shankland, 1981. The ● is from Stacey, 1977b; the ▲ are from Baumgardner and Anderson, 1981; the ○ are from Anderson, 1982.

The model is checked by observing that the resulting density distribution matches that given by seismic models. This approach produced the results for α listed in Table 4.5. The value of γ, computed from (1.19), rises steadily from 0.98 at a depth of 2771 km to 1.2 at 771 km, which, when extrapolated to $P = 0$, yields $\gamma_0 = 1.4$. An interesting aspect of these calculations is the result that $\gamma\rho =$ constant throughout the lower mantle, in agreement with the expectation that at high T and low η, $\delta_T \rightarrow K'$ and $q \rightarrow 1$, as shown in Section 3.5.

4.12. α vs. ρ in the lower mantle from three thermal models

The values of α versus ρ taken from α versus z curves of the lower mantle for three thermal models given in Tables 4.3, 4.4. and 4.5 are plotted in a log-log plot in Fig. 4.10. According to (3.18) and (4.19), the slope of the curve is $-\delta_T$. The curve shows that all the average values of δ_T lie between 3.5 and 4, in spite of the temperature differences of the thermal models. This range of values in δ_T can thus be considered as arising from the seismic model PREM. We see that it is the same as that found for δ_T for MgO in the compression range of the mantle (see Figs. 3.5 and 3.6). This verifies the assumption by Chopelas and Boehler (1992a) that the value of m in their equation for $\delta_T(\eta)$ (3.48) for perovskite is close to the value they measured for MgO. Hence the average δ_T over the P, T range of the lower mantle for all lower mantle minerals should be between 3.5 and 4.

4.13. Thermal expansivity of silicate perovskite at high *P* and *T*

4.13.1. *The thermoelastic parameters*

In Section 4.4.3, $\alpha(\eta, T)$ was found for MgO. There the experimental data give $\alpha(T)$ at $P = 0$, the upper curve in Fig. 4.2. Using that data, $\alpha(T)$ at $\eta = 1$ was found, giving the lower curve in Fig. 4.2, to which (4.17) was applied, yielding the data plotted in Fig. 4.3. For perovskite, we do not have data on $\alpha(T)$ over a sufficiently large T range to proceed as we did for MgO. We do, however, have information on the basic thermoelastic parameters from which $\Delta V/V_0$ at $P = 0$ can be found by the Suzuki equation (4.32). Thermoelastic parameters from four sources are shown in Table 4.6.

Table 4.6. Thermoelastic parameters for silicate perovskite

Material	$(Mg, Fe)SiO_3$	$(Mg, Fe)SiO_3$	$(Mg, Fe)SiO_3$	$(Mg, Fe)SiO_3$
Source	Knittle et al. (1986)*	Hemley et al. (1992)	Wang et al. (1994)	Anderson & Masuda (1994)
Θ (K)	825	1017	(1017)	1017
K_{T_0} (GPa)	260	263	261	261
K'	4.0	3.9	4.0	4.0
V_0 (cc mol^{-1})	24.39	24.46	24.46	24.46
γ_0	2.20	1.98	1.3	1.5
δ_{T_0}		7.0‡	4.3	5.0
κ		0.8*		1.5

‡From Stixrude et al., 1992a. *The high temperature model.

4.13.2. *Isobars showing $\Delta V/V_0$ and α versus T at $P = 0$*

$\Delta V/V_0$ versus T ($P = 0$) at constant γ_0 for the uncompressed lower mantle according to Anderson and Sumino (1980)'s set of thermoelastic parameters is found in Fig. 4.8. This figure shows that $\Delta V/V_0$ is about 5% for $T = 1900$ K using $\gamma_0 = 1.5$. $\gamma_0 = 1.5$ is chosen because γ_{ac} of the lower mantle at $P = 0$ is 1.5 using PREM data (see Figs. 1.41 and 5.16). If we take ρ_{oc} (cold) $= 4.22$ and ρ_0(hot) $= 4.00$ (Bukowinski and Wolf, 1990) for the lower mantle, taking T from 300 K to 1900 K at $P = 0$, then $\Delta V/V_0$ is on average 5%; compare this value with the $\gamma_0 = 1.5$ solution in Fig. 4.8. $\alpha(T)$ versus T for the uncompressed lower mantle ($\gamma_0 = 1.5$) is shown in Fig. 4.9: α at 800 K is close to 2.5×10^{-5} K^{-1}.

If the thermoelastic parameters of silicate perovskite proposed by Anderson and Masuda (1994) are used (except that γ_0 is arbitrary), the pattern for $\Delta V/V_0$ versus T is as shown in Fig. 4.11. In that figure, the solution for $\Delta V/V_0$ at $T = 1900$ K is 4.5% for $\gamma_0 = 1.5$, and 5% for $\gamma_0 = 1.6$. These values are not significantly different from those found for the lower mantle from the seismic data (Fig. 4.8).

In subsequent discussions, we will favor $\gamma_0 = 1.5$ for silicate perovskite. This choice is consistent with the thermoelastic properties of MgO and the lower mantle. Table 1.2 shows that for MgO, $\gamma_{ac} = 1.5$, in close agreement with $\gamma = 1.54$. Thus γ_{ac} is representative of γ for this solid. For the earth's uncompressed lower mantle (Anderson, 1979a), $\gamma_{ac} = 1.5$. Since the earth's lower mantle is composed mostly of perovskite and some magnesiowüstite, it is apparent that γ for silicate perovskite should be close to 1.5.

As shown in Fig. 4.11, the thermoelastic parameters from Knittle et al. (1986) yield $\Delta V/V_0 = 7.5\%$ for $T = 1900$ K. This is 2.8% more than the $\Delta V/V_0$ of the uncompressed lower mantle from 300 K to 1900 K. It led them to conclude that the lower mantle had to have more iron than the upper mantle. The Hemley et al. (1992) choice of $\gamma_0 = 1.98$ leads to 6% for $\Delta V/V_0$, which is higher by about 1% than $\Delta V/V_0$ of silicate perovskite from ambient to 1900 K using Bukowinski and Wolf's (1990) values of uncompressed ρ_0. This suggests a small change in iron content.

The choices of Θ and γ_0 are crucial for the solutions of α and $\Delta V/V_0$ for lower mantle silicates. The value of Θ selected by Knittle et al. (1986) is too low for perovskite (see Fig. 1.3), and when it is used, the Suzuki equation will predict a relatively high $\Delta V/V_0$. In Fig. 4.12, α versus T is plotted from the integration of $\Delta V/V_0$ (Fig. 4.11), where the Anderson and Masuda thermoelastic data (except for γ_0, which is adjustable) are used. The intercepts on the y axis show the value of α at $T = 800$ K and $P = 0$ according to the choice of γ_0: 3.7×10^{-5}K^{-1} for $\gamma_0 = 1.98$ (Hemley et al., 1992); 2.75×10^{-5}K^{-1} for $\gamma_0 = 1.5$; and 2.4×10^{-5}K^{-1} for $\gamma_0 = 1.3$ (Wang et al., 1994). For comparison we observe that $\alpha = 4.3 \times 10^{-5}$K^{-1} from the Knittle et al. choice of thermoelastic parameters. Their choice of thermoelastic parameters leads to high values of α and $\Delta V/V_0$.

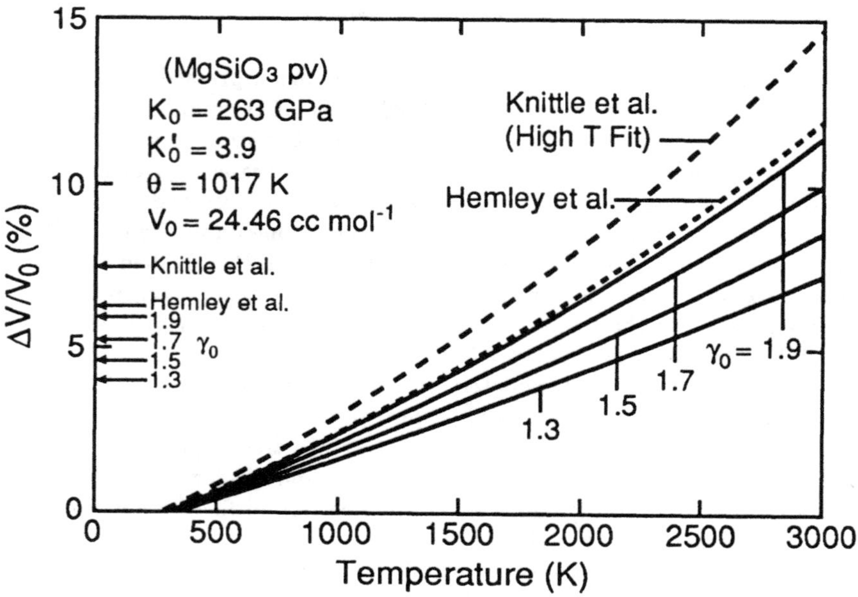

Fig. 4.11. $\Delta V/V_0$ versus T at $P = 0$ from (4.32) using the thermoelastic parameters of Anderson and Masuda (1994). The plot of the Suzuki equation using the Knittle et al. (1986) thermoelastic data is shown by the large-dashed line. The plot of the Suzuki equation using the Hemley et al. (1992) thermoelastic data is shown by the small-dashed line. Arrow intercepts on the y axis are for $\Delta V/V_0$ at 1900 K, depending on γ_0 (modified from Anderson and Masuda, 1994).

4.13.3. *Isochores showing α versus T (constant η)*

To find α at higher compression, we first need to change from constant P conditions to constant V conditions. In Section 4.4.3, this problem was solved for MgO using (3.4), but we do not know αK_T versus T for silicate perovskite. Instead, we use the calculus identity (3.14) and evaluate the term $\alpha K_T(\partial\alpha/\partial T)_P$ in (3.14) by using (3.3) and (1.63), yielding

$$\left(\frac{\partial \alpha}{\partial T}\right)_V = \left(\frac{\partial \alpha}{\partial T}\right)_P - \delta_T \alpha^2. \tag{4.38}$$

In Section 2.9 it was shown that if $\alpha(T)$ (constant P) is linear in T at high T, then

$$\left(\frac{\partial \alpha}{\partial T}\right)_V = 0. \tag{2.51}$$

Placing this in (4.38) and integrating the resulting $(\partial\alpha/\partial T)_P$, we find

$$\frac{\alpha}{\alpha_{T^*}} = \left[1 - \delta_T \alpha_{T^*}(T - T_0)\right]^{-1}, \tag{4.39}$$

where T^* represents the beginning of the integration, $T^* > \Theta$.

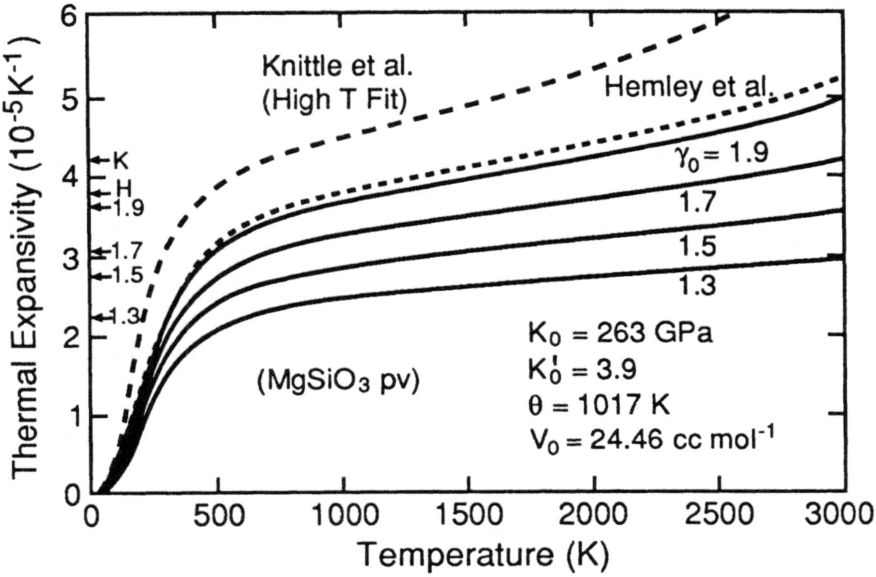

Fig. 4.12. α versus T at $P = 0$ according to the Suzuki equation when the thermoelastic parameters of Anderson and Masuda (1994) are used. γ_0 is variable. The large-dashed lines are from the thermoelastic parameters proposed by Knittle et al. (1986). The intercepts on the y axis show values of α at $T = 800$ K according to the value of γ_0 (modified from Anderson and Masuda, 1994).

Expanding the right side of (4.39) out as a series and truncating after the first term, we find (2.52), which agrees with the data. Using (2.52), we guarantee that $\alpha(T)$ is parallel to the T axis. For T below Θ, we evaluate (4.38) from the data on $\alpha(T)$. Doing these two operations on the appropriate $P = 0$ curve in Fig. 4.12 yields $\alpha(T)$ at $\eta = 1$, which is shown in Fig. 4.13, along with the appropriate $\alpha(T)$ at $P = 0$ curve, selected from Fig. 4.12 for the $\gamma_0 = 1.5$ case. From the $\eta = 1$ curve, other isochores are found using (4.17), as shown in Fig. 4.13. We assume $\delta_{T_0} = 5.0$ and $\kappa = 1.5$. The value of δ_{T_0} is chosen considering the value of δ_{T_0} arising from the rate of change of α with ρ for the lower mantle (see Fig. 4.10) and the report by D.L. Anderson (1987) that δ_S at deep mantle conditions is 1.8 (and therefore $\delta_T = 2.8$). Calculations show that $\alpha(T)$ is insensitive to the exact value of κ for $1.1 < \kappa < 1.9$. The value $\kappa = 1.5$ was assumed.

4.13.4. *Isotherms of α versus P for silicate perovskite*

The data in Fig. 4.13 can be presented in several forms. To change variables so as to find $\alpha(P,T)$, we must find the appropriate value of P for a given V and T, $P(V,T)$, or the EoS. Since $P(V,T) = P(V,T=0) + P_{TH}(V,T)$, we need to find the isothermal EoS and the thermal pressure for silicate perovskite. This is handled in Chapters 6 and 10. We note, however, that to determine the thermal pressure, the value of $\alpha(0,\Theta)$, which is the subject of this chapter, and $K_T(0,\Theta)$, are needed. From Fig. 4.12, $\alpha(0,1000)$ is $2.9 \times 10^{-5}\mathrm{K}^{-1}$. The value of $K_T(0,1000)$ requires $(\partial K_T/\partial T)_P$. From the definition of δ_T, (1.47), $(\partial K_T/\partial T)_P$ can be found, since δ_T and $K_T(0,300)$ are known. We find $(\partial K_T/\partial T)_P = 0.035$ GPa/K, and $K_T(0,1000) = 239$ GPa to be self-consistent with $\delta_T = 5.0$ and $K_T(0,300) = 263$ GPa. Thus $\alpha K_T = .00692$ GPa/K. In Section 10.10, the EoS of perovskite, based on these numbers, is presented in Table 10.6. Using the EoS, values of $\alpha(V,T)$ in Fig. 4.13 are transformed into $\alpha(P,T)$ and plotted as isotherms of α versus P in Fig. 4.14.

The value of α at 1900 K and $P = 23.8$ GPa (corresponding to the 670 km depth seismic discontinuity) from Fig. 4.14 is $2.4 \times 10^{-5}\mathrm{K}^{-1}$. This value is slightly higher than the values computed from actual mantle seismic profiles: $2 \times 10^{-5}\mathrm{K}^{-1}$ by Brown and Shankland (1981) and $1.7 \times 10^{-5}\mathrm{K}^{-1}$ by Baumgardner and Anderson (1981) (see Section 4.10). It is slightly lower, however, than the value computed by Chopelas and Boehler (1992a) (3×10^{-5} K^{-1}). In their 1992a paper Chopelas and Boehler acknowledged that their earlier paper (1989) on α was in error because δ_T had been assumed to be independent of P.

The value of α at 2900 K and $P = 135$ GPa corresponding to the region of the mantle near the mantle-core boundary ($0.9 \times 10^{-5}\mathrm{K}^{-1}$) is reasonably close to the finding of Chopelas and Boehler (1992a), $1 \times 10^{-5}\mathrm{K}^{-1}$. These two estimates are remarkably close to those determined by the inversion of seismic profiles: $0.97 \times 10^{-5}\mathrm{K}^{-1}$ by Stacey (1977b), $1 \times 10^{-5}\mathrm{K}^{-1}$ by Baumgardner and Anderson (1981), and $1 \times 10^{-5}\mathrm{K}^{-1}$ by Brown and Shankland (1981).

A rather remarkable result shown by Fig. 4.14 is the convergence of the isotherms at high P. Thus α is independent of T at high P, provided $T > \Theta$. Apparently α depends only slightly on pressure, providing the pressure range is higher than about 1/3 K_{T_0}. This convergence, when considered with the remarkable agreement of values at 135 GPa described in the paragraph above, suggests that many different thermoelastic models will yield similar values of α at deep mantle-like conditions.

4.13.5. *Isobars of $\Delta V/V_0$ versus T: the experimental cross-check*

The α versus P isotherm curve can be transformed into a $\Delta V/V_0$ versus P isotherm by integration. The isotherms of $\Delta V/V_0$ versus P can then be cross-plotted to obtain isobars of $\Delta V/V_0$ versus T. Such isobars are the way in which actual experimental data of $\Delta V/V_0$ versus T are often presented. Comparing such isobars, derived from the Suzuki equation (4.32) and the $\alpha(\eta)$ equation (4.17), with the experimental data is a way of checking the validity of the assumed thermoelastic parameters.

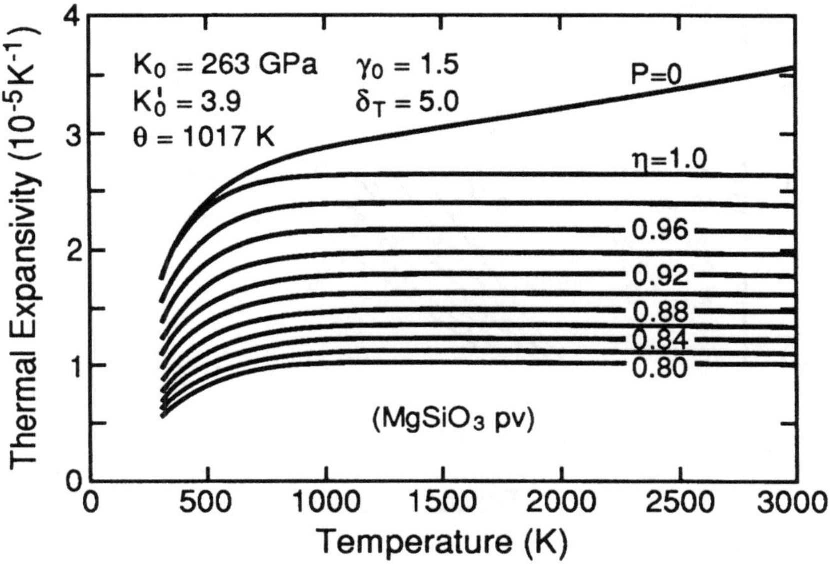

Fig. 4.13. Isochores of α versus T. The curve for $P = 0$ from Fig. 4.12 ($\gamma_0 = 1.5$) is shown for reference (modified from Anderson and Masuda, 1994).

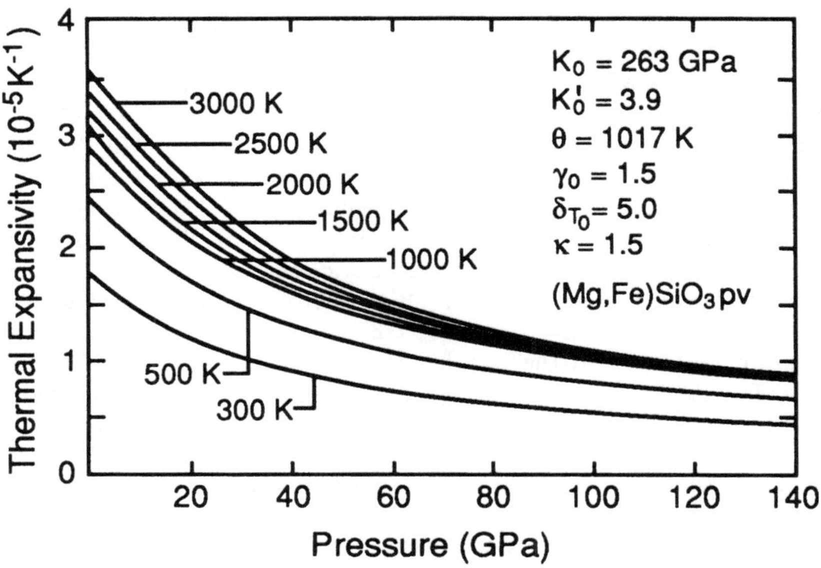

Fig. 4.14. Isotherms of α versus P using the thermoelastic parameters shown in the inset and the equation of state. High T isotherms converge above 100 GPa (modified from Anderson and Masuda, 1994).

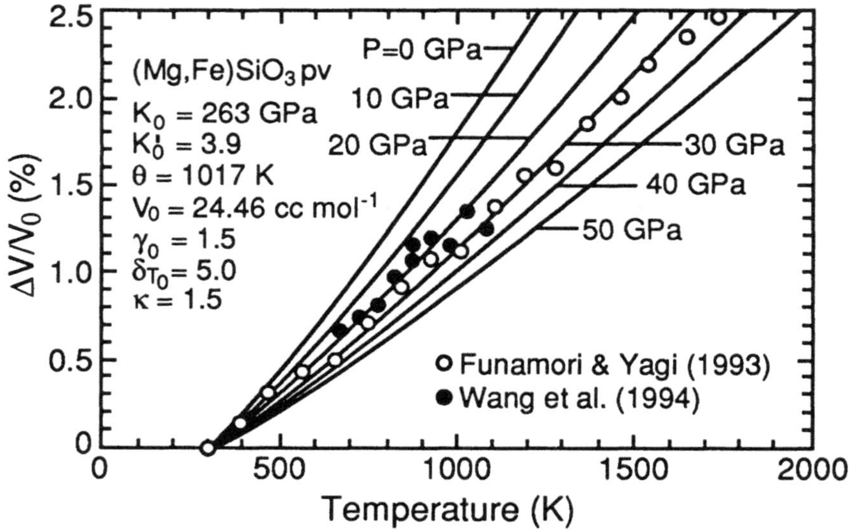

Fig. 4.15. Isobars of expansivity, $\Delta V/V_0$, versus T. Solid circles represent the isobar experimental data of Wang et al. (1994) (10.8 GPa), open circles those of Funamori and Yagi (1993) (36 GPa) (modified from Anderson and Masuda, 1994).

Such a comparison is made in Fig. 4.15, which shows plot of the isobaric 36 GPa experimental points of Funamori and Yagi (1993): some of their points fall between our calculated 30 and 40 GPa isobars. Also plotted are the isobaric 10.8 GPa experimental points of $\Delta V/V_0$ versus T of Wang et al. (1994). Some of these lie close to the 10 GPa isobar. In Fig. 4.16, the 36 GPa isobar calculated from Anderson and Masuda's thermoelastic constants is compared to the experimental data of Funamori and Yagi. The 36 GPa isobars calculated from the Hemley et al. thermoelastic constants and from the Wang et al. thermoelastic constants are also presented. There is a fair consistency between all solutions and the data. The effect of increasing γ_0 is to shift the calculated isobar to the left. The effect of increasing δ_{T_0} is to shift the calculated isobar to the right. Thus the high values of δ_{T_0} and γ_0 of Hemley et al. produce an isobar not too far removed from that using much lower values of γ_0 and δ_{T_0} reported by Wang et al.

4.13.6. *Isotherms of ρ versus P: comparisons with PREM*

There is geophysical interest in comparing the density distribution of silicate perovskite with the density distribution from the PREM seismic model. As $\rho/\rho_0 = (1 - \Delta V/V_0)^{-1}$, a solution for the isotherms of ρ versus P only requires a simple manipulation of the data behind the solid curves in Fig. 4.15. The results are shown in Fig. 4.17.

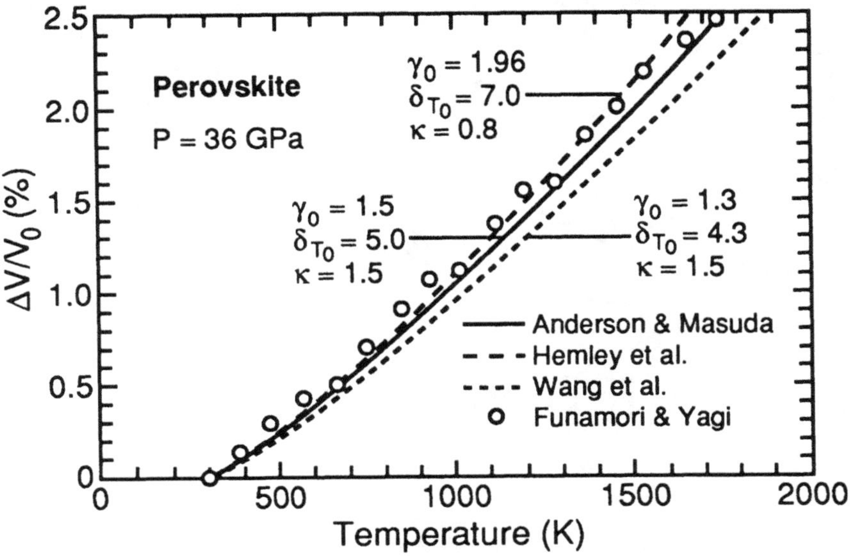

Fig. 4.16. The 36 GPa isobar of perovskite calculated from sets of thermoelastic parameters, as indicated in inset, all compared with the Funamori and Yagi (1993) experimental data (modified from Anderson and Masuda, 1994).

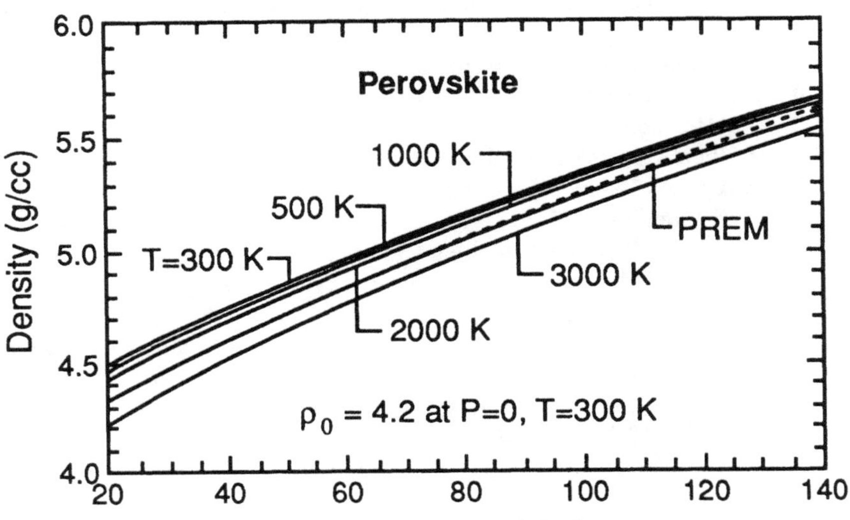

Fig. 4.17. Isotherms of ρ versus P for perovskite using $\gamma_0 = 1.5$ and $\delta_{T_0} = 5.0$ (modified from Anderson and Masuda, 1994). Dashed line represents PREM.

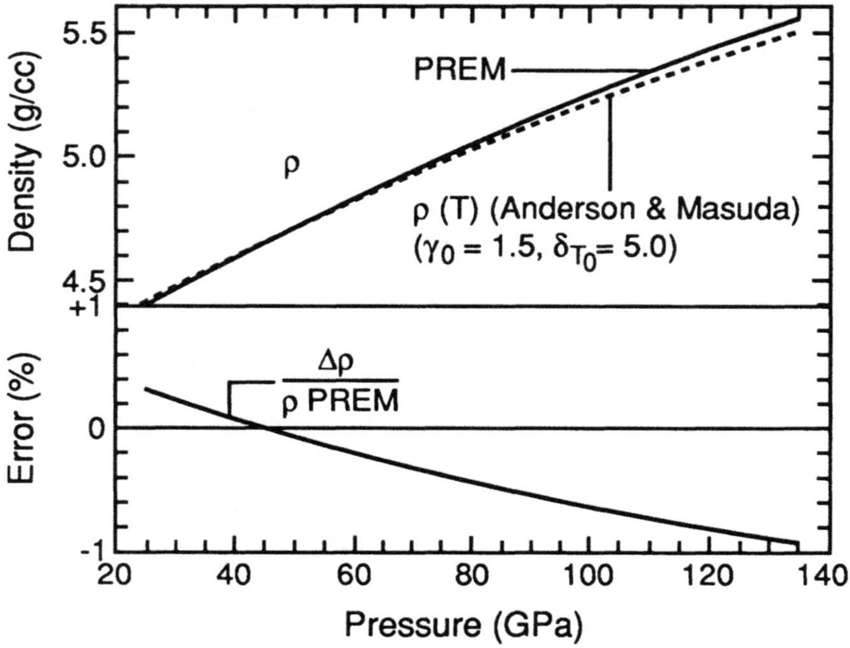

Fig. 4.18. Adiabat of ρ versus P computed from the thermoelastic parameters of Anderson and Masuda (1994) (modified from Anderson and Masuda, 1994).

The 2000 and 3000 K isotherms in Fig. 4.17 are close to PREM. The PREM solution, however, corresponds to mantle conditions that are adiabatic. A better comparison to PREM is made by using the appropriate adiabat of silicate perovskite. The calculation requires that P_{TH} change with T along the adiabat. We use the adiabatic temperature distribution of Brown and Shankland (1981) with $T_0 = 1900$ K (28 GPa). The result is shown in Fig. 4.18. There is less than 1% difference between the calculated adiabatic ρ of PREM and that of silicate perovskite using the parameters $\gamma_0 = 1.5$ and $\delta_{T_0} = 5.0$. Thus we find that a ($\gamma_0 = 1.5$, $\delta_{T_0} = 5.0$) perovskite lower mantle agrees with PREM. Using a larger value of γ_0, 1.96, Stixrude et al. (1992a) also reported agreement between their computed ρ and PREM. The larger value of γ_0 by Hemley et al. (1992), 1.98, however, leads to $\alpha = 3.8 \times 10^{-5}\text{K}^{-1}$ at 800 K, considerably larger than the α value ($2.8 \times 10^{-5}\text{K}^{-1}$), from the recommended value of γ_0, 1.5 (see Fig. 4.12).

4.14. Problems

1. Verify that (4.24) follows from (4.23).
2. Verify that (4.37) follows from (4.35) and (4.36).
3. Assuming $P(\eta, T) = K_T(T) f(w)$, prove $\alpha/\alpha_0 = 1 - (P/K_T)\delta_{T_0}$.

5

OXIDES THAT ARE DEBYE-LIKE SOLIDS

5.1. Introduction

By physicists' standards, the materials of planets are not well characterized. The material in the mantle of terrestrial planets is rocky and composed of many minerals that belong to a few phase diagrams. The large potential number of minerals in these phase diagrams is reduced and constrained by the density of seismological models. Even so, uniqueness is not possible because many minerals or combinations of minerals have the same density at zero pressure. Density discrimination at high pressure and high temperature could reduce the ambiguity further if the temperature of the planet were well characterized (which it is not), and if the equation of state were universally agreed upon (which it is not, as will be shown in Part II). Minerals that survive as candidates for planet interiors are of interest to ceramic science.

The description of the physical properties of planet materials must be general. A unique solution to composition cannot be made. There is little possibility of assuring crystallographic class, precise chemical composition, or mechanical history, much less purity and defect structure.

The corresponding physical theories applied to these planetary materials are also necessarily simplified, as the material parameters needed for a more generalized theory are usually not obtainable. For this reason simple theories found in the earliest stages of solid state science and promulgated early in this century keep cropping up in theories of planet interiors.

A good example is the Debye theory of specific heat. Some such theory is needed, because a means must be had to provide a transition between the measured mechanical properties of a planet's interior (e.g., its seismic velocities) and its calculated thermal properties (which cannot be measured except at the surface).

The question of whether the application of such a simple theory leads to meaningful results arises, and it is precisely this we wish to examine.

5.2. Packing fraction and coordination

Consider the packing of equal spheres. Let *P.F.* be the packing fraction of a structure, equal to the ratio of the volume of the spheres forming the cell to the volume of the crystallographic cell itself, or

$$P.F. = \frac{\sum \mathbf{n}_i V_i}{V_c}, \tag{5.1}$$

where $\mathbf{n}_i$ is the number of atoms of type i in the unit cell; V_i is the volume of a single atom; and V_c is the volume of the unit cell (Fairbairn, 1943). The summation extends over all atoms in the cell. For identical spheres of radius r, the above equation becomes

$$P.F. = \frac{4\pi}{3}\left(\frac{q r^3}{V_c}\right), \tag{5.2}$$

where q is the number of spheres in the unit cell.

The coordination number M is a function of the crystal class, and the packing fraction ($P.F.$) depends on the crystal class through M. Table 5.1 shows the results for various packing fractions. As M decreases, so does the packing fraction. Even in the most densely packed solids, about one-fourth of the structure is occupied by voids. The volume of voids in a simple cubic structure is about 50% and 60% for diamond.

Birch (1961) showed that the ambient sound velocity v_p is linear with ρ_0 and inverse with the mean atomic weight $\mathbf{M}/p$. This is called Birch's law. Thus the systematics of v_p versus ρ_0 scaled approximately as ρ_0/μ, the inverse of which is the average volume of the atom in the unit cell,

$$V_{\text{ion}} = \frac{\mu}{\rho_0}. \tag{5.3}$$

Equation (5.3) is also a measure of the packing of the ions in oxides and silicates. The values of $P.F.$ from (5.2), calculated using Pauling's (1960) ionic radii and the average volume per ion using (5.3),, are presented in Table 5.2 and correlated in Fig. 5.1.

Table 5.1. Degree of close packing of spheres

Lattice types	Coord. no., M	Packing fraction q	(%)	Example types
close packing of equal spheres	12	$\frac{\sqrt{2}}{6}\pi$	74.0	Zr (hex) Fe (hcp at high P)
tetragonal	10	$\frac{2}{9}\pi$	69.8	protoctinium
body-centered cubic	8	$\frac{\sqrt{3}}{8}\pi$	68.1	CsCl (Fe, ambient)
face-centered cubic	6	$\frac{\pi}{8}$	52.3	NaCl (Fe at high T)
diamond	4	$\frac{\sqrt{3}}{16}\pi$	34.0	C, Si, Ge

Table 5.2. Properties of silicates and oxides leading to packing fraction (*P.F.*) and V_r values (**M**, molecular weight; *p*, number of atoms in **M**)

Mineral name	Ideal formula	ρ_0^*	μ^* $\mathbf{M}/p$	V_c	V_{ions} μ/ρ_0	V_r	*P.F.*†
Stishovite	SiO_2	4.28	20.03	23.40	23.28	4.7	.99‡
Corundum	Al_2O_3	4.00	20.39	42.48	35.34	5.1	.84
Kyanite	Al_2SiO_3	3.60	20.25	75.02	58.83	5.6	.81
Periclase	MgO	3.58	20.16	18.74	12.65	5.6	.68
Spinel	$MgAl_2O_4$	3.58	20.32	65.87	48.09	5.7	.67
Pyrope	Mg_2Al_3-Si_3O_{12}	3.51	20.29	192.71	142.69	5.8	.76
Pigeonite	$MgSiO_3$	3.42	20.08	48.93	35.93	5.9	.73
Jadeite	$NaAlSi_2O_6$	3.43	20.21	98.23	73.68	5.9	.75
Rutile	TiO_2	4.86	26.60	27.40	24.30	6.0	.69
Hematite	Fe_2O_3	5.27	31.98	50.46	38.17	6.0	.76
Clino-enstatite	$MgSiO_3$	3.28	20.08	110.05	74.78	6.1	.69
Forsterite	Mg_2SiO_4	3.22	20.10	72.84	48.59	6.3	.72
Silimanite	Al_2SiO_5	3.23	20.25	33.62	58.83	6.4	.70
Jeffersonite	$MgCaSi_2O_6$	3.39	21.65	106.48	74.78	6.4	.70
Andalusite	Al_2SiO_5	3.15	20.25	85.74	58.83	6.4	.69
Mullite	$Al_6Si_2O_{13}$	3.13	20.28	226.81	153.19	6.5	.68
Diopside	$CaMgSi_2O_6$	3.28	21.65	110.05	74.78	6.6	.68
Enstatite	$MgSiO_3$	2.98	20.08	55.47	35.93	6.6	.65
Fayalite	Fe_2SiO_4	4.39	29.12	77.36	49.96	6.6	.65
Schefferite	$MgCaSi_2O_6$	3.39	23.07	91.24	51.79	6.8	.57
Coesite	SiO_2	2.92	20.03	34.30	23.28	6.5	.68
Sarcolite	$Ca_3Al_2SiO_2$	2.93	21.67	283.54	152.08	7.4	.54
Quartz	SiO_2	2.65	20.03	37.79	23.28	7.6	.62
Akerman-ite	Ca_2Mg-SiO_{12}	2.94	22.70	154.35	90.34	7.7	.59
Anorthite	$CaAl_2Si_2O_8$	2.77	21.40	167.40	96.67	7.7	.58
Albite	$NaAlSi_3O_8$	2.61	20.17	167.47	96.73	7.7	.57
Carnegieite	$NaAlSiO_4$	2.57	20.29	92.13	50.39	7.9	.52
Anortho-clase	$KNaAl_2$-Si_6O_{16}	2.58	20.79	349.22	197.12	8.1	.58
Orthoclase	$KAlSi_3O_8$	2.55	21.41	181.93	103.23	8.4	.57
Cristo-balite	SiO_2	2.34	20.03	42.80	23.28	8.6	.55
Leucite	$KAlSi_2O_6$	2.47	21.88	142.27	79.97	8.8	.55
Vitreous silica	SiO_2	2.20	20.03	45.52	23.28	9.1	.51

†*P.F.* is the ratio V_{ions}/V_c. *M and ρ_0 data from Deer et al., 1962. ‡For discussion of this unrealistic *P.F.*, see Section 5.3.

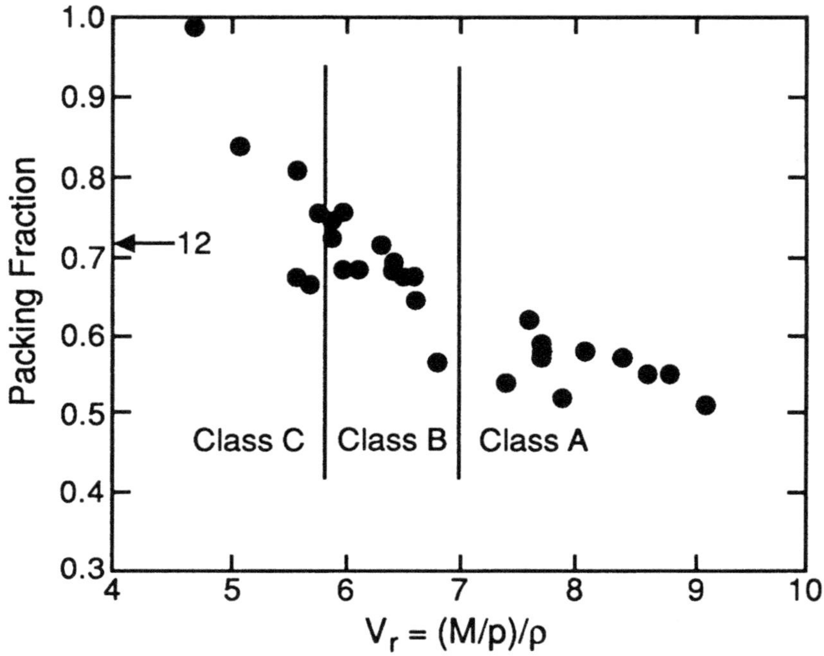

Fig. 5.1. Packing fraction versus average volume per atom, $V_r = (\mathbf{M}/p)/\rho$, for 29 oxides and silicates as listed in Table 5.2. The number 12 on the ordinate refers to the packing fraction of equal spheres for the coordination number 12. The minerals are sorted into three classes (see text). MgO, Mg_2SiO_4, and α-SiO_2 are representatives, respectively, of Class C, Class B, and Class A.

Table 5.2 illustrates the point that a great many minerals have a value of μ close to 20.5. Exceptions to this are minerals containing Fe or Ca; in those cases μ does not depart significantly from 23–31.

A plot of V_r versus the *P.F.* is shown in Fig. 5.1. This plot indicates that there is a regular correspondence between these two variables. The purpose of this curve is to demonstrate that a high *P.F.* corresponds to a high density in the class of oxides and ceramics where μ has about the same value.

We shall show that the position a rock-forming mineral has on the curve shown by Fig. 5.1 determines the applicability of the Debye theory in representing thermal and mechanical properties. The field of the curve is divided into three zones. Class A consists of the open structures, which have smaller density and smaller packing fractions. Class C consists of densely packed structures, and Class B is intermediate. The center of Class B corresponds approximately to a 68% packing fraction (i.e., body center cubic packing of equal spheres).

This division is not altogether arbitrary. Pauling (1960), assuming spherical anions, pointed out that anions are in contact whenever the cation–anion radius is less than 0.414. He listed the following as the radius ratios with respect to the oxygen anion: Be^{2+}, 0.25; Li^{+}, 0.34; Si^{4+}, 0.37; Al^{3+}, 0.41; Mg^{2+}, 0.47. This means that insofar as mechanical properties are concerned, minerals in Class C can be represented as closely packed oxygen spheres, with each oxygen sphere having a mean atomic weight of near 21. Included in this class are stishovite, silicate perovskite, corundum, periclase, and perhaps kyanite. For Class C, $V_r < 5.8$.

In Class A are found all the low-density SiO_2 polymorphs, calcite, feldspars, anorthite, and various open structure oxides like ZnO. In Class A, $V_r > 6.8$. The borderline cases, which we shall lump together as Class B minerals, all have a *P.F.* near 68%. These include some important oxides and silicates—olivine, pyroxene, CaO, rutile, and the calcium garnets. For Class B, $5.8 < V_r < 6.8$.

For silicates and oxides of Class A, volume changes can occur by the change of the angle between the tetrahedra without changing the cation–anion radius significantly. Thus the *P.F.* may change with pressure or temperature, even though the cation–anion distance remains relatively constant (Hazen and Finger, 1982).

The classification of silicates here into three classes, based on packing fraction, is close to the classification of silicates made by D.L. Anderson (1969) on the basis of the bulk sound velocity and cell volume per SiO_2 unit.

5.3. Polyhedral groups in crystal chemistry and V_r

The primary classification scheme used here is based not on the *P.F.*, but on the average volume V_r. The *P.F.* scheme used is based on the assumption that the radius of the oxygen ion is invariant with structure and density (Fairbairn, 1943). Very likely the radius of the oxygen ions is less for Class C than for Class A. We see from Table 5.1 that the *P.F.* for stishovite is 0.99, based on $r_0 = 1.32$ Å for the oxygen ion. This value leads to an unreasonably high *P.F.*, and so the r_0 of the oxygen ion for this mineral must be smaller than Pauling's value. The significance of the classification by V_r is that it identifies those silicates and oxides in which the oxygen–oxygen repulsion is a dominant feature of the potential energy.

The classification of oxides and silicates by the magnitude of V_r is analogous to the classification usually found in crystal chemistry (for example, Hazen and Finger, 1979). A central concept of silicate chemistry is the identification of the polyhedral of the cation (typically tetrahedral when surrounded by four oxygens and octahedral when surrounded by eight oxygens). Two-cation polyhedra may be linked by a shared face, a shared edge, a shared corner, or by van der Waals forces.

The most dense oxides and silicates share faces or edges in three dimensions. Fully edge-linked structures include stishovite, corundum, periclase, spinels, and garnets. Thus structures with shared cation polyhedra edges are typically Class C, but there are some edge-linked polyhedral structures found in Class B. In Class C, the oxygens are in contact, and the primary factor in compression is oxygen–oxygen compression, although cation–anion compression also plays a small role.

The least dense silicates, such as α-quartz, feldspar, and zeolites, have primarily corner–linked tetrahedra. These are often called framework structures. In these minerals, volume change can occur by the change of angle between the tetrahedra without significant change in the cation–anion distance. These structures are found in Class A. In addition, the less dense 6 fold coordinated oxides (CaO) and all the 4 fold coordinated oxides (ZnO) are in Class A. For these solids, resistance to compression involves bending of the cation–anion bond.

In some silicate structures, polyhedra share edges with some adjacent polyhedra and link corners with others. These structures belong in Class B. But also included in Class B are the garnets and spinels, for reasons that will be discussed later. In Class B, volume compression arises from a mix of oxygen–oxygen compression and oxygen–anion compression, although some structures have bond angle changes under compression as well.

5.4. Comparing Θ_{ac} with Θ from calorimetry, Θ_{cal}

We now propose to show that the Debye theory can be used to approximate the properties of the Class C silicates, but not the Class A silicates. In the classical theory of a thermal solid, Debye approximated the phonon spectrum by assuming that all phonons are acoustic and the frequency of the phonons is controlled by the sound velocities of an isotropic body, v_p and v_s (see Section 1.6). This theory led to good approximations of specific heat and entropy versus T at high T for metals by using the so-called Debye function. In his derivation, Debye defined the relationship between the Debye frequency ω_D and the mean sound velocity v_m. Now ω_D can be converted to a temperature (the Debye temperature) by $\omega_D = (k/h)\Theta$.

The Debye temperature (see Chapter 1) as a function of ρ and sound velocities is given by (1.65), where the fundamental constants are evaluated.

$$\Theta_{ac} = 251.2 \left(\frac{\rho}{\mu}\right)^{1/3} v_m \quad \text{K}. \tag{1.65}$$

Using (1.65) the Debye temperature for any solid can be estimated from data on sound velocities (Tables 1.9, 1.10, and 1.11 and Appendix A–7).

A test of the applicability of (1.65), shown above, is to compare it with the value of Θ derived by matching specific heat curves to the Debye function. Θ_{ac} versus T can be calculated by using the measured velocity versus T, as in (1.65). Θ_{cal} can also be calculated from measured specific heat data by matching the specific heat data with a standard specific heat

Debye function. The comparison of these two methods of finding Θ has been well documented by Kieffer (1980, 1982), where the two values of Θ are close for oxides of Class C, like MgO, but where a large discrepancy exists between the two Θ's for silicates of Class A, such as α-quartz.

It is of some importance to know more about the conditions when

$$\Theta_{cal} \cong \Theta_{ac}. \tag{5.4}$$

For this we examine briefly the moments of the vibrational frequency spectrum, $g(\omega)$, often called the density of states.

5.5. The moments of the vibrational density of states

Detailed comparison of experimental data with reference to the Debye temperature is best understood in terms of the moments of the frequency distribution as defined by

$$\left\langle \omega^n \right\rangle = \sum \omega_j^n / 3pN, \quad n > -3, \tag{5.5}$$

where the sum is taken over all the modes in the vibrational density of states, $g(\omega)$; N is the number of atoms; and $3pN$ is the number of vibrational degrees of freedom. In the Debye theory for very low temperatures, the expression for ω_D is defined in terms of reciprocal frequencies (or sound velocities) to the third power (see (1.65)). This corresponds to $n = -3$ in (5.5).

The Debye moments $\Theta(n)$ are defined in terms of the frequency moments by

$$\Theta_D(n) = \frac{\hbar}{k}\left[\frac{1}{3}(n+3)\left\langle \omega^n \right\rangle\right]^{1/n} \quad \text{for } n > -3,\ n \neq 0 \tag{5.6}$$

and

$$\Theta_D(0) = \lim_{n \to 0} \Theta_0(n). \tag{5.7}$$

At $T = 0$, the value of Θ is found from $\Theta(-3)$. It turns out (Barron et al., 1980) that

$$\Theta(-3) = \Theta^S_{T=0} = \Theta^{C_P}_{T=0}, \tag{5.8}$$

where the Θ^S and Θ^{C_P} are the values of Θ found at $T = 0$ from the experimental values of entropy and specific heat.

The high temperature Θ, Θ_∞, found from specific heat is related to the second moment of the spectrum (Barron et al., 1980)

$$\Theta^{ht}_{cal}(+2) = \Theta^{C_P}_\infty. \tag{5.9}$$

Considering oxides and silicates where $\Theta > 300$ K, the sound velocities at room temperature differ little from sound velocities at absolute zero (Section 1.52). Therefore the use of (1.65) to find Θ from measured velocities corresponds closely to the case $\Theta(-3)$. From Table 1.9, $\Theta(-3) = 942$ K for MgO. The high temperature specific heat for MgO leads to $\Theta(2) = 779$ K (Barron et al., 1980). Thus $\Theta(-3)/\Theta(+2) = 1.2$. We should not demand better agreement from the experiments than observed from the calculated $\Theta(n)$ curve. Barron et al. calculated this curve, shown in Fig. 5.2, and we see that $\Theta(-3)/\Theta(+2) \approx 1.21$, showing that the experiments follow the theory fairly well. If we evaluate Θ_{ac} at high T, there is good agreement between Θ_{ac} and Θ_{cal} for MgO because sound velocities decrease with T. On that account, we could expect the Debye theory to give meaningful values of many physical properties of MgO, and we say the solid is Debye-like.

For silicate perovskite, it has been found (Stixrude et al., 1992b) from moments of the calculated density of states that $\Theta(-3) = 950$ K, and $\Theta(+2) = 1095$ K, which are reasonably close in value. (This closeness is also seen in the case of MgO, as shown above). The theoretical numbers for Θ check out well with experiments, which yield $\Theta_{cal} = \Theta(+2) = 920 \pm 50$ and 1000 ± 50 from infrared and Raman frequencies (Stixrude et al., 1992b). The low T Θ_{ac} turns out to be 1090 K (see Fig. 1.3), but Θ_{ac}^{ht} extrapolated to 1700 K yields 990 K in agreement with Θ_{cal} at high T (see Fig. 1.3). The $\Theta(n)$ curve for silicate perovskite was calculated by Wolf (1992) and is also shown in Fig. 5.2.

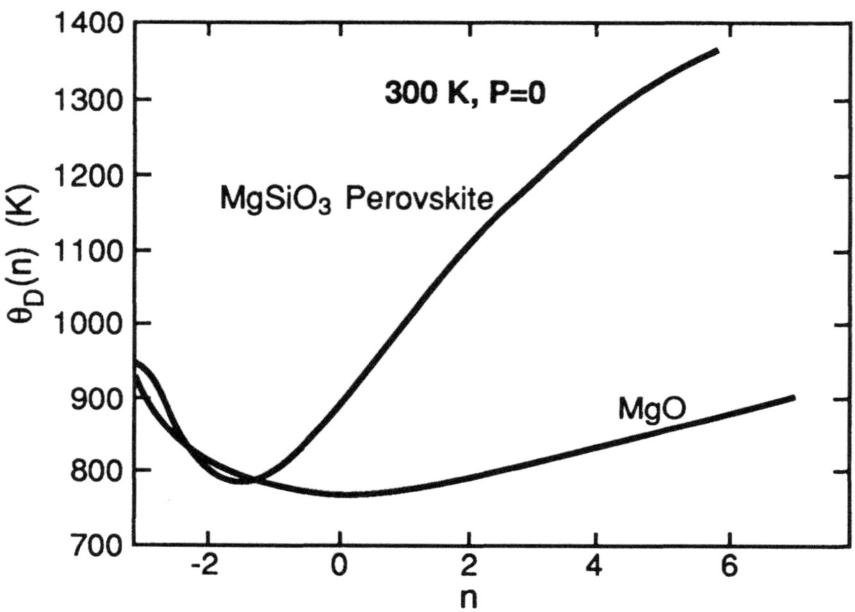

Fig. 5.2. The curve of moments of Θ_D plotted versus the moment n, $\Theta_D(n)$, for MgO and $MgSiO_3$ perovskite. Wolf (1992) presented the perovskite curve, and Barron et al. (1980) presented the MgO curve.

Again we might expect many of the physical properties of silicate perovskite to be calculable from a Debye model. Many materials do not have reasonable agreement between $\Theta(-3)$ and $\Theta(+2)$. Bailey and Yates (1970) show that $\Theta(-3)/\Theta(+2) = 400/900 = 0.44$ for pyrolitic graphite.

$\Theta_D(-3)$ is weighted heavily by low frequency modes, and $\Theta_D(+2)$ is weighted heavily by high frequency modes. In a pure Debye spectrum, where $g(\omega)$ is quadratic up to the cutoff at ω_D, $\Theta_D(-3) = \Theta_D(+2)$. If there is a substantial fraction of modes in $g(\omega)$ higher than ω_D, or if $g(\omega)$ does not equal the Debye spectrum at low frequencies, then $\Theta_D(-3)$ cannot be close to $\Theta_D(+2)$. If these numbers are close, however, say within 15%, then $\Theta_D(-3) = \Theta_{ac}$ used in a Debye function $D(\Theta/T)$ gives a good approximation to the measured specific heat at high T. A Debye-like solid requires that only a small fraction of the total modes lie above ω_D (see Figs. 5.3 and 5.4).

If a very intense peak occurs in the middle of $g(\omega)$, as in the case of CaO, then $\Theta_D(0)$ is not a very good approximation to $\Theta_D(+2)$.

The volume derivatives of $\Theta_D(n)$ define the Grüneisen parameters in their limits for two values of η. In general,

$$\gamma(n) = -d\,\ell\text{n}\,\Theta_D(n)/\partial\,\ell\text{n}\,V. \tag{5.10}$$

For very low temperatures,

$$\gamma(-3) = \gamma_{T=0}, \tag{5.11}$$

and for high temperatures,

$$\gamma(0) = \gamma_{T=\infty}. \tag{5.12}$$

$\gamma(-2)$ has a significant physical meaning. This moment gamma arises from the root mean square determination of the amplitude of vibration (Wolf and Jeanloz, 1984) and is related to melting (see Chapter 11).

5.6. The vibrational spectra (density of states) $g(\omega)$

For many problems in oxides and silicates, and especially so in geophysics, sound velocity data at high T are much more plentiful than specific heat or spectroscopic data at high T. It is very helpful, therefore, to know when sound velocity data at high T can be safely used to find thermal properties.

5.6.1. *Class C density of states*

In this section we compare the density of states with the simple Debye theory, to show when one might expect the simple Debye theory to apply to a solid. The Debye theory is a good approximation to some physical properties whenever the vibrational density of states, $g(\omega)$, is such that the optical modes bunch together and are close to the ω_0 peak found from Θ_{ac}.

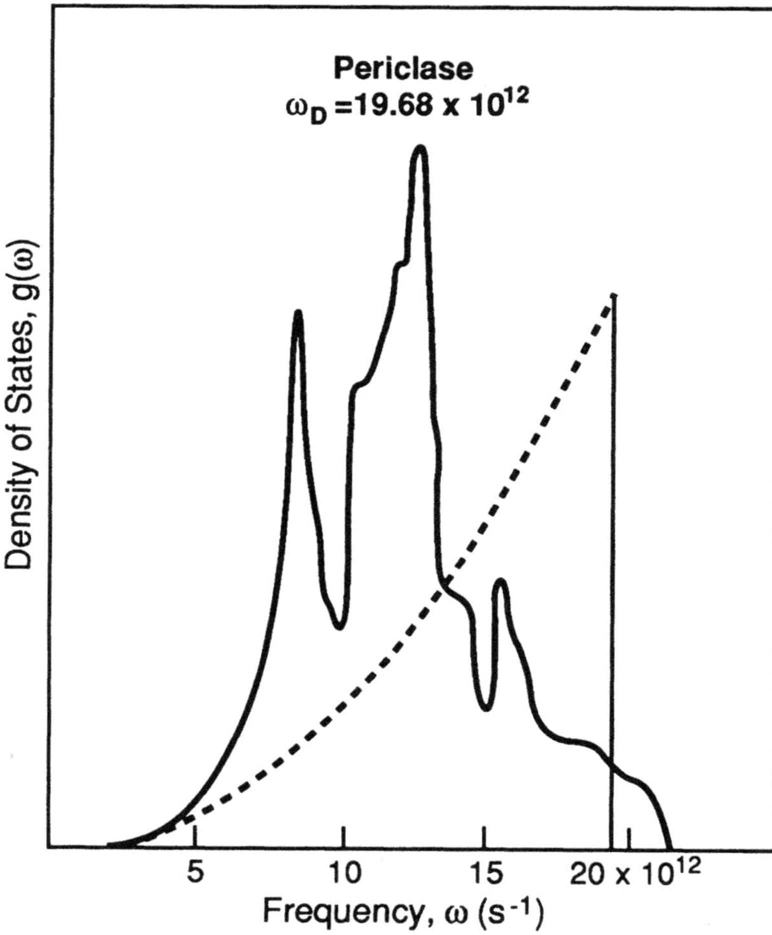

Fig. 5.3. The computed frequency spectrum or density of states for periclase, MgO, as calculated by Sangster et al. (1970). It is compared with the Debye spectrum, where ω_D is computed by sound velocity measurements. Note that the optical spectrum is mostly contained within the Debye spectrum, and that the spectra converge at low frequencies—a Class C correlation.

The density of states for periclase (shown in Fig. 5.3) has been calculated by Sangster et al. (1970). The Debye spectrum is also given as a parabola, where the Debye frequency, $\omega_D = 19.68 \times 10^{12}$ cps, is determined from acoustic experiments according to (1.65). Agreement is good at low frequencies, and the fraction of the spectrum lying to the right of ω_D is small. This agreement between the Debye spectrum and the density of states at low frequency is considered to be fairly good.

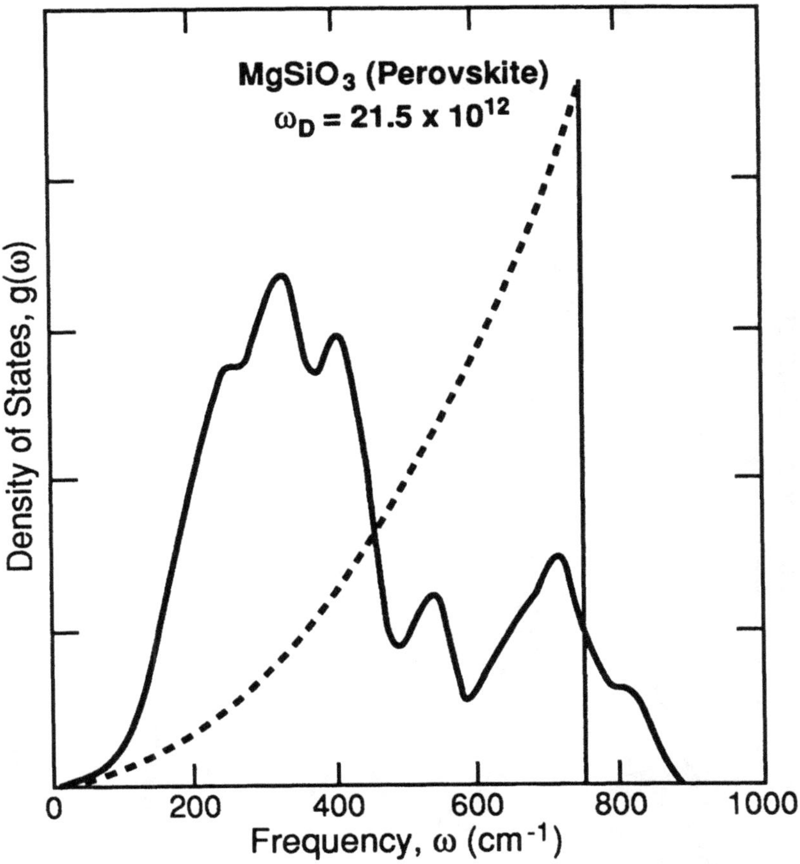

Fig. 5.4. The computed frequency spectrum or density of states for silicate perovskite, as calculated by Choudhury et al. (1988). It is compared with the Debye spectrum computed by sound velocity measurements.

The density of states of silicate perovskite has been presented by Choudhury et al. (1988). In Fig. 5.4 the density of states is compared with the Debye spectrum determined by $\omega_D = 21.5 \times 10^{12}$ (from the sound velocity measurements of Yeganeh-Haeri et al., 1989). Only a small fraction of the optical modes lie outside ω_D, and there is good agreement between the two optical spectra at low frequencies. It is expected, therefore, that the Debye theory will be very useful for computing thermal properties of this solid. To be noted is the similarity of the silicate perovskite spectrum to the MgO spectrum (Fig. 5.3). In Section 5.5, we found for perovskite, $\Theta_D(-3)/\Theta_D(+2) = 0.87$, and $\Theta_{ac}/\Theta_{cal} = 1.18$, and that at 1000° Θ_{ac} approaches Θ_{cal}, indicating Debye-like behavior. Altogether there is good evidence that silicate perovskite is to be classified as a Class C solid.

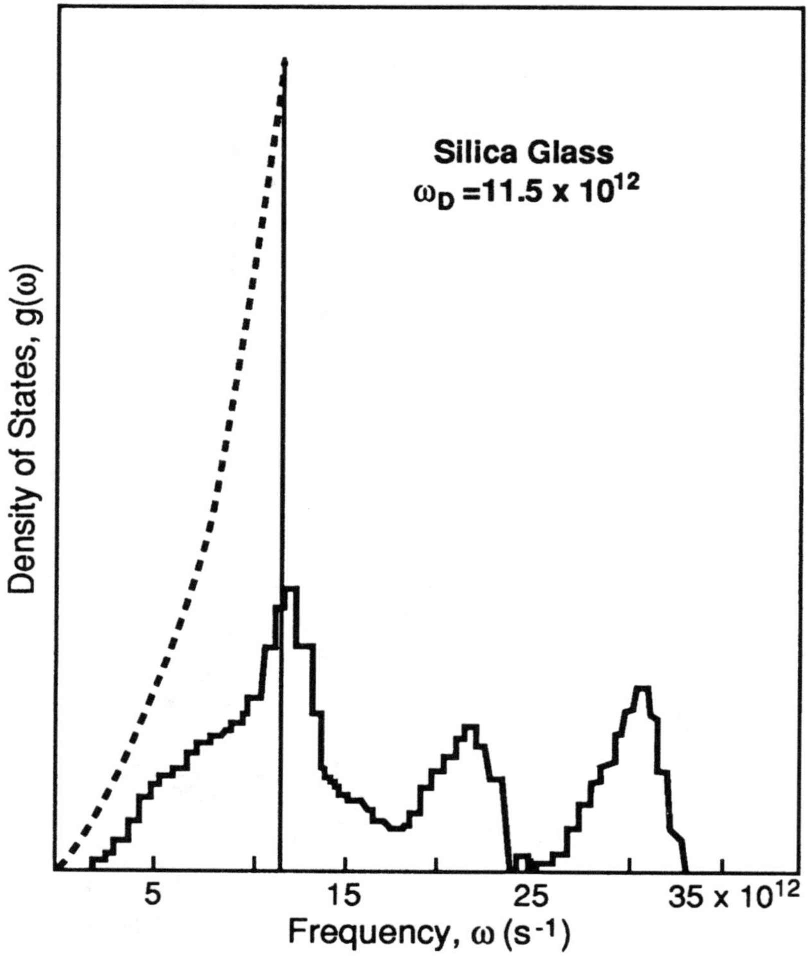

Fig. 5.5. The computed density of states for silica glass calculated by Bell (1976) compared with the Debye spectrum, where ω_D is computed from sound velocity measurements. The frequency spectra match poorly at both high frequencies and low frequencies.

5.6.2. *Class A density of states*

The density of states of silica glass has been calculated by Bell (1976) and is reproduced in Fig. 5.5. The Debye spectrum, calculated by the Debye approximation, (1.65), by means of ω_D, is indicated by the parabola. Here it is seen that the Debye approximation is very poor, and a great deal of the spectrum lies at frequencies higher than ω_D. Also note that the low frequencies of the density of states do not match the low Debye frequencies.

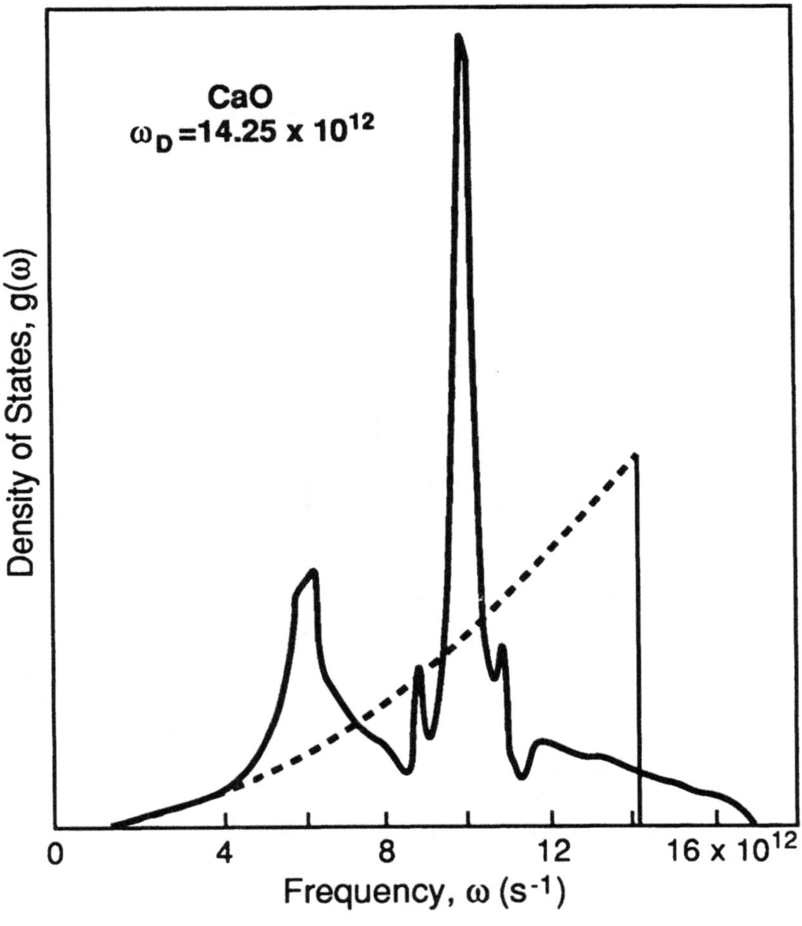

Fig. 5.6. The density of states for CaO calculated by Bilz and Kress (1979), where ω_D is computed from sound velocity measurements. The agreement is judged to be poor because of the large optical band in the mid-spectrum and the large distribution of modes above ω_D. The spectra favor high frequencies over low frequencies.

5.6.3. *Class B density of states*

The calculated density of states for CaO (illustrated in Fig. 5.6) has been presented by Bilz and Kress (1979). The Debye spectrum calculated by means of $\omega_D = 14.25 \times 10^{12}\ \mathrm{s}^{-1}$ is a poor approximation because of the strong optical peak at about $\omega = 10 \times 10^{12}\ \mathrm{s}^{-1}$ and the large high frequency tail, which make the high frequencies weigh heavily in the calculation of $\Theta_D(+2)$. This poor correlation between the two spectra is reflected in

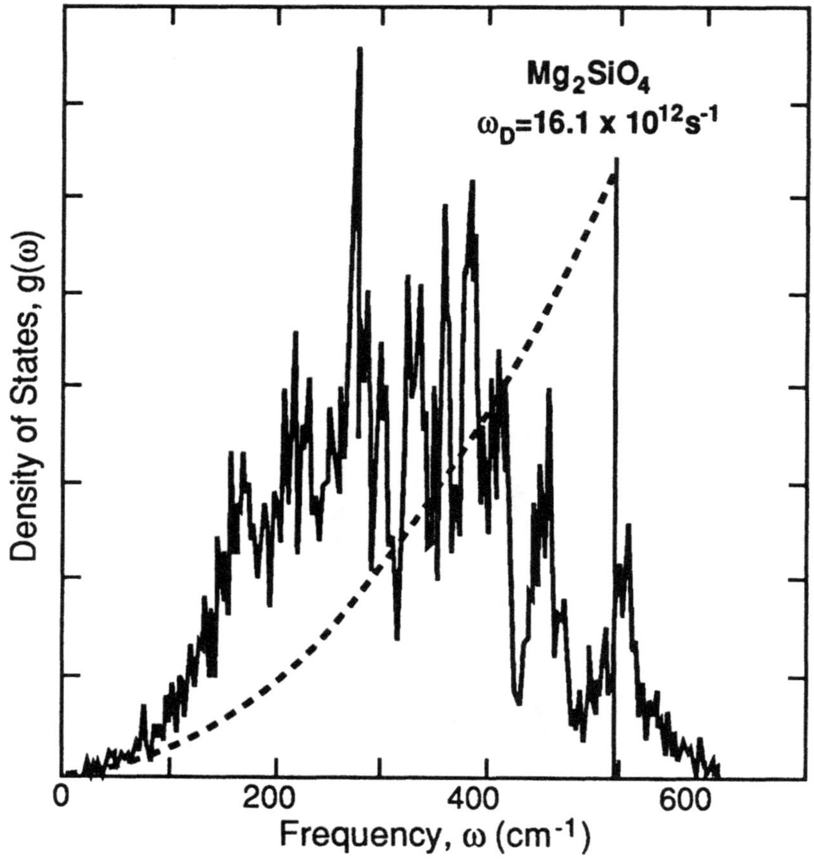

Fig. 5.7. The calculated density of states for forsterite (Rao et al., 1988), where $\omega_D = 16.1 \times 10^{12}$ s^{-1} is calculated from sound velocity measurements. The correlation of the two spectra is similar to that found for MgO, but there is a longer high frequency tail in forsterite.

various poor correlations between measured properties and those derived from simple Debye theory (for example, $\gamma_{ac} < \gamma$; see Table 5.3).

The calculated density of states for Mg_2SiO_4, presented by Rao et al. (1988), is illustrated in Fig. 5.7. The correlation is similar to that for MgO, but there are unplotted internal SiO_4 modes between 800 cm^{-1} and 1000 cm^{-1} (Chopelas, 1990b). This contributes to $\omega_D(+2)$, but not $\omega_D(-3)$. Mg_2SiO_4 is marginally close to Class C, and so correlations between measured properties and the Debye (calculated) properties are sometimes good, and sometimes not good.

Another representative of Class B is rutile, TiO_2. The density of the states of rutile has been calculated by Traylor et al. (1971). This is shown

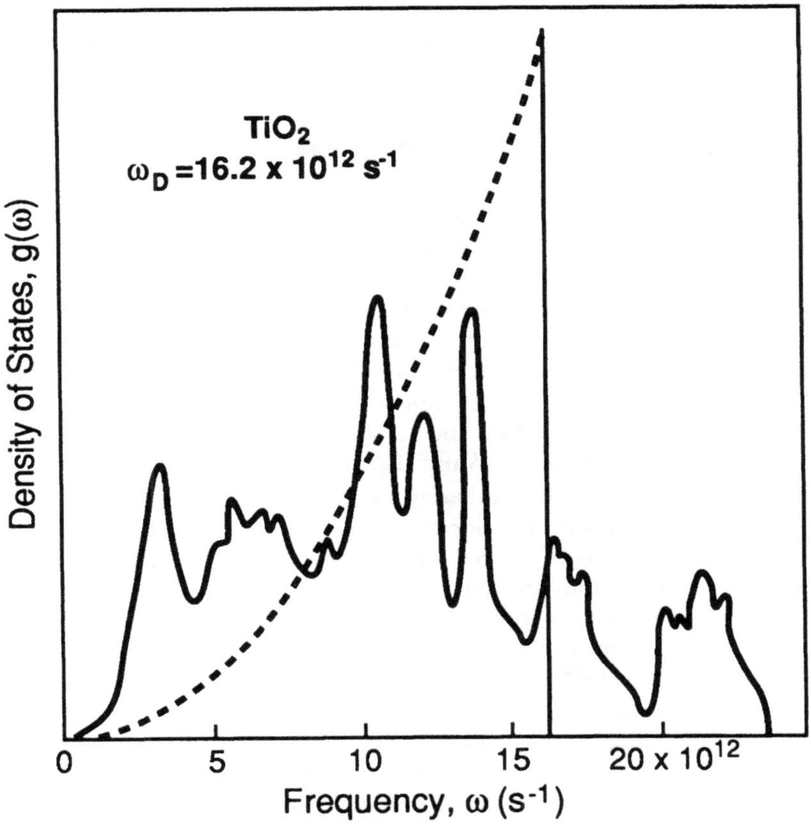

Fig. 5.8. The computed density of states for rutile, compared with the Debye spectrum, where $\omega_D = 16.2 \times 10^{12}\ \mathrm{s}^{-1}$ is computed from sound velocity measurements. Too much of the high frequency spectrum lies outside of ω_D, and the convergence at low frequency is only fair.

in Fig. 5.8, along with the Debye spectrum using $\omega_D = 16.2 \times 10^{12}$. Here we see that the match is not good, but not bad either. A substantial part of the spectrum lies above ω_D. Further, the match at the low frequency end is not quite right.

These results reinforce the assertions made in the above sections. Solids with a high *P.F.* (e.g., for rock-forming minerals at high density) will tend to have a compressed vibrational density of states analogous to that of a monatomic closely packed crystal.

5.7. Velocity systematics

We now show good correlations between ambient sound velocities from one mineral to another in Class C. But the overall correlations are diminished if members of Class A are added to the group. We also demonstrate that the good correlation found in Class C is to be expected as a result of Debye theory.

5.7.1. *Velocity at constant mean atomic mass, μ*

The Debye theory has been used to explain the relationship between the elastic properties of mantle minerals and their densities (Shankland, 1972) that has come to be called velocity-density systematics. Shankland proposes that the form of an equation relating sound velocity to density at constant mean atomic mass (e.g., $\mu = \mathbf{M}/\rho$) is constant from mineral to mineral. Data on the subject accumulated in the past 20 years have induced authors to show that the velocity-density systematics concept is valid only in the statistical sense, and that the number of exceptions make it less than satisfactory as a general rule (Wang, 1978).

Nevertheless, we will emphasize in this section that the exceptions to the original proposal for velocity-density systematics are the minerals to which the Debye theory is not applicable (i.e., Class A). The velocity-density systematics rules are reliable guides if one restricts the application of velocity systematics to minerals in Class C. The derivations following Debye's theory are most instructive. We review and follow the derivations of Shankland (1971, 1972), which were presented in part earlier by Anderson and Nafe (1965) and later amplified by Wang (1978).

There is a long history of velocity systematics starting with Birch (1960, 1961). He demonstrated empirically that for minerals and rocks, ambient sound velocity is linear in ambient density comparing mineral to mineral. Variations of Birch's equations have appeared in D.L. Anderson (1967), Anderson (1965c, 1972), Shankland (1971) and Wang (1978). Shankland brought Debye theory to confirm the empirical results for velocity systematics found by previous authors.

In the Debye theory examined by Shankland, the beginning was focused on a well known result in lattice dynamics; namely, that at low values of the wave number $\mathbf{k}$, the dispersion relationship between frequency and wave number is linear,

$$\omega_k = v_k \mathbf{k}, \tag{5.13}$$

where $k = 1, 2, 3$ for the two shear and longitudinal acoustic branches, and v_k is the sound velocity for each branch. At the maximum cut-off frequency in the vibrational spectrum (i.e., at the Debye limit), we have the relationship

$$\mathbf{k}_{\max} = \frac{\pi}{r_0},$$

where r_0 is the lattice spacing constant at $P = 0$. In the Debye theory, $\mathbf{k}_{\max}$ is defined by the Debye frequency ω_D. Because the Debye theory treats only acoustical vibrations, the sound velocity, $v_k (k = 1, 2, 3)$ is easily solved. Using the above relationship between r_0 and $\mathbf{k}_{\max}$, (5.13) becomes

$$v_k = \omega_{k_{\max}} \left(\frac{r_0}{\pi} \right). \tag{5.14}$$

Now $V_0 = \mu/\rho$, where μ is the mean atomic mass. Taking the derivative of (5.14) with respect to ρ (at constant mean atomic mass) gives

$$\left(\frac{\partial \ln v_k}{\partial \ln \rho} \right)_\mu = \left(\frac{\partial \ln \omega_{k_{\max}}}{\partial \ln \rho} \right)_\mu - \frac{1}{3}. \tag{5.15}$$

The first term on the right is the Grüneisen parameter, arising from Debye theory (Grüneisen, 1926). The subscript k takes on three values appropriate for the longitudinal and two sound velocities. Thus

$$\left(\frac{d \ln v_k}{d \ln \rho} \right)_\mu = \gamma_k - \frac{1}{3}, \tag{5.16}$$

where γ_k is a quantity arising from an acoustic mode of vibration. Equation (5.16) can be written as

$$v_k = a_k \rho^{\gamma_k - \frac{1}{3}}. \tag{5.17}$$

Thus we expect that when oxides and minerals are sorted by constant μ and restricted to those whose density of states can be approximated by a Debye spectrum, the variation of sound velocity with density from solid to solid (all at ambient conditions) should obey (5.17). Using a slightly differing approach relying on Grüneisen's (1926) theory, Anderson and Nafe (1965) derived (5.17) for the shear velocities.

Equation (5.17) is an exponential version of Birch's law, given by

$$v_k = a_k + b_k \rho. \tag{5.18}$$

Plotting (5.17) and (5.18) over the (ambient) density range found in oxides and silicates shows that these two equations produce virtually identical results. In (5.17) and (5.18) k can be s, ℓ, or b, representing the shear, longitudinal, or bulk sound velocity modes. A plot of v_b versus ρ_0 for constant μ is shown in Fig. 5.9 (taken from Shankland, 1972). For review articles on the application of these equations, see Anderson (1972), Shankland (1977), and Wang (1978).

At this point, I comment that many rock-forming oxides and silicates that are exceptions to the rule given by (5.18) or its equivalent, (5.17), have involved calcium-rich compounds (Simmons, 1964a,b) or silicates like α-quartz with low Poisson's ratio values (Mao, 1974). Such solids, lying off the curve given by (5.18), are not members of Class C and are characterized by having low packing fractions or high values of V_r. They have densities of states far too complicated to be represented by a Debye law, so (5.17) or (5.18) is not appropriate. From Table 1.9, μ for CaO is high, and because Ca is nearly the same size as oxygen, its V_r value is high, giving support to Simmons's findings.

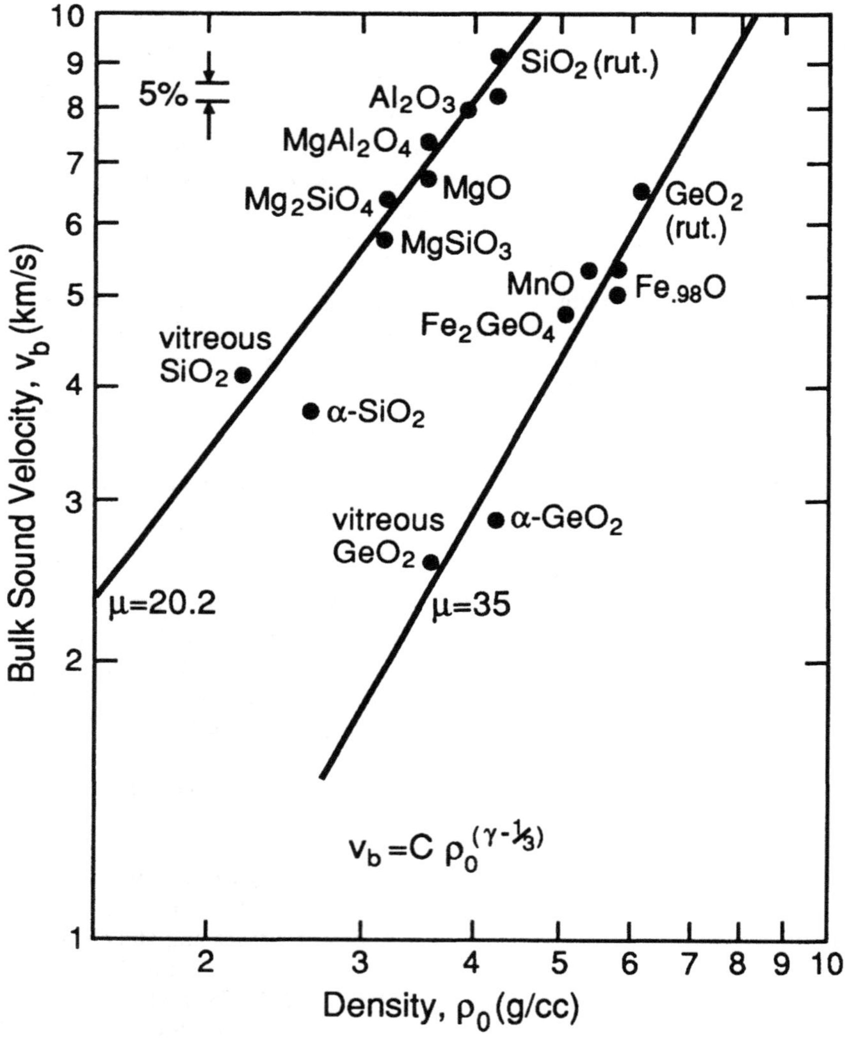

Fig. 5.9. The correlation of bulk sound velocity v_b with ρ_0 at constant mean atomic weight on a log-log plot. The straight lines represent the equation $v_b = \alpha_i \rho_0^{\gamma_i - 1/3}$ (5.17) (Plotted from data of Shankland, 1972).

Some specific exceptions to (5.17) and (5.18) will now be noted. A garnet's sequence may not all fall on a common line because the large Ca atoms in some garnets would increase the value of V_r. Pyrope, with no Ca and with a $V_r = 5.66$, is on the border of Class C, but almandine, having a significant amount of Ca, has $V_r = 6.26$ (Babuska et al., 1978) and is a Class B solid (see Fig. 5.1).

Some oxides and glasses have coordination numbers M of 4 or 2, producing optical modes of rotation that greatly broaden their densities of states (Barron, 1937; Sato and Anderson, 1980). For Zn0, $M = 4$, which places the oxide in Class A. In general, failure of (5.17) and (5.18) will often be noted for sequences where crystal class and μ are maintained constant but a high coordination number is not (see data of Liebermann and Ringwood, 1973; Davies, 1974; Shaw, 1976).

Many oxides and silicates have complicated vibrational spectra that are unsuitable for comparison with a simple Debye spectrum. Placing velocity data of these minerals in velocity-density correlations will lead to a distortion of the correlation. That is why points representing CaO, TiO_2, and calcite are not plotted in Fig. 5.9. Chrisoberyl (Al_2BeO_4), with a low value of V_c, (4.89), does not fit these ideas and is an unexplained aberration to the pattern (Wang et al., 1975). This aberration has been attributed to Be. For planet interiors, the chrisoberyl deviant can be ignored because Be is a scarce element.

5.7.2. *Velocity at constant crystal structure, variable μ*

In solid solutions where, for example, iron is substituted for magnesium, the crystal structure is held constant, and the mass μ is varied. Some simple rules resulting from the Debye approximation apply to this case also. Following Shankland (1972), we introduce μ into the formula for frequency. Consider a vibrating one-dimensional system with a spring constant $\mathbf{g}$ and mass μ. We find

$$\omega = \left(\frac{\mathbf{g}}{\mu}\right)^{1/2} . \tag{5.19}$$

Replacing ω by $v_b \mathbf{k}_{\max}$ according to (5.13), we have

$$\left(\frac{\pi}{r_0}\right) v_b = \left(\frac{\mathbf{g}}{\mu}\right)^{1/2} .$$

From the above,

$$\frac{d \ln v_b}{d \ln \mu} = -\frac{1}{2} + \frac{d \ln \left(\mathbf{g} r_0^2\right)^{1/2}}{d \ln \mu} . \tag{5.20}$$

It has been shown (Shankland, 1972) that the term on the right is close to zero for minerals, because $\mathbf{g} r_0^2$ is independent of mass. For ionic solids the spring constant, being the second derivative of the force holding atoms together, is, using the Born-Mie repulsion (Shankland, 1972),

$$\mathbf{g} r_0^2 = \frac{\mathbf{A} Z_1 Z_2 e^2 (n-1)}{r_0} ,$$

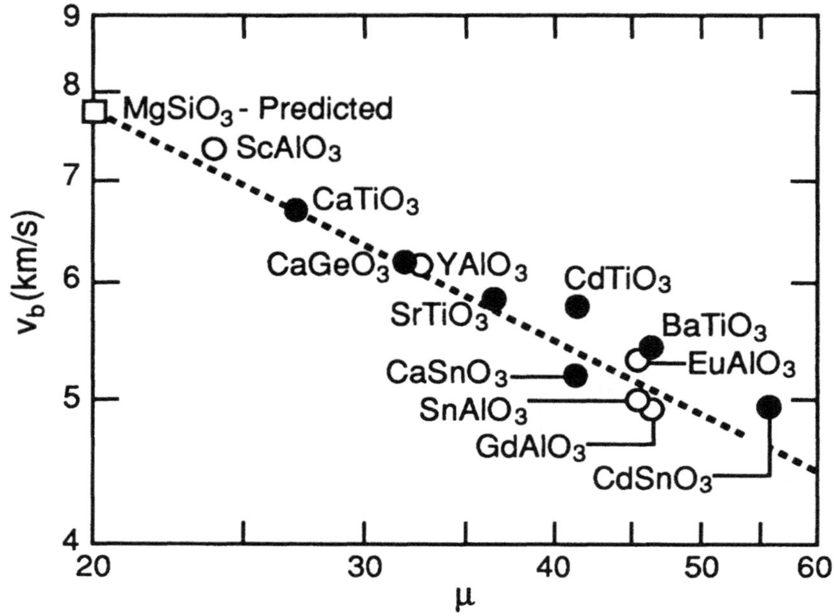

Fig. 5.10. v_b versus μ for a sequence of perovskite solids using Liebermann's data (1982). The □ is the predicted value of v_b for silicate perovskite (1982), using the $v_b\mu^{1/2}$ = constant best fit for the pervoskite data, represented by the dashed line (figure modified from Liebermann, 1982).

where n is the repulsion coefficient (see (8.8)). For a sequence in the same crystal class (isostructural), gr_0^2 will not change, resulting in

$$(K_{T_0} V_M)_x = \text{constant}, \tag{5.21}$$

where V_M is the molecular volume; K_{T_0} is the bulk modulus; and x indicates constant crystal class. Equation (5.21) is called the law of corresponding states (Anderson, 1965c) and has been verified for many oxides (D.L. Anderson and Anderson, 1970) and generalized to monatomic solids (Anderson, 1972). That is to say, (5.21) holds for a sequence in a solid solution. Liebermann (1975) showed that (5.21) holds for olivines and spinels (see his Table 8), and also for a series of perovskites (Liebermann, 1982).

Because of (5.21), gr_0^2 is independent of μ, and (5.20) becomes simply

$$\left(\frac{d\,\ell\text{n}\,v_b}{d\,\ell\text{n}\,\mu}\right)_x = -\frac{1}{2}. \tag{5.22}$$

An equation equivalent to (5.22) is (Shankland, 1972)

$$v_b\mu^{1/2} = \text{constant}. \tag{5.23}$$

Liebermann (1982) tested (5.23) over a wide range in μ by considering a sequence of perovskite solids (see Fig. 5.10).

The square in Fig. 5.10 representing Mg silicate perovskite represents an extrapolation using the mean value of $v_b\mu^{1/2}$ = constant. This extrapolation leads to a predicted value of v_b that is about 7.8 km/s, as shown by the location of the □ in the figure. Using $\rho_0 = 4.2$ gm/cc for perovskite, $v_b^2\rho_0$ yields about $K_{S_0} = 255$ GPa, which is close to the experimental values of K_{T_0} measured about a decade later, around 260 GPa, as listed in Table 4.6. All the perovskites fit into the pattern of Fig. 5.10, which demonstrates that perovskites satisfy not only (5.22) and (5.23), but also the assumption implicit in (5.19); that is, their density of states can be represented by a single Debye frequency. They are, therefore, Debye-like solids.

The details of applying the concepts behind (5.21) to framework silicates were explored by Hazen and Finger (1979, 1982). They found

$$\frac{K_{T_0} r_a^{-3}}{s^2 Z_1 Z_2} = \text{constant}, \tag{5.24}$$

where s is the ionicity; Z_1 and Z_2 are the ionic charges; and r_a is the average cation-anion bond distance.

5.7.3. *Velocity at constant crystal structure, variable ρ*

Keeping the crystal class constant but varying ρ, we have

$$\left(\frac{\partial \ln v_b}{\partial \ln \rho}\right)_x = \left(\frac{\partial \ln v_b}{\partial \ln \mu}\right)_x \left(\frac{\partial \ln \mu}{\partial \ln \rho}\right)_x, \tag{5.25}$$

where x stands for constant crystal class. But, because $\rho = \mu/V_M$,

$$\left(\frac{\partial \ln \rho}{\partial \ln \mu}\right)_x = 1 - \frac{\partial \ln V_M}{\partial \ln \mu}.$$

Using (5.22) in (5.25),

$$\left(\frac{\partial \ln v_b}{\partial \ln \rho}\right)_x = -\frac{1}{2}\left(1 - \frac{\partial \ln V_c}{\partial \ln \mu}\right)^{-1} = -\frac{1}{2}\left(1 - c'\right)^{-1}. \tag{5.26}$$

The term c' in (5.26) is a crystal structural factor determined by the way the substitution of an ion of a differing mass affects the molecular volume. Thus, c' is essentially a correction for the size of the ion and tends to be nontrivial for solid solutions where calcium is substituted for iron or magnesium. For sequences where the c' correction is small, (5.26) becomes

$$\left(\frac{\partial \ln v_b}{\partial \ln \rho}\right)_x = -\frac{1}{2}, \tag{5.27}$$

similar to the situation found for v_b versus μ at constant x, (5.22).

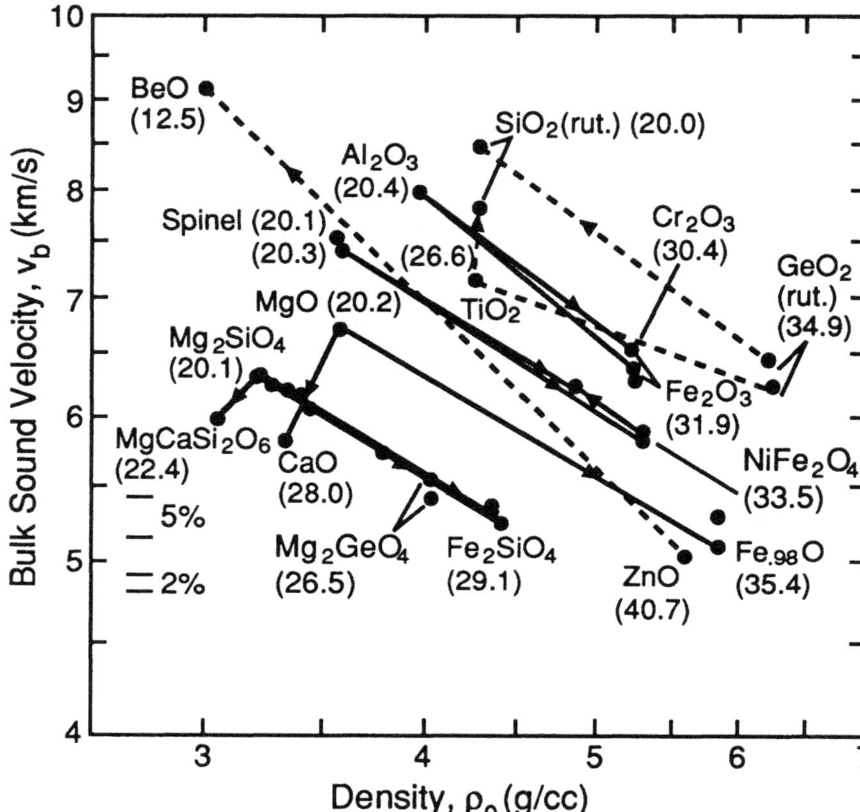

Fig. 5.11. Bulk sound velocity v_b versus ρ_0 at constant crystal structure (modified from Fig. 2, Shankland, 1972). Values of μ are in parentheses. On a log-log plot, (5.27) requires straight lines with a slope of $-1/2$. Lines connecting oxides and silicates of Classes A, B, and C do not all show slopes of $-1/2$ on this plot. If solids of Class A and Class B (ZnO, BeO, CaO, TiO_2, $MgCaSi_2O_6$) are removed from the plot, the remaining lines (C to C) have slopes close to $-1/2$.

Equation (5.27) is derived directly from (5.19), a model in which there is a vibration of one atom responding to a linear spring; that is, the solid is monatomic. Then Debye theory comes in by means of (5.13), so that ω is replaced by sound velocity v_b. Thus (5.27) should be valid for Class C solids, but should not necessarily hold for Class B solids. We should not expect the slope of the $\ell n\ v_b$ versus $\ell n\ \rho_0$ curve to be $-1/2$ for Class A solids. This is clearly demonstrated in Fig. 5.10, which is a plot of $\ell n\ v_b$ versus $\ell n\ \rho_0$ at constant crystal structure.

From this figure we see that the slope for many solid pairs is $-1/2$. Disturbance of this simple pattern is found by the lines connecting CaO with

MgO, BeO with ZnO, stishovite with TiO_2, and $MgCaSiO_4$ with Mg_2SiO_4. If BeO, ZnO, TiO_2, and $MgCaSiO_4$—all members of Class A—are removed from the graph, and if CaO is also removed, the remaining pattern is regular and follows (5.27).

These variants to the pattern predicted by (5.27) have been interpreted as evidence of the corrections arising from the size effect given by c' in (5.26) (Shankland, 1977; Wang, 1978). They can also be interpreted, as we do in this chapter, as arising because certain solids do not abide by the Debye approximation due to strong optical modes in the frequency spectrum, so (5.19) does not apply. For calcium-rich silicates, both effects are important.

5.7.4. *The isostructural variation of K_{T_0} with ρ_0 in a solid solution*

Consider a solid solution series of dense minerals for which (5.27) arising from Debye theory is appropriate. One can write (5.18) in the exponential form

$$v_b = \mathbf{b}\rho^{-1/2}, \tag{5.28}$$

where $\mathbf{b}$ is an integrating constant, different for each crystal class.

Because $v_b = (K_S/\rho)^{1/2}$, (5.28) predicts that K_{S_0} is independent of ρ_0 as the average atomic mass μ changes from mineral to mineral in a solid solution. D.L. Anderson (1976) noted that experimental evidence showed that K_{S_0} depends little on density as iron content increases within a solid solution. This, of course, could have been easily derived from (5.22), as presented by Shankland (1972). D.L. Anderson found this relationship for pyroxenes, olivines, dense garnets, and spinels. However, the magnitude of $\mathbf{b}$ is sensitive to the crystal structure considered, as experimentalists have confirmed repeatedly. For example, Weidner et al. (1982b) found that for the orthopyroxenes the bulk modulus is virtually independent of iron content (Fig. 5.12).

The bulk modulus is also virtually independent of iron content for the olivines [Isaak, 1992 (see also Jackson et al., 1978)], as shown in Fig. 5.13. K_S is nearly independent of iron for the Mg/Fe solid solution of spinel. In garnets, going from pyrope ($Mg_3Al_2Si_3O_{12}$) to almandine ($Fe_3Al_2Si_3O_{12}$), there is virtually no increase in K_S where density varies from 3.56 to 4.32. Babuska et al. (1978) reported that K_S = 173 GPa for pyrope and 178 GPa for almandine.

For the solid solution periclase-wüstite, MgO-FeO, the value of K_S is reported to vary with density. K_S for MgO (Spetzler, 1970) is 163 GPa, 8% less than the value reported by Mizutani et al. (1972). Jeanloz and Sato-Sorenson (1982), however, reported K_S = 157 GPa for wüstite. Measurements of intermediate compositions of polycrystalline samples by Jackson et al. (1978) indicate that the value of K_S may rise as iron content increases. It may be that the samples of single crystals used have variable defect structures.

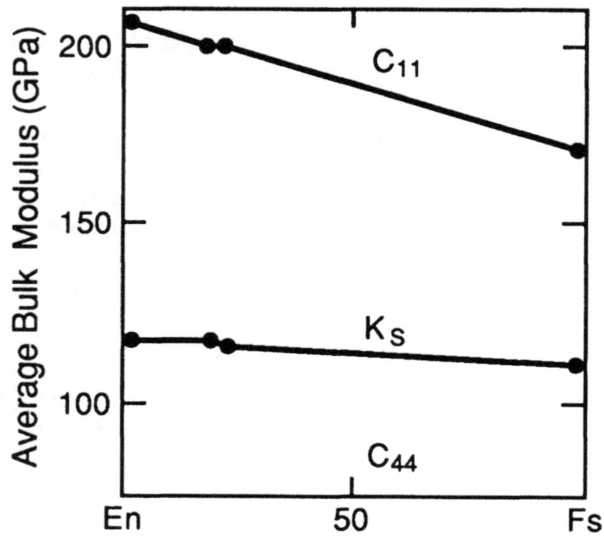

Fig. 5.12. Bulk modulus versus fractional composition for the solid solution of orthopyroxenes. Note that adiabatic bulk modulus is virtually independent of composition and also that C_{11} decreases with increasing iron content (modified from Weidner et al., 1982b).

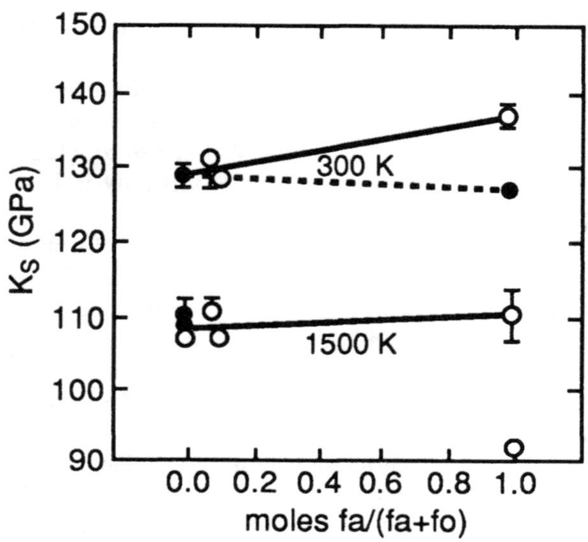

Fig. 5.13. Bulk modulus versus fractional fayalite component at 300 K and 1500 K (modified from Isaak, 1992) (see Isaak for data references). Note that bulk modulus is virtually independent of iron content, especially at high T.

Wüstite is a complex nonstoichiometric oxide that exists in a variety of types characterized by differing ratios of octahedral to tetrahedral ferric iron and by varieties of clustered defects (Hazen and Jeanloz, 1984). As a consequence, there is a large reported range in bulk modulus for wüstite, and the subject still open.

The issue of how K_S varies in the MgO-FeO solid solution is not yet decided; the answer will be determined by future measurements on defect-controlled single crystals in the iron-rich end of the (Mg,Fe)O solution. Acoustic experiments on FeO strengthen the proposition that K_S is independent of iron content in the MgO-FeO solid solution (Graham and Kim, 1986).

5.7.5. *The isostructural variation of shear constants with ρ in a solid solution*

Although the Debye theory is a guide to the variation of K_S in a solid solution, it is of little help in predicting how the isotropic shear modulus changes in such a solid solution. For that, we need equations of lattice dynamics. In a solid solution the bulk modulus is controlled by (5.22) and (5.27) in response to simple rules concerning the change of atomic potential with lattice spacing. But the shear constants depend critically on the details of the lattice sums (Anderson, 1970; Anderson and Liebermann, 1970). Thus, we probably cannot have both the ambient shear constants and the ambient bulk modulus independent of ambient density along a solid solution line. This can be illustrated, for example, by the equations for C_{44} and C_S for the CsCl structure deduced from lattice sums (Anderson, 1970) using a simple power law repulsive potential.

Using the central force approximation in CsCl crystal structure, Anderson and Liebermann (1970) found that

$$C_{44} = K_S - \frac{4}{3}P - \frac{0.6887\mathbf{A}Z_1Z_2e^2}{V_0 r_0}\left(\frac{\rho}{\rho_0}\right)^{4/3}. \tag{5.29}$$

In a solid solution series, $\mathbf{A}Z_1Z_2e^2$ is nonvariant, and for $P = 0$, (5.29) becomes

$$C_{44} = K_{S_0} - \text{constant} \cdot V_0^{-4/3}. \tag{5.30}$$

For CsCl at $P = 0$ (Anderson, 1970), the other shear constant is

$$C_S = \text{constant} \cdot V_0^{-4/3}. \tag{5.31}$$

As the density increases with substitution of heavier atoms, V_0 usually increases (Shankland, 1972). If so, according to (5.30), C_{44} must increase if K_{S_0} remains constant, and according to (5.31), C_S must decrease as the Fe/Mg ratio increases. The isotropic shear modulus will be intermediate between C_{44} and C_S. Whether the isotropic shear modulus will rise or fall with iron substitution depends very much on the actual equations for C_{44} and C_S, and for each crystal structure the equations of the shear constants will be different from (5.30) and (5.31). For most structures, the isotropic shear modulus decreases as the iron content increases.

For orthopyroxenes, in contrast to the CsCl structure, both C_{44} and C_S decrease with increasing iron content, so the isotropic shear modulus must also decrease (Fig. 5.14). The opposite case is found for the garnet sequence, pyrope to almandine (Babuska et al., 1978; Leitner et al., 1980), where the isotropic shear constant rises with increasing iron content even though the bulk modulus is independent of iron content.

For the olivine solid solution, forsterite to fayalite, the shear modulus drops with increasing iron content, both at low temperatures and high temperatures, as shown in Fig. 5.15. This has been confirmed by Suzuki et al. (1992), who found some interior points at about Fa/(Fa + Fo) $\cong$ 1/2.

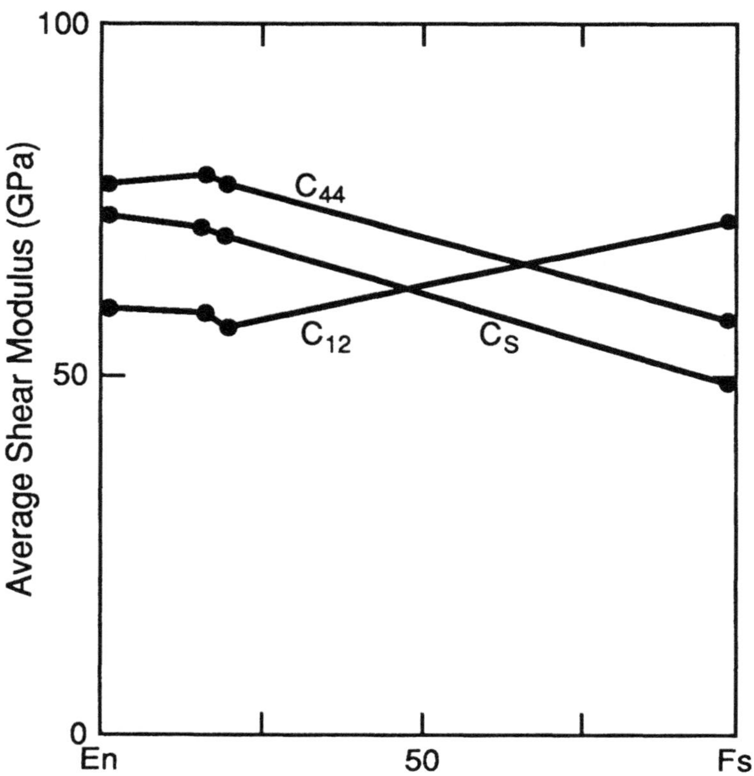

Fig. 5.14. For the solid solution of orthopyroxenes, both C_{44} and C_S decrease as the iron content is increased (modified from Weidner et al., 1982b).

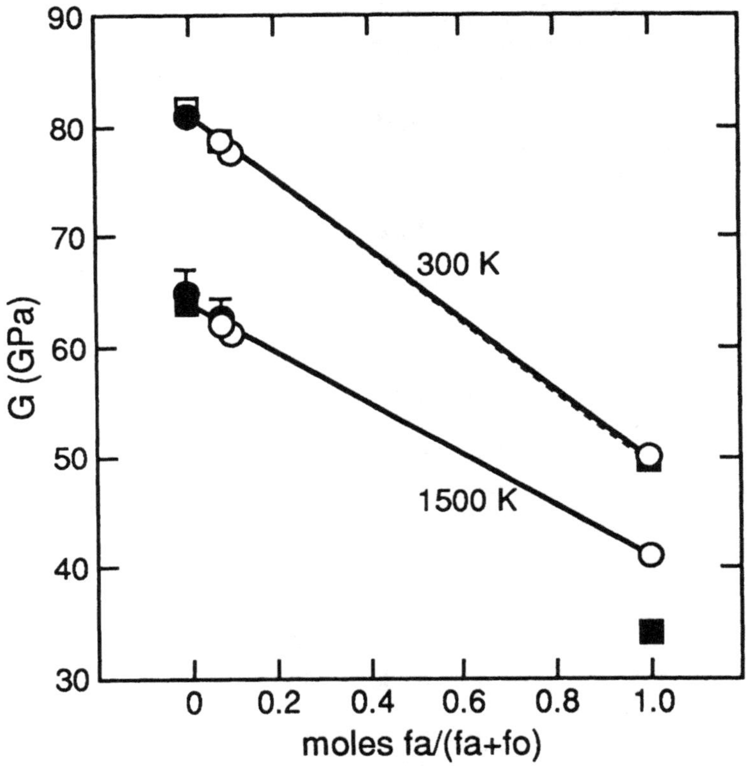

Fig. 5.15. The variation of shear modulus G with fractional fayalite component at 300 K and 1500 K (modified from Isaak, 1992).

The amount of change of the isotropic shear modulus or Poisson's ratio with density depends on the lattice sum details of the atomic potential, and in general these will not be known for materials that actually comprise planet interiors. Thus, variation of the shear constants with the iron content for assumed mixtures of silicates in planet interiors is presently unknown. Consequently, there is no simple rule governing v_s versus ρ in a solid solution such as is given for v_b versus ρ by (5.28).

5.8. The Grüneisen ratio γ and γ_{ac}

Equations for the Grüneisen parameters were presented in Section 1.4. We also showed by (5.12) that $\gamma_D(-3) = \gamma_{ac}$ and that $\gamma_D(0) = \gamma$. We therefore expect that γ_{ac} will be close in value to γ for the same solids for which Θ_{ac} is close in value to Θ_{cal}. In Table 5.3 we show that for materials in Class C, γ_{ac} is close in value to the measured γ. On the contrary, for materials in Class A, we find that γ_{ac} is not close in value to the measured γ. For members of Class B, γ_{ac} is only somewhat close to γ, an intermediate case.

Table 5.3. Values of γ_{ac} estimated from sound velocities (at room temperature) (1.41) compared with γ_0 found from (1.24) and (1.30)

	γ_{ac} (1.41)	γ_0 (1.24)	(1.30)	
Class **C** Solids				
α-iron[c]	1.81[a]	1.81[c]		Earth's lower mantle,
Al_2O_3[a]	1.31	1.32		uncompressed: $\gamma = 1.5$[d]
MgO[a]	1.50	1.54	1.39[b]	Earth's inner core,
				compressed: $\gamma = 1.5$[d]
Class **B** Solids				
Forsterite[a]	1.47	1.15	1.15[b]	
Olivine[a]	1.33	1.21		
TiO_2[a]	1.07	1.28		
Spinel[a]	0.79	1.13		
Garnet (alm.-spes.)[a]	1.38	1.01		
Garnet (pyr.-alm.)[a]	1.28	1.37		
CaO	1.57	1.72		
Class **A** Solids				
ZnO[a]	−0.44	0.81		
SiO_2 (quartz)[a]	0.46	0.66	0.57[b]	
SiO_2-glass[a]	−2.46	0.036	0.03[f]	
Soda-lime glass[f]	−0.33		0.61	
Lead-silica glass[f]	0.09		0.51	
$CaCO_3$[a]	−1.004	0.567		
Alkali halides				
NaCl[a]	1.33	1.56	1.62[e]	
KCl[a]	1.50	1.54		

Sumino and Anderson, 1984[a]; Boehler, 1982[b], Guinan and Beshers, 1968[c]; Anderson, 1979a[d]; Boehler, 1981[e]; Sato and Anderson, 1980[f].

Table 5.3 is another example showing that Debye theory—here in the form of γ_{ac}—can be applied usefully to oxides and silicates of Class C (γ_{ac} and γ are quite close in value) but not to Class A.

5.9. dK_S/dP for closely packed oxides and silicates

In densely packed oxides, where cation radii are small compared with those of oxygen, the oxygen ion plays a major role in compression. Anderson (1965b) suggested that the elasticity of a wide group of rock-forming minerals is controlled primarily by the volumes and packing fractions of the oxygen ions. Following this idea, Runcorn (1976) suggested that for dense silicates the oxygen ion network could be approximated as a densely packed monatomic cubic solid, and that silicon, iron, and magnesium ions, being small, play little part in determining elasticity. As an example, he pointed out that for spinel, the oxygen ions are arranged in a face-centered cubic lattice.

It is well known that the less densely packed oxides have relatively high values of dK_S/dP (Anderson et al., 1968; Anderson, 1972). Further, it is generally found that K_0' decreases across a phase change going from a low-pressure phase to a high-pressure phase. For example, K_0' is 5.2 for α-iron (bcc) (Guinan and Beshers, 1968), but is 4.37 for ϵ-iron (hcp) (Jephcoat et al., 1986).

Runcorn (1976) went on to suggest that for densely packed solids, whether iron or silicates, the repulsive potential between the atoms would take a simple form, and we could expect the value of K_0' to be close to $(n+7)/3$, where n is the exponent on the interatomic relative spacing giving a power-law repulsion. Thus all densely packed oxides, as well as iron, should have a value near 4 for K_0', as n is close to 10 and $K_0' = (n+7)/3$ for this repulsion law. It has long been noted that K_0' for closely packed silicates is 4 or less (D.L. Anderson and Anderson, 1970).

Runcorn's suggestion seems to be borne out by experimental results (Fig. 5.16). K_0' is 3.99 for Al_2O_3 and 4.2 for MgO (Table 1.8). Wolf and Jeanloz (1984) reported $K_0' = 3.79 \pm 0.4$ for Mg_2SiO_4 perovskite.

As pointed out above, $K_0' = 4.23$ for hcp iron. The stishovite data for K_0' are reported variously by different authors. Early static high-pressure measurements indicated that $K_0' > 6$ (Liebermann et al., 1976), but it was noted that the value of K_0' inferred from shock waves was considerably lower. The definitive measurements of K_{S_0} made by Weidner et al. (1982a) using Brillouin scattering led them to suggest that K_0' for stishovite was substantially lower than for other rutile structure oxides. Thus K_0' for stishovite may be close to 4.0, following the pattern in Fig. 5.16.

5.10. The Grüneisen ratio of the earth's lower mantle

The high temperature acoustic gamma γ_{ac}^{ht} is derived in Chapter 1.

$$\gamma_{ac}^{ht} = (1/3)\,(2\gamma_s + \gamma_p)\,, \tag{1.41}$$

where the equivalent equations to (1.39) and 1.40) are

$$\gamma_s = (1/3) + (\partial\,\ell\text{n}\;v_s/\partial\,\ell\text{n}\;\rho)_T \tag{5.32}$$

$$\gamma_p = (1/3) + (\partial\,\ell\text{n}\;v_p/\partial\,\ell\text{n}\;\rho)_T\,. \tag{5.33}$$

Because we have found from Section 5.6 that silicate perovskite is a Debye-like solid, $\gamma_{ac}^{ht} = \gamma$, it follows that in the lower mantle, we probably will have $\gamma_{ac}^{ht} = \gamma$. There is the problem of FeO in magnesiowüstite to consider. There are no data on the iron effect of FeO in magnesiowüstite, but there are some data on the iron effect in (Mg, Fe)SiO_4. Sumino and Anderson's (1984)

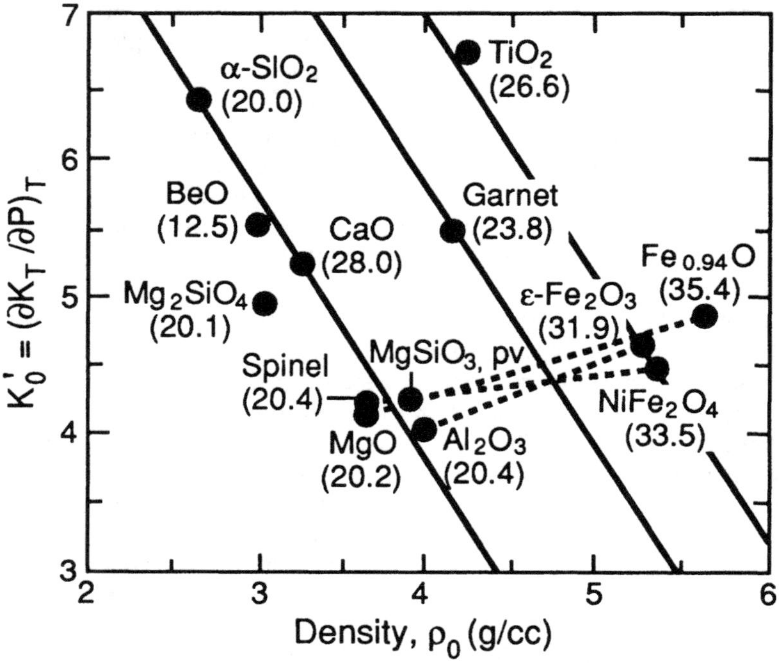

Fig. 5.16. $K_0' = (\partial K_S/\partial P)_T$ at $P = 0$ versus ambient density ρ_0. At constant $\mathbf{M}/p$, a large value of ρ_0 is correlated with a smaller value of K_0'. The dashed lines show isostructural correlation, suggesting that K_0' rises slightly for solid solutions. For high values of ρ_0 and low values of $\mathbf{M}/p$, $K_0' \approx 4$ (most data from Sumino and Anderson, 1984; $Fe_{0.94}0$ data from Jackson et al., 1990).

summary shows that γ_{ac}/γ_{cal} is 0.86 for pure Mg_2SiO_4 and drops to 0.8 for $(Mg_{0.93}Fe_{0.07})_2SiO_4$. Thus the effect of Fe lowers γ_{ac} by about 7% and is probably even less discernible in the value of γ_{ac}/γ_{cal} in magnesiowüstite. The concentration of magnesiowüstite is small in the mantle rocks, however, so the effect of iron on γ_{ac}/γ_{cal} is probably not discernible for the average rock of the lower mantle. This means that γ_{ac}^{ht} calculated by (5.33) is a good representation of γ in the lower mantle. That is to say, γ can be closely estimated by finding γ with the sound velocity and density distribution.

We use the PREM data from Dziewonski and D.L. Anderson (1981) with 5 significant figures to evaluate (1.41), (5.32), and (5.33). Values of γ_{ac}^{ht} for the lower mantle are listed in Table 5.4, which shows that γ_{ac}^{ht} is a monotonically decreasing function (going from 1.40 to 1.0 from the top to the bottom of the mantle by extrapolation).

To a good approximation, $\ell n\ \gamma$ versus $\ell n\ \rho$ data of the lower mantle define a straight line with a slope of about 1.35 (Fig. 5.16), or $q = 1.32$, as found by (1.32).

Table 5.4. Calculation of γ_{ac} for the lower mantle using the PREM model*

Depth (km)	v_p (km s^{-1})	v_s (km s^{-1})	ρ (g/cc)	γ_P	γ_S	γ_{ac}
871	11.245	6.311	4.504	1.50	1.15	1.26
971	11.416	6.378	4.563	1.47	1.14	1.24
1071	11.578	6.442	4.621	1.43	1.13	1.22
1171	11.734	6.504	4.678	1.40	1.10	1.20
1271	11.882	6.563	4.735	1.37	1.08	1.18
1371	12.024	6.619	4.790	1.35	1.06	1.16
1471	12.161	6.673	4.844	1.32	1.05	1.14
1571	12.293	6.726	4.898	1.30	1.04	1.13
1671	12.421	6.776	4.951	1.29	1.03	1.11
1771	12.545	6.825	5.003	1.27	1.02	1.10
1871	12.665	6.873	5.055	1.26	1.01	1.09
1971	12.784	6.920	5.106	1.25	1.00	1.09
2071	12.900	6.965	5.157	1.25	1.00	1.08
2171	13.016	7.011	5.207	1.25	1.00	1.08
2271	13.131	7.055	5.257	1.25	1.00	1.08
2371	13.245	7.100	5.307	1.26	1.00	1.09
2471	13.361	7.144	5.357	1.27	1.00	1.09
2571	13.477	7.189	5.407	1.28	1.01	1.10

*Calculations at depths < 871 km omitted because of roughness of the seismic data near the 671 km discontinuity.

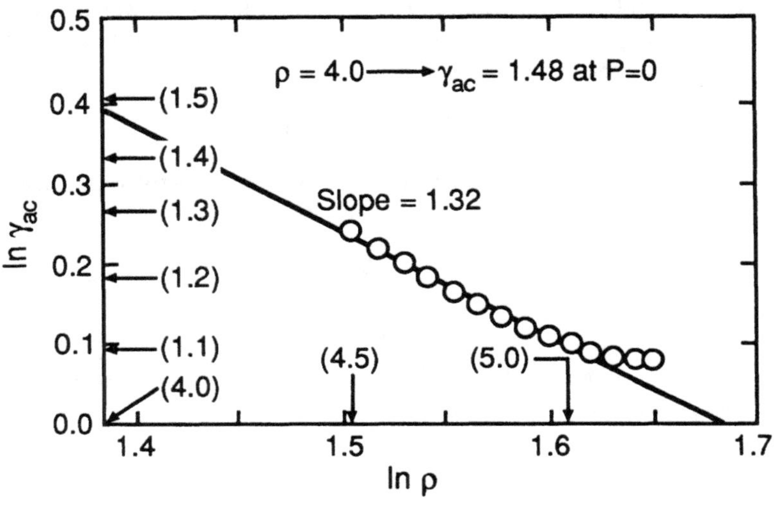

Fig. 5.17. The value of γ_0, 1.48, for the lower mantle determined from plotting $\ell n\ \gamma_{ac}$ versus $\ell n\ \rho$ from Table 5.4.

Equation (2.52) demonstrates the relationship between q, δ_S, and K_0' at high T:

$$\delta_T^{ht} = q^{ht} + K_0' - 1. \tag{5.34}$$

The value of K_0' is 3.34 for the lower mantle (Section 7.7); the value of q from Fig. 5.17 is 1.35. Anderson (1979b) showed that, when coupled with the requirements of $\Theta(z)$ in the lower mantle, the seismic data are satisfied for $0.5 < q^{ht} < 2.2$, so that $q^{ht} = 1$ is a reasonable solution (see also Fig. 10.11). In Section 3.8, it was shown that q is virtually independent of T and η at high T. Thus we expect $\delta_T \cong 3.3 \pm 0.2$ at deep mantle conditions, where $\delta_S = \delta_T - \gamma$ (see Isaak et al., 1990). Taking $\gamma \cong 1.1$ (Fig. 3.10), (5.34) yields $\delta_S \approx 2.1 \pm 0.3$. D.L. Anderson (1987) found $\delta_S \approx 1.8$ for the deep lower mantle from considerations of mantle tomography, which is reasonably close.

From Fig. 5.17 q in the lower mantle is 1.35, somewhat higher than unity even though it has been repeatedly stressed that at extreme conditions, q is near 1 (see, for example, Figs. 3.12 and 3.13). However, the determination of q by Fig. 5.17 is not isobaric, but under conditions corresponding to the geotherm (perhaps this is close to adiabatic). The adiabatic correction is given by the calculus equation

$$\left(\frac{V}{\gamma}\frac{\partial\gamma}{\partial V}\right)_e = \left(\frac{V}{\gamma}\frac{\partial\gamma}{\partial V}\right)_T - \left(\frac{\partial\gamma}{\partial T}\right)_P \frac{K_T}{\gamma}\left(\frac{\partial T}{\partial P}\right)_e, \tag{5.35}$$

where e refers to the earth's geotherm. The first term on the right of (5.35) is the q calculated isothermally. The factors K_T/γ and $(\partial T/\partial P)_e$ are positive, so the sign of $(\partial\gamma/\partial T)_P$ determines whether q_e is larger or smaller than q. For the case of MgO, $(\partial\gamma/\partial T)_P$ is slightly negative at high T (Fig. 3.10 and (2.2)). Assuming this to hold for the lower mantle, then $(\partial \ln \gamma/\partial \ln V)_e$ is slightly larger than $(\partial \ln \gamma/\partial \ln V)_T$. Thus $q_e^{ht} > q^{ht}$, and Fig. 5.17 is reconciled with Fig. 3.12.

5.11. The seismic equation of state

D.L. Anderson (1967) presented an important and useful relationship between seismic velocity and density, which is

$$\rho = A\left(\frac{\mathrm{M}}{p}\right)\Phi^{\mathrm{r}}; \tag{5.36}$$

$$\mathrm{r} = 1/4 - 1/3,$$

where A is a constant and r is related to the Grüneisen constant. Φ is the same as v_b^2 in (5.17). Using (5.17), we have

$$\rho = \text{const}\left(\frac{\mathrm{M}}{p}\right)\Phi^{(2\gamma-2/3)^{-1}}, \tag{5.37}$$

so that $\mathrm{r} = (2\gamma - 2/3)^{-1}$, a typical value of which is 0.3–0.5. Thus the seismic EoS (5.37) has a foundation in Debye theory.

5.12. Entropy of closely packed minerals

Brown and Shankland (1981) showed that the entropy of closely packed polyatomic minerals is more closely represented by a Debye model than is entropy of minerals with open structures. Low coordinated solids (open structures) tend to have acoustic entropies in excess of thermodynamic measurement. Examples of silicates that do not follow the Debye theory for entropy include framework silicates (quartz and albite), coesite, GeO_2, calcite, and ZnO.

5.13. The Kieffer model for density of states $g(\omega)$

It is clear from this chapter that Class C solids for which the Debye theory is applicable for finding thermal properties are few in number compared to those of Class A and Class B (Fig. 5.1). Something different is needed to represent the vibrational density of states for Classes A and B.

In Class C, the vibrational density of states $g(\omega)$ is replaced by the quadratic Debye frequency distribution in which all information on optical modes is submerged. In the Debye model, it is assumed that all $3pN$ modes can be represented by acoustic modes, and that the acoustic sound velocities determine the frequency of the highest mode, ω_D. To find the physical properties for solids of Classes A and B requires the use of the actual $g(\omega)$, or a good approximation to it. Kieffer (1980, 1982) found a good and useful approximation to $g(\omega)$, a model in which $g(\omega)$ is replaced by delta functions and bands when the exact $g(\omega)$ is unknown.

If the primitive cell has p atoms, $3p$ is the number of degrees of freedom (or modes) of the cell, and of these only 3 are acoustic modes. Each of the three acoustic modes appears as a separate branch in $g(\omega)$ and is approximated by a sine wave. Low symmetry silicates have a large number of atoms in the unit cell, so p is large, and the optic modes number $3p - 3$ modes. They often occupy a large amount of frequency space. Kieffer's (1980) model for forsterite is shown in Fig. 5.18.

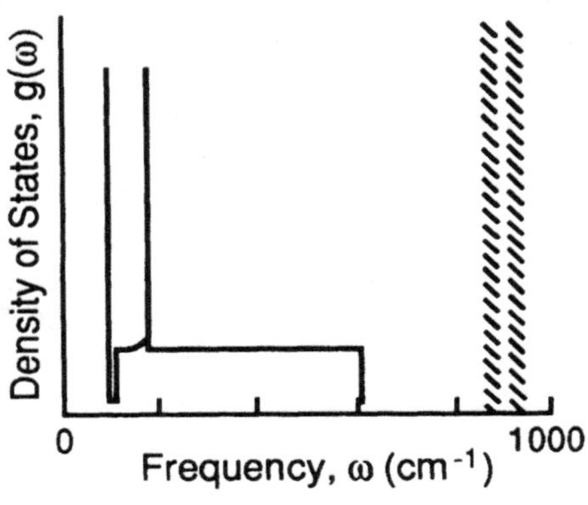

Fig. 5.18. Kieffer's model of $g(\omega)$ for forsterite, where $p = 80$, showing the positions of the 3 acoustic curves (two of which overlap), the optical spectrum (less than 620 cm^{-1}), and two Einstein functions (after Kieffer, 1980).

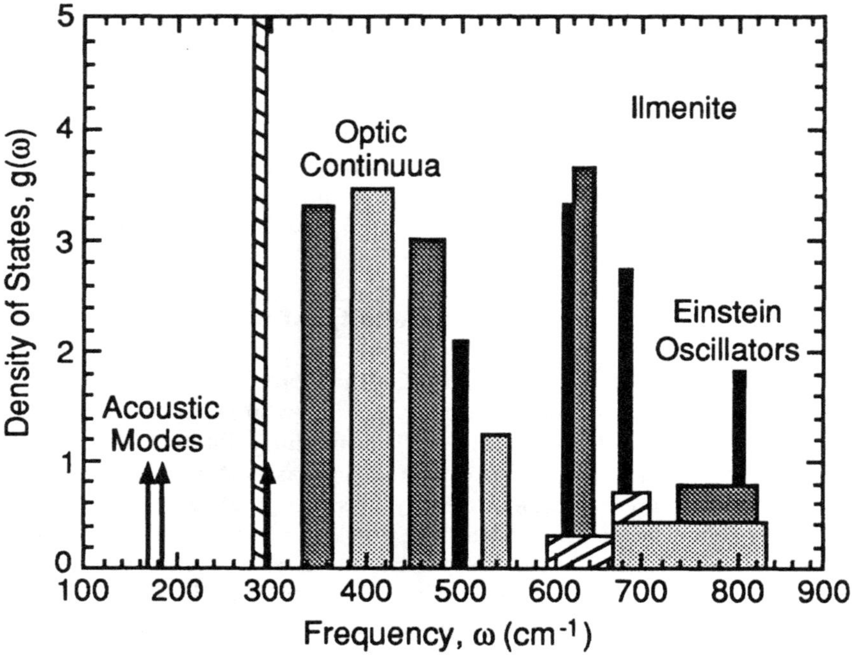

Fig. 5.19. Model density of states for ilmenite, $MgSiO_3$. The three arrows are the three acoustic modes. The boxes are optic continua, each with appropriate height. The Einstein oscillators are represented as four thick, solid lines with heights proportional to the number of modes represented. The overlapping boxes, distinguished from one another by degrees of shading, represent continua, and the tall box represents a narrow continuum. The scale on density of states is arbitrary (after Hofmeister and Ito, 1992).

Kieffer's method was applied to ilmenite by Hofmeister and Ito (1992). They took account of all measured optical modes and assigned continua (boxes) to regions of the model spectrum where there are clusters of optic modes, as shown in Fig. 5.19. The optical modes are classified as rotation, translation, Raman, and infrared, requiring ten continua. There are four Einstein functions and three acoustic Debye spectra, as shown in Fig. 5.19. The continua limits were fixed by measured optical band limits.

They were able to determine entropy $\mathcal{S}$ and C_V of ilmenite using $g(\omega)$ in Fig. 5.19 along with (1.51). They found excellent agreement between their calculated C_P and measured C_P and between calculated $\mathcal{S}$ and measured $\mathcal{S}$ over the experimental range, 100 K to 700 K. Thus the $g(\omega)$ found by Kieffer's method can be quite good and comparable in accuracy with $g(\omega)$ found from theory (such as given in Fig. 5.3 for MgO).

PART I REFERENCES

Akaogi, M., and Akimoto, S., 1979, "High pressure equilibria in a garnet lherzolite with special reference to Mg^{2+}-Fe^{2+} partitioning among constituent minerals," *Phys. Earth Planet. Inter.* **19**, 31–51.

Akimoto, S., 1972, "The system MgO-FeO-SiO_2 at high pressures and temperatures—phase equilibria and elastic properties," *Tectonophysics* **13**, 161–187.

Anderson, D.L., 1967, "A seismic equation of state," *Geophys. J. Roy. Astr. Soc.* **13**, 9–30.

Anderson, D.L., 1969, "Bulk modulus-density systematics," *J. Geophys. Res.* **74**, 3857–3864.

Anderson, D.L., 1976, "The 650 km mantle discontinuity," *Geophys. Res. Lett.* **3**, 347–349.

Anderson, D.L., 1983, "Chemical composition of the mantle," *J. Geophys. Res.* Suppl. 88, B41–B52.

Anderson, D.L., 1987, "A seismic equation of state. II. Shear properties and thermodynamics of the lower mantle," *Phys. Earth Planet. Inter.* **45**, 307–327.

Anderson, D.L., 1988, "Temperature and pressure derivatives of elastic constants with application to the mantle," *J. Geophys. Res.* **93**, 4688–4700.

Anderson, D.L., and Anderson, O.L., 1970, "The bulk modulus-volume relationship for oxides," *J. Geophys. Res.* **75**, 3494–3500.

Anderson, O.L., 1965a, "An approximate method of estimating shear velocity from specific heat," *J. Geophys. Res.* **70**, 4726–4728.

Anderson, O.L., 1965b, "Determination and some uses of isotropic elastic constants of polycrystalline aggregates using single crystal data," *Physical Acoustics: Principles and Methods*, Vol. 3, Part B (W. Mason, Ed.). New York: Academic Press, pp. 43–95.

Anderson, O.L., 1965c, "Bulk modulus-volume relationship for oxide compounds and related geophysics problems," *J. Geophys. Res.* **70**, 3951–3963.

Anderson, O.L., 1967, "Equation for thermal expansivity in planetary interiors," *J. Geophys. Res.* **72**, 3661–3668.

Anderson, O.L., 1970, "Elastic constants of the central force model for three cubic structures: Pressure derivatives and equations of state," *J. Geophys. Res.* **75**, 2719–2740.

Anderson, O.L., 1972, "Patterns in elastic constants of minerals important to geophysics," *The Nature of the Solid Earth* (E.C. Robertson, J.F. Hayes, and L. Knopoff, Eds.). New York: McGraw-Hill, pp. 575–613.

Anderson, O.L., 1974, "The determination of the volume dependence of the Grüneisen parameter γ," *J. Geophys. Res.* **79**, 1153–1155.

Anderson, O.L., 1979a, "The high temperature acoustic Grüneisen parameter in the Earth's interior," *Phys. Earth Planet. Inter.* **18**, 221–231.

Anderson, O.L., 1979b, "Evidence supporting the approximation $\gamma\rho =$ constant for the Grüneisen parameter of the earth's lower mantle," *J. Geophys. Res.* **84**, 3537–3542.

Anderson O.L., 1981, "Temperature profiles of the Earth," *Evolution of the Earth* (R.J. O'Connell and W.S. Fyfe, Eds.). Washington, D.C.: American Geophysical Union, pp. 19–27.

Anderson O.L., 1982, "The earth's core and the phase diagram of iron," *Phil. Trans. Roy. Soc. London* **A**, 21–35.

Anderson O.L., and Isaak, D., 1993, "The dependence of the Anderson-Grüneisen parameter δ_T upon compression at extreme conditions," *J. Phys. Chem. Solids* **54**, 221–227.

Anderson, O.L., and Liebermann, R.C., 1970, "Equations for the elastic constants and their pressure derivatives for three cubic lattices and some geophysical applications," *Phys. Earth Planet. Inter.* **3**, 61–85.

Anderson, O.L., and Masuda, K., 1994, "A thermodynamic method of computing thermal expansivity of silicate perovskite at high *P* and *T*," *Phys. Earth Planet. Inter.*, in press.

Anderson, O.L., and Nafe, J., 1965, "Bulk modulus-volume relationship for oxide compounds and related geophysical problems," *J. Geophys. Res.* **70**, 3951–3963.

Anderson, O.L., and Sumino, Y., 1980, The thermodynamic properties of the earth's lower mantle," *Phys. Earth Planet. Inter.* **23**, 314–331.

Anderson, O.L., and Yamamoto, S., 1987, "The interrelationship of thermodynamic properties obtained by the piston-cylinder high pressure experiments and RPR high temperature measurements for NaCl," *High Pressure Research in Mineral Physics*. Geophys. Monogr. Ser., vol. 39 (M.H. Manghnani and Y. Syono, Eds.). Washington, D.C.: American Geophysical Union, pp. 289–298.

Anderson, O.L., Schreiber, E., Liebermann, R., and Soga, N., 1968, "Some elastic constant data on minerals relevant to geophysics," *Rev. Geophys.* **6**, 491–524.

Anderson, O.L., Isaak, D.G., and Yamamoto, S., 1989, "Anharmonicity and the equation of state for gold," *J. Appl. Phys.* **65**, 1534-1543.

Anderson, O.L., Chopelas, A., and Boehler, R., 1990, "Thermal expansivity versus pressure at constant temperature: A re-examination," *Geophys. Res. J.* **17**, 685–688.

Anderson, O.L., Isaak, D.G., and Oda, H., 1991, "Thermoelastic parameters for six minerals at high temperature," *J. Geophys. Res.* **96**, 18037-18046.

Anderson O.L., Isaak, D.G., and Oda, H., 1992a, "High-temperature elastic constant data on minerals relevant to geophysics," *Rev. Geophys.* **30**, 57–90.

Anderson O.L., Oda, H., and Isaak, D.G., 1992b, "A model for thermal expansivity at high compression and high temperature: MgO as an example," *Geophys. Res. Lett.* **19**, 1987–1990.

Anderson, O.L., Oda, H., Chopelas, A., and Isaak, D.G., 1993, "A thermodynamic theory of the Grüneisen ratio at extreme conditions: MgO as an example," *Phys. Chem. Miner.* **19**, 369–380.

Babuska, V., Fiala, J., Kumazawa, M., Ohno, I., and Sumino, Y., 1978, "Elastic properties of garnet solid-solution series," *Phys. Earth Planet. Inter.* **16**, 157–176.

Bailey, H.C., and Yates, B., 1970, "Anisotrope thermal expansion of pyrolitic graphite at low temperatures," *J. Appl. Phys.* **41**, 5088–5091.

Barron, T.H.K., 1955, "Thermal expansion of solids at low temperature," *Philos. Mag.* **7**, 720–777.

Barron, T.H.K., 1957, "Grüneisen parameters for the equation of state of solids," *Ann. Phys.* **1**, 77–89.

Barron, T.H.K., 1979, "A note on the Anderson-Grüneisen functions," *J. Phys. C* **12**, 155–159.

Barron, T.H.K., Collins, J.G., and White, G.K., 1980, "Thermal expansion of solids at low temperatures," *Adv. Phys.* **29**, 609–730.

Barsch, G.R., and Achar, B.N.N., 1970, "Shell model calculation of the second Grüneisen parameter and pressure dependence of the first Grüneisen parameter for alkali halides," *Proc. Conf. Am. Inst. Phys.* **3**, 221–229.

Barton, M.A., and Stacey, F.D., 1985, "The Grüneisen parameter at high pressure: A molecular dynamical study," *Phys. Earth Planet. Inter.* **39**, 167–177.

Bassett, W.A., Takahashi, T., Mao, H., and Weaver, J.S., 1968, "Pressure induced transformation in NaCl," *J. Appl. Phys.* **34**, 319–325.

Baumgardner, J.R., and Anderson, O.L., 1981, "Using the thermal pressure to compute physical properties of terrestrial planets," *Adv. Space Res.* **1**, 159–176.

Bell, H., 1976, "Vibrational properties of amorphous solids," *Meth. Comput. Phys.* **15**, 215–255.

Bilz, H., and Kress, W., 1979, *Phonon Dispersion Relations in Insulators.* New York: Springer-Verlag, p. 51.

Birch, F., 1952, "Elasticity and constitution of the Earth's interior," *J. Geophys. Res.* **57**, 227–286.

Birch, F., 1960, "The velocity of compressional waves in rocks to 10 kilobars," Part 1, *J. Geophys. Res.* **65**, 1083–1102.

Birch F., 1961, "The velocity of compressional waves in rocks to 10 kilobars," Part 2, *J. Geophys. Res.* **66**, 2199–2224.

Birch, F., 1968, "Thermal expansion at high pressures," *J. Geophys. Res.* **73**, 817–819.

Birch, F., 1986, "Equation of state and thermodynamic parameters of NaCl to 300 kbar in the high-temperature domain," *J. Geophys. Res.* **91**, 4949–4954.

Boehler, R., 1981, "Adiabats $(\partial T/\partial P)_S$ and the Grüneisen parameter of NaCl up to 50 kilobars and 800°C," *J. Geophys. Res.* **86**, 7159–7162.

Boehler, R., 1982, "Adiabats of quartz, coesite and magnesium oxide at 50 kbar and 1000 K, and the adiabatic gradient in the earth's mantle," *J. Geophys. Res.* **81**, 5501–5506.

Boehler, R., 1983, "Melting temperature, adiabats and Grüneisen parameter of lithium, sodium and potassium versus pressure," *Phys. Rev. B* **27**, 6754–6762.

Boehler, R., and Kennedy, G.C., 1980, "Pressure dependence of the thermodynamic Grüneisen parameter of fluids," *J. Appl. Phys.* **48**, 4183–4186.

Boehler, R., and Ramakrishnan, J., 1980, "Experimental results on the pressure dependence of the Grüneisen parameter: A review," *J. Geophys. Res.* **85**, 6996–7002.

Boness, D.A., Brown, J.M., McMahon, A.K., 1986, "The electronic thermodynamics of iron under earth core conditions," *Phys. Earth Planet. Inter.* **42**, 227–240.

Born, M., and Huang, K., 1954, *Dynamical Theory of Crystal Lattices.* London: Oxford University Press.

Brennan, B.J., and Stacey, F.D., 1979, "A thermodynamically based equation of state for the lower mantle," *J. Geophys. Res.* **84**, 5535–5539.

Brillouin, L., 1946, *Wave Propagation in Periodic Structures.* New York: McGraw-Hill Book Co., Inc., p. 246.

Brown, J.M., and Shankland, T.J., 1981, "Thermodynamic parameters in the Earth as determined from seismic profiles," *Geophys. J. Roy. Astr. Soc.* **66**, 579–596.

Bukowinski, M.S.T., 1977, "A theoretical equation of state for the inner core," *Phys. Earth Planet. Inter.* **14**, 333–347.

Bukowinski, M.S.T., and Wolf, G.H., 1990, "Thermodynamically consistent decompression: Implications for lower mantle composition," *J. Geophys. Res.* **95**, 12583–12593.

Chhabildas, L.C., and Ruoff, A.L., 1976, "Isothermal equation of state for sodium chloride by the length-change-measurement technique," *J. Appl. Phys.* **47**, 4182–4187.

Chopelas, A., 1990a, "Thermal expansion, heat capacity, and entropy of MgO at mantle pressure," *Phys. Chem. Miner.* **17**, 142–148.

Chopelas, A., 1990b, "Thermal properties of forsterite at mantle pressures derived from vibrational spectroscopy," *Phys. Chem. Miner.* **17**, 149–156.

Chopelas, A., and Boehler, R., 1989, "Thermal expansion measurements at very high pressures, systematics and the case for a homogeneous mantle," *Geophys. Res. Lett.* **16**, 1347–1350.

Chopelas, A. and Boehler R., 1992a, "Thermal expansivity of the lower mantle," *Geophys. Res. Lett.* **19**, 1983–1986.

Chopelas, A., and Boehler, R., 1992b, "Raman spectroscopy of high pressure $MgSiO_3$ phases synthesized in a CO_2 laser heated diamond anvil cell: Perovskite and clinopyroxene," *High Pressure Research: Application to Earth and Planetary Sciences* (Y. Syono and M.H. Manghnani, Eds.). Tokyo: Terra Scientific Publishing Co., pp. 101–108.

Choudhury, N., Chapalet, S.L., Rao, K.R., and Ghose, S., 1988, "Lattice dynamics of $MgSiO_3$ perovskite," *Pramãna J. Phys.* **30**, 423–428.

Davies, G.F., 1974, "Elasticity, crystal structure and phase transitions," *Earth Planet. Sci. Lett.* **22**, 339–346.

Deer, W.A., Howie, R.A., and Zussman, J., 1962, *Rock-Forming Minerals*, Vols. I, II, and IV. New York: John Wiley & Sons.

Dugdale, J.S., and MacDonald, D.K.C., 1953, "The thermal expansion of solids," *Phys. Rev.* **89**, 832–834.

Dziewonski, A.M., and Anderson, D.L., 1981, "Preliminary reference earth model," *Phys. Earth Planet. Inter.* **25**, 297–356.

Dziewonski, A.M., Hales, A.I., and Lapwood, E.R., 1975, "Parametrically simple Earth models consistent with geophysical data," *Phys. Earth Planet. Inter.* **10**, 12–48.

Einstein, A., 1911, "Eine besiehung zwischen dem elastichen verhalten und der spezifischen wärme bei festen korpen mit einatomizen molecule," *Ann. d. Physik* **34**, 170–184.

Fairbairn, H.W., 1943, "Packing in ionic minerals," *Geol. Soc. Am. Bull.* **54**, 1305–1374.

Fritz, J.N., Marsh, S.P., Carter, W.J., and McQueen, R.G., 1971, "The Hugoniot equation of state of sodium chloride in the sodium chloride structure," *Accurate Characterization of the High Pressure Environment* (E.C. Lloyd, Ed.). NBS Spec. Publ. **326**, 201–208.

Funamori, N., and Yagi, T., 1993, "High pressure and high temperature in situ x-ray observation of $MgSiO_3$ perovskite under lower mantle conditions," *Geophys. Res. Lett.* **20**, 387–390.

Gilvarry, J.J., 1956, "Grüneisen parameter for a solid under finite strain," *Phys. Rev.* **102**, 331–340.

Graham, E.K., and Kim, H.S., 1986, "The bulk modulus of wüstite (FeO)," (abstract), *EOS Trans. AGU* **67**, 1240.

Grüneisen, E., 1912, "Theorie des festen zustandes einatomizer elemente," *Annelen der Physics (Leipzig)* **39**, 257–306.

Grüneisen, E., 1926, "The state of a solid body," *Handbuch der Physik*, Vol. **10**. Berlin: Springer-Verlag, pp. 1–52; Engl. transl. NASA RE 2-18-59W, 1959.

Guinan, M.W., and Beshers, O.N., 1968, "Pressure derivatives of the elastic constants of α-iron to 10 kbar," *J. Phys. Chem. Solids* **29**, 541–549.

Hardy, R.J., 1980, "Temperature and pressure dependence of intrinsic anharmonicity and quantum corrections to the equation of state," *J. Geophys. Res.* **85**, 7011-7015.

Hazen, R.M., and Finger, L.W., 1979, "Bulk modulus-volume relationship for cation-anion polyhedra," *J. Geophys. Res.* **84**, 6723–6727.

Hazen, R.M., and Finger, L.W., 1982, High-temperature and high-pressure crystal chemistry, *High Pressure Researches in Geoscience* (W. Schreyer, Ed.). New York: Academic Press, pp. 151–176.

Hazen, R.M., and Jeanloz, R., 1984, "Wüstite ($Fe_{1-x}O$): A review of its defect structure and physical properties," *Rev. Geophys. Space Phys.* **22**, 37–46.

Hemley, R.J., Cohen, R.E., Yeganeh-Haeri, A., Mao, H.K., and Ito, E., 1989, "Raman spectra and lattice dynamics of $MgSiO_3$-perovskite at high pressure," *Perovskite* (A. Navrotsky and D.J. Weidner, Eds.). Washington, D.C.: American Geophysical Union, pp. 35-43.

Hemley, R.J., Stixrude, L., Fei, Y., and Mao, H.K., 1992, "Constraints on lower mantle composition from *P, V, T* measurements of (Fe,Mg)SiO_3-perovskite and (Fe,Mg)O," *High Pressure Research: Application to Earth and Planetary Sciences*, American Geophysical Union Monographs, no. 67. Tokyo: Terra Scientific Publishing Co., p. 183–189.

Hofmeister, A.M., and Chopelas, A., 1991, "Thermodynamic properties of pyrope and grossular from vibrational spectroscopy," *Am. Mineral.* **76**, 880–891.

Hofmeister, A.M., and Ito, E., 1992, "Thermodynamic properties of $MgSiO_3$ ilmenite from vibrational spectra," *Phys. Chem. Miner.* **18**, 423–432.

Irvine, R.D., and Stacey, F.D., 1975, "Pressure dependence of the thermal Grüneisen parameter with application to the earth's lower mantle and outer core," *Phys. Earth Planet. Inter.* **11**, 157–165.

Isaak, D.G., 1992, "High-temperature elasticity of iron-bearing olivines," *J. Geophys. Res.* **97**, 1871–1885.

Isaak, D.G., 1993, "The mixed *P, T* derivatives of elastic moduli and implications on extrapolating throughout Earth's mantle," *Phys. Earth Planet. Inter.* **80**, 37–48.

Isaak, D.G., Anderson, O.L., and Goto, T., 1989a, "Elasticity of single-crystal forsterite measured to 1700 K," *J. Geophys. Res.* **94**, 5895–5906.

Isaak, D.G., Anderson, O.L., and Goto, T., 1989b, "Measured elastic moduli of single-crystal MgO up to 1800 K," *Phys. Chem. Miner.* **16**, 703–704.

Isaak, D.G, Cohen, R.E., and Mehl, M.J., 1990, "Calculated elastic and thermal properties of MgO at high pressures and temperatures," *J. Geophys. Res.* **95**, 7055–7067.

Isaak, D.G., Anderson, O.L., and Cohen, R.E., 1992, "The relationship between shear and compressional velocities at high pressures: Reconciliation of seismic tomography and mineral physics," *Geophys. Res. Lett.* **19**, 741-744.

Jackson, I., Liebermann, R.C., and Ringwood, E.A., 1978, "The elastic properties of $Mg_xFe_{1-x}0$ solid solutions," *Phys. Chem. Miner.* **3**, 11–31.

Jackson, I., Khanna, S.K., Revcolevschi, A., and Berthon, J., 1990, "Elasticity, shear-mode softening and high-pressure polymorphism of wüstite ($Fe_{1-x}0$)," *J. Geophys. Res.* **95**, 21671–21685.

Jamieson, J.C., Demarest, H.H., and Schiferl, D., 1978, "A re-evaluation of the Grüneisen parameter for the Earth's core," *J. Geophys. Res.* **83**, 5929–5935.

Jamieson, J.C., Fritz, J.N., and Manghnani, M.H., 1982, "Pressure measurement of high temperatures in x-ray diffraction studies: Gold as a primary standard," *High Pressure Research in Geophysics* (S. Akimoto and M.H. Manghnani, Eds.). Tokyo: Center for Academic Publishing, pp. 27–48.

Jeanloz R., 1979, "Properties of iron at high pressures and the state of the core," *J. Geophys. Res.*, **84**, 6059–6069.

Jeanloz R., and Sato-Sorenson, Y., 1982, "Wüstite (Fe_xO): Compression, nonstoichiometry and bulk viscosity," (abstract), *EOS Trans. AGU* **63**, 1097–1098.

Jephcoat, A.P., Mao, H.K., and Bell, P.M., 1986, "The static compression of iron to 78 GPa with rare gas solids as pressure-transmitting media," *J. Geophys. Res.* **91**, 4677–4684.

Kieffer, S.W., 1980, "Thermodynamics and lattice vibrations of minerals: 4. Application to chain and sheet silicates and orthosilicates," *Rev. Geophys. Space Phys.* **18**, 862–886.

Kieffer, S.W., 1982, "Thermodynamics and lattice vibrations of minerals: 5. Applications to phase equilibria, isotopic fractionation, and high-pressure thermodynamic properties," *Rev. Geophys. Space Phys.* **20**, 827–849.

Kittel, C., 1956, *Introduction to Solid State Physics*, 2nd Ed. New York: John Wiley & Sons, pp. 134–135; 250–251.

Knittle, E., and Jeanloz, R., 1987, "Synthesis and equation of state of (Mg,Fe)SiO_3 to over 100 GPa," *Science* **235**, 668–670.

Knittle, E., Jeanloz, R. and Smith, G.L., 1986, "Thermal expansion of silicate perovskite and stratification of the earth's mantle," *Nature* **319**, 214–215.

Knopoff, L., and Shapiro, J.N., 1969, "Comments on the interrelationship between Grüneisen's parameter and shock and isothermal equations of state," *J. Geophys. Res.* **74**, 1439–1450.

Landau, L.D., and Lifshitz, E.M., 1958, *Statistical Physics*. London: Pergamon Press, Ltd., pp. 48–53, 482.

Leitner, B.J., Weidner, D.J., and Liebermann, R.C., 1980, "Elasticity of single crystal pyrope and implications for garnet solid solution series," *Phys. Earth Planet. Inter.* **22**, 111–121.

Liebermann, R.C., 1975, "Elasticity of olivine (α), beta (β) and spinel (γ) polymorphs of germanates and silicates," *Geophys. J. Roy. Astr. Soc.* **42**, 899–929.

Liebermann, R.C., 1982, "Elasticity of minerals at high pressure and temperature," *High Pressure Researches in Geosciences* (W. Schreyer, Ed.). Stuttgart: Schweizerbart'sche, pp. 1–14.

Liebermann, R.C., and Ringwood, A.E., 1973, "Birch's law and polymorphic phase transformations," *J. Geophys. Res.* **78**, 6926–6932.

Liebermann, R.C., Ringwood, A.E., and Major, A., 1976, "Elasticity of polycrystalline stishovite," *Earth Planet. Sci. Lett.* **32**, 127–140.

Mao, H., 1974, "Velocity-density systematics and its implications for the iron content of the mantle," *J. Geophys. Res.* **79**, 5447-5452.

Mao, H.K., Hemley, R.J., Fei, Y., Shu, J.F., Chen, L.C., Jepthcoat, A.P., and Wu, Y., 1991, "Effect of pressure, temperature and composition on lattice parameters and density of (Fe,Mg)SiO_3-perovskite to 30 GPa," *J. Geophys. Res.* **96**, 8069–8080.

McQueen, R. G., Marsh, S. P., Taylor, J. W., Fritz, J. N., and Carter, W. J., 1970, "The equation of state of solids from shock wave studies," *High Velocity Impact Phenomena* (R. Kinslow, Ed.). San Diego: Academic Press, pp. 249–419.

Mizutani, H., Hamano, Y., and Akimoto, S., 1972, "Elastic wave velocities of polycrystalline stishovite," *J. Geophys. Res.* **77**, 3744–3749.

Nakamura, Y., Latham, G., Lammlein, D., Ewing, M., Duennebier, F., and Dorman, J., 1974, "Deep lunar interior inferred from recent seismic data," *Geophys. Res. Lett.* **1**, 147–151.

Navrotsky, A., 1989, "Thermochemistry of Perovskites," in *Perovskite*, Geophys. Monogr. Ser., vol. 45 (A. Navrotsky and D.J. Weidner, Eds.). Washington, D.C.: American Geophysical Union, pp. 67–80.

Palciauskas, V.V., 1975, "Volume dependence of the vibrational free energy," *Geophys. Res. Lett.* **2**, 317–319.

Parise, J.B., Wang, Y., Yeganeh-Haeri, A., Cox, D.E., and Fei, Y., 1990, "Crystal structure and thermal expansion of (Mg,Fe)SiO_3 perovskite," *Geophys. Res. Lett.* **17**, 2089–2092.

Pauling, L., 1960, "The sizes of ions," *Nature of the Chemical Bond*. Ithaca, New York: Cornell University Press, pp. 511–519.

Quareni, F., and Mulargia, F., 1988, "The validity of the common approximations for the Grüneisen parameter," *Geophys. J.* **93**, 505–519.

Rao, K.R., Chaplot, S.L., Choudhury, N., Ghose, S., and Price, D.L., 1988, "A lattice dynamical calculation of phonon density states and specific heat of forsterite, Mg_2SiO_4," *Phys. Chem. Miner.* **16**, 83–97.

Ross, N.L., and Hazen, R.M., 1989, "Single crystal x-ray diffraction study of Mg_2SiO_3-perovskite from 77 to 440 K," *Phys. Chem. Miner.* **16**, 415–420.

Runcorn, S.K., 1976, "A physical interpretation of Bullen's compressibility-pressure hypothesis," *Physics and Chemistry of Minerals and Rocks* (R.G.J. Strens, Ed.). New York: John Wiley & Sons, pp. 245–247.

Sangster, M.J.L., Peckham, G., and Saunderson, D.H., 1970, "Lattice dynamics of magnesium oxide," *J. Phys. Chem.* **3**, 1026–1036.

Sato, Y., and Anderson, O.L., 1980, "A comparison of the acoustic and thermal Grüneisen constants for three glasses," *J. Phys. Chem. Solids*, **41**, 401–410.

Shankland, T.J., 1971, "Velocity-density systematics," (abstract), *EOS Trans. AGU* **52**, 920.

Shankland, T.J., 1972, "Velocity-density systematics: Derivation from Debye theory and the effect of ionic size," *J. Geophys. Res.* **77**, 3750–3758.

Shankland, T.J., 1977, "Elastic properties, chemical composition, and crystal structures of minerals," *Geophys. Surv.* **3**, 69–100.

Shankland, T.J., and Brown, J.M., 1985, "Homogeneities and temperatures in the lower mantle," *Phys. Earth Planet. Sci.* **38**, 51–58.

Shaw, G.H., 1976, "Calculation of entropies of transition and reaction and slopes of transition and reaction lines using Debye theory," *J. Geophys. Res.* **81**, 3031–3035.

Simmons, G., 1964a, "Velocity of compressional waves in various minerals at pressures to 10 kilobars." *J. Geophys. Res.* **69**, 1117–1121.

Simmons, G., 1964b, "Velocity of shear waves in rocks to 10 kilobars. 1." *J. Geophys. Res.* **69**, 1123–1130.

Slater, J.C., 1939, *Introduction to Chemical Physics*, 1st edition. New York: McGraw-Hill Book Co., pp. 26, 239.

Spetzler, H., 1970, "Equation of state of polycrystalline and single crystal MgO to 8 kilobars and 800 K," *J. Geophys. Res.* **75**, 2073–2087.

Spetzler, H., Sammis, C.G., and O'Connell, R.J., 1972, "Equation of state of NaCl: Ultrasonic measurements to 8 kbar and 800°C and static lattice theory," *J. Phys. Chem. Solids* **33**, 1727–1750.

Stacey, F.D., 1977a, "Applications of thermodynamics to fundamental earth physics," *Geophys. Surv.* **3**, 175–204.

Stacey, F.D., 1977b, "A thermal model of the earth," *Phys. Earth Planet. Inter.* **15**, 341–348.

Stacey, F.D., 1993, "Properties of a harmonic lattice," *Phys. Earth Planet. Inter.* **78**, 19–22.

Stixrude, L., Hemley, R.J., Fei, Y., and Mao, H.K., 1992a, "Thermoelasticity of silicate perovskite and magnesiowüstite and stratification of the earth's mantle," *Science* **257**, 1099–1101.

Stixrude, L., Hofmeister, A., Anderson, O.L., and Wolf, G., 1992b, "The Debye temperature of orthorhombic $MgSiO_3$ perovskite and of the lower mantle," (abstract), *EOS Trans. AGU* **73**, 317.

Sumino, Y., and Anderson, O.L., 1984, "Elastic constants of minerals," *CRC Handbook of Physical Properties of Rocks* (R.S. Carmichael, Ed.). Boca Raton, Fl.: CRC Press, Inc., pp. 39–138.

Sumino, Y., Nishizawa, 0., Goto, T., Ohno, I., and Ozima, M., 1977, "Temperature variation of elastic constants of single-crystal forsterite between $-190°$ and $400°$C," *J. Phys. Earth* **25**, 377–392.

Sumino, Y., Anderson, O.L., and Suzuki, I., 1983, "Temperature coefficients of elastic constants of single crystal MgO between 80 and 1300 K," *Phys. Chem. Miner.* **9**, 38–47.

Suzuki, I., 1975, "Thermal expansion of periclase and olivine and their anharmonic properties," *J. Phys. Earth* **23**, 145–159.

Suzuki, I., Ohtani, E., and Kumazawa, M., 1980, "Thermal expansion of modified spinel, β-Mg_2SiO_4, *J. Phys. Earth* **28**, 273–280.

Suzuki I., Seya K., Tokai H., and Sumino Y., 1979, "Thermal expansion of single-crystal manganosite," *J. Phys. Earth* **27**, 63–69.

Suzuki, I., Takei, H., and Anderson, O.L., 1984, "Thermal expansion of forsterite," *Thermal Expansion 8* (T.A. Hahn, Ed.). New York: Plenum, pp. 79–88.

Suzuki, I., Inouye, Y., Hirao, J., Oda, H., Saito, T., and Seya, K., 1992, "The resonant sphere technique for measurements of elasticity and anelasticity of a small mineral specimen," *J. Seism. Soc. Jpn.* **45**, 213–228 (in Japanese).

Thomsen, L., 1970, "On the fourth-order anharmonic equation of state of solids," *J. Phys. Chem. Solids* **31**, 2003–20016.

Toksöz, M.N., 1979, "Planetary seismology and interiors," *Rev. Geophys. Space Phys.* **17**, 1541–1655.

Traylor, J.G., Smith, H.G., Nicklow, R.M., and Wilkinson, M.K., 1971, "Lattice dynamics of rutile," *Phys. Rev. B* **3**, 3457–3472.

Usselman, T.M., 1975, "Experimental approach to the state of the core. Part II. Composition and thermal regime," *Am. J. Sci.* **275**, 291–303.

Vashchenko, V.Y., and Zubarev, V.N., 1963, "Concerning the Grüneisen constant," *Sov. Phys.-Solid State* (Engl. transl.) **5**, 653–655.

Wallace, D.C., 1972, *Thermodynamics of Crystals*. New York: John Wiley and Sons, Inc.

Wang, C.Y., 1978, "Elastic constant systematics," *Phys. Chem. Miner.* **3**, 251–261.

Wang, H., Gupta, M.C., and Simmons, G., 1975, "Chrysoberyl (Al_2BeO_4): Anomaly in velocity-density systematics," *J. Geophys. Res.* **80**, 3761–3764.

Wang, Y., Weidner, D.L., Liebermann, R.C., and Zhao, Y., 1994, "*P-V-T* equation of state of (Mg,Fe)SiO_3 perovskite: Constraints on the composition of the lower mantle," *Phys. Earth Planet. Inter.* **83**, 13–41.

Wang, Y., Weidner, D.L., Liebermann, R.C., Lui, X., Ko, J., Vaughn, M.T., Zhao, Y., Yeganeh-Haeri, A., and Pacalo, R.E.G., 1991, "Phase transformation and thermal expansion of $MgSiO_3$ perovskite," *Science* **251**, 410–413.

Watt, J.P., Shankland, T.J., and Mao, N., 1975, "Uniformity of mantle composition," *Geology* **3**, 92–97.

Weidner, D.J., Bass, J.D., Ringwood, A.E., and Sinclair, W., 1982a, "The single-crystal elastic moduli of stishovite," *J. Geophys. Res.* **87**, 4740–4746.

Weidner, D.J., Bass, J.D., and Vaughn, M.T., 1982b, "The effect of crystal structure and composition on elastic properties of silicates," *High Pressure Research in Geophysics* (S. Akimoto and M.H. Manghnani, Eds.). Tokyo: Center for Academic Publishing, pp. 125–134.

White, G.K., and Anderson, O.L., 1966, "Grüneisen parameter of magnesium oxide," *J. Appl. Phys.* **37**, 430–432.

Williams, Q., Jeanloz, R., and McMillan, P., 1987, "Vibrational spectrum of $MgSiO_3$ perovskite: Zero pressure Raman and mid-infrared spectra to 27 GPa," *J. Geophys. Res.* **92**, 9116–9129.

Wolf, G.H., 1992, Private correspondence about data found in work of (but deleted from publication of) Wolf and Bukowinski, 1987.

Wolf, G.H., and Bukowinski, M, 1987, "Theoretical study of the structural properties and equations of state of $MgSiO_3$ and $CaSiO_3$ perovskites: Implications for lower mantle compositions," *High Pressure Research in Mineral Physics* (M.H. Manghnani and Y. Syono, Eds.). Tokyo: Terra Scientific Publishing Co., pp. 313–331.

Wolf, G.H., and Jeanloz, R., 1984, "Lattice dynamics and structure distortions of $CaSiO_3$ and $MgSiO_3$ perovskites," *Geophys. Res. Lett.* **12**, 413–416.

Yagi, T., 1978, "Experimental determination of the thermal expansivity of several alkali halides at high pressure," *J. Phys. Chem. Solids* **39**, 563–571.

Yamamoto, S., Ohno, I., and Anderson, O.L., 1987, "High temperature elasticity of sodium chloride," *J. Phys. Chem. Solids* **48**, 143–151.

Yeganeh-Haeri, A., Weidner, D.J., and Ito, E., 1989, "Single crystal elastic moduli of magnesium metasilicate perovskite," *Perovskite* (A. Navrotsky and D.J. Weidner, Eds.). Washington, D.C.: American Geophysical Union, pp. 13–26.

Zemansky, M.W., 1943, *Heat and Thermodynamics.* New York: McGraw-Hill Book Co., p. 73.

Zharkov, V. N., and Kalinin, V.A., 1971, *Equations of State for Solids at High Pressures and Temperatures* (Engl. transl.). New York: Consultants Bureau, pp. 5–31.

PART II
ISOTHERMAL EQUATIONS OF STATE

...we never have perfectly clean-cut knowledge of anything, but all our experience is surrounded by a twilight zone, a penumbra of uncertainty... the penumbra is to be penetrated by improving the accuracy of measurement.

—Nobelist Percy Bridgman

In the beginning of the science of the equation of state (EoS), pressure was related to volume by two measurable parameters, K_{T_0} and V_0, and an empirical constant, ξ. Birch (1938, 1952) showed that ξ is determined by the value of K_0'. At first, K_0' could not be measured independently, and so uncertainty in K_0' defined the edge of the penumbra. Since the early 1960's, K_0' has been accurately determined by ultrasonic measurements under pressure, providing researchers with a valuable three-parameter EoS. This, however, only moved the penumbra, because researchers then desired a four-parameter EoS that had an empirically chosen parameter, $K_0'' = \left(\partial^2 K/\partial P^2\right)_T$. This parameter has defined the edge of the twilight zone for many years. Today new, accurate experiments in physical acoustics are beginning to disclose the value of K_0'' as an experimental value independent of the P, V measurement, and the penumbra of uncertainty is about to be penetrated and shifted again.

The fourth order isothermal EoS is a function of the four measurable parameters (all at $P = 0$), so that $P = f((V/V_0), K_{T_0}, K_0', K_0'')$. Three chapters in this part are concerned with what the functional relationship f might be in both third and fourth order EoS's. Assuming that the required parameters have been measured, the EoS can be represented by $f(V,T)$. Emphasizing that the isothermal EoS is strictly determinable only at $T = 0$, we often use $P = f(V,0)$ in Part II. We are concerned with the region of variables where f$(V,0)$ is single-valued and continuous, with definable derivatives, over a wide range of V, and so the effect of phase boundaries on the EoS is not discussed.

The EoS's considered can be lumped into three classes: (1) those based on the solid mechanics definition of finite strain; (2) those based on assumed relationships between the variables within the EoS, such as K'; and (3) those based on interatomic potentials. These three classes, some of which are given in terms of quantum mechanics parameters, will be discussed separately in Chapters 6, 7, and 8. I will show that for most applications of the EoS, there is little justification for choosing one EoS over another strictly on the basis of fundamentals. The choice is made on the basis of convenience and tradition.

Chapter 9 concerns shear constants and shear velocities. The theory of lattice dynamics is applied to obtain the shear elastic constants for a cubic single crystal. This is done in terms of certain derivatives of the potential. Finite strain also can be used to define the shear constants in terms of third order elastic constants. The variation of the shear constants with P and T depends greatly upon the crystallographic structure, whereas the variation of K_T with P and T is insensitive to the crystallographic details.

A large number of equations in the literature are suitable for expressing the isothermal pressure–volume relationship. It is now becoming recognized that all such equations are essentially empirical. Equation of state calculations based on first principles methods using numerical methods (the *ab initio* approach) are now being published, and these will be the course of much future work. (A full treatment of the first principles methods belongs in a theoretical textbook).

Ab initio calculations have been developed to calculate static and dynamic properties of solids, including ceramic solids and oxides important to solid earth geophysics (see, for example, Hemley and Gordon, 1985; Mehl et al., 1986; Cohen, 1987; Isaak et al., 1990). *Ab initio* theory is at the state in which calculations of elastic moduli and their pressure and temperature derivatives and melting boundaries are possible, where no experimental data are used in the theory other than values of universal constants, such as Planck's constant; atomic variables, such as mass; and structural information, such as the point group representation. Excellent agreement between calculated and computed properties has been demonstrated in a number of cases. (Principles of *ab initio* calculations applied to thermal physics were discussed lightly in Part I, Chapter 5). Even with this pending triumph in EoS theory, there is still a need for analytical expressions, although largely empirical, to represent and analyze the measured data, and for extrapolation and interpolation. That is the justification for these four chapters.

6

FINITE STRAIN

6.1. Introduction

In this chapter we consider a few EoS's based on finite strain. The classical theory of infinitesimal elasticity is based on two assumptions: 1) the strains are uniquely determined by the stresses and are reversible, and 2) the strains are limited to 'linear' elasticity; that is, they are so small that squares and products are negligible. In finite strain, the first assumption is retained, but the restriction to small strains is removed (Murnaghan, 1937). In generalized elasticity theory, finite strain representations go back to Love (1927), but practical expressions for finite strain were developed by Birch (1938, 1952). He used Murnaghan's basic theory but restricted it to the case where the initial stress is a large hydrostatic pressure. The crucial assumption in finite strain theory is in the formal relationship between compression and coordinate displacement (see Stacey et al., 1981). One form is

$$\frac{V}{V_0} = \eta = (1 + 2\epsilon)^{3/2}\,, \tag{6.1}$$

where ϵ is the coordinate displacement (strain).

In Chapter 1, I defined E_{ST} as the lattice energy at $P = 0$ and $T = 0$. The Helmholtz energy will increase with strain, so the total energy due to compression will be $\mathcal{F} = E_{ST} + E(\epsilon)$, where $E(\epsilon) = E(\eta, 0)$; that is, at $T = 0$ where $\mathcal{F}_{VIB}$ vanishes (1.2), there still may be $E(\epsilon)$. The strain energy, $E(\epsilon)$, will be some assumed function of ϵ, and the pressure at $T = 0$ is $P(\eta) = \eta\,(\partial E/\partial\eta)_T$, where $P = 0$ when $E(\eta) \rightarrow 0$. We shall see that a large variety of paths can be used to obtain the desired EoS, $P(\eta)$.

The relationship between η and strain is not unique (Knopoff, 1963). An equally plausible alternative to (6.1) is

$$\eta = (1 - 2\epsilon)^{-3/2}\,. \tag{6.2}$$

The difference between (6.1) and (6.2) arises from different reference states; the former is the Lagrangian formulation, whereas the latter is the Eulerian formulation (see Section 6.7). Other formulations, including a linear combination of (6.1) and (6.2), are possible; the number of different constructions between η and ϵ in finite strain is endless (Knopoff, 1965).

6.1.1. *The third order EoS*

Because K_{T_0}, K_0', and ρ_0 can be measured precisely in the laboratory, they are available to define the isothermal EoS. Thus, any isothermal EoS with only three arbitrary constants can be defined completely. Because the pressure function P vanishes at $V = V_0$, these three parameters define four related functions of the EoS, $\mathcal{F} - E_{ST}$, P, K_T, and K_0', at all pressures. Higher order derivatives of K_{T_0} taken beyond K_0' have rarely been measured in the laboratory, and reported measurements have large error bars. Thus a fourth order equation of state (four arbitrary constants) usually involves K_0'', whose experimental determination is not very robust. The product $K_{T_0}K_0''$ is a dimensionless number near the value -10. A three-parameter EoS has constants that are expressed in the constants in the potential. The potential could include three terms in a polynominal expansion in strain

$$E(\epsilon) = \mathcal{F} - E_{ST} = b_1\epsilon + b_2\epsilon^2 + b_3\epsilon^3, \tag{6.3}$$

where $\epsilon = \mathrm{f}(\eta)$ when $\eta = V/V_0 = \rho_0/\rho$. Now $\mathcal{F} \rightarrow \mathcal{U}$ as $T \rightarrow 0$, (1.2).

In a strict sense, the isothermal EoS requires setting T to absolute zero so that E_{VIB} in (1.14) is zero. Also both $\mathcal{U}(V_0, 0)$ and $P(1,0)$ are zero. Thus the parameters are K_{T_0}, K_0', and ρ_0 (through η), used in the EoS to represent conditions at $T = 0$. Using (6.3) and taking $T = 0$, the three-parameter isothermal EoS is now symbolically defined as:

$$\begin{aligned}
\mathcal{U}(\eta,0) &= \mathrm{f}_1(\eta,\ K_{T_0},\ K_0') + E_{ST}\\
P(\eta,0) &= \mathrm{f}_2(\eta,\ K_{T_0},\ K_0')\\
K_T(\eta,0) &= \mathrm{f}_3(\eta,\ K_{T_0},\ K_0')\\
K_0'(\eta,0) &= \mathrm{f}_4(\eta,\ K_{T_0},\ K_0').
\end{aligned}$$

Now $K_T = K_{T_0}$ at $\eta = 1$, so $K_T(\eta, 0)$ is proportional to K_{T_0}, the other variables being dimensionless. Also P has the same dimensions as K_{T_0}. The parameter η includes the parameter V_0. $\mathcal{U} - E_{ST}$ has dimensions of energy, the same dimensions as $K_{T_0}V_0$. Therefore, the above can be replaced by

$$\begin{aligned}
\mathcal{U}(\eta,0) &= K_{T_0}V_0\mathrm{f}_1(\eta, K_0') + E_{ST}\\
P(\eta,0) &= K_{T_0}V_0\mathrm{f}_2(\eta, K_0')\\
K_T(\eta,0) &= K_{T_0}V_0\mathrm{f}_3(\eta, K_0')\\
K_0'(\eta,0) &= K_{T_0}V_0\mathrm{f}_4(\eta, K_0').
\end{aligned} \tag{6.4}$$

The function $E(\epsilon)$ does not have to be a series expansion in a second variable such as found in (6.3). Many atomic theories regard the potential as a specific function of volume or interatomic radius. These potentials, often called intermolecular pair potentials ϕ used in place of ϵ in (6.3), are important to EoS's and will be discussed in Chapter 8.

We do not have to restrict $E(\epsilon)$ to three elastic parameters. To define the isothermal EoS, we need as many experimental parameters as $E(\eta)$ parameters. Let us suppose that we can identify certain terms in the potential with the dielectric constant ε_0 and the index of refraction n_0. Then we use the internal energy $\mathcal{U}$ to include ε and n_0^2 (zero-pressure values) as well as E_{ST}, giving a five-parameter isothermal EoS. In that case, there are six associated functions:

$$\begin{aligned}
\mathcal{U}(\eta,0) &= \mathrm{f}_0(\eta, K_{T_0}, K_0', \varepsilon_0, n_0^2) + E_{ST} \\
P(\eta,0) &= \mathrm{f}_1(\eta, K_{T_0}, K_0', \varepsilon_0, n_0^2) \\
K_T(\eta,0) &= \mathrm{f}_2(\eta, K_{T_0}, K_0', \varepsilon_0, n_0^2) \\
K_0'(\eta,0) &= \mathrm{f}_3(\eta, K_{T_0}, K_0', \varepsilon_0, n_0^2) \\
\varepsilon(\eta,0) &= \mathrm{f}_4(\eta, K_{T_0}, K_0', \varepsilon_0, n_0^2) \\
n^2(\eta,0) &= \mathrm{f}_5(\eta, K_{T_0}, K_0', \varepsilon_0, n_0^2).
\end{aligned}$$

Most of the equations in this chapter deal with the set of equations given by (6.4). The Decker equation of state (Decker, 1971) for NaCl, however, is based on the set of six equations above, as we shall see in Section 8.9.

6.1.2. *Parameters in the isothermal EoS*

Though the isothermal EoS, $P(\eta,0)$, is strictly applicable only at absolute zero, it is often used to describe the EoS of geophysical materials at room temperature or to duplicate the experimental data of materials taken at the room temperature isotherm. The justification for this rests on the similarity of the bulk modulus values of oxides and silicates at room temperature and at absolute zero, i.e., for minerals with a high Debye temperature

$$K_T(1, 273^\circ) \rightarrow K_T(1,0)$$

(see Table 1.5). Such a statement cannot be made for soft solids like the alkali metals; however, for the hard solids of geophysics and ceramic science, room temperature is well below the Debye temperature, and K_T changes little between 0 K and 273 K.

There is little point in insisting that only the zero degree bulk modulus data be used for the EoS, because the variation of the measured value of K_T $(1, 273^\circ)$ from author to author is often as large as the difference between K_T $(1, 273^\circ)$ and K_T $(1, 0^\circ)$ reported by the same author. An example of this experimental uncertainty for NaCl, which is one of the most carefully measured solids, is shown in Table 1.6. The experimental uncertainty in $K_T(1, 273^\circ)$ is even greater for solids less emphasized.

The pressure equation is affected more by K_{T_0} than K_0'. Once K_{T_0} has been identified, the major features of the EoS are predictable. In this chapter, $P(\eta, 0) = P(\eta)$ will be used to identify the isothermal EoS.

Pressure experiments near absolute zero are difficult and therefore data are scarce, but a great deal of room temperature isothermal data exist. Because most of the applications of EoS's made here are to hard solids, found in geophysics and ceramics, $P(\eta)$ shall be understood to mean $P(\eta, 273)$ in this chapter. When the isothermal EoS is applied to softer solids (i.e., when Θ is near room temperature, as, for example, with gold), the longer nomenclature is used (see details in Chapter 2).

In the experimental dimensionless EoS plot, $P(\eta)$ versus η does not change significantly from solid to solid. Here $K_T \rightarrow K_{T_0}$ as $\eta \rightarrow 1$ and is not an adjustable parameter. Dimensionless plots of EoS data tend to overlap each other at low pressure because they have a common slope, -1. At higher compressions they diverge but the divergence is not marked whenever the values of K_0' are close for solids under comparison. In Table 1.8 it was shown that values of K_0' generally lie between 4.0 and 6.5 for dense oxides and silicates, with $K_0' = 5$ a good representative value.

The experimental data for four halides (LiF, NaF, NaCl, and NaI) are shown in Fig. 6.1, which also shows the remarkable overlap of the data on a dimensionally reduced EoS curve η versus P/K_{S_0}. The values of K_0' are 5.4 ± 0.15 for the four compounds. This figure demonstrates that a simple theoretical EoS to match the data can be attained. A theoretical isothermal EoS of the third order, (6.4), may be suitable to match the experimental data of minerals at comparable compressions. From Fig. 6.1 we deduce that

$$\frac{P(\eta)}{K_{T_0}} = \mathrm{f}(K_0', \eta),$$

where $\mathrm{f}(K_0')$ varies slowly with K_0' at constant η.

The question is, what theoretical isothermal EoS is best? Unfortunately, an arbitrary assumption about the potential, or equivalently the strain energy function $E(\epsilon)$, is behind the specification of $\epsilon(\eta)$. Thus we shall see that often a presumed theoretical isothermal EoS rests on unproven assumptions.

6.2. Basic assumption: a series in strain (ϵ) for the energy function $E(\epsilon)$

6.2.1. *A series expansion for* $E(\epsilon)$

The beginning of finite strain formulations discussed here is the relationship between the compression η and the strain variable ϵ; that is, $\epsilon = f(\eta)$. Several functions of η have been successfully used in the EoS. It is often convenient to expand $E(\epsilon)$ as a series in ϵ. This series expansion must be truncated after a few terms, as the experimental information required by the coefficients of the terms in the expansion is limited. The experimental information consists of the bulk modulus and density at zero pressure, K_{T_0}

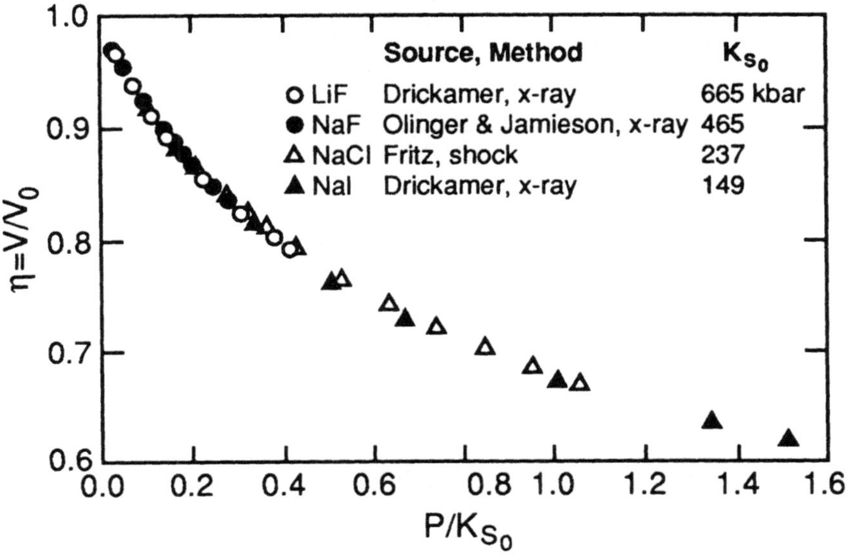

Fig. 6.1. The dimensionless experimental data for V/V_0 versus P/K_{S_0} for four alkali halides at room temperature (modified from Smith and McLean, 1973).

and ρ_0, and the higher volume derivatives, K_0', K_0'', etc. The accuracy of K_0'', K_0''', etc., based on experimental evidence is severely limited.

Let the symbol ϵ' denote $d\epsilon/dV$, $\epsilon'' = d^2\epsilon/d^2V$, etc. Take the expansion in the potential as a Taylor's series in the equation

$$E(\epsilon) = K_{T_0} V_0 \sum_{\mathbf{p}=2}^{\infty} \left(\frac{C_{\mathbf{P}}}{\mathbf{p}!}\right) \epsilon^{\mathbf{P}}, \tag{6.5}$$

where $\mathbf{p}$ denotes a power, not a derivative, in ϵ. The condition $P(1) = P_0 = 0$ is satisfied by ignoring the term for $\mathbf{p} = 1$. The factor $K_{T_0} V_0$ has the dimensions of energy so that all coefficients C_{P} and ϵ are dimensionless.

From (2.1) and (2.5) and dropping the subscript T_0 for P and K in (6.4) so that isothermal conditions are assumed for the remainder of this section, we have $E' = P$, where E' is $(\partial E/\partial V)_T$, and further

$$E'' = -\frac{K}{V}; \tag{6.6}$$

$$E''' = -\frac{K}{V^2}\left(\frac{VK'}{K} + 1\right). \tag{6.7}$$

The derivatives of (6.5), where the subscript 0 refers to $\epsilon = 0$, are

$$E' = K_{T_0} V_0 \epsilon \epsilon' C_2 \left\{ 1 + \sum_{\mathbf{p}=3} \frac{C_P}{C_2} \left[\frac{\epsilon^{\mathbf{p}-2}}{(\mathbf{p}-1)!} \right] \right\}; \tag{6.8}$$

$$E'' = K_{T_0}V_0C_2\left((\epsilon'\epsilon' + \epsilon\epsilon'') + \sum_{\mathbf{p}=3} \frac{C_\mathbf{p}}{C_2}\left\{\left[\frac{\epsilon'\epsilon'}{(\mathbf{p}-2!)} + \frac{\epsilon''\epsilon}{(\mathbf{p}-1)!}\right]\right\}\epsilon^{\mathbf{p}-2}\right); \qquad (6.9)$$

$$E''' = K_{T_0}V_0C_2\left(3\epsilon'\epsilon'' \sum_{\mathbf{p}=3} \frac{C_\mathbf{p}}{C_2}\left\{\left[\frac{\epsilon'\epsilon'\epsilon'}{(\mathbf{p}-3)!} + \frac{3\epsilon'\epsilon''\epsilon}{(\mathbf{p}-2)!} + \frac{\epsilon'''\epsilon^2}{(\mathbf{p}-1)!}\right]\right\}\epsilon^{\mathbf{p}-3}\right). \qquad (6.10)$$

For $P = 0$, $\epsilon = \epsilon_0 = 0$. Since

$$E'' = \frac{K_T}{V},$$

then

$$E_0'' = \frac{K_{T0}}{V_0}. \qquad (6.11)$$

Another expression for E_0'' is found from (6.9). For $n = 2$ and $\epsilon_0 = 0$, the terms behind the summation sign in (6.9) vanish, and E_0'' becomes

$$E_0'' = K_{T_0}V_0C_2(\epsilon_0')^2. \qquad (6.12)$$

Equating (6.11) and (6.12),

$$C_2 = \frac{1}{(V_0\epsilon_0')^2}. \qquad (6.13)$$

Similarly,

$$E_0''' = -\frac{K_T}{V^2}(1 + K_0'), \qquad (6.14)$$

where

$$K_0' = \left(\frac{\partial K_T}{\partial P}\right)_T \quad \text{at } P = 0.$$

For $n = 3$ and $\epsilon_0 = 0$, the $\epsilon\epsilon'''\epsilon^2$ term in (6.10) vanishes, and all but the $(\epsilon')^3$ term behind the summation sign in (6.10) vanish, leaving

$$E_0''' = K_{T_0}V_0C_2\left[3\epsilon_0'\epsilon_0'' + \frac{C_3}{C_2}(\epsilon_0')^3\right]. \qquad (6.15)$$

Equating (6.14) and (6.20), C_3 becomes

$$C_3 = -\left[\frac{1}{(\epsilon_0' V_0)^3}(1 - K_0') + 3C_2 \frac{\epsilon_0'''}{(\epsilon_0')^2}\right]. \tag{6.16}$$

The pressure equation to third order is:

$$\begin{aligned} P = E' =& K_{T_0} V_0 \epsilon \epsilon' \left(C_2 + \frac{1}{2} C_3 \epsilon\right) \\ =& K_{T_0} V_0 \epsilon \epsilon' \frac{1}{(\epsilon_0 V_0)^2}\left\{1 - \frac{\epsilon}{2}\left(\frac{1 + K_0'}{\epsilon_0' V_0} + \frac{3\epsilon_0''}{(\epsilon_0')^2}\right)\right\}. \end{aligned} \tag{6.17}$$

The energy equation to third order is

$$E(\epsilon) = \frac{1}{2} K_{T_0} V_0 E^2 \left(C_2 + \frac{1}{3} C_3 \epsilon\right) \tag{6.18}$$

$$\begin{aligned} =& \frac{1}{2} K_{T_0} V_0 \epsilon^2 \left\{\frac{1}{(\epsilon_0' V_0)^2} - \frac{\epsilon}{3}\left[\frac{1 + K_0'}{(\epsilon_0 V_0^3)} + \frac{3}{(\epsilon_0 V_0)^3} \frac{\epsilon_0''}{(\epsilon_0')^2}\right]\right\} \\ =& \frac{1}{2} \frac{K_{T_0} V_0 \epsilon^2}{(\epsilon' V_0)^2}\left\{1 - \frac{\epsilon}{3}\left[\frac{1 + K_0'}{\epsilon_0 V_0} + 3\frac{\epsilon_0}{(\epsilon_0')^2}\right]\right\}. \end{aligned} \tag{6.19}$$

As an example, consider Eulerian strain for $\epsilon(\eta)$. From (6.2)

$$\epsilon(\eta) = \frac{1}{2}\left(1 - \eta^{-2/3}\right) = \frac{1}{2}\left[1 - \left(\frac{V_0}{V}\right)^{2/3}\right]. \tag{6.20}$$

We have

$$\begin{aligned} \epsilon' &= \frac{1}{3V_0}\eta^{-5/3} : & \epsilon_0' &= \frac{1}{3V_0} \\ \epsilon'' &= -\frac{5}{9V_0^2}\eta^{-8/3} : & \epsilon_0'' &= -\frac{5}{9V_0^2} \\ \frac{\epsilon'}{\epsilon_0'} &= \eta^{-5/3}; & \frac{\epsilon_0''}{(\epsilon_0')^2} &= -5 \\ \epsilon''' &= \frac{40}{27}\frac{1}{V_0^3}\eta^{-11/3} \\ \frac{\epsilon_0'''}{(\epsilon_0')^3} &= 40, \end{aligned}$$

where $\epsilon' = (\partial\epsilon/\partial V)$; $\epsilon'' = (\partial E'/\partial V)$; and $\epsilon''' = (\partial\epsilon''/\partial V)$. The third order coefficient, C_3, found by using $\epsilon_0''/(\epsilon_0')^2 = -5$ and $\epsilon_0' V_0 = 1/3$ in (6.16), is

$$C_3 = -27\,(K_0' - 4). \tag{6.21}$$

Thus, to the third degree in Eulerian strain (at $T = 0$):,

$$E(\epsilon) = \mathcal{U} - E_{ST} = \frac{9}{2} K_{T_0} V_0 \epsilon^2 \left[1 - \frac{3}{2}(K_0' - 4)\epsilon\right], \qquad (6.22)$$

and the EoS is

$$P(\epsilon, 0) = -3K_{T_0}(1 - 2\epsilon)^{-5/2}\epsilon\left[1 - \tfrac{3}{2}(K_0' - 4)\epsilon\right]. \qquad (6.23)$$

Equations (6.22) and (6.23) are commonly called the Birch-Murnaghan equations of state (B-M EoS) of third degree written in terms of the strain function. When incorporated into the above, (6.2) redefines these equations in terms of the compression parameter η.

There is no analytical obstacle to finding formulae for C_4, C_5, etc. However, C_4 requires knowledge of K_0'' in addition to K_0' and K_{T_0}. K_0'', the second derivative of K with respect to V, is difficult to determine experimentally, so C_4 is mostly of interest in terms of convergent problems in the series expansion of $\mathcal{U} - E_{ST}$. The formula for C_4 is

$$C_4 = 81\left[(K_0' - 8)(K_0' + 1) + K_0 K_0'' + \frac{215}{9}\right]. \qquad (6.24)$$

To those who are accustomed to dealing with stress calculations that are linear and infinitesimal, the definition of strain such as given by (6.1) may seem strange. Such formulae for ϵ arise because of the requirements that the strain energy be invariant under rotation. In finite strain theory, the strain energy has terms arising from second order terms in strain that cannot be neglected when ϵ is about 0.1 and above. One reason why there are a number of possible definitions of $\epsilon(\eta)$ that can possibly be used in place of (6.1) is that the reference strain is arbitrary. The precise way in which ϵ is defined becomes significant when the strain is large.

The Lagrangian definition of strain, in contrast to (6.25), is

$$\epsilon = \frac{1}{2}\left(\eta^{2/3} - 1\right) = \frac{1}{2}\left[\left(\frac{V}{V_0}\right)^{2/3} - 1\right]. \qquad (6.25)$$

Evaluating $\epsilon_0''/(\epsilon_0')^2$ as shown on the previous page, we find -1, so that $C_2 = 9$, and by (6.16)

$$C_3 = -27K_0',$$

which is quite different from the Eulerian case seen in (6.21). Thus, C_3 can only be zero if K_0' is zero. This means that the Lagrangian formula has fewer convergence problems at high P than the Eulerian formula for any value $0 < K_0' < 4$. A value of $K_0' = 0$ is appropriate, however, for incompressible flow in liquids.

An EoS in the third order is found from (6.22):

$$P(\epsilon, 0) = -9K_{T_0}V_0\epsilon\epsilon'(1 - 3K_0'\epsilon).$$

In terms of the compression η, it is (Thomsen, 1970)

$$P(\eta, 0) = -\tfrac{3}{2}K_{T_0}\left(\eta^{1/3} - \eta^{-1/3}\right)\left[1 - \tfrac{3}{4}K_0'\left(\eta^{2/3} - 1\right)\right]. \tag{6.26}$$

In terms of the density ratio, $w = \rho/\rho_0 = 1/\eta$, (6.26) is

$$P(w, 0) = \tfrac{3}{2}K_{T_0}(w^{1/3} - w^{-1/3})\left[1 + \tfrac{3}{4}K_0'(1 - w^{-2/3})\right]. \tag{6.27}$$

6.3. Finite strain equations of state based on $\epsilon(\eta)$

6.3.1. *The Birch-Murnaghan isothermal equation of state*

The most widely used isothermal EoS in solid geophysics is known as the Birch-Murnaghan EoS, given by (6.23). The long use and wide application of this EoS has engendered for it a certain authority in the literature. Nevertheless, this EoS, like all other isothermal EoS's, is based on an unproven assumption. In this case, the EoS rests on the assumption of Eulerian strain (6.20). Other definitions of strain and their consequences will be presented in subsequent sections. Knopoff (1965) has discussed the variety of possibilities in assuming a strain function and the resulting ambiguity. There is no *a priori* reason to assume (6.20) and the resulting (6.23), other than for reasons of convenience, reasonability, and utility (see Sections 6.6 and 8.15).

Birch's original derivations were based on Murnaghan's theory of finite elasticity. The EoS presented by Birch in his classic 1952 paper is the same as (6.26) except that K_0' was not explicitly represented. Instead, a parameter ξ was used, defined by Birch as

$$\xi = \tfrac{3}{4}(4 - K_0'). \tag{6.28}$$

In the body of his derivation, Birch was very careful about the physical meaning of ξ. However, some early authors who applied the B-M EoS to a particular geophysical problem allowed ξ to be a floating constant or simply assumed a reasonable value of ξ. In the early days of this theory, such approximations were unavoidable because K_0' had not yet been carefully measured. However, because K_0' has now either been measured or constrained to narrow limits by associated measurements, the application of the B-M EoS presently requires strict attention to (6.28).

In terms of the compression η and the experimentally determined constants K_{T_0} and K_0', the isothermal B-M EoS to third degree is

$$\frac{P(\eta,0)}{K_{T_0}} = -\frac{3}{2}\left(\eta^{-5/3}-\eta^{-7/3}\right)\left[1-\frac{3}{4}(K_0'-4)\left(1-\eta^{-2/3}\right)\right], \tag{6.29}$$

$$\frac{K}{K_{T_0}} = \eta^{-5/3}\left\{1+\frac{1}{2}\left(1-\eta^{-2/3}\right)\left[5-3K_0' -\frac{27}{4}(4-K_0')(1-\eta^{-2/3})\right]\right\}. \tag{6.30}$$

In terms of the density ratio, w,

$$\frac{P(w,0)}{K_{T_0}} = \frac{3}{2}\left(w^{7/3}-w^{5/3}\right)\left[1+\frac{3}{4}(K_0'-4)(w^{2/3}-1)\right]. \tag{6.31}$$

The B-M EoS in the second degree is ($C_3 = 0$)

$$\frac{P\left(\eta,0\right)}{K_{T_0}} = \frac{3}{2}\left(\eta^{-7/3}-\eta^{-5/3}\right). \tag{6.32}$$

The third order B-M EoS reduces to (6.32) when $K_0' = 4$. Dense oxides, such as Al_2O_3, FeO, stishovite, and $MgSiO_3$ perovskite, have values of K_0' close to 4 (Table 1.8). Consequently, the B-M second order EoS is often a close approximation to the B-M third order EoS, and in the past it has been used to estimate high-compression data when K_0' is unknown.

6.3.2. *The Bardeen equation of state*

Bardeen's (1938) EoS arises from an assumption of an interatomic potential between atoms. The potential giving rise to the Bardeen EoS, which could have been treated by the methods found in Chapter 8, is

$$\phi(r) = A\left(\frac{r_0}{r}\right)+B\left(\frac{r_0}{r}\right)^2+C\left(\frac{r_0}{r}\right)^3. \tag{6.33}$$

This well-known potential and its resulting EoS have been applied to alkali metals. The treatment of this EoS logically belongs in Chapter 8, but we include it in this chapter to illustrate a point.

The EoS that results from (6.33) can be derived from a particular strain function, $\epsilon(\eta)$. I therefore present it in this section to illustrate that deriving an EoS from a strain function is a general approach. The strain function appropriate to the Bardeen potential is

$$\epsilon(\eta) = \left[1-\left(\frac{V_0}{V}\right)^{1/3}\right] = \left(1-\eta^{-1/3}\right). \tag{6.34}$$

This is often called the infinitesimal definition of strain.

From (6.34) we have $\epsilon_0''/(\epsilon_0')^2 = -4$, and from (6.16),

$$C_3 = -27\,(K_0' - 3)\,. \tag{6.35}$$

C_3 vanishes at $K_0' = 3$ in this EoS, just as C_2 vanishes for $K_0' = 4$ in the B-M EoS. For many solids the third order term in this EoS will be small due to the closeness of K_0' to 4. Thus, the second order EoS is of special interest.

The third order EoS for the Bardeen potential is

$$P\,(\epsilon, 0) = -3K_T\eta^{-4/3}\epsilon\left[1 - \tfrac{3}{2}\,(K_0' - 3)\,\epsilon\right]. \tag{6.36}$$

In terms of η,

$$P\,(\eta, 0) = -3K_{T_0}\left(\eta^{-4/3} - \eta^{-5/3}\right)\left[1 - \frac{3}{2}\,(K_0' - 3)\left(1 - \eta^{-1/3}\right)\right]. \tag{6.37}$$

For $K' = 3$, we have the second order Bardeen EoS (see also Section 8.15)

$$P\,(\eta, 0) = -3K_{T_0}\left(\eta^{-5/3} - \eta^{-4/3}\right). \tag{6.38}$$

6.3.3. *The Ullman-Pan'kov equation of state*

As shown by (6.3), the potential $E(\epsilon)$ is expanded into a Taylor's series in which ϵ has been interpreted thus far as a finite strain function. However it is not necessary that ϵ, in some of its various forms, be strain in its traditional sense. ϵ can be any convenient dimensionless function of volume. We have seen that the third order term is large or small depending on the magnitude of K_0'. Because the parameter K_0' is important in defining the higher order terms, the function ϵ can be chosen to include K_0' in such a way as to facilitate convergence of the series. Ullman and Pan'kov (1976) defined such an $\epsilon(\eta)$ function with the result that C_3 always vanishes and with the feature that K_0' is in the exponent of η in the pressure equation, i.e., they chose an ϵ that is dependent on both η and K_0'.

$$\epsilon = \tfrac{1}{u}\left(\eta^{u/3+1} - 1\right), \tag{6.39}$$

from which

$$C_3 = -27\left[\frac{u^3}{(u+3)^3}\right]\left[1 + K_0' + u\right]. \tag{6.40}$$

Thus $C_3 = 0$ if $u = -K_0' - 1$, which holds for any value of K_0'.

To the third order, the Ullman-Pan'kov equation of state (U-P EoS) is

$$P\,(\epsilon, 0) = -3K_{T_0}\epsilon\,(1 + u\epsilon)^{(u-1)/u}\,.$$

In terms of the compression η, it is

$$\frac{P(\eta,0)}{K_{T_0}} = \frac{-3}{K_0' - 2}\left(\eta^{((-K_0'+1)/3)+3} - 1\right)\eta^{-(K_0'+1)/3}. \tag{6.41}$$

My purpose in placing the U-P EoS in this chapter is not necessarily to recommend the use of (6.41) for oxides and silicates, but to illustrate the point that $\epsilon(\eta)$ can be chosen to facilitate convergence of the series expansion. We see that K_0' may be an exponent of η in finite strain theories, similar to the kinds of EoS derived from atomic theories of solids. The Eulerian and Langrangian functions $\epsilon(n)$ are based on sound solid mechanics derivations (see Section 6.7), whereas the Ullman-Pankov derivation is chosen to insure a mathematical implication.

6.4. Problems with truncation of the series

We have emphasized that the third order EoS is used because there is often insufficient experimental information to make generalized statements about fourth order and higher terms. This amounts to a truncation of the Taylor's series expansion of the energy after the third term by making $C_4 = 0$. There is a danger that the sum of all higher power terms is sufficiently large that they are not negligible compared with the first three terms of the series.

Equation (6.23) should be replaced by

$$E(\epsilon) = \mathcal{U} - E_{ST} = \frac{1}{2}K_{T_0}\left(\frac{\epsilon^2}{\epsilon_0'}\right)\left(1 + \frac{C_3}{3}\epsilon\right) + R_{n+4}, \tag{6.42}$$

where R_{n+4} is the remainder of the $n+4$ and higher terms. We cannot evaluate R_{n+4} if the experimental value of K_0'' is not sufficiently robust, which is often the case. We can, however, find the conditions necessary to find the vanishing C_4 coefficient. If C_4 is sufficiently small, we may assume

$$R_{n+4} = \left(\frac{K_{T_0}}{\epsilon_0'}\right)\left(\frac{C_4}{4}\right)\epsilon^4. \tag{6.43}$$

Equation (6.24) evaluates C_4 in terms of K_0', K_0'', and the function $\epsilon(\eta)$. The equations for the four functions of C_4 considered thus far are:

Birch-Murnaghan EoS

$$\begin{aligned} C_4 &= (K_0'+1)(K_0'-8) + \tfrac{215}{9} + K_{T_0}K_0'' \\ &= (3-K_0')(4-K_0') + \tfrac{35}{9} + K_{T_0}K_0'' \end{aligned} \tag{6.44}$$

Thomsen EoS

$$C_4 = (K_0'+1)(K_0'+1) - \tfrac{1}{9} + K_{T_0}K_0'' \tag{6.45}$$

Bardeen EoS

$$\begin{aligned} C_4 &= (K_0'+1)(K_0'-6)+\tfrac{128}{9}+K_{T_0}K_0'' \\ &= (2-K_0')(3-K_0')+\tfrac{20}{9}+K_{T_0}K_0''; \end{aligned} \tag{6.46}$$

Ullman-Pan'kov EoS

$$C_4 = \tfrac{1}{9}(2K_0'-1)(K_0'+1)+K_{T_0}K_0''. \tag{6.47}$$

We see that if C_4 is zero, then K_0'' is a function of K_0' and K_0. This is unlikely from a physical point of view because the interatomic pair potential has great variety when comparing from one compound with another.

Considering that the range K_0' varies from 3.5 to 6.5, we find the values of $K_{T_0}K_{T_0}''$ shown in Table 6.1 for the case $C_4 = 0$. The value of C_4 in the Ullman-Pan'kov EoS is remarkably close to that in the Birch-Murnaghan EoS, especially in the range ($3.5 < K_0' < 5.0$) that is appropriate for planetary interiors. $K_T K_0''$ is exactly the same for the two EoS's in the case $K_0' = 4$.

For C_4 to be sufficiently small that R_{n+4} is negligible, K_{T_0}'' must have a negative value. If it is positive, then C_4 is too large for the third order EoS to approximate the correct P. Because $C_3 = 0$ for $K_{T_0}' = 4$ in the B-M EoS, the fourth order term will be larger than the third order term for solids where $K_0' = 4 \pm \epsilon$. This means that the B-M EoS in the third order is no better, and perhaps even worse, than in the second order when K_0' is close to 4.

We see that $K_0'' = 0$ or $C_4 = 0$ are two very different assumptions and should not be confused. The consequences of $C_4 = 0$ are different with each of the strain functions we have considered (See Table 6.1).

6.5. The fourth order isothermal equation of state

The fourth order EoS requires that terms for which $n > 4$ vanish in (6.5). The EoS is

$$P(\epsilon, 0) = -K_{T_0}\left(\frac{\epsilon_0}{\epsilon_0'}\right)\epsilon\left[1+\left(\frac{C_3}{2}\right)\epsilon+\left(\frac{C_4}{6}\right)\epsilon^2\right]. \tag{6.48}$$

Table 6.1. Values of $K_{T_0}K_0''$ for $C_4 = 0$

EoS	K_0' : 3.5	5.0	6.5
Birch-Murnaghan	− 3.639	− 5.889	−12.633
Thomsen	−15.639	−29.889	−29.889
Bardeen	− 2.972	− 8.222	−17.977
Ullmann-Pan'kov	− 3.000	− 6.000	−10.000

Once C_3 and C_4 are evaluated from the assumed strain function, P can be written out explicitly. For example, the fourth order Lagrangian EoS is

$$P(\eta,0) = \tfrac{3}{2}K_{T_0}\left(\eta^{1/3}-\eta^{-1/3}\right)\left[1-\tfrac{3}{4}\left(\eta^{2/3}-1\right)\right]$$

$$\left\{\tfrac{3}{8}\left[\left(K_0'+1\right)K_0'+K_{T_0}K''\right]-\tfrac{1}{24}\right\}\left(\eta^{2/3}-1\right)^2. \quad (6.49)$$

In general, P involves four material property parameters—K_{T_0}, K_0', K_0'', and V_0—involved in the compression ratio η. In general the fourth order EoS can be written:

$$P = -K_{T_0}f\left(\eta K_0', K_0''\right). \quad (6.50)$$

Although a fourth order EoS requires four material property parameters, they need not necessarily be those given in (6.50). The material property parameters shown in (6.50) are a direct consequence of expanding the energy in a Taylor's series (6.3).

The Birch-Murnaghan EoS of fourth degree is (Stacey et al., 1981)

$$P(\eta,0) = \tfrac{3}{2}K_0\left(\eta^{-7/3}-\eta^{-5/3}\right)\left\{1+\tfrac{3}{4}\left(K_0'-4\right)\left(\eta^{-2/3}-1\right)\right.$$

$$\left.+\tfrac{3}{8}\left[K_0K_0''+\left(K_0'-4\right)\left(K_0-3\right)+\tfrac{35}{9}\right]\left(\eta^{-2/3}-1\right)^2\right\}. \quad (6.51)$$

The equations for the isothermal bulk modulus of the Birch-Murnaghan EoS K_T and its pressure derivative K_0' are

$$K' = \frac{A+B}{D}; \quad (6.52)$$

$$A = K_0' + \tfrac{1}{2}\left\{3K_0K_0' + (3K_0'+7)(K_0'-4)+28\right\}(\eta^{-2/3}-1) \quad (6.53)$$

$$B = \tfrac{27}{8}\left\{K_0K_0'' + K_0'(K_0'-4)+\tfrac{35}{9}\right\}(\eta^{-2/3}-1)^2$$

$$D = 1 + \tfrac{1}{2}(3K_0'-5)(\eta^{-2/3}-1)$$

$$+\tfrac{9}{8}\left\{K_0K_0''+K_0'(K_0'-4)+\tfrac{35}{9}\right\}(\eta^{2/3}-1)^2, \quad (6.54)$$

and where the bulk modulus

$$K_{TB} = K_0\eta^{-5/3}D. \quad (6.55)$$

6.6. On the instability of the Eulerian equation of state

Stacey et al. (1981) pointed out that in the fourth order B-M EoS, the value of K_0' is negative at densities exceeding

$$\frac{\rho}{\rho_0} = \left[\frac{1+2K_0'}{(-3K_{T_0}K_0'')}\right]^{\frac{3}{2}}.$$

They said, "Depending on the value of $K_0 K_0''$, [the range where K_0' is negative] may not be much beyond the terrestrial [range of pressure] and is therefore a serious defect of the Eulerian equations." For K_0' lower than 4 this defect arises well within terrestrial pressures.

Hofmeister (1993) has analyzed the stability of the Eulerian formulation of finite strain in terms of equivalent repulsive interatomic potentials. She finds that an instability in the potential exists for $K_0' < 4$, and no minimum exists for $K_0' >$ about 7. Thus, stable results exist for applying the B-M third order EoS for $3.9 < K_0' < 7.1$. This is the usable region and is satisfactory for most solids to which it is applied. However, for some solids, like orthopyroxene, where $K_0' \approx 11$ (Webb and Jackson, 1990) and liquids and glasses where $7 < K_0' < 23$ (Bottinga, 1985), the B-M EoS is not reliable. For ionic solids, when $K_0' < 3$ (Anderson and Liebermann, 1970), the shear moduli become negative, and the structure is not stable.

6.7. More on the volume strain relation, $\epsilon = f(\eta)$

In Section 6.1 it was pointed out that there is a non-uniqueness in the definition of finite strain as evidenced by the choice between (6.1) and (6.2). We now show how these equations may be derived.

Let $\mathbf{r}$ be a vector locating the final state and $\mathbf{r}_0$ a vector defining the initial strain position, then the vector strain is ε, where

$$\mathbf{r} - \mathbf{r}_0 = \varepsilon \mathbf{r}_0. \tag{6.56}$$

This is the accepted Lagrangian approach, where the deformation is in the framework of the initial state. Similarly, the vector strain $\mathbf{f}$ defines the deformation in the framework of the final state (the Eulerian approach):

$$\mathbf{r} - \mathbf{r}_0 = \mathbf{f}\,\mathbf{r}. \tag{6.57}$$

Define $\bar{\boldsymbol{\epsilon}}$ as the symmetric finite strain tensor in relation to ε as

$$\bar{\epsilon}_{ij} = \frac{1}{2}\left(e_{ij} + e_{ji} + e_{mj}e_{mi}\right). \tag{6.58}$$

It is known from the most general foundations in finite strain theory (Truesdell, 1966) that the stress tensor $\mathbf{S}$ is related to strain ε when starting with (6.56), by

$$\mathbf{S} = (1+\varepsilon)\left(\frac{\partial \mathcal{F}}{\partial \bar{\boldsymbol{\epsilon}}}\right)_T (1+\varepsilon^{\dagger})\frac{1}{V}, \tag{6.59}$$

where $\varepsilon^{\dagger}$ is the transpose of ε. This is the Lagrangian formulation. Using (6.57), the Eulerian approach, the general stress and strain relationship is (Truesdell, 1966)

$$\mathbf{S} = (1-2\boldsymbol{\epsilon})\left(\frac{\partial \mathcal{F}}{\partial \boldsymbol{\epsilon}}\right)_T \frac{1}{V}, \tag{6.60}$$

where the components of the strain tensor ϵ are

$$\epsilon_{ij} = \frac{1}{2}\left(f_{ij} + f_{ji} - f_{mj}f_{mi}\right). \tag{6.61}$$

Independent of these equations, the relation between strain and volume change for the Lagrangian case (Thomsen, 1970) is

$$\eta^2 = \left(\frac{V}{V_0}\right)^2 = \left|\det\left(1 + 2\epsilon\right)\right|, \tag{6.62}$$

and for the Eulerian case, is

$$\eta^2 = \left(\frac{V}{V_0}\right)^2 = \left|\det\left(1 - 2\epsilon\right)\right|. \tag{6.63}$$

For the case of hydrostatic stress, $\mathbf{S} \rightarrow P$, there is great simplification, and both (6.59) and (6.60) reduce to

$$P = -\left(\frac{\partial \mathcal{F}}{\partial V}\right)_T, \tag{6.64}$$

which is the beginning point of this chapter. But, on the other hand, (6.62) and (6.63) reduce to different equations when $\mathbf{S} \rightarrow P$:

$$\bar{\epsilon} = \frac{1}{2}\left(\eta^{2/3} - 1\right) \tag{6.65}$$

$$\epsilon = \frac{1}{2}\left(1 - \eta^{-2/3}\right), \tag{6.66}$$

which are the same as (6.1) and (6.2) (except that early in this chapter we do not distinguish between the two kinds of strain). Holding to identifying separately the two kinds of strain, we have

$$\bar{\epsilon} = \eta^{2/3}\epsilon.$$

The essential difference between the two kinds of strain is that the Lagrangian approach uses the original and unstrained state as the reference, whereas the Eulerian approach uses the deformed state as the reference. There is a potential incompatibility when a Eulerian strain-based EoS, such as the B-M EoS, is used in conjunction with theories based on the Lagrangian framework. Lattice dynamics-based derivations of C_{ij}, such as described in Chapter 9, are Lagrangian based.

6.8. Problems

1. Derive $C_3 = -27K_0'$ for the Langrangian strain.

2. Find the analytical expression for K_0' for the Birch-Murnaghan EoS of third degree.

3. Prove (6.28) and (6.39).

4. Derive (6.51) and (6.52).

7

CONSTRAINING PARAMETERS TO GET THE EQUATION OF STATE

7.1. Introduction

In this chapter I present some examples showing EoS's that arise with the assumption of special characteristics of EoS parameters. The majority of EoS's in the literature arise from special and arbitrary assumptions about the parameters. Of the many possibilities, we review three illustrative cases: (1) an assumption about K' at high P, (2) an assumption about q, and (3) an assumption about $K_T(P)$.

The first case is attributed to Keane (1954), who suggested that dK/dP is a monotonically decreasing function with pressure, eventually reaching a limiting value of $K'_\infty = (\partial K/\partial P)_T$ as $P \to \infty$. K'_∞ can be considered as a floating parameter that is used to adjust the curve to pass through any specified high pressure point in the $P-\eta$ plane or the $P-w$ plane. The Keane EoS is useful for interpolation between shock wave data and lower pressure P-V data.

The second case is a thermodynamic approach (Brennan and Stacey, 1979), where the Grüneisen parameter is written as a special function of pressure (known as the "free volume" approximation), yielding a differential equation in P. With the help of the assumption relating γ and $y = V_0/V = \rho/\rho_0$, the differential equation is solved, yielding

$$P = K_{T_0}\mathrm{f}\,(\gamma_0, w)\,.$$

The third case is attributed to Murnaghan (1944), who defined it as a special case in his finite strain theory; K_T is taken to be linear in P (i.e., $K' = K'_0$).

7.2. The Keane EoS: $dK_T/dP \to K'_\infty$ at high P

Keane's boundary value assumption is that K'_T is a monotonically decreasing function passing from K'_0 at $y = 1$ to $K'_\infty(P = \infty)'$ for $\rho >> \rho_0$. More exactly,

$$\frac{dK_T}{dP} = (K'_0 - K'_\infty)\left(\frac{K_{T_0}}{K_T}\right) + K'_\infty. \tag{7.1}$$

Thus K' is a monotonically decreasing function descending from K'_0 at $P = 0$ to a lower limit K'_∞ at large pressure.

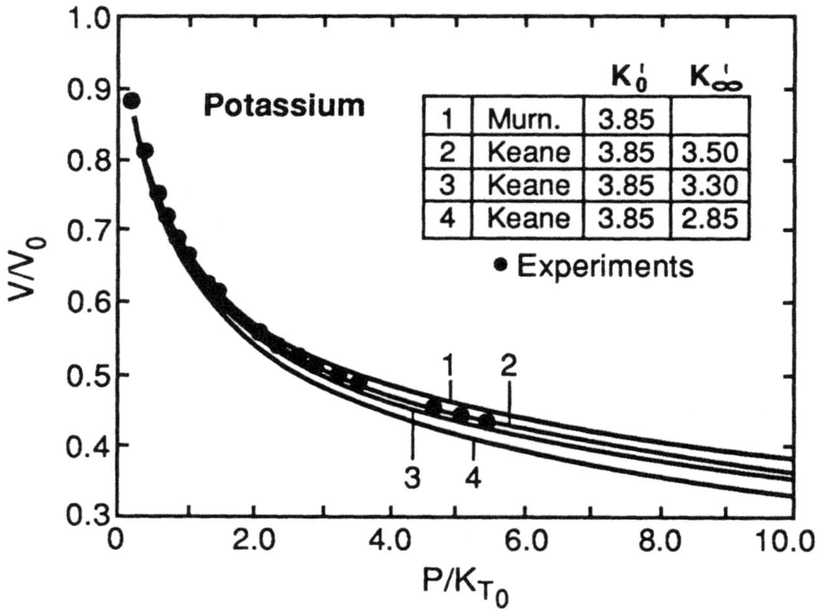

Fig. 7.1. The Murnaghan and Keane equations of state for potassium. The values shown are from Monfort and Swenson (1965). The best value of K'_∞ appears to be 3.50, this value being sensitive to the precision of the highest pressure (isothermal) shock wave data such as reported by Rice (1965).

Keane integrated (7.1) and obtained the following isothermal expressions that constitute the Keane EoS:

$$\frac{P}{K_{T_0}} = \frac{K'_\infty}{(K'_\infty)^2}\left(w^{K'_\infty} - 1\right) - \frac{K'_0 - K'_\infty}{K'_\infty}\,\ell\text{n}\, w \tag{7.2}$$

and

$$K = K_{T_0}\left[1 + \frac{K'_0}{K'_\infty}\left(y^{K'_\infty} - 1\right)\right]. \tag{7.3}$$

Equation (7.2) is useful for describing both the pressure determined from low pressure acoustic data and shock wave data measured on the same substance (Anderson, 1966). K_0 and K'_0 are taken from the acoustically determined laboratory data, and the value K'_∞ is adjusted so that the pressure–density curve fits shock wave data points. Equation (7.3) is a good interpolation formula for intermediate pressures (see Fig. 7.1).

The value of K'_∞ can be evaluated from the interatomic pair potential. For many pair potentials, $(K'_0 - K'_\infty) = 1/3$. K'_∞ is the value of the bulk modulus derivative at a very high pressure in which the crystallographic structure remains intact and in which there is no phase change. For a Fermi-Dirac gas, composed of nucleons, $K'_\infty = 5/3$.

7.3. The Brennan-Stacey EoS and the Barton-Stacey EoS

In Chapter 2, we demonstrated that for some materials, $(\partial K_T/\partial T)_V = 0$, or close to zero. This requires that $q \approx 1$. Brennan and Stacey (1979) showed that this condition leads straight to the Vashchenko-Zubarev (1963) or "free-volume" formula for γ (see Section 1.4.8).

$$\gamma_{vz} = \frac{\frac{1}{2}\left(\frac{dK_T}{dP}\right) - \frac{5}{6} + \frac{2}{9}\left(\frac{P}{K_T}\right)}{1 - \frac{4}{3}\left(\frac{P}{K_T}\right)}. \tag{7.4}$$

The zero pressure value of (7.4) is

$$\gamma_0 = \frac{1}{2}\left(\left(\frac{\partial K}{\partial P}\right) - \frac{5}{3}\right). \tag{7.5}$$

Since K' and K are differentials of P, (7.4) is a second order differential equation in P. Integration of (7.4) yields K_T of the Brennan-Stacey (1979) EoS:

$$K_T(w) = \frac{K_{T_0}}{3\gamma_0} w^{1/3}\left[(2w + 3\gamma_0)\, e^{2\gamma_0}\left(1 - w^{-1}\right) - 2w\right]. \tag{7.6}$$

Integration of (7.6) yields the pressure,

$$P_{Br-S} = \frac{K_0}{2\gamma_0} w^{4/3}\left[e^{2\gamma_0(1-w^{-1})} - 1\right]. \tag{7.7}$$

Brennan and Stacey (1979) called (7.7) the thermodynamic formulation of the equation of state because it arises from the thermodynamic condition $(\partial K_T/\partial T)_V = 0$. For the general case of q, (7.7) is replaced by form called the incomplete gamma function, which requires a numerically based solution (Brennan and Stacey, 1979). Because $\gamma\rho =$ constant is consistent with the thermodynamic properties of NaCl, the calculated pressure from (7.7) can be tested against the 300 K measured isotherm data of NaCl (for example, the isotherm deduced from shock wave measurements reported by Fritz et al. (1971)). We use $K_T\,(300) = 23.84\,\mathrm{GPa}$ and $K' = 5.35$ from Spetzler et al.'s (1972) acoustic measurements to evaluate (7.7), P_{Br-S}. The assignment of a value for γ_0 requires a choice: using (7.5), γ_0 is 1.84, but the measurements of Yamamoto et al. (1987) yield $\gamma_0 = 1.614$. Both values of γ_0 are used for P_{Br-S}, to be compared with the Fritz et al. (1971) experimental isotherm, P_F, as well as to the third order Birch-Murnaghan EoS, P_{B-M}. The comparisons are made in Table 7.1.

Table 7.1. Comparison of analytically determined pressure with experiment for NaCl

V/V_0	Pressure Experimental GPa P_F at 300 K	Pressure Analytical Equation of State GPa P_{Br-S} $\gamma_0 = 1.84$	P_{Br-S} $\gamma_0 = 1.614$	P_{B-M}
1.0	0.000	0.00	0.00	0.000
0.95	1.403	1.42	1.41	1.398
0.90	3.468	3.34	3.30	3.289
0.85	5.996	5.95	5.73	5.838
0.80	9.395	9.51	9.04	9.266
0.75	14.080	14.37	13.47	13.871
0.7	20.350	21.05	19.43	20.05

P_F from Fritz et al., 1971. P_{Br-S} from the Brennan-Stacey EoS (7.6). P_{B-M} from the Birch-Murnaghan EoS (6.23).

We see that P_{Br-S} follows the isotherm deduced from shock data, P_F, rather well. It appears that the Brennan-Stacey formulation is quite satisfactory for the isothermal EoS, provided that the solid under question approximately satisfies the condition $(\partial K_T/\partial T)_{V=0}$.

Barton and Stacey (1985), using a molecular dynamics approach, computed γ for an fcc crystal with central force atomic potential function and found correlation of motions of neighboring atoms beyond what was observed in the Brennan-Stacey model of γ, γ_{vz}. This led to a correction term f in the expression for γ_{vz} given by

$$\gamma_{ba-s} = \left[\left(\frac{1}{2}\frac{dK}{dP} - \frac{1}{6}\right)\right] - \tfrac{1}{3}f\left(\frac{1}{3}\frac{P}{K}\right)\left(1 - \tfrac{2}{3}f\frac{P}{K}\right)^{-1}. \qquad (7.8)$$

Barton and Stacey (1989) showed that (7.8) yields the following differential equation, where $w = \rho/\rho_0$, and where f is considered as an arbitrary parameter.

$$9w^2P'' - \left[6(f-1)w^2 + 18\gamma_0 x\right]P' + f\,(2w + 12\gamma_0)\,P = 0. \qquad (7.9)$$

Taking $q = 1$ and the boundary conditions $P = 0$ and $K'_{T_0} = wP'$ at $w = 1$, the solution of (7.9) becomes

$$P_{Ba-S} = K_0 \exp(2\gamma_0) w^{2f/3} \sum_{k=0}^{\infty} \frac{(-2\gamma_0)^k}{k!\,[(2f/3) - (1/3) + k]} \times \left[1 - w^{[(2f/3)-(1/3)+k]}\right], \qquad (7.10)$$

where the subscript $Ba\text{-}S$ represents the Brennan-Stacey approximation to P. At $P = 0$, γ is constrained only by K_0'. Equation (7.4) is replaced by

$$\gamma_0 = \frac{K_0' - \frac{1}{3}}{2}. \tag{7.11}$$

7.4. The Murnaghan EoS

Murnaghan (1937, 1944) noted in his general theory of finite strain that K is linear in P to a good approximation up to high compressions.

$$K = K_{T_0} + K_0' P, \tag{7.12}$$

which can be written as

$$-\frac{w\,dP}{dw} = K_{T_0} + K_0' P \tag{7.13}$$

and can be integrated to yield the isothermal Murnaghan EoS,

$$\frac{P}{K_{T_0}} = \frac{1}{K_0'}\left(w^{K_0'} - 1\right). \tag{7.14}$$

In spite of its simplicity, (7.14) is accurate up to surprisingly high pressures.

Anderson (1966) demonstrated that (7.14) duplicated experimental data for a wide variety of materials up to pressures of the order of $K_{T_0}/2$. Many experimentalists find the Murnaghan EoS very convenient to express their results. Mao and Bell (1979) measured the compression of MgO up to 94.1 GPa, which is about $P/K_{T_0} = 0.62$, using a diamond anvil pressure cell. They found the data fit a Murnaghan EoS where $K_{T_0} = 156$ GPa and $K_{T_0}' = 4.7$. The parameters are in reasonable agreement with values determined from other methods. A plot of their data is given in Fig. 7.2. The dashed line is from shock wave data.

7.5. The $K_{T_0} K_0''$ parameter

The EoS's described so far in this chapter have an implied value of $K_{T_0} K_{T_0}''$. For the Keane EoS, differentiation of (7.1) with respect to P yields

$$K_{T_0} K_{T_0}'' = \frac{K_\infty}{K_0' - 1}. \tag{7.15}$$

For the Brennan-Stacey EoS, differentiation of (7.9) yields

$$K_{T_0} K_{T_0}'' = -\frac{7}{3} K_0' + \frac{31}{9}. \tag{7.16}$$

For the Murnaghan EoS,

$$K_{T_0}'' = 0.$$

In these three cases, $C_4 = 0$. For the Brennan-Stacey EoS, K_0'' is negative, just as we found for the cases of finite strain (Chapter 6).

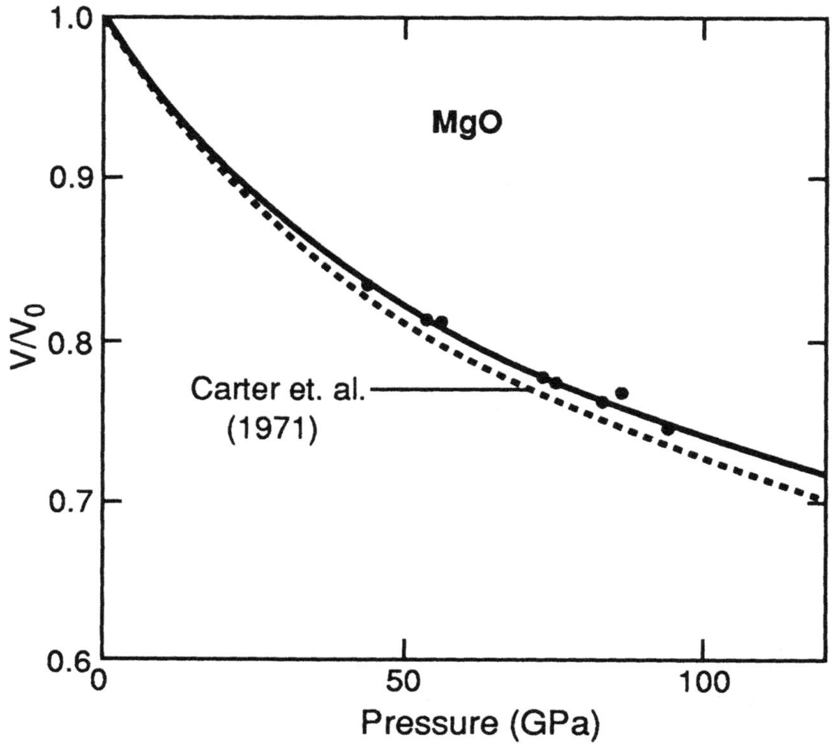

Fig. 7.2. *V-P* data for MgO, after Mao and Bell (1979). Circles on the graph represent their measurements. Shock wave isotherm by Carter et al. (1971) is shown by the dashed line. Parameters of the Murnaghan EoS —the solid line— are: $K_{T_0} = 159.9$ GPa and $K'_{T_0} = 4.56$, acoustically measured by Anderson and Andreatch (1966) (modified from Mao and Bell, 1979).

7.6. Other relationships leading to an EoS

We have by no means exhausted the possibilities for EoS's using reasonable relationships between the thermodynamic parameters. Five more examples are noteworthy. N. Mao (1970) suggested that $K\,(\partial P/\partial K)$ is linear in P. Grover et al. (1973) suggested a relationship between the logarithm of the bulk modulus and volume. Slater (1939) defined pressure as a quadratic polynomial in $\Delta V/V_0$, where V_0 refers to zero pressure volume; but for this case, Stacey et al. (1981) pointed out that dK/dP becomes negative at $\rho/\rho_0 = 1.7$. If V_0 is replaced by the instantaneous V in the quadratic polynomial, dK/dP remains positive at all P, and the resulting EoS describes mercury (Davis and Gordon, 1967). The Tait EoS (MacDonald, 1969) arises by taking V as proportional to the logarithm of P.

7.7. Compression in the earth's lower mantle

Bullen (1968, 1969) found that the seismic data of the earth, when transformed into K_S and P data, can be represented by linear functions. He reported that to an accuracy of two percent, the two linear relations,

$$K_S = 229 + 3.16P \text{ GPa} \quad (7.17)$$

and

$$K_S = 184 + 3.44P \text{ GPa}, \quad (7.18)$$

represent the results for the lower mantle and the outer core separately.

These empirical relations show that the earth's lower interior obeys approximately the Murnaghan condition (6.11). However, the situation is not quite comparable to that described in Section 6.4, because the earth's geotherm is not isothermal. The temperature is steadily increasing with pressure, approximately adiabatically, throughout the lower mantle. The EoS derived from (7.17), of the form given by (7.14), will not be an isothermal EoS, but will closely approximate an adiabatic EoS. Bullen's discovery has not been pursued because an adiabatic EoS does not contribute to the understanding of temperature in the general EoS given by (1.17).

More recent seismic models verify that K_S is linear in P, at least through the upper portion of the lower mantle. Fig. 7.3 is a plot of K_S versus P for the lower mantle using the seismic solution of Dziewonski et al. (1975).

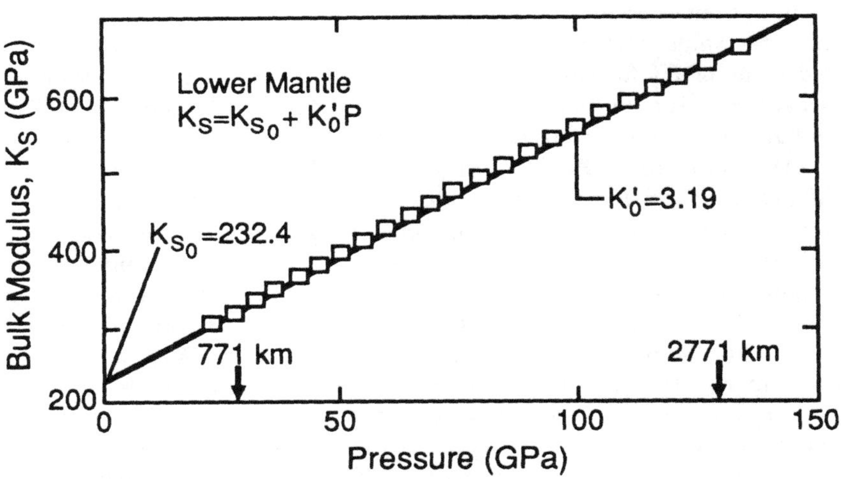

Fig. 7.3. The variation of the adiabatic bulk modulus with pressure in the lower mantle as determined from the PEM model seismic solution (Dziewonski et al. 1975). The data fall on a linear curve (after Anderson and Sumino, 1980).

Up to $P = 100$ GPa, very good linearity also exists for the PREM seismic model. Thus the slope and intercept are well defined. The (hot) values of K_0' and K_{S_0} determined by K_S versus P using PREM data are (Anderson and Sumino, 1980) 3.19 and 232.4 GPa.

The value of $K_0' = 3.19$ for the lower mantle in Fig. 7.3 is the adiabatic derivative of the adiabatic modulus. This value is less than the typical values of K_0' for silicates and mantle material candidates, as shown in Table 1.8. Part of this discrepancy arises because of different thermodynamic conditions.

Representing conditions in the earth by the subscript e and isothermal conditions by T, we have from calculus

$$\left(\frac{\partial K_S}{\partial P}\right)_e = \left(\frac{\partial K_S}{\partial P}\right)_T + \left(\frac{\partial K_S}{\partial T}\right)_P \left(\frac{\partial T}{\partial P}\right)_e . \tag{7.19}$$

The last term of (7.19) is negative because K_S diminishes with T, whereas T increases with P in the earth.

For many compounds, $(\partial K_S/\partial T)_P$ is of the order of -0.015 GPa/deg. The $\partial T/\partial P$ gradient across the earth's upper mantle is near 10 deg/GPa, so that the correction term is about -0.15. We therefore may expect that the uncompressed lower mantle will have a value of K_0' (hot) $= 3.34$.

In Sections 3.4 and 3.5, we have shown that K' increases with T, where $\partial^2 K_T/\partial T \partial P$ is about $0.4 \times 10^{-3}\text{K}^{-1}$, for MgO and olivine. Assuming this value holds for silicate perovskite and the temperature for the uncompressed lower mantle is 1900 K, K_0' should be greater by about 0.07 at 300 K than at mantle temperatures. Thus K_0' (1900) $= 3.34$ becomes K_0' (300) $= 3.41$. The latter value is still lower than the Knittle and Jeanloz (1980) measured value of K_0' for silicate perovskite, 3.9. However, the result for K_0' is sensitive to the method of extrapolation. Bukowinski and Wolf (1990) found K_0' (hot) $= 3.7$ using the third order B-M EoS, and K_0' (hot) $= 4.2$ using the fourth order Birch-Murnaghan EoS. If applied to the third order B-M EoS result on K_0', the corrections above would make K_0' (300) almost exactly coincide with the measured value of K_0' for silicate perovskite.

The uncompressed value of K_{S_0} is temperature dependent and will be higher at room temperature than at high T. Typical mantle materials have a value of $(dK/dT) = -0.015$ GPa/deg. Going from K_{S_0} (1900) to K_{S_0} (300) adds about 30 GPa, yielding an ambient $K_{S_0} = 255$ GPa. Changing K_{S_0} to K_{T_0} will require an increase of about 1.5 %. Thus the final K_{T_0} for the cold uncompressed mantle will be close to the measured value of K_{T_0} for silicate perovskite. Solving (7.14) for w, we find from the data shown in Fig. 7.3 that ρ_0 is near 4.00 ± 0.01 for the hot uncompressed lower mantle, and applying a thermal expansion correction, we find $\rho_0(300)$ close to 4.25 g/cc (as also found by Bukowinski and Wolf, 1990). Therefore, the cold uncompressed lower mantle has EoS parameters that are quite close to the laboratory measurements of silicate perovskite.

8

EQUATIONS OF STATE FROM THE INTERATOMIC POTENTIAL

8.1. Introduction

In principle, the pressure can be exactly determined if the interatomic potential can be exactly determined. Recently much progress has been made to in realizing adequate interatomic potentials for geophysically important solids having low crystal symmetry and large cell sizes. This has yielded reasonably good EoS's starting from first principles. A brief summary of these advances was covered in Chapter 5.

Here we will approach the subject of interatomic potentials heuristically, using approximations appropriate for simple cubic solids. This approach is appealing because of its simplicity and because aspects of the interatomic potential expose parameters in the EoS that have physical meaning.

Consider a simple structure in which the volume V of a mole of atoms is determined by the separation of an adjacent pair of atoms, r. Then the volume V is related to r by

$$V = DNr^3,$$

where D is a structural factor accounting for the efficiency of packing. Changing the variable V to r, the interatomic distance,

$$\eta = \frac{V}{V_0} = \left(\frac{r}{r_0}\right)^{1/3}. \tag{8.1}$$

We are concerned with the change of $\mathcal{F}$ beyond E_{ST} due to a change of r, similar to the change of $\mathcal{F}$ beyond E_{ST} due to a change of strain in Chapter 7. In Chapter 7 we defined $\mathcal{F} - E_{ST}$ as the change due to η, $E(\eta)$.

Here we define $\mathcal{F} - E_{ST}$ as the change in interatomic potential ϕ due to r, $\phi(r)$. Thus

$$P(V, 0) = \frac{d\phi}{dV} = -\left(\frac{1}{r^2}\right)\frac{d\phi}{dr}. \tag{8.2}$$

Further,

$$K_T(V, 0) = -\left(\frac{1}{r}\right)\frac{d^2\phi}{dr^2} - \left(\frac{2}{r^2}\right)\frac{d\phi}{dr}. \tag{8.3}$$

The attractive component of ϕ is sufficiently different from the repulsive component that it is useful to separate it into two components: ϕ_a and ϕ_v. Below we neglect many-body terms, which are important in the general approach and are often included for oxide EoS's. Further, we neglect consideration of lattice strain as it arises from relative motion of tetrahedra and octahedra in network silicates (Hazen and Finger, 1979, 1982). The parameter ϕ is a consequence of the force between two atoms arising in the presence of the surrounding crystal field, and not just the force between two isolated atoms.

8.2. The attractive interatomic potential ϕ_a

In the definition of the interatomic potential for a solid, assumptions must be made about the chemical bonds holding the solid together and about the range of effective force of the bonds. Most structures are made up of mixtures of bond types, but here we make simplifying assumptions. For minerals in geophysical problems, the ionic bond is by far the most important. In an ionic solid, the electrons are bound to the nucleus, and the solid is nonconducting or only slightly conducting. Visible light is transmitted in thin sections; the solid has good cleavage; and it exhibits strong infrared absorption. These and other physical properties, as found in many oxides and silicates of interest to geophysics and ceramic science, justify the assumption of ionicity in the attractive potential.

It is assumed that the structure is an ionic point lattice (i.e., all lattice points are composed of charges whose attraction is given by a coulombic potential). For the attraction of one ion pair separated by the ionic distance r_i, the attractive potential is

$$\phi_{a_i} = -\frac{Z_1 Z_2 e^2}{r_i},$$

where Z_1 and Z_2 are the valences of the two ions; e is the electric charge; ϕ_{a_i} designates the attractive potential for one ion pair; and r_i is the interatomic spacing between an arbitrary ion pair.

The above equation must be summed over all pairs in the lattice. When this is done, we have an expression for the attractive potential energy:

$$\phi_a = \sum \phi_{a_i} = -\frac{\mathbf{A} Z_1 Z_2 e^2}{r}, \tag{8.4}$$

where r is the nearest-neighbor interatomic distance, and $\mathbf{A}$ is the Madelung constant, which is dependent on the crystallographic structure. $\mathbf{A}$ can be found by the appropriate Ewald sum and is tabled in many references. It is usually defined on the basis of a mole of the chemical formula of a substance. In this case, $\mathbf{A}$ is a small number for a diatomic substance (1.74576 for NaCl) and a much larger number for complicated unit cells such as found for Al_2O_3, as shown by Table 8.1.

Table 8.1. Modified Madelung constant for a variety of structures using two definitions

Structure	Examples of Solids	Coord. No.	Energy/ atom pair	Energy/ valence bond
Rock salt	NaCl, CaO, AgCl	6	1.748	1.748
Cs chloride	CsCl, TlCl	8	1.763	1.763
Rutile	MgF_2	3,6	2.408	1.605
Corundum	Al_2O_3, Fe_2O_3	4,6	4.172	1.668

8.3. A simplification in the Madelung constant

For iron-free oxides and silicates, the "average" mean atomic weight μ is close to about 20.5, as shown in Table 1.9. This gets close to 22 when the Fe/(Fe + Mg) ratio gets as high as 20%, but for the earth's mantle itself, μ is close to 21 (Watt et al., 1975). For this reason it is often useful to convert specific volume V to μ/ρ or constant/ρ. The Madelung constant of minerals can then be converted to units of energy per valence bond instead of energy per atom pair. In the former case the value of the Madelung constant is virtually independent of structure, as Table 8.1 shows.

Thus the coulombic energy of many oxides and silicates can be well approximated, defining M as the product of $Z_a Z_c$ (averaged) multiplied by M corresponding to NaCl. Here $Z_a Z_c$ (ave) $= \sum x_j\, Z_j\, \bar{p}\, Z_a/(x_j + \bar{p})$, where x_j is the number of cations; Z_a is the valence of the anion; and $\bar{p}$ is the number of anions in the cell.

8.4. The repulsive energy term

The interatomic potential ϕ must have a repulsive energy term or terms; otherwise the lattice would collapse under the attractive forces. Calling the repulsive energy of a particular ion pair ϕ_v, the total repulsive energy of a mineral is

$$\phi_v = \sum \phi_{v_i}. \tag{8.5}$$

Because the repulsive energy is very short ranged, the terms in (8.5) become negligible after a few ion pair distances. In simple structures, the repulsive energy can be represented satisfactorily by one, two, or three terms:

$$\phi_v = M v_{nn} + M' v_{nnn} + \dots \tag{8.6}$$

where the subscript nn stands for nearest neighbors; the subscript nnn stands for next nearest neighbors; and M and M' represent the coordination number. In a biatomic ionic structure where the physical properties of anion and cation are quite different, two terms are required for the *nnn* repulsive energy, and in that case the repulsion is well approximated by three terms:

$$\phi_v = M v_{nn} + M' v\,(a-a)_{nnn} + M'' v\,(c-c)_{nnn}\,, \tag{8.7}$$

where $\phi_v\,(a-a)_{nnn}$ stands for anion–anion interaction and $\phi_v\,(c-c)_{nnn}$ stands for cation–cation interaction.

For many minerals of interest in geophysics and ceramics, the attractive component of ϕ is taken to be (8.4). Though this is common practice, there are a great variety of proposals for the repulsive energy, ϕ_v. Authors have differing opinions on the functional form of v_{nn} and on the number of terms in (8.6) needed to approximate ϕ_v.

8.5. The Born-Mie equation of state

This case has the simplest repulsive energy term. No *nnn* terms are considered, and the *nn* term is given by an inverse power law:

$$\phi_v = \frac{Mb}{r^n}, \tag{8.8}$$

where b is an arbitrary constant ($M = 6, 8, 4$ for NaCl, CsCl, and ZnS structures, respectively, for example).

Now assume that ϕ is the sum of $\phi_a + \phi_v$, by neglecting many-body terms, magnetic terms, electronic terms, and internal strain. That is, ϕ is the sum of (8.4) and (8.8).

Now $\eta = V/V_0 = (r/r_0)^3$, so that (see also Fürth, 1944)

$$\begin{aligned}\phi(\eta) &= -\left(\frac{\mathbf{A} Z_1 Z_2 e^2}{r_0}\right)\eta^{-1/3} + \left(\frac{Mb}{r_0^n}\right)\eta^{-n/3} \\ &= -\,\bar{a}\eta^{-1/3} + \bar{b}\eta^{-n/3},\end{aligned} \tag{8.9}$$

where the thermal energy is neglected ($T = 0$).

The three arbitrary parameters of $\phi(\eta)$ are $\bar{a}$, $\bar{b}$, and n. Replacing the parameter V with $\eta = V/V_0$ in (8.1) and (8.2), $P(V_0, 0)$ becomes $P(1, 0)$. The boundary condition $P(1,0) = 0$ eliminates one parameter, and the others are found in terms of K_{T_0} and K'. At equilibrium,

$$\frac{d\phi}{dV} = \left(\frac{1}{V_0}\right)\frac{d\phi}{d\eta} = 0, \quad \text{so that } \bar{b} = \frac{\bar{a}}{n}, \tag{8.10}$$

and

$$P(\eta) = \left(\frac{\bar{a}}{3}\right) V_0^{4/3}\left[\eta^{-(n+3)/3} - \eta^{-4/3}\right]. \tag{8.11}$$

Using (2.1),

$$K_T(\eta) = \left(\frac{\bar{a}}{9}\right) V_0^{4/3}\left[(n+3)\,\eta^{-(n+3)/3} - 4\eta^{4/3}\right], \tag{8.12}$$

where again the condition $T = 0$ is implied by $P(\eta)$ and $K_T(\eta)$. When $\eta \to 1$,

$$K_{T_0} = \frac{a\,(n-1)}{9} V_0^{4/3}, \tag{8.13}$$

so

$$P(\eta) = \frac{3K_{T_0}}{(n-1)} \left[\eta^{(n+3)/3} - \eta^{-4/3}\right]. \tag{8.14}$$

Using (2.5) on the above,

$$K'(\eta) = -\left(\frac{\eta}{K}\right)\frac{dK}{d\eta} = \frac{n+3}{3} + \frac{4}{3}\left[\left(\frac{n-1}{n+3}\right)\eta^{-(n-1)/3} - 1\right].$$

As $\eta \to 1$,

$$K' = \frac{n+7}{3}. \tag{8.15}$$

Using (8.13) and (8.15) in the expressions for P, K, and K' (Anderson, 1970),

$$\phi(\eta) = \frac{3K_{T_0}}{3K_0' - 8}\left[\frac{\eta^{-(K'_{T_0}-7/3)}}{K_0' - \frac{7}{3}} - \frac{\eta^{-1/3}}{\frac{1}{3}}\right] \tag{8.16}$$

$$P(\eta) = \frac{3K_{T_0}}{3K_0' - 8}\left[\eta^{(-K'_{T_0}-4/3)} - \eta^{-4/3}\right] \tag{8.17}$$

$$K_T(\eta) = \frac{K_{T_0}}{3K'_{T_0} - 8}\left[(3K_0' - 4)\,\eta^{-(K'_{T_0}-4/3)} - 4\eta^{-4/3}\right] \tag{8.18}$$

$$K'(\eta) = \frac{\left(\frac{K_{T_0}-4}{3}\right) + \left(\frac{4}{3}\right)3K'_{T_0} - 8}{(3K_0' - 4)\,\eta^{-\left(3K'_{T_0}-8\right)/3} - 4}. \tag{8.19}$$

Note that the EoS is completely defined in terms of the parameters K_{T_0} and K_0'. This gives a physical meaning to the atomic parameters a, b, and n in terms of EoS parameters.

The Born-Mie EoS, (8.17), blows up at $K_{T_0} = 8/3$, but this is of little physical significance, as most oxides and silicates have $K'_{T_0} > 3$. The Born-Mie EoS, (8.18), has some similarity in its form to the U-P EoS (6.36). In fact, at $K'_{T_0} = 3$ the Born-Mie EoS is identical to both the U-P EoS and the Bardeen EoS of the second degree.

Mammone (1980) applied the Born-Mie EoS to his measurements of CaO. Choosing $K_{T_0} = 1149$ and $K'_{T_0} = 4.04$, he found that the Born-Mie EoS made a close fit to his data taken up to 64.1 GPa ($P/K_{T_0} = 0.56$). This is shown in Table 8.2. The values chosen for K_{T_0} and K'_{T_0} agree well with other data in the literature.

Table 8.2. Comparison of experimental compression for CaO with theory using the Born-Mie EoS and the parameters $K_{T_0} = 1149$ and $K'_{T_0} = 4.04$.

Measured		Theory		
P(GPa)	V/V_0	V/V_0	%	P/K_0
0.9	0.9897	0.9924	+0.27	0.0078
5.9	0.9640	0.9544	−1.00	0.0513
9.0	0.9362	0.9341	−0.22	0.0783
9.1	0.9339	0.9334	−0.05	0.0792
10.9	0.9273	0.9226	−0.51	0.0949
13.2	0.9073	0.9069	−0.04	0.1192
15.8	0.8916	0.8959	+0.48	0.1375
18.7	0.8750	0.8817	+0.77	0.1628
18.8	0.8824	0.8812	−0.12	0.1636
21.0	0.8676	0.8711	+0.40	0.1828
21.5	0.8659	0.8689	+0.35	0.1871
26.0	0.8591	0.8496	+1.11	0.2263
29.3	0.8362	0.8373	+0.13	0.2550
34.4	0.8208	0.8193	−0.18	0.2994
36.9	0.8132	0.8110	−0.27	0.3211
41.0	0.7980	0.7983	+0.04	0.3565
46.1	0.7794	0.7836	+0.54	0.4012
51.0	0.7710	0.7706	−0.05	0.4439
54.5	0.7631	0.7618	−0.17	0.4743
60.3	0.7507	0.7483	−0.32	0.5248
62.8	0.7425	0.7427	+0.03	0.5466
64.1	0.7410	0.7399	−0.15	0.5579

From Mammone, 1980.

8.6. The Born-Meyer equation of state: the method of potentials

Like the Born-Mie approximation in the previous subsection, the *nnn* terms in ϕ_v are neglected, but the *nn* term is given by a simple exponential, commonly known as the Meyer repulsion term,

$$\phi_v = Me^{-b(r/r_0)}. \tag{8.20}$$

The potential energy at absolute zero is given by

$$\phi(\eta) = -\left[\frac{\mathbf{A}Z_1Z_2e^2}{r_0}\right]\eta^{-1/3} + Mbe^{(-b\eta^{1/3})}. \tag{8.21}$$

The EoS solutions are found by employing the same calculus given by (8.10) to (8.15) (Kalinin, 1960),

$$P(\eta) = \left(\frac{3K_{T_0}}{a-2}\right)\left[\eta^{2/3}e^{\left(a-a\eta^{1/3}\right)} - \eta^{4/3}\right] \tag{8.22}$$

and

$$K(\eta) = \left(\frac{K_{T_0}}{a-2}\right)\eta^{-1/3}\left(\eta^{-1/3}+a\right)e^{\left(a-a\eta^{1/3}\right)} - 4\eta^{-4/3}, \tag{8.23}$$

where

$$a = nr_0 = \frac{3}{2}\left(K_T' - 1\right) + \frac{1}{2}\left[\left(3K_T' - 7\right)^2 + 8\right]^{1/2}. \tag{8.24}$$

These equations have been referred to as the Method of Potentials by many authors in the USSR and are in common use by workers in the Schmidt Institute of Physics of the Earth's Interior. Equation (8.22) has been used to represent experimental data on many materials up to high pressure (Zharkov and Kalinin, 1971; Zharkov et al., 1972; Zharkov and Trubitsyn, 1978).

8.7. The Demarest equation of state for NaCl: *nnn* repulsion

Demarest used (8.21) for ϕ, but he added additional terms for the *nnn* components of ϕ_v. There were two papers in which the *nnn* terms were defined differently. In one of these (Demarest, 1974), he assumed that the *nnn* repulsion was anion–anion, and he used the inverse power law, similar to (8.8). In the other paper (Demarest, 1975), he assumed the *nnn* repulsion was an exponential law, similar to (8.20).

The extra constants introduced by the *nnn* potential required extra boundary values. He used the boundary conditions of the elastic constants and their pressure derivatives by invoking not only K_{T_0} but also $(C_{44})_0$, the elastic shear constant, and by invoking $(dC_{44}/dP)_0$, as well as K'_{T_0}. Simple algebraic expressions for $P(\eta)$ and $K(\eta)$ are not possible, and the computer is needed for the final results. Demarest found that at high compression $P(\eta)$ and $K(\eta)$ are not sensitive to the choice of the *nnn* potential. However, he found that at high compression the elastic shear constants are quite sensitive to the choice of the *nnn* potential. Demarest's main contribution is that he showed that the next nearest neighbor terms are needed to make fine adjustments to the shear elastic constants and their pressure derivatives in the EoS.

8.8. Van der Waals bonds in the potential ϕ_a

A few authors have included the van der Waals interaction in ϕ_a. This is a weak attraction arising from the interaction of the dipole moments of two neighboring atoms. A dipole moment arises because the electron cannot be everywhere at once, and the electron shell's center of charge makes the interionic distance different from the nuclear distance. The electrostatic energy due to two interacting dipoles varies as r^{-6}. Even weaker interactions exist for quadrapole–quadrapole interactions. These energy terms vary as r^{-8}, r^{-12}, etc.

The van der Waals interaction energy is very weak, often about 1/100 of the strength of an ionic bond. In most minerals important to geophysics and ceramics research, the van der Waals energy is so weak compared with the coulombic attraction that it can be neglected altogether, although it has been included in the case for NaCl discussed below.

8.9. The Decker equation of state for NaCl

Decker (1971) assumed (8.20) for three repulsion components of ϕ_v, and for ϕ_{a_i} he took (8.4) plus a term in r^{-6} and a term in r^{-8}, altogether three attractive components. The full explanation of all these six terms is found in Seitz's (1940) book. Decker assumed that he needed all these terms to find the EoS for the NaCl and CsCl structures. The Decker potential represented as the internal energy at absolute zero is

$$\phi(\eta) = \left(\frac{\mathbf{A}e^2}{r_0}\right)\eta^{1/3} - c\eta^{-3} - \bar{d}\eta^{8/3} + Mbe^{\left(-bx^{1/3}\right)} + M'b^- e^{\left(-b^+ d\eta^{1/3}\right)} + M'b^+ e^{\left(-b^- d\eta^{1/3}\right)}. \quad (8.25)$$

Here $\bar{d}$ is the ratio of the *nnn* distance to the *nn* distance, and M' is the *nnn* coordination number.

There are a large number of unknowns in (8.25). Some of them, like c and $\bar{d}$, were numbers taken from evaluations from previous papers, and some of them, like b^+ and b^-, were computed by Decker from polarizabilities using the measured dielectric constant and the index of refraction. The parameters c and d are essentially determined by using K_{T_0} and the lattice parameter. However, K_0' is not used as a boundary condition to find c, and so Decker's analyses departed substantially from methods common in most EoS analyses.

The value of $P(\eta)$ is determined by computer calculations. Decker added a term for the thermal energy, using the Debye theory covered in Chapter 1. Decker published his EoS as a table of numbers listing $P(\eta, T)$. His tabled values, especially $P(\eta, 273)$, have been used by many experimentalists as a secondary pressure standard. The measured compression of NaCl can be used as a pressure calibration through $P(\eta, T_0)$ in the Decker table. Thus the relative compression of the measured sample to a standard NaCl sample in parallel is enough to find $P(\eta, T_0)$ for the unknown sample.

8.10. Equations of state for metals

In solid earth geophysics, when considering properties of the earth's core, interatomic potentials appropriate to iron are often of interest. In the core, the compression is of the order $\eta \simeq 0.4$. The Morse (1929) interatomic pair potential has been favored by Welch et al. (1977) and Fazio et al. (1979) and has been discussed by Zharkov and Kalinin (1971).

This potential consists of two purely covalent bonding terms:

$$\phi = \phi_0 \left[e^{2f(1-r/r_0)} - 2e^{f(1-r/r_0)} \right], \tag{8.26}$$

where ϕ_0 is the binding energy at $r = r_0$. Applying the formula for P and substituting back as before,

$$P = \left(\frac{3K_{T_0}}{f} \right) w^{2/3} \left[e^{2f\left(1-w^{-1/3}\right)} - e^{f\left(1-w^{-1/3}\right)} \right], \tag{8.27}$$

where $w = V_0/V = \rho/\rho_0 = 1/\eta$. It can be shown (Stacey et al., 1981) that

$$f = K'_{T_0} - 1, \quad \text{and } K_{T_0} K''_{T_0} = -\frac{1}{9}\left(2f^2 + 9f + 2\right). \tag{8.28}$$

Formulas for K and K' are found in Stacey et al. (1981).

A variation of the Morse potential is the Rydberg potential,

$$\phi(r) = \phi_0 \left[1 - f\left(1 - V/V_0\right)\right] e^{f(1-r/r_0)}. \tag{8.29}$$

The resulting pressure equation and the K_{T_0} (K''_0) equation are

$$P = 3K_{T_0} w^{2/3} \left(1 - w^{1/3}\right) e^{\left[(3/2)\left(K'_{T_0}-1\right)\left(1-w^{1/3}\right)\right]} \tag{8.30}$$

$$K_{T_0} K''_0 = -\frac{1}{4}\left[\left(K'_0\right)^2 + 2K'_0 - \frac{19}{9}\right]. \tag{8.31}$$

Stacey et al. (1981) presented, apparently for the first time, the derivation of the pressure from the Davydov potential, which was mentioned by Zharkov and Kalinin (1971).

$$\phi = \left(A\frac{r_0}{r} - B\right) e^{f(1-r/r_0)} \tag{8.32}$$

$$P = \frac{3K_{T_0}}{f+2} \left[w^{4/3} + fw - (f+1)e^{2/3}\right] e^{f\left(1-w^{-1/3}\right)}, \tag{8.33}$$

from which

$$f = \frac{3}{4}\left\{\left(K'_0 - 3\right) + \left[\left(K'_0 + 1\right)\left(K'_0 - \frac{5}{3}\right)\right]^{1/2}\right\} \tag{8.34}$$

and

$$K_{T_0} K_0'' = -\frac{f^4 + 12f^3 + 44f^2 + 62f + 32}{2(f+2)^2}. \tag{8.35}$$

Now, taking $K_{T_0}' = 5$ as representative for iron in the high temperature phases and solving for K_0', we find

$$K_0'' = -7.78, \quad \text{Morse}$$
$$K_0'' = -8.82, \quad \text{Rydberg}$$
$$K_0'' = -7.80, \quad \text{Davydov,}$$

which are similar to those values found for other potentials. All suggest K_0'' must be negative. That is, K' decreases gradually with pressure.

8.11. EoS parameters for iron at core pressures

Many attempts have been made to find a phase diagram of iron valid up to core pressures (Birch, 1972; Liu, 1975; Anderson, 1986; Boehler, 1986; Ross et al., 1990; Williams et al., 1991; Anderson, 1992; Boehler, 1994). The phase diagram must be known to show which phase of iron is present at the inner core pressure.

There are now six phases of iron and four triple points. Birch (1972) knew of four phases: α (bcc), γ (fcc), ϵ (hcp), and a nonmagnetic δ (also bcc) sandwiched between α and γ, and restricted to very low pressures. Boehler (1986) proposed a new high temperature phase, which he called θ, that was required on the basis of his experiments. This high T phase was also suggested by the theory of Young and Grover (1984), who indicated that it is nonmagnetic bcc and called it α'. We shall designate this phase as θ. Ross et al. (1990) proposed the phase boundary to be near 5000 K. There is also a new phase called β (Saxena et al., 1993) found near 90 GPa.

One of Birch's (1972) proposed phase diagrams (which lacks the phases β and θ) is shown in Figure 8.1. Here the upper triple point is at 100 GPa, making the core entirely within the ϵ (hcp) phase. The Ross et al. (1990) proposal for a high temperature, high pressure bcc phase (called by them α') would put the inner and part of the outer core within the phase θ. There are theoretical arguments against a stable bcc phase at high pressure. Hence we follow Boehler and call it θ. Thus, the EoS appropriate for pure iron at inner core pressures and temperatures does not appear to be appropriate for a bcc or an hcp structure, whereas that appropriate for the upper part of the outer core might arise from the new β phase. The phase diagram as of 1993 (Anderson, 1993) including θ and β is shown in Fig. 8.2.

Brown and McQueen (1980) measured the longitudinal wave velocity set up in a solid after the compression shock wave had passed and before the tensile shock wave had arrived, so they were able to record the longitudinal sound velocity at a value of P and compare it with the bulk sound velocity at that same P (recorded in Table 12.1). From this they found two transitions

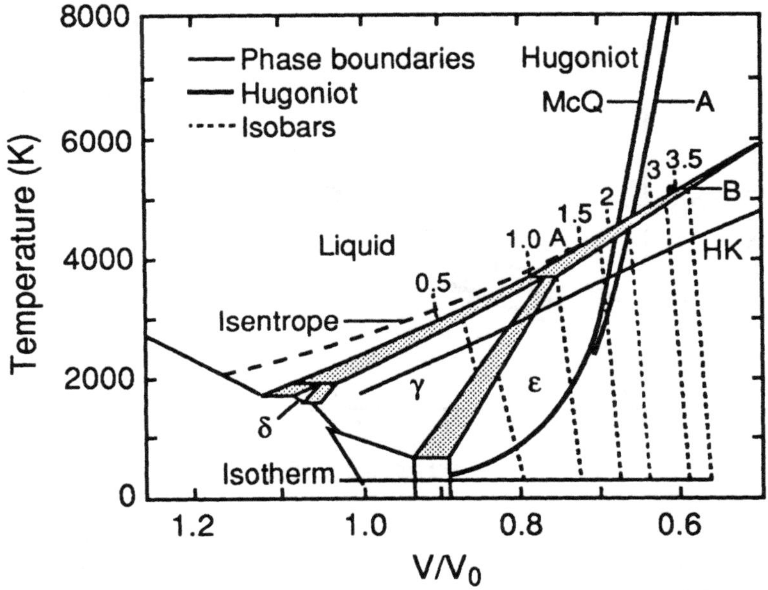

Fig. 8.1. The phase diagram of iron in T versus V/V_0 space according to Birch (1972). A and B represent pressure at the core–mantle boundary and the inner–outer core boundary. The convergence of the γ-ϵ boundary and the ϵ-liquid boundary defines the ϵ-γ-ℓ triple point at 100 GPa. Dotted lines are isobars. The lines marked A and McQ are the Hugoniot from Al'tshuler et al. (1968) and McQueen et al. (1970), respectively (modified from Birch, 1972).

indicated by a sharp decrease in v_p: $P = 200$ GPa and $P = 243$ GPa. At the higher pressure transition and higher P's, v_p became exactly v_b, so that $v_s \to 0$. Thus the onset of melting in iron was found to be at $P = 243$ GPa along the Hugoniot. The plot of v_p versus P is shown in Fig. 8.3. The melting temperature T_m at 243 GPa is then the calculated T of the Hugoniot at the pressure (see also Chapter 12). Although Brown and McQueen (1986) found $T_m = 5800 \pm 500$ K for the 243 GPa transition, Williams et al. (1987) reported 6700 ± 400 K.

Parameters used in the EoS of iron are listed in Table 8.3. We consider three phases of iron as possible candidates for the inner core, α(bcc), ϵ(hcp), and γ(fcc). Values of ρ_0, K_0, and K_0' of these three phases are listed in Table 8.3 (Anderson, 1986, 1993). (The parameters for the nonmagnetic bcc phase are not the same as those for the α phase). There are four sets of values of K_0 and K_0' for ϵ-iron in Table 8.3, but we see the tradeoff between K_0 and K_0' typical of P-V compression measurements in Fig. 8.4. The value found for K_0 from the experiments would be better determined if K' could be constrained. The recommended approach is to assume $C_4 = 0$.

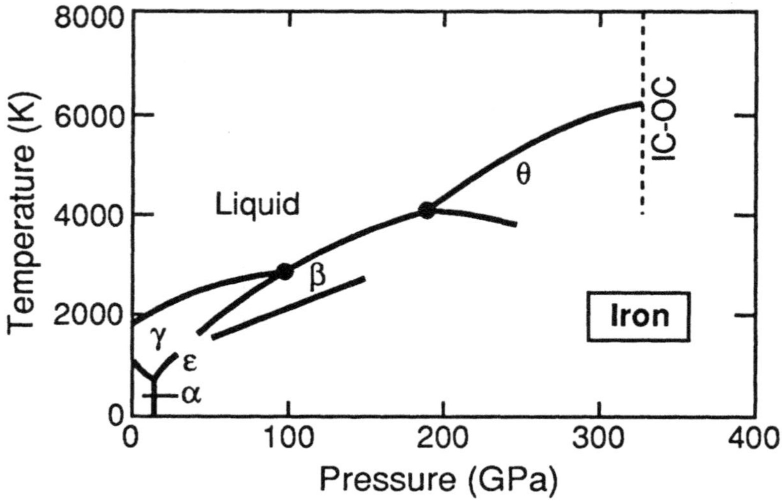

Fig. 8.2. A version of the phase diagram of iron (Anderson, 1993), including the θ phase (Boehler, 1986) and the new phase β (Saxena et al., 1993). This is one of several interpretations (modified from Anderson, 1993).

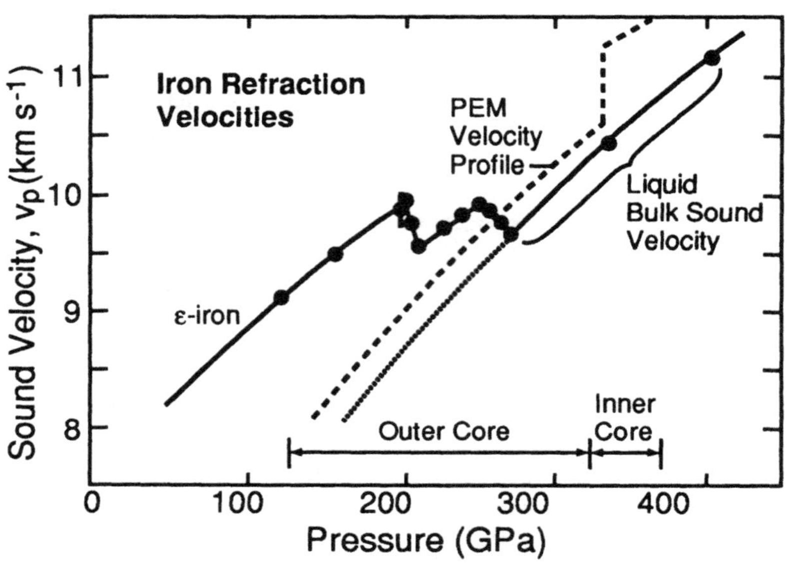

Fig. 8.3. Elastic wave velocities (solid line) versus pressure along the Hugoniot of iron (modified from Brown and McQueen, 1982). The dotted line is the calculated bulk sound velocity. The first break represents a solid-solid transition; the second break represents a solid-liquid transition, both on the Hugoniot.

Table 8.3. Equation of state parameters for phases of iron at room temperature pressures

Iron phase ($P = 0$, $T = 300$ K)	ρ_0 g/cc	K_0 GPa	K_0' GPa	Source
α	7.873	166.6	5.29	Guinan & Beshers, 1968
	7.873	166.6	5.97	Rotter & Smith, 1966
	7.87	167	4.9	Mao & Bell, 1979
$\alpha(nm)^{**}$	8.37	284	5.3 – 6.0	Isaak and Masuda, 1993
ϵ	8.28	178.2	5.15	Brown & McQueen, 1982 (reduced shock wave data)
ϵ	8.28	156.2	5.4	Mao & Bell, 1979 (static compression data)
ϵ	8.28	173.2	4.37	Jephcoat et al., 1986
ϵ	8.28	182.7	4.4*	Andrews, 1973
γ	8.0	180†	5.5	Zaretsky & Stassis, 1987 Stassis 1994 (at 1400 K)
Liquid, 1900 K	7.0	136.0	5.0	Birch, 1972; Filipov et al., 1966 (phase diagram data)

*Obtained from Andrews, Fig. 5, using $K_0 = 182.7$ GPa. **nonmagnetic values (Isaak and Masuda, 1993). †Extrapolated to 300 K from 1400 K

The chosen values for ϵ(hcp) iron at 300 K are the centroid of Fig. 8.4: $\rho_0 = 8.28$, $K_{T_0} = 172$ GPa, and $K_0' = 4.8$. To compute ρ at the inner core P, an isothermal equation of state, that is, $P_0(V)$, must first be chosen.

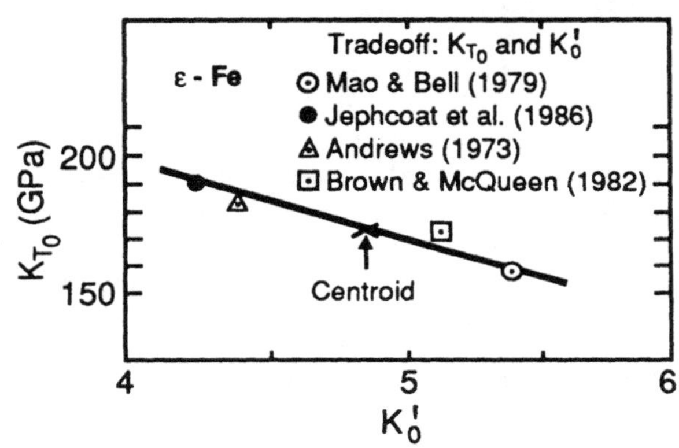

Fig. 8.4. The tradeoff between K_0' and K_{T_0} for ϵ-iron, typical of static and dynamic compression studies. We use the centroid given by the mark X.

It is sufficient here to show that two well known semi-theoretical isothermal EoS's can be used to test for the presence of impurities in the inner core when P_{TH} is considered. These EoS's, as applied below, are not entirely empirical, nor are they used here as curve-fitting devices with floating values of the parameters. The three parameters determined by experiment or theory, ρ_0, K_0, K_0', fix the density–pressure trajectory once the EoS has been selected. We choose the third order Birch-Murnaghan EoS ((6.15), (6.18), and (6.45)) and the Morse EoS (8.26). The use of the Morse equation for compressed iron has been recommended by a number of authors (for example, Fazio et al., 1979; Welch et al., 1977). The Birch-Murnaghan EoS is widely used among geoscientists.

8.12. Equations of state of iron at inner core pressures

The trajectories of the EoS curves for the ϵ phase and the γ phase calculated by the B-M third order EoS and α-iron calculated by both the Morse potential and the Birch-Murnaghan third order EoS for core pressures (T = 300 K) are graphed in Fig. 8.5. Also included is the shock experimental ρ-P curve of Brown and McQueen (1982) corrected to 300 K. It is clear that ϵ-iron has a density greater than 14 g/cc at inner core conditions, somewhat greater than that of the inner core as determined by the seismic PREM model (12.85 g/cc) and greater than that of γ-iron. γ-iron appears to have a density somewhat greater than the PREM value and α-iron. Since the calculations of $P(V)$ do not include the effect of T, the values of ρ appear to be higher than the PREM values of ρ. To correct for T, we must add the thermal pressure term, P_{TH}, as shown in (1.17). This will shift the calculated values of ρ in Fig. 8.5 to the right, improving the agreement with PREM. (The value of P_{TH} for the inner core will be treated in Chapter 10). Curves 4 and 5 show the variance induced by changing the EoS. The error bar on the γ curve (3) shows the variance induced by changing the value of K' over a reasonable range.

8.13. How to choose the best EoS: a general discussion

A number of formulations for analytical expressions for the EoS have been presented in Chapters 6, 7, and 8, and I have stressed that all of them arise from an unchecked and unprovable assumption concerning an assumed interatomic potential, an assumed strain function, or an assumed boundary condition that is not testable. I have also emphasized that all converge at low pressures ($P < 1/2K_{T_0}$). Divergence of the various theories from one another unfortunately occurs in the pressure region where the experimental data begin to have large error bars. If a favored EoS does not duplicate the experimental data at very high pressure, does one blame the data, the EoS, or both? If one is confident of the experimental data or wants to statistically treat the high-pressure experimental values of $P(\eta, T_0)$, the best EoS for that solid can be determined (MacDonald, 1969).

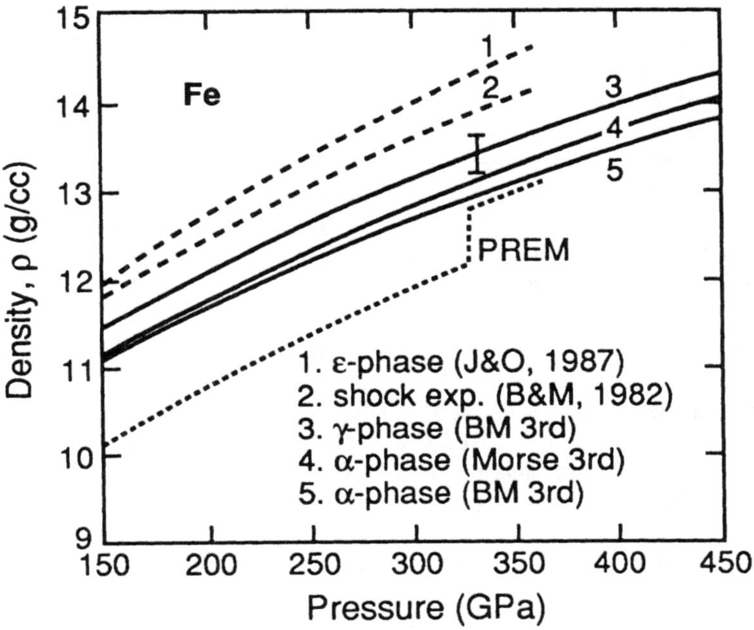

Fig. 8.5. The 300 K density trajectories of ϵ (hcp) iron, γ(fcc) iron, and α (bcc) iron arise from the Birch-Murnaghan and Morse third order EoS. They are compared to PREM data of the earth's core. The same trajectories corrected to core temperatures are presented later (in Fig. 10.14). Curves 4 and 5 demonstrate the difference between the B-M third order and the Morse EoS. The error bars on 3 demonstrate the variance as a result of changing the value of K_0' by $\pm 10\%$.

For rough approximations the Murnaghan EoS (6.18) is quite useful because it is simple. Whenever the pressure is less than about $(K_{T_0})/3$, all EoS's give the same result, so the Murnaghan EoS is good for low compression. The Born-Mie EoS has distinct advantages if the problem at hand has atomic variables, because it (8.11) is defined in terms of mass, valence, and structure of the solid. The Born-Meyer EoS (8.12) is also defined in terms of crystallographic variables. This EoS is well ingrained in the Soviet literature, where it is called the Method of Potentials, and many published tabular data are based on it. Further, as we shall see in Chapter 9, the Born-Mie and the Born-Meyer EoS each can be related directly to the shear elastic constants at high P, and thus help clarify the physics of shear velocity at high pressure.

The Birch-Murnaghan EoS (8.13) is well ingrained in the geophysical literature of the Western world, so its use is familiar to many readers. It is also quite compatible with seismological data and is easily related to the shear and longitudinal velocities of isotropic media. Furthermore, it has been shown to be compatible with the linear shock velocity–particle velocity curve of shock wave analysis (Jeanloz, 1989). However, as we showed in Chapter 6, it is not reliable except for the range in K_0' given by $3.4 < K_0' < 7$. Fortunately this covers most of the range in K_0' found in minerals of the earth's interior (an exception is orthopyroxene). It is not valid for liquids or composites where K_0' is larger than 7.

This book is not intended to treat the EoS of liquids, but they are characterized as having large values of K_0', where the Birch-Murnaghan EoS is not reliable. For these cases, the student is referred to EoS proposed by Vinet et al. (1987), which is

$$P(\eta) = K_{T_0}\eta^{-2/3}(1-\eta^{-1/3})\exp\left[\tfrac{3}{2}(K_0'-1)\left(1-\eta^{-1/3}\right)\right]. \tag{8.36}$$

An excellent discussion of the advantages of this EoS, as well as a critique of the shortcomings of the B-M third order EoS for $K_0' < 2.9$, $K_0' > 7$ is given by Hofmeister (1993). The EoS given by (8.36) is derived by differentiating the binding energy of the solid (Rose et al., 1983).

8.14. The virial theorem equation of state

The virial theorem has identical form in classical and quantum-mechanical physics (Landau and Lifshitz, 1958; Slater, 1968; McClellan, 1974). It can be used to find the pressure and bulk modulus from the interatomic potential or from quantum mechanical based calculations, as was done by Bukowinski (1976) and Bukowinski and Knopoff (1976) for fcc iron.

The virial theorem is derived from an expression for the kinetic energy, which for solids is split into two parts: one for the particles and one for the external forces on the nucleons. For the classical case, the first part is the internal potential energy $\mathcal{V}$. The second is the external force (pressure) acting on the volume (McMillan and Latter, 1958), yielding

$$KE = -\tfrac{1}{2}\mathcal{V} + \tfrac{3}{2}PV. \tag{8.37}$$

The kinetic energy is also $KE = \mathcal{U} - \mathcal{V}$, so that (8.37) is $3PV = 2\mathcal{U} - \mathcal{V}$.

McMillan and Latter (1958) argued that the potential energy was the interatomic potential ϕ, so that $3PV = 2\mathcal{U} - \mathcal{V}$ becomes

$$3PV = 2\mathcal{U} - \phi. \tag{8.38}$$

Replacing $\mathcal{U}$ with (1.2), (8.38) becomes $3PV = 2(\mathcal{F} + T\mathcal{S}) - \phi$.

Differentiating the above and using (1.9) and (1.11), we have

$$3V\left(\frac{\partial P}{\partial V}\right)_T + 5P = -\left(\frac{\partial \phi}{\partial V}\right)_T + \alpha K_T T. \tag{8.39}$$

The last term in (8.39) is the high temperature approximation for thermal pressure P_{TH} in the quasiharmonic approximation. Multiplying both sides by $(1/3)V^{2/3}$, the left side becomes a differential, and

$$\left(PV^{5/3}\right)' = -\frac{V^{2/3}}{3}\left(\frac{\partial\phi}{\partial V}\right)_T + \frac{V^{2/3}}{3}\alpha K_T T.$$

Changing from V to η and integrating the above, we have

$$P\eta^{5/3} = \frac{-1}{3V_0}\int_1^\eta \eta^{2/3}\left(\frac{\partial\phi}{\partial\eta}\right)_T d\eta + \frac{T}{3}\int_1^\eta \eta^{2/3}\alpha K_T\, d\eta. \qquad (8.40)$$

Taking αK_T as independent of η (justified in Chapter 10), P_{TH} is

$$P_{TH} = (\tfrac{1}{5})\alpha K_T T. \qquad (8.41)$$

A case where P_{TH} does depend on η, a rare exception, is gold.

Libby and Libby (1972) suggested that ϕ be given by

$$\phi = \phi_0\left(\frac{V}{V_0}\right)^{1/3} = \phi_0\eta^{1/3}, \qquad (8.42)$$

which is the ϕ_a (see (8.4)) in this chapter: in other words, ϕ is assumed to be the coulombic potential with no repulsion term. Using (8.42) in (8.40) and taking $T = 0$, upon integration the Libby and Libby EoS is

$$P = \frac{-\mathcal{U}_0}{3V_0}\left(\eta^{-5/3} - \eta^{-4/3}\right). \qquad (8.43)$$

The first derivative of the above yields $K_{T_0} = \mathcal{U}_0/9V_0$, so that

$$P = -3K_{T_0}\left(\eta^{-5/3} - \eta^{-4/3}\right). \qquad (6.39)$$

This is also the second order Bardeen EoS, (6.39), controlled by one parameter, K_{T_0}. While K_{T_0} will vary from material to material, (6.39) requires that K_0' have a constant value, 3, for all solids to which (6.39) is applied. Libby and Libby (1972) compared data of metals with (6.39) with some success, but the K_0' value of most solids (including most metals) is larger than 3.

We note that for oxides and silicates, there is a strong repulsive term in ϕ, so that for these solids, so we must replace ϕ in (8.40) with $\phi_a + \phi_v$, say (8.4) and (8.5), and then integrate. This requires a numerical computation, but the result is that K_0' takes on the same values as in the Born-Mie EoS, where $K_0' = (n+7)/3$, $6 < n < 10$ (Anderson, 1970).

8.15. Choosing an EoS for the earth's lower mantle

Some equations presented in Chapters 5–8 are useful for extrapolations into the earth's interior. Since the core–mantle boundary pressure is about

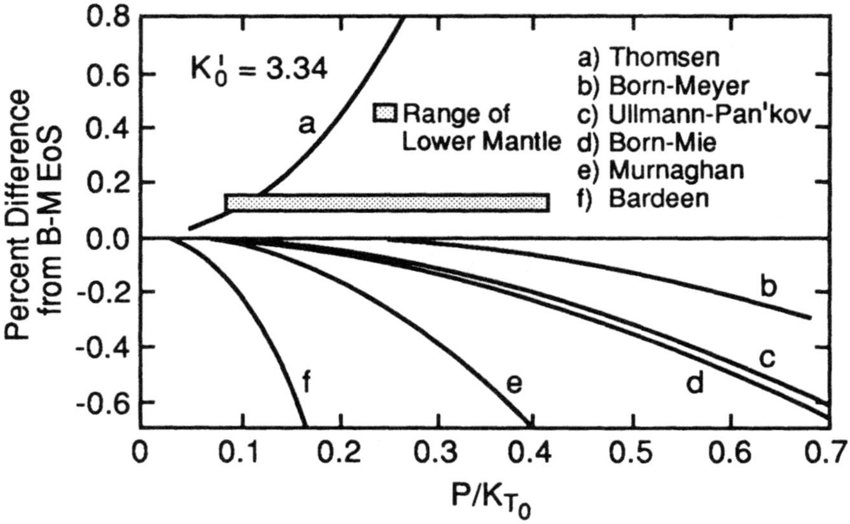

Fig. 8.6. Percent difference in density for six EoS's compared to the B-M EoS at $K_0' = 3.34$ (for the earth's lower mantle). The Murnaghan, the Born-Mie, the Born-Meyer, and the Ullman-Pan'kov EoS's are equal to ρ of the Birch-Murnaghan EoS to within one part in the third significant figure.

129 GPa, and candidate materials for the lower mantle have K_0 near 250 GPa, extrapolations involve $P/K_{T_0} \leq 0.5$. Comparison between the various isothermal EoS's at these values of P/K_{T_0} is appropriate.

The third order B-M EoS is taken as a standard and six other third order EoS's are compared with it. Percentage differences are plotted in Fig. 8.6. All but the Thomsen EoS have lower values of ρ/ρ_0 than found from the B-M EoS. The Born-Mie EoS and the U-P EoS are very close to each other at $K_{T_0}' = 3.3$, and they depart only slightly from the B-M EoS (0.2%) at the core–mantle boundary pressure. The Born-Meyer EoS, or Method of Potentials, is the closest to the Birch-Murnaghan EoS.

Within the uncertainties found between the various seismic models of the mantle reported in the literature, one cannot choose between the four isothermal EoS's—Birch-Murnaghan, Born-Mie, Born-Meyer, and Ullman-Pan'kov. However, the Murnaghan EoS and the Thomsen (Lagrangian formula) EoS have distinctly different behavior from that of the Birch-Murnaghan EoS at lower mantle conditions. The value of K_0', 3.34, in the lower mantle (see Section 2.8) puts the B-M EoS of third degree (see Section 6.6) on the edge of reliability. However, as so many EoS's have had the same behavior as the B-M third degree EoS for lower mantle parameters ρ_0, K_0, and K', the Eulerian formulation is satisfactory for the lower mantle, although marginally so.

9

SHEAR VELOCITY AT HIGH PRESSURE

9.1. Introduction

A significant phenomenon in high pressure elasticity is that the behavior of the shear elastic constants does not follow the simple rules found in the EoS, which are restricted to relationships of the bulk modulus in P, V, T space. The behavior of the EoS, such as found in Chapters 6, 7, and 8, is not very structure dependent. On the other hand, the shear constants under pressure are very dependent on structure even to the extent of being different from point group to point group in the same crystal class. Two generalizations stand out. (1) Whereas the bulk modulus is linear in pressure, at least to values of $P/K_0 \simeq 1/2$, the shear elastic constants are, by comparison, quite nonlinear. (2) The Poisson ratio increases with pressure for oxides and ceramics that are densely packed.

In seismic models of the earth, it is found that the Poisson ratio σ increases with depth. This interesting phenomenon is interpreted as resulting from the decrease of v_s/v_p as P increases. It can be shown from a number of different approaches (continuum elasticity, lattice dynamics, and atomic physics) that the shear velocity associated with the shear elastic constants increases with P at a slower rate than the longitudinal velocity increases with P. This is the primary reason why σ for an isotropic body increases with P. For single crystals, as pressure increases, the spread between the highest and lowest shear velocity increases.

For heuristic purposes the subsequent sections discuss the pressure dependence of the three elastic constants found in a cubic solid. In the last sections the discussion is extended to solids of lower symmetry.

9.2. Elastic constant relationships in cubic solids (centrosymmetry)

9.2.1. *The shear constants versus pressure in the low pressure range*

The following relationships of a cubic solid are useful. We have the bulk modulus,

$$K_S = \frac{C_{11} + 2C_{12}}{3}, \tag{9.1}$$

where all elastic constants are functions of P. If we consider isothermal conditions close to absolute zero, then $K_S = K_T = K$. For the remainder of this chapter, but only here, K_S is replaced by K (meaning conditions at low temperature). We have the Cauchy relationship,

$$C_{12} - C_{44} = 2P + \epsilon_r, \tag{9.2}$$

where ϵ_r is a number measuring the residual departure from the Cauchy condition, $C_{12} = C_{44}$. If $\epsilon_r = 0$, the Cauchy condition is maintained. Here we assume that ϵ_r is small and independent of pressure. (The pressure independence of ϵ_r is often assumed in lattice dynamic calculations). We have the definition of the second shear constant,

$$C_S = \frac{1}{2}(C_{11} - C_{12}). \tag{9.3}$$

The two shear constants can be written in terms of C_{11} and K:

$$C_S = \frac{3}{4}(C_{11} - K), \tag{9.4}$$

and

$$C_{44} = \frac{3}{2}K - \frac{1}{2}C_{11} - 2P - \epsilon_r. \tag{9.5}$$

From (9.4) we note that $C_{11} > K$ for C_S to be non zero and non-vanishing. From (9.6) we note that $C_{11} < 3K + 2\epsilon_r - 2P$ for C_{44} to be non zero. These are strict outer limits on K, because the lattice will most certainly become unstable if either C_S or C_{44} vanishes. The lattice may in fact become unstable if either of these constants becomes small but non zero (Demarest, 1974). We find that C_{11} is contained in the limits

$$K < C_{11} < 3K - 2\epsilon_r - 4P. \tag{9.6}$$

Because $2\epsilon_r$ is small compared to $3K$ and $2P$ when P is large, it will be ignored.

C_{11} can vary between the lower limit of K and the upper limit of $3K - 4P$ without incurring mechanical instability. By expanding the elastic constants, (9.4) and (9.6), out as polynomials in pressure, it is easily shown that the pressure derivatives of C_S and C_{44} are sensitive to both the pressure derivatives, K_0' and $(C_{11})_0'$. We note that if we take $K_0'' = 0$; that is, if we assume that the bulk modulus is linear with pressure, the other elastic constants may still be quadratic with pressure because C_{11} may be quadratic with pressure.

Thus we see that although it is well known that up to large pressure $(P \simeq (\frac{1}{2}) K_0)$ the bulk modulus is linear with pressure, either of the shear constants can be nonlinear with pressure, even small pressure. Whether the shear constants are linear depends very much on the behavior of C_{11} as a function of pressure. It can be further demonstrated that the magnitude of the pressure dependence of the shear elastic constants depends very much on the magnitude of the pressure dependence of C_{11}.

If we set $(C_{11})'_0$ close to a minimum allowable value, K'_0, it is easily shown that dC_S/dP has a large positive value, whereas dC_{44}/dP is a small number. On the contrary, if $(C_{11})'_0$ is close to the maximum allowable value, $3K'_0 - 2$, then $(C_{44})'_0$ has a large positive value, $(3/2)\ (K'_0 - 1)$, whereas C'_S has a very small pressure derivative.

9.2.2. *Noncentrosymmetric lattices: lattice dynamic equations*

The above holds for centrosymmetric cubic lattices. For cubic lattices in which the atoms are not all on centrosymmetric sites, as for example, in diamond or zinc sulfide, the expressions for the elastic constants have a correction term. This term arises because a net dipole is set up, in the absence of centrosymmetric sites, by the fields of the atoms in certain distortions. If this induced dipole is coupled with the stress, an additional term is subtracted from the expressions for one or more of the elastic constants.

It turns out from calculations made using lattice dynamical techniques that the noncentrosymmetric cubic lattices correspond to the case where $(C_{11})'_0$ is at a minimum value. Thus for ZnS it is shown by Born (1923) that C_{44} is replaced by

$$\mathcal{C}_{44} = C_{44} - \frac{\mathcal{C}^2}{\mathcal{D}} = C_{44} - F(P), \tag{9.7}$$

where $\mathcal{C}$ and $\mathcal{D}$ are pressure dependent operators, and $F(P)$ is positive. In all cases, the effect of this additional term is to decrease the shear elastic constant below the value for a centrosymmetric lattice. In fact, for four-fold coordinated structures, dC_{44}/dP is often negative. Examples where dC_{44}/dP is negative are diamond, ZnO, ZnS, and TiO_2.

The pressure dependence of C_{11} and of the shear elastic constants is greatly dependent on the crystallographic structure and varies from point group to point group, even in the same crystal class. The elastic constants are functions of the derivatives of the potential, and the operators on the potential producing the elastic constants are often expressed in terms of spherical coordinates.

The operators we discuss below are given by (Born, 1923)

$$\mathcal{P} = \left(\frac{1}{r}\right)\left(\frac{\partial\phi}{\partial r}\right); \tag{9.8}$$

$$\mathcal{Q} = \left(\frac{1}{4}\right)\left(\frac{\partial}{\partial r}\right)\left(\frac{1}{r}\right)\left(\frac{\partial\phi}{\partial r}\right), \tag{9.9}$$

where ϕ is the interatomic potential between atom pairs. The elastic constants of a cubic crystal are found by the operators $\mathcal{P}$ and $\mathcal{Q}$ by the following

(Born, 1923; Anderson and Liebermann, 1970):

$$C_{11} = \frac{1}{V}\sum_{\ell}\left[\mathcal{Q}\left(x^{\ell}\right)^{4} + \mathcal{P}\left(x^{\ell}\right)^{2}\right], \tag{9.10}$$

$$C_{44} = \frac{1}{V}\sum_{\ell}\left[\mathcal{Q}\left(x^{\ell}\right)^{2}\left(y^{\ell}\right)^{2} - \mathcal{P}\left(x^{\ell}\right)^{2}\right], \tag{9.11}$$

and

$$C_{S} = \frac{1}{V}\sum_{\ell}\left\{\mathcal{Q}\left[\left(x^{\ell}\right)^{2} - \left(x^{\ell}\right)^{2}\left(y^{\ell}\right)^{2}\right] + \mathcal{P}\left(x^{\ell}\right)^{2}\right\}, \tag{9.12}$$

where ℓ is summed over the lattice coordinates $(x,\ y)$ in the cell. For non-centrosymmetric lattices, like diamond,

$$\mathcal{C}_{44} = C_{44} - \frac{\mathcal{C}^{2}}{\mathcal{D}}, \tag{9.13}$$

$$\mathcal{C} = \frac{1}{V}\sum_{\ell} x^{\ell} y^{\ell} z^{\ell} \mathcal{Q}, \tag{9.14}$$

$$\mathcal{D} = \frac{1}{V}\sum_{\ell}\left[\mathcal{P} + \mathcal{Q}\left(x^{\ell}\right)^{2}\right], \tag{9.15}$$

where V is the volume per cell (two atoms), and ℓ is the index on lattice sites.

At high pressures, the repulsive term of the potential is far more important to the value of the elastic constants than the attractive (or coulombic) term, so for the many purposes described here, ϕ can be replaced by $\phi_v(r)$, the repulsive potential. To show how the operators are affected by lattice sums, we ignore the attractive (or coulombic) component of the potential and consider only the repulsive component of the potential. The operators are then identified by the superscript R (meaning repulsive):

$$\mathcal{P}^{R} = \left[\left(\frac{1}{r}\right)\left(\frac{\partial \phi_v}{\partial r}\right)\right]_r; \quad \mathcal{Q}^{R} = \left[\left(\frac{1}{r}\right)\left(\frac{\partial}{\partial r}\right)\left(\frac{1}{r}\right)\left(\frac{\partial \phi_v}{\partial r}\right)\right]_r. \tag{9.16}$$

The summation of ℓ in (9.11) on the repulsive component of C_{11}^{R} is not taken beyond the M nearest neighbors because it is assumed $\phi_v(r)$ varies so rapidly that it can be ignored at distances $2r$ and greater.

The question is then: How does the following vary with crystallographic structure?

$$C_{11}^{R} = \frac{1}{V}\sum_{\ell=1}^{M}\left[\mathcal{Q}^{R}\left(x^{\ell}\right)^{4} + \mathcal{P}^{R}\left(x^{\ell}\right)^{2}\right], \tag{9.17}$$

where C_{11}^R stands for the repulsive contribution to the value of C_{11}. In the NaCl structure, the $M = 6$ nearest neighbors are located at $[x^\ell/r, y^\ell/r, z^\ell/r]$ positions given by $[(1,0,0), (0,1,0), (-1,0,0), (0,-1,0), (0,0,1), (0,0,-1)]$, so that only two positions are non zero in the sum taken by (9.17). Thus the lattice sum arising from (9.17) is

$$VC_{11}^R\,(\mathrm{NaCl}) = 2\mathcal{Q}^R r^4 + 2\mathcal{P}^R r^2. \tag{9.18}$$

Further, the lattice sum taken in (9.14) yields

$$V\mathcal{C}^R\,(\mathrm{NaCl}) = 0 \tag{9.19}$$

because there are no atomic positions that have x^ℓ, y^ℓ, z^ℓ all non zero. This means the $\mathcal{C}$ term in (9.13) is zero.

In the CsCl structure, the $M = 8$ nearest neighbors are located, in terms of $r/3^{1/2}$, at $[(1,1,1), (-1,-1,-1), (-1,-1,1), (1,1,-1), (1,-1,1), (-1,1,-1), (-1,1,1), (1,-1,-1)]$. The lattice sum arising from (9.17) is therefore

$$VC_{11}^R(\mathrm{CsCl}) = \tfrac{8}{9}\left(\mathcal{Q}^R r^4 + 3\mathcal{P}^R r^2\right). \tag{9.20}$$

There are eight atoms with x^ℓ components, but the relation between x^ℓ and r involves the factor $1/3^{1/2}$. In the sum given by (9.14), there are four positive terms and four negative terms that cancel. This leads to

$$V\mathcal{C}^R(\mathrm{CsCl}) = 0. \tag{9.21}$$

Thus, as for NaCl, the $\mathcal{C}^2/\mathcal{D}$ in (9.13) is zero.

In the ZnS structure, the $M = 4$ nearest neighbors are located, in terms of $r/3^{1/2}$, at $[(1,1,1), (-1,-1,1), (1,-1,-1), (-1,1,-1)]$. The lattice sum arising from (9.17) is

$$VC_{11}^R(\mathrm{ZnS}) = \tfrac{4}{9}\left(\mathcal{Q}^R r^4 + 3\mathcal{P}^R r^2\right). \tag{9.22}$$

There are four atoms with x^ℓ components, but as in CsCl the relation between x^ℓ and r involves the factor $1/3^{1/2}$. In the sum given by (9.14), there are four positive terms so that the resulting sum is non zero. This leads to

$$V\mathcal{C}^R(\mathrm{ZnS}) = \frac{4(3)^{1/2}}{9}\mathcal{Q}^R r^3, \tag{9.23}$$

$$V\mathcal{D}^R(\mathrm{ZnS}) = 4\left(\mathcal{P}^R + \tfrac{1}{3}\mathcal{Q}^R r^2\right). \tag{9.24}$$

We have now determined the operators for three crystallographic structures, which, when used on the repulsive potential, give us formulas for the repulsive part of the elastic constants. Using (9.9) and (9.10) we can write the repulsive contributions to C_{11} and K in terms of derivatives of the repulsive potential $\phi_v(r)$, as in Table 9.1.

The pressure P has a repulsive and a coulombic component. The repulsive component is

$$P^R = -\left(\frac{M}{3V}\right) r \left(\frac{d\phi_v}{dr}\right). \tag{9.25}$$

Using (9.25) and Table 9.1, we compute the values of C_{11} and dC_{11}/dP in terms of K and P, and ignoring the contribution of the coulombic term so that $P^R = P$, we have for high pressure

$$\text{NaCl}: \quad C_{11} = 3K - 2P, \tag{9.26a}$$

$$\text{CsCl, ZnS CaF}_2: \quad C_{11} = K - \frac{4}{3}P. \tag{9.26b}$$

The pressure derivatives are

$$\text{NaCl}: \quad \left(\frac{\partial C_{11}}{\partial P}\right) = 3\left(\frac{\partial K}{\partial P}\right) - 2; \tag{9.27a}$$

$$\text{CsCl, ZnS CaF}_2: \quad \left(\frac{\partial C_{11}}{\partial P}\right) = \left(\frac{\partial K}{\partial P}\right) - \frac{4}{3}. \tag{9.27b}$$

To evaluate the above at low pressure, the full expressions for the elastic components must also include the coulombic terms that are given in Tables 9.2 and 9.3. The contribution arising from coulombic terms changes slowly with pressure (Anderson and Demarest, 1971), but the main contributions to C_{11} at high pressure are given by (9.26) and (9.27). For many high pressure physics problems, the coulombic term is ignored. A procedure similar to that leading to (9.27*a*) and (9.27*b*) will produce the repulsive contribution to C_{44} and C_{12} in terms of the derivative of the repulsive potential $\phi_v(r)$ (Anderson, 1970).

Table 9.1. The derivative of the elastic constant parameters in terms of the repulsive potential, $\phi_v(r)$ for C_{11} and K

Structure	C_{11}^R	K^R
NaCl $(M = 6)$	$\frac{M}{3V}r^2\left(\frac{d^2\phi_v}{dr^2}\right)$	$\frac{M}{9V}\left(r^2\frac{d^2\phi_v}{dr^2} - 2r\frac{d\phi_v}{dr}\right)$
CsCl $(M = 8)$	$\frac{M}{9V}r^2\left(\frac{d^2\phi_v}{dr^2} + 2r\frac{d\phi_v}{dr}\right)$	$\frac{M}{9V}r^2\left(\frac{d^2\phi_v}{dr^2} - 2r\frac{d\phi_v}{dr}\right)$
ZnS $(M = 4)$	$\frac{M}{9V}r^2\left(\frac{d^2\phi_v}{dr^2} + 2r\frac{d\phi_v}{dr}\right)$	$\frac{M}{9V}\left(r^2\frac{d^2\phi_v}{dr^2} - 2r\frac{d\phi_v}{dr}\right)$

Source: Anderson, 1970.

The differentials in Table 9.1 can be changed into variables involving P, K, K_0, and ρ_0, which are useful to show the trend of elastic constants with pressure. This is done by expressing the derivative of the C_{ij} in terms of the derivative for K. Assuming that ϕ_v consists of only interactions arising from next nearest neighbors, the elastic constants are defined in Table 9.2. To observe the qualitative effects of P on the C_{ij}, ignore the last term involving the factor $(\rho/\rho_0)^{4/3}$, which is weak compared to the first two terms. To be noted is the strong contrast between C'_{44} for the NaCl and CsCl structures.

Table 9.2. The elastic constants for NaCl and CsCl structures, where the potential is coulombic plus one central repulsive term $M\phi_v$ (no next-nearest neighbor potentials) (after Anderson, 1970)

	NaCl	CsCl
	Elastic constants	
C_{11}:	$3K - 2P - \frac{4}{5}\left(\frac{K_0}{q}\right)\left(\frac{\rho}{\rho_0}\right)^{4/3}$	$K - \frac{4}{3}P + 1.3774\left(\frac{K_0}{q}\right)\left(\frac{\rho}{\rho_0}\right)^{4/3}$
C_{12}:	$P + \frac{2}{5}\left(\frac{K_0}{q}\right)\left(\frac{\rho}{\rho_0}\right)^{4/3}$	$K + \frac{2}{3}P - 0.6887\left(\frac{K_0}{q}\right)\left(\frac{\rho}{\rho_0}\right)^{4/3}$
C_{44}:	$-P + \frac{2}{5}\left(\frac{K_0}{q}\right)\left(\frac{\rho}{\rho_0}\right)^{4/3}$	$K - \frac{4}{3}P - 0.6887\left(\frac{K_0}{q}\right)\left(\frac{\rho}{\rho_0}\right)^{4/3}$
C_S:	$\frac{3}{2}(K - P) + \frac{3}{5}\left(\frac{K_0}{q}\right)\left(\frac{\rho}{\rho_0}\right)^{4/3}$	$-P + 1.033\left(\frac{K_0}{q}\right)\left(\frac{\rho}{\rho_0}\right)^{4/3}$
	Pressure derivatives	
C'_{11}:	$3K' - 2 - \left(\frac{1.0613}{q}\right)\frac{(\rho/\rho_0)^{4/3}}{K/K_0}$	$K' - \frac{4}{3} + \left(\frac{1.8365}{q}\right)\frac{(\rho/\rho_0)^{4/3}}{K/K_0}$
C'_{12}:	$1 + \left(\frac{1.5037}{q}\right)\frac{(\rho/\rho_0)^{4/3}}{K/K_0}$	$K' - \frac{2}{3} - \left(\frac{0.9183}{q}\right)\frac{(\rho/\rho_0)^{4/3}}{K/K_0}$
C'_{44}:	$-1 + \left(\frac{1.5037}{q}\right)\frac{(\rho/\rho_0)^{4/3}}{K/K_0}$	$K' - \frac{4}{3} - \left(\frac{0.9183}{q}\right)\frac{(\rho/\rho_0)^{4/3}}{K/K_0}$
C'_S:	$\frac{3}{2}(K'-1) - \left(\frac{0.796}{q}\right)\frac{(\rho/\rho_0)^{4/3}}{K/K_0}$	$\left(\frac{1.3778}{q}\right)\frac{(\rho/\rho_0)^{4/3}}{K/K_0} - 1$

Note: $\frac{K_0}{q} = A_m \frac{Z_1 Z_2 e^2}{v_0 r_0}$. For $\phi_v = \frac{b}{r^n}$, $q = \frac{3K'_0 - 8}{9K_0}$.

Table 9.3. Elastic constants for the ZnS and CaF_2 structures (the potential is coulombic plus one central repulsive term $M\phi_v$) (Anderson, 1970)

	ZnS	CaF_2
	Elastic constants	
C_{11}:	$K - \frac{4}{3}P + 0.06546\left(\frac{K_0}{q}\right)\left(\frac{\rho}{\rho_0}\right)^{4/3}$	$K - \frac{4}{3}P + 1.5733\left(\frac{K_0}{q}\right)\left(\frac{\rho}{\rho_0}\right)^{4/3}$
C_{12}:	$K + \frac{3}{2}P - 0.03273\left(\frac{K_0}{q}\right)\left(\frac{\rho}{\rho_0}\right)^{4/3}$	$K + \frac{2}{3}P - 0.786\left(\frac{K_0}{q}\right)\left(\frac{\rho}{\rho_0}\right)^{4/3}$
$\mathcal{C}_{44}$:	$K - \frac{4}{3}P - 0.03273\left(\frac{K_0}{q}\right)\left(\frac{\rho}{\rho_0}\right)^{4/3}$	$K - \frac{4}{3}P - 0.786\left(\frac{K_0}{q}\right)\left(\frac{\rho}{\rho_0}\right)^{4/3}$
C_S:	$-P + 0.04910\left(\frac{K_0}{q}\right)\left(\frac{\rho}{\rho_0}\right)^{4/3}$	$-P + 1.17965\left(\frac{K_0}{q}\right)\left(\frac{\rho}{\rho_0}\right)^{4/3}$
C_{44}:	$\mathcal{C}_{44} - \frac{\mathcal{C}^2}{\mathcal{D}}$	$\mathcal{C}_{44} - \frac{2\mathcal{C}^2}{\mathcal{D} + 2\mathcal{D}'}$
$\mathcal{C}$:	$\frac{3^{1/2}}{r}\left(K - \frac{P}{3}\right) + \frac{0.75}{r}\left(\frac{K_0}{q}\right)\left(\frac{\rho}{\rho_0}\right)^{4/3}$	$\frac{3^{1/2}}{3r}\left(K - \frac{P}{3}\right) + \frac{1.59}{r}\left(\frac{K_0}{q}\right)\left(\frac{\rho}{\rho_0}\right)^{4/3}$
$\mathcal{D}$:	$\frac{3}{r^2}\left(K - \frac{4}{3}P\right) - \frac{1.32692}{r^2}\left(\frac{K_0}{q}\right)\left(\frac{\rho}{\rho_0}\right)^{4/3}$	$\mathcal{D} + 2\mathcal{D}' = \frac{3}{2r^2}\left(K - \frac{4}{3}P\right)$
	Pressure derivatives	
C'_{11}:	$K' - \frac{4}{3}\left(\frac{0.08728}{q}\right)\frac{(\rho/\rho_0)^{4/3}}{K/K_0}$	$K' - \frac{4}{3} + \left(\frac{0.69923}{q}\right)\frac{(\rho/\rho_0)^{4/3}}{K/K_0}$
C'_{12}:	$K' + \frac{2}{3} - \left(\frac{0.04364}{q}\right)\frac{(\rho/\rho_0)^{4/3}}{K/K_0}$	$K' + \frac{2}{3} - \left(\frac{0.34961}{q}\right)\frac{(\rho/\rho_0)^{4/3}}{K/K_0}$
$\mathcal{C}'_{44}$:	$K' - \frac{4}{3} - \left(\frac{0.04364}{q}\right)\frac{(\rho/\rho_0)^{4/3}}{K/K_0}$	$K' - \frac{4}{3} - \left(\frac{0.34961}{q}\right)\frac{(\rho/\rho_0)^{4/3}}{K/K_0}$
C'_S:	$-1 + \left(\frac{0.06546}{q}\right)\frac{(\rho/\rho_0)^{4/3}}{K/K_0}$	$-1 + \left(\frac{0.52441}{q}\right)\frac{(\rho/\rho_0)^{4/3}}{K/K_0}$
C'_{44}:	$\mathcal{C}'_{44} - \frac{d}{dP}\left(\frac{\mathcal{C}^2}{\mathcal{D}}\right)$	$\mathcal{C}'_{44} - 2\frac{d}{dP}\left(\frac{\mathcal{C}^2}{\mathcal{D} + 2\mathcal{D}'}\right)$

Table 9.4. Ratio of pressure derivatives of elastic constants to K_0'

	NaCl Structure						
Measured Pressure Derivative	LiF	NaF	NaCl	KCl	KBr	KI	MgO
$(C_{11})_0'/K_0'$	1.938	2.333	2.217	2.395	2.42	2.74	2.10
$(C_{44})_0'/K_0'$	0.269	0.039	0.074	−0.074	−0.053	−0.031	0.257
$(C_S)_0'/K_0'$	0.704	0.973	0.9315	1.04	1.06	0.93	0.829

	CsCl Structure		
Measured Pressure Derivative	CsCl	CsBr	CsI
$(C_{11})_0'/K_0'$	1.216	1.169	1.209
$(C_{44})_0'/K_0'$	0.631	0.628	0.700
$(C_S)_0'/K_0'$	0.160	0.123	0.157

Data from Anderson, 1970.

Table 9.2 shows that the pressure derivative of C_{11} is much larger in an NaCl structure than in a CsCl structure. In the former case, dC_{11}/dP should be a little smaller than $3K_0'$, and in the latter case, a little smaller than K_0'. Putting this information in (9.2) and (9.5), we find dC_{44}/dP to be very low, approaching zero, in the NaCl structures, and much larger in the CsCl structure. Conversely, we find dC_S/dP to be small for the CsCl structure and larger for the NaCl structure. Also we see from Table 9.2 that the pressure derivative of C_{44} for the NaCl structure is small, near zero, while the pressure derivative of the CsCl structure is large, near K_0'. Further, the situation is reversed for the pressure derivative of C_S. The confirmation of these general relationships by experimental results is shown in Table 9.4.

The validity of the general conclusions above is independent of the choice of the potential function, $\phi_v(r)$. To make quantitative predictions of C_{11} and C_{44} at high pressure, $\phi_v(r)$ has to be specified. But the sound velocities associated with the shear constants will have more curvature with pressure than the bulk modulus irrespective of the choice of $\phi_v(r)$.

The general expression for the sound velocity associated with C_{ij} is

$$v_{ij} = \left(\frac{C_{ij}}{\rho}\right)^{1/2}, \qquad i \neq j. \tag{9.28}$$

Consider the shear elastic constant C_{ij}, which has a low value compared to K. It either increases slowly with P or decreases with P, as shown in Table 9.4. Because ρ increases steadily with P, it is possible that the velocity associated with some shear mode can decrease with pressure or can pass through a maximum and then decrease with pressure.

The maximum elastic constant in a solid (C_{11}) increases rapidly with pressure, and the minimum elastic constant increases slowly, or decreases. Thus, the spread between the maximum and minimum sound velocities increases with pressure in solids. As a consequence, the value of Poisson's ratio increases steadily with pressure.

9.3. Pressure derivatives for the repulsion model, $v = b/r^n$

The pressure derivatives of the elastic constants (including the coulombic terms) are now easily established by using the relationship $dK/dP = (\rho/K)(dK/d\rho)$.

We saw in (8.15) that if $\phi_v(r) = b/r^n$, then $K_0' = (n+7)/3$. Because $\mathrm{q} = (n-1)/9$, we have for this potential $\mathrm{q} = (3K_0' - 8)/9$. Using this definition of q, the ratio $(C_{ij}')_0$ to K_0' is found for NaCl and CsCl and is presented in Table 9.5 (Anderson, 1970).

The equations in Table 9.5 compare qualitatively well with measurement for those solids where the Cauchy relation is closely obeyed (KCl and KBr). As the departure from $C_{12} = C_{44}$ gets worse, the agreement between the measured and predicted pressure derivatives gets worse. Nevertheless, we can deduce the reason why dC_{44}/dP is low and sometimes negative.

Table 9.5. Formulas for (dC_{ij}/dP) at $P = 0$ for two cubic lattices using the repulsive potential $v(r) = b/r^n$

Computed Pressure Derivative	NaCl Lattice	CsCl Lattice
$\dfrac{(C_{11})_0'}{K_0'}$	$3 - \dfrac{1}{K_0'}\left(2 + \dfrac{9.60}{3K_0' - 8}\right)$	$1 - \dfrac{1}{K_0'}\left(\dfrac{4}{3} - \dfrac{16.53}{3K_0' - 8}\right)$
$\dfrac{(C_{12})_0'}{K_0'}$	$\dfrac{1}{K_0'}\left(1 + \dfrac{4.8}{3K_0' - 8}\right)$	$1 + \dfrac{1}{K_0'}\left(\dfrac{2}{3} - \dfrac{8.265}{3K_0' - 8}\right)$
$\dfrac{(C_{44})_0'}{K_0'}$	$\dfrac{1}{K_0'}\left(-1 + \dfrac{4.8}{3K_0' - 8}\right)$	$1 - \dfrac{1}{K_0'}\left(\dfrac{4}{3} + \dfrac{8.265}{3K_0' - 8}\right)$
$\dfrac{(C_S)_0'}{K_0'}$	$\dfrac{3}{2} - \dfrac{1}{K_0'}\left(\dfrac{3}{2} + \dfrac{7.2}{3K_0' - 8}\right)$	$\dfrac{1}{K_0'}\left(-1 + \dfrac{12.4}{3K_0' - 8}\right)$

From Anderson, 1970.

From Table 9.5, we see that $dC_{44}/dP < 0$ for $K_0' < 4.2$. See Fig. 9.2 for theoretical predictions of this effect. It is clear from Tables 9.2, 9.3, and 9.5 that any theory describing how the elastic constants (apart from K) vary with pressure must account for the symmetry properties of the lattice in detail. The elastic constants are different for each point group. All four lattices described are cubic crystals, yet the two shear constants (C_{44} and C_S) vary with pressure in quite different ways from one cubic lattice to another.

In the NaCl lattice, C_{44} goes down with pressure (slightly), whereas C_S goes up; in the CsCl lattice, C_{44} goes up sharply with pressure, whereas C_S goes up for a small range of pressure and then down; and in the ZnS lattice, both C_{44} and C_S go down with pressure. Yet for all three lattices, K varies with P in an identical manner.

Following a procedure corresponding to that outlined in Section 9.2.2., Sammis (1970) calculated the elastic constants for NaCl and spinel $MgAl_2O_4$ using the repulsive potential known as the Born-Mayer potential, (8.20), which he wrote in this form:

$$\phi(v) = \lambda_1 \exp\left(\frac{-r_{ij}}{d_i}\right).$$

He found, as did Anderson and Liebermann (1970), who used the Born-Mie form of $\phi(v)$, (8.15), that dC_{44}/dP for NaCl is close to zero at $P = 0$, and that dC_{44}/dP becomes negative as P increases.

In addition, Sammis applied his theory to spinel, $MgAl_2O_4$. He found that although dC_{44}/dP is positive at $P = 0$, it passes through zero at about 50 GPa and becomes negative, e.g., the lattice becomes unstable, at higher pressure (see Fig. 9.1). One sees this same qualitative behavior in C_{44} for NaCl and C_S for CsCl and ZnS (see Fig. 3 of Anderson and Liebermann, 1970).

9.3.1. *The shear elastic constants for two phases of NaCl*

A theoretical study of NaCl in its two structures, B1 (NaCl) and B2 (CsCl), by Hemley and Gordon (1985) illustrates the large effect structure has on the shear constants. Hemley and Gordon calculated elastic constants from the binding energy, which included the long-range repulsive interaction, a short-range coulombic interaction using an electron gas density functional, and finally a term for the self energies of the component ions (the Hartrel-Fock calculation of self energies). The variation of the elastic constants for NaCl is shown in Fig. 9.2, and that for the CsCl structure of NaCl in Fig. 9.3. Here we see the reversed roles of C_S and C_{44} between the two phases, in agreement with the results of Tables 9.2 and 9.4.

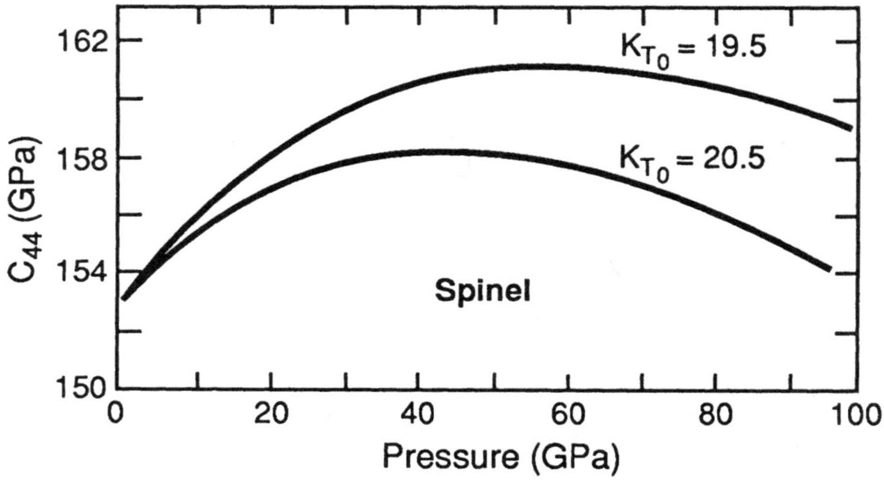

Fig. 9.1. Theoretical pressure dependence of the shear constant C_{44} for the spinel structure under the assumption of two different values for the zero-pressure bulk modulus (modified from Sammis, 1970).

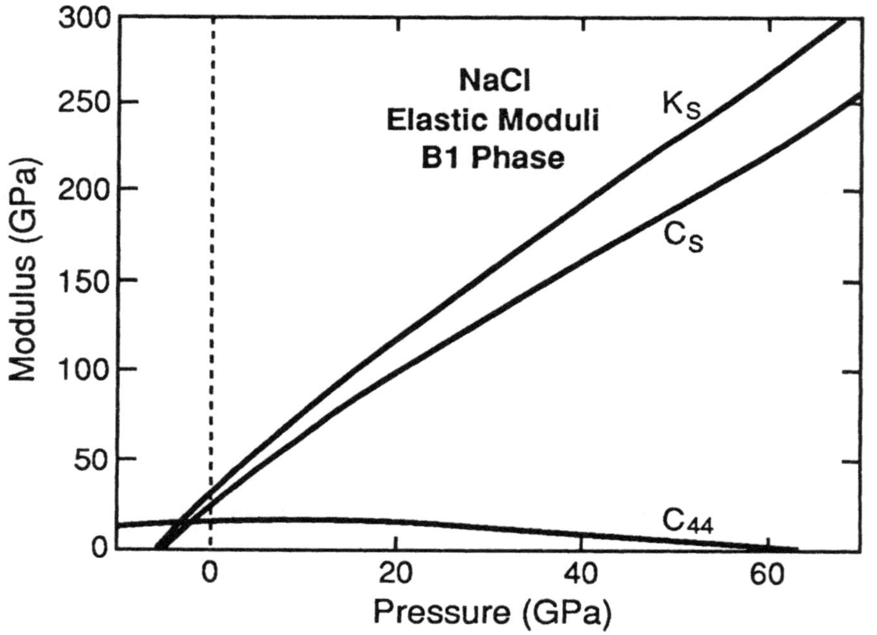

Fig. 9.2. Calculated pressure dependence of the elastic moduli for solid NaCl in the B1 (or NaCl) structure (after Hemley and Gordon, 1985).

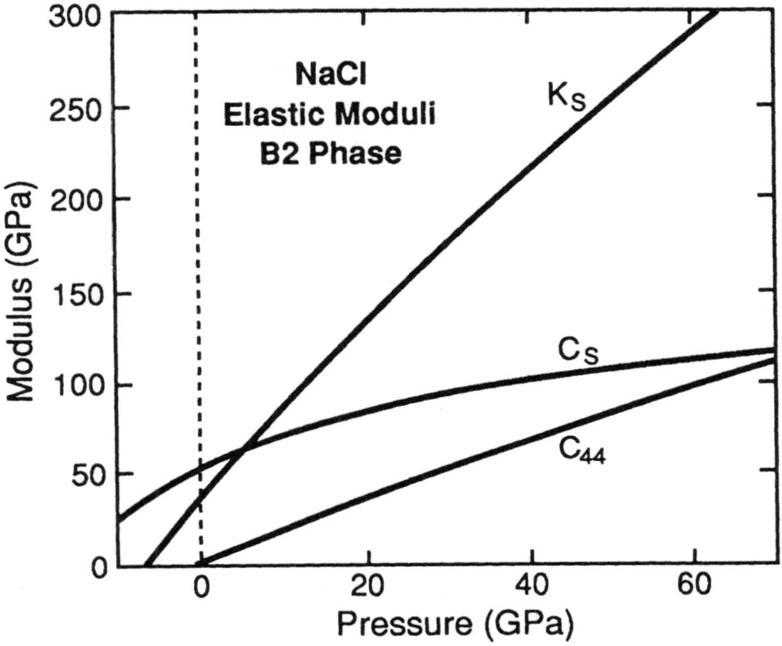

Fig. 9.3. Calculated pressure dependence of the elastic moduli for solid NaCl in the B2 (or CsCl) structure (after Hemley and Gordon, 1985).

9.4. Averaging to get isotropic moduli and velocity

9.4.1. *The Voigt-Reuss-Hill averaging scheme*

Figure 9.2 and Table 9.2 indicate that the two shear elastic constants of the NaCl structure diverge with pressure, whereas the shear constants of the CsCl structure (B_2) converge with pressure. What is needed is the isotropic equivalent shear constant—some sort of average of C_S and C_{44}. Two averaging schemes, the Hill (1952) method and the Hashin-Shtrikman (1962) method, are commonly used to find the isotropic equivalent of C_{ij}. The average values between the two methods are very close even though their upper and lower limits are different. Here we use the Hill method, the average of the upper bound (superscript V) and the lower bound (superscript R), because it is less complicated. The average isotropic shear modulus G is found from G^V and G^R, where

$$G^V = \frac{1}{5}(2C_S + 3C_{44}) \tag{9.29}$$

$$G^R = \left[\frac{1}{5}\left(\frac{2}{C_S} + \frac{3}{C_{44}}\right)\right]^{-1} \tag{9.30}$$

$$\left(\frac{dG}{dP}\right)^V = \frac{1}{5}\left(2\frac{dC_S}{dP} + 3\frac{dC_{44}}{dP}\right) \tag{9.31}$$

$$\left(\frac{dG}{dP}\right)^R = \frac{\left(G^2\right)^2}{5}\left[\frac{2}{C_S^2}\left(\frac{dC_S}{dP}\right) + \frac{3}{C_{44}^2}\left(\frac{dC_{44}}{dP}\right)\right]. \tag{9.32}$$

Barsch (1968) gives different equations for the R limit, defining (9.30) in terms of the thermodynamic rather than the effective elastic constants, but I shall use the effective elastic constants in this chapter. The value of G used is the arithmetic mean called the Hill average (Hill, 1952),

$$G^H = \frac{1}{2}\left(G^V + G^R\right). \tag{9.33}$$

Kumazawa (1969) has argued for the use of the geometric mean,

$$\bar{G} = \left(G^V G^R\right)^{1/2}. \tag{9.34}$$

According to Knopoff and Shapiro (1969), it is dangerous to assume mean values for the pressure derivatives. If the average value of G is allowed to oscillate between the two bounds as pressure is increased, the calculated value of dG/dP can exceed the bounds given by (9.33) and (9.34). In general, the average value of G will not behave erratically, and the true value of dG/dP will be bounded by the derivatives of (9.33) and (9.34).

9.4.2. *The shear velocities*

Of special interest are the three isotropic velocities given by

$$v_p = \left(\frac{K + \frac{4}{3}G}{\rho}\right)^{1/2} \tag{9.35}$$

$$v_s = \left(\frac{G}{\rho}\right)^{1/2} \tag{9.36}$$

$$v_b = \left(\frac{K}{\rho}\right)^{1/2}. \tag{9.37}$$

It is useful to use the dimensionless forms of the above

$$\frac{v_p}{(v_b)_0} = \left[\frac{K/K_0 + (4/3)(G/K_0)}{\rho/\rho_0}\right]^{1/2} \tag{9.38}$$

$$\frac{v_s}{(v_b)_0} = \left[\frac{G/K_0}{\rho/\rho_0}\right]^{1/2} \tag{9.39}$$

$$\frac{v_b}{(v_b)_0} = \left[\frac{K/K_0}{\rho/\rho_0}\right]^{1/2}. \tag{9.40}$$

At $P = 0$ the dimensionless pressure derivatives of the isotropic sound velocities are

$$\left[\left(\frac{1}{v_p}\right)\frac{dv_p}{d(P/K_0)}\right]_0 = \frac{1}{2}\left[\left(\frac{3K_0}{1+4G_0}\right)\left(K_0' + \frac{4}{3}G_0' + \frac{4/3G_0}{K_0} - 1\right)\right] \quad (9.41)$$

$$\left[\left(\frac{1}{v_s}\right)\frac{dv_s}{d(P/K_0)}\right]_0 = \frac{1}{2}\left[\left(\frac{K_0}{G_0}\right)G_0' - 1\right] \quad (9.42)$$

$$\left[\left(\frac{1}{v_b}\right)\frac{dv_b}{d(P/K_0)}\right]_0 = \frac{1}{2}(K_0' - 1). \quad (9.43)$$

9.4.3. *Poisson's ratio in the Voigt limit for centrosymmetric lattices*

The value of Poisson's ratio is given by

$$\nu = \frac{3 - (2G/K)}{2(3 + (G/K))}. \quad (9.44)$$

The pressure derivative of ν, in dimensionless form, is

$$\frac{d\nu}{d(P/K_0)} = \left(\frac{9}{2}\right)\frac{K'(G/K_0) - G'(K/K_0)}{\left[3(K/K_0) + (G/K_0)\right]^2}. \quad (9.45)$$

The Cauchy relation for centrosymmetric lattices tells us that $C_{12} - C_{44} = 2P$. From this and (9.1) and (9.3) it follows directly that

$$G^V = \frac{3}{5}(K - 2P). \quad (9.46)$$

By using (9.46) in (9.44), the Voigt-Poisson ratio for a Cauchy solid is

$$\nu^V = \frac{3K + 4P}{12K - 4P}. \quad (9.47)$$

At $P = 0$, $\nu_0^V = 0.25$ for all values of K_0' for any centrosymmetric lattice. That is, the value of ν^V is not affected by the choice of point group. Note that, as $K = K_0 + K_0'P$, and K_0' is in general 3.9–7, then ν^V must increase with pressure.

Similarly we find

$$\frac{d\nu^V}{d(P/K_0)} = \left(\frac{15}{4}\right)\frac{1}{\left[3 - (P/K)\right]^2}. \quad (9.48)$$

At $P = 0$, we have

$$\left[\frac{d\nu^V}{d(P/K_0)}\right]_0 = \frac{5}{12}.$$

Thus, both Poisson's ratio and its pressure derivatives have one value at ambient conditions for all values of K_0' in the Voigt approximation for the NaCl and CsCl lattices, a severe limit on applications of this simple theory.

The Voigt limit for the ZnS and CaF_2 lattices is not so restricted because C_{44} has the factor $\mathcal{C}^2/\mathcal{D}$ subtracted from the central force calculations. Thus G/K is less for the ZnS lattice than for the CsCl lattice; consequently, using (9.47), ν^V for the ZnS central force model is greater than (or equal to) 0.25.

The value of ν^R will be greater than or equal to ν^V. This results from the fact that in the averaging scheme, $G^V \geq G^R$, as proven by Hill (1952). Thus we conclude that the central force model has the following properties for the four cubic lattices: the ambient value of Poisson's ratio is greater than or equal to 0.25, the exact value depending on the hardness of the pair potential (reflected by K_0') and the appropriate averaged result between the Voigt and Reuss limits.

It is known that the value of ν_0 for some oxides is less than 0.25; for example, in MgO it is 0.17. Noncentral components of the potential very likely account for these low values. It has been shown (Anderson, 1970) from data on alkali halides that the noncentral correction affects the value of G much more than the value of K, and that the fraction G/K increases as noncentral effects become more important. This, in turn, means that the computed value of ν decreases as noncentral effects become more important.

9.5. dv_s/dP can be negative

The dimensionless pressure derivative of the shear velocity at zero pressure is found by using the appropriate computed value of G_0/K_0. There are four structures and the Reuss and Voigt limits to solve. The most significant point emerging is that there are different classes of solutions for the different cubic lattices. For example, all possible solutions in the ZnS lattice require that $(dv_s/dP)_0$ be negative. All possible solutions in the CsCl lattice require that $(dv_s/dP)_0$ be positive. In contrast to these two lattices, for NaCl and CaF_2 some solutions allow a positive value of $(dv_s/dP)_0$ and some allow a negative value. Within the assumption of central forces, the exact value of $[dv_s/d(P/K_0)]_0$ depends on K_0' and the average chosen between the limiting solutions.

These results are of significance to geophysical theories involving the properties of matter at high pressure. They show that there is no single solution for the elastic behavior of isotropic shear elasticity at finite strain. Crystalline symmetry of the crystallites comprising the isotropic material substantially affects the variation of shear velocity with pressure.

It is clear that to predict exactly the experimental values of BeO, CaO, and MgO, an exact theory from fundamentals that can account for the Cauchy condition is required. Nevertheless we see that a great many oxides and silicates should have negative values (or low positive values near zero) for dv_s/dP, especially for isotropic aggregates where the coordination of the atoms is low. For the earth's deep interior, where the mineral structures are closely packed and their coordination numbers are high, we therefore suggest that $[dv_s/d(P/K_0)]_0$ is normal, such as the behavior suggested for CsCl. Negative pressure derivatives of the shear velocity could be found in crustal rocks, but not in deep mantle rocks.

9.6. Finite strain

Birch's (1938, 1952) early equations from second order finite strain theory do not allow $[d\,\ell n\, v_s/d(P/K_0)]_0$ to be negative or even take on a low positive value such as has been measured for spinel (Sammis, 1970) and garnet. Birch's traditional equation is:

$$\left[\frac{d\,\ell n\, v_s}{d(P/K_0)}\right]_0 = \frac{3+4(G/K_0)_0}{6(G/K_0)_0} = \frac{(4-3\nu)}{3(1-2\nu)}. \tag{9.49}$$

Note that as $\nu < 0.5$ according to (9.49), the pressure derivative of the shear velocity must always be positive.

Sammis et al. (1970) corrected (9.49) by including some terms required by the Eulerian expansion to second order in strain. Their solution adds another term to (9.49) with two new independent parameters,

$$\left[\frac{d\,\ell n\, v_s}{d(P/K_0)}\right]_0 = \frac{3+4(G/K_0)_0}{6(G/K_0)_0} + \frac{3m_1+m_2}{12G_0}. \tag{9.50}$$

The second term added to (9.50) involves the third order elastic constants independent of K_0 and $(G/K_0)_0$. Equation (9.50) is therefore consistent with the implications discussed above: because m_1 and m_2 are independent of K and G and can be negative, adjustments can be made in their magnitude so that a negative derivative for (9.50) is allowed.

This means that a low (or negative) pressure derivative of v_s may arise because of the influence of third order elastic constants.

9.7. Shear velocity versus pressure

In Figs. 9.4 and 9.5, Equations (9.38), (9.39), and (9.40) versus pressure are plotted for three values of K_0', where the Born-Mie repulsion is used for $Q(v)$. The Hill arithmetic mean (9.33) of $(G^H/\rho)^{1/2}$ is used in computing

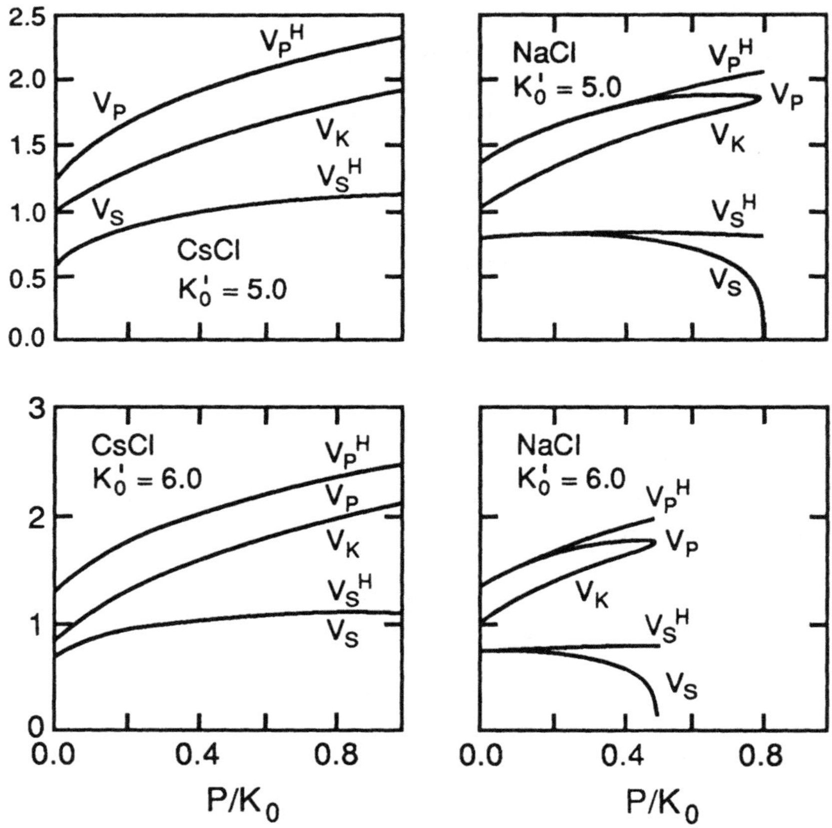

Fig. 9.4. Compression, bulk, and shear velocities versus pressure for the CsCl and NaCl lattices ($K_0' = 5.0$ and 6.0). The graphs terminate at a lattice instability due to the vanishing of a shear constant (after Anderson and Demarest, 1971).

the sound velocities, v_s^H; v_s without a superscript indicates that the geometric mean (9.34) was used. The sound velocities are terminated at the pressure at which either C_S or C_{44} vanishes.

Generally the two methods used here to compute the velocity of an aggregate predict similar results except in the neighborhood of a lattice instability. When either C_{44} or C_S vanishes at a high P, according to our equations, a lattice instability is predicted to occur. At such an instability, the geometric mean predicts that the shear velocity will plunge to zero, where the P-wave velocity will approach the bulk sound velocity, v_ϕ. On the other hand, the Hill arithmetic mean predicts a leveling off of the shear velocity and a slight decrease in dv_p^H/dP at the point of lattice instability. Thus for detection of an instability of a lattice, the geometric mean shear velocity is preferred.

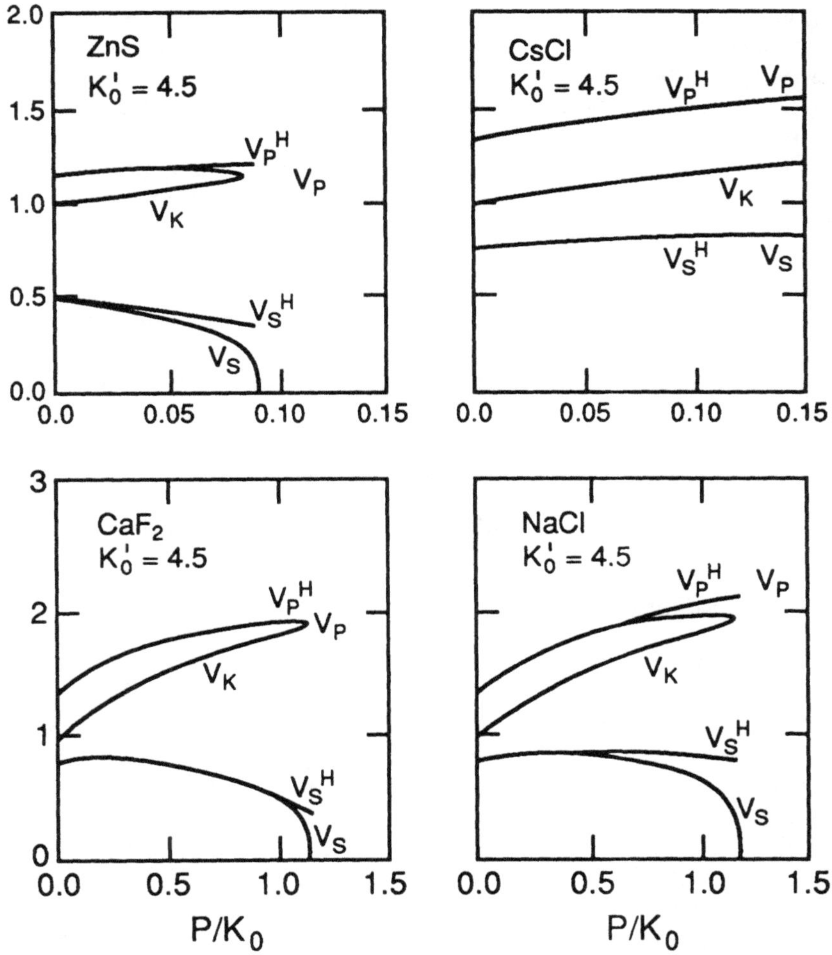

Fig. 9.5. Compression, bulk, and shear velocities versus pressure for the ZnS, CsCl, CaF_2, and NaCl lattices for $K_0' = 4.5$. The graphs are terminated at a lattice instability due to the vanishing of a shear constant (after Anderson and Demarest, 1971).

Although v_b depends only on K_0', v_p and particularly v_s are quite sensitive to the crystallographic structure. A comparison is made between the velocities of four cubic lattices for equal values of K_0' in Fig. 9.5. In the NaCl structure, v_s initially increases with P, then flattens out, but soon decreases with P. In the CsCl structure, the leveling off of v_s occurs at a much higher pressure. For the CaF_2 structure, the behavior of v_s is similar to that of NaCl, and a negative slope in v_s is predicted. For the ZnS structure, v_s strongly decreases with pressure.

From these figures it is easily seen that the pressure variation of v_s is quite unlike the pressure variation of v_b. v_b is always a slowly monotonically increasing function of P. It is also apparent that the curves of v_s vs. P for NaCl and CsCl can be quite nonlinear. This nonlinearity has significant implications for geophysical theories of the earth's mantle. There are geophysical applications of the result of the C_{44} behavior of spinel (Fig. 9.1).

Sammis (1970) pointed out that, as the upper mantle region between 400 km and 600 km is assumed to be mainly olivine in the spinel structure, dv_s/dz may approach a low and possibly negative value in this region. This is the seismic region of the "low velocity zone." However, more recent measurements by Gwanmesia et al. (1990) yield normal values of dG/dP for β and α-Mg_2SiO_4. In all these calculations the behavior of v_b is normal for all structures, whereas that of v_s has anomalies for solids with low coordination numbers. We note that the v_s versus P behavior of a low pressure phase may not be duplicated in the high pressure phase (Figs. 9.2 and 9.3).

9.8. Poisson's ratio in closely packed cubic metals at high pressure

9.8.1. *Introduction*

The previous discussion is appropriate to ionic structures and could possibly apply to silicate structures comprising the earth's mantle and crust. The core of the earth is metallic, however, and ionic potentials may not be appropriate. Falzone and Stacey (1980) analyzed the compression of a generalized central force model of atomic interactions in the approximation that ignores the intrinsic rigidity of bond angles (i.e., a closely packed metal). Their resulting equations account for the pressure dependences of the elastic constants in an hcp solid. It was presumed that this model would be ideal for consideration of the elasticity of the earth's core.

Their approach was to find the energy appropriate to the deformations leading to C_{44}, C_S, and K in terms of the shortenings (Δr) of the interatomic distance, r.

9.8.2. *The energy of deformation*

The interatomic energy, given by Falzone and Stacey (1980), is

$$\begin{aligned}\phi(r_i) =&\phi(r_0) + \phi'(r_0)(r_i - r_0) + \frac{1}{2}\phi''(r_0)\big[(r_i - r_0)^2 \\ &+ \alpha_2(r_i - r_0)\sum_{j=1}^{4}(r_j - r_0) + \beta_2(r_i - r_0)(r_k - r_0)\big], \qquad (9.51)\end{aligned}$$

where α_2 and β_2 represent bond interactions depending on different angles of bonds. Each bond (i) is considered to have 4 neighboring (j) bonds at 60° and one (k) bond at 90°. Equation (9.51) represents the unit cell.

Table 9.6. Values of changes in the bond length in a face-centered cubic crystal deformed for shear strain ϵ_S or volume strain $\Delta V/V_0$; the unstrained bond length is r_0.

Bond Orientations	[100] Shear	[110] Shear
1,1,0; -1,-1,0	$r_0\left(\frac{\epsilon_s}{2}+\frac{\epsilon_s^2}{8}\right)$	$r_0\frac{\epsilon_s^2}{4}$
1,0,1; -1,0,-1 1,0,-1; -1,0,1	$r_0\frac{\epsilon_s^2}{8}$	$r_0\left(-\frac{\epsilon_s}{4}+\frac{3}{32}\epsilon_s^2\right)$
0,1,1; 0,-1,-1; 0,1,-1; 0,-1,1	$r_0\frac{\epsilon_s^2}{8}$	$r_0\left(\frac{\epsilon_s}{4}+\frac{3}{32}\epsilon_s^2\right)$
1,-1,0; -1,1,0	$r_0\left(-\frac{\epsilon_s}{2}+\frac{\epsilon_s^2}{8}\right)$	$r_0\frac{\epsilon_s^2}{4}$

Volume change for all cases: $r_0\left[1/3(\Delta V/V_0) - 1/9(\Delta V/V_0)^2\right]$
Source: Falzone and Stacey, 1980.

Keeping the second order term in calculating the effect of volume compression gives the following for the volume per atom:

$$V = \frac{r^3}{2^{1/2}} = \frac{(r_0+\Delta r)^3}{2^{1/2}}. \tag{9.52}$$

The change in V is therefore

$$\begin{aligned}\Delta V = & V - V_0 = \frac{1}{2^{1/2}}\left[(r_0+\Delta r)^3 - r_0^3\right]\\ & \sim \frac{1}{2^{1/2}}\left[3r_0^2\Delta r = 3r_0^2\Delta r + 3r_0(\Delta r)^2\right],\end{aligned} \tag{9.53}$$

which gives the expression for ΔV at the bottom of Table 9.6.

The total energy of deformation is obtained by applying (9.51) to each of the bond-length changes in Table 9.6. We count six bonds per atom with 12 neighbors. The orientations of all 12 bonds are listed in Table 9.6. The strain energies per atom (volume $V_0 = r_0^3/2^{1/2}$) are, for the three types of deformation,

$$E_{100} = \frac{1}{2}G_{100}\epsilon^2\left(\frac{r_0^3}{2^{1/2}}\right) = \frac{3}{4}r_0\phi'\epsilon^2 + \frac{1}{4}(1-\beta_2)\,r_0^2\phi''\epsilon^2, \tag{9.54}$$

$$E_{110} = \frac{1}{2} G_{110} \epsilon^2 \left(\frac{r_0^3}{2^{1/2}} \right) = \frac{7}{8} r_0 \phi' \epsilon^2 + \frac{1}{8} (1 - 2\alpha_2 + \beta_2)\, r_0^2 \phi'' \epsilon^2, \quad (9.55)$$

$$E_V = -PV \left(\frac{\Delta V}{V_0} \right) + \frac{1}{2} K V_0 \left(\frac{\Delta V}{V_0} \right)^2 = 2 r_0 \phi' \left(\frac{\Delta V}{V_0} \right)$$
$$+ \left[\frac{1}{3} (1 + 4\alpha_2 + \beta_2) r_0^2 \phi'' - \frac{2}{3} r_0 \phi' \right] \left(\frac{\Delta V}{V_0} \right)^2, \quad (9.56)$$

where the pressure P is

$$P = - \left(\frac{2(2)^{1/2}}{r_0^2} \right) \phi'. \quad (9.57)$$

9.8.3. *Elastic constants versus P*

Taking the second derivative of the appropriate energy with respect to V and isolating the pressure term, (9.57), we have

$$C_{44} = \left(\frac{1 - \beta_2}{2^{1/2}} \right) \left(\frac{\phi''}{r_0} \right) - \frac{3}{4} P, \quad (9.58)$$

$$C_S = \frac{1}{2} (C_{11} - C_{12}) = \left(\frac{1 - 2\alpha_2 + \beta_2}{2^{3/2}} \right) \left(\frac{\phi''}{r_0} \right) - \frac{7}{8} P, \quad (9.59)$$

$$K = \frac{1}{3} (C_{11} + 2C_{12}) = \left(\frac{2(2)^{1/2}}{3} \right) (1 + 4\alpha_2 + \beta_2) \left(\frac{\phi''}{r_0} \right) + \frac{2}{3} P. \quad (9.60)$$

These equations satisfy the Cauchy relation if $\alpha_2 = \beta_2 = 0$.

Eliminating ϕ'' from (9.58) and (9.60), we can express each of the rigidity moduli in terms of K and P:

$$C_{44} = \frac{3}{4} \left(\frac{1 - \beta_2}{1 + 4\alpha_2 + \beta_2} \right) \left(K - \frac{2}{3} P \right) - \frac{3}{4} P \quad (9.61)$$

$$C_S = \frac{3}{8} \left(\frac{1 - 2\alpha_2 + \beta_2}{1 + 4\alpha_2 + \beta_2} \right) \left(K - \frac{2}{3} P \right) - \frac{7}{8} P. \quad (9.62)$$

Equations (9.61) and (9.62) yield a conclusion similar to that for the case of the central force ionic models discussed in the previous sections. Regardless of the dependence of K on P (disallowing very strong variations of the small parameters α_2 and β_2 with P), both of the principal shear moduli decrease with pressure relative to K. This solution is approximately the same as the central force model, which we found for CsCl. Thus Poisson's ratio increases strongly with pressure.

The plausible range of values for the parameters α_2 and β_2 are found by expressing them in terms of the crystal constants; at zero pressure, the relationships are:

$$\alpha_2 = \frac{2C_{12} - C_{11}}{4C_{44} + 3C_{11} - 2C_{12}} \tag{9.63}$$

$$\beta_2 = \frac{3C_{11} - 2C_{12} - 4C_{44}}{4C_{44} + 3C_{11} - 2C_{12}}. \tag{9.64}$$

Calculations using (9.63) and (9.64) show that α_2 and β_2 are a few percent, but are as large as 0.2 for some metals.

Calculating the Hill average of the shear modulus G taken from (9.61) and (9.62), Falzone and Stacey found the Hill average of Poisson's ratio to be

$$\nu^H = \frac{3 - 2\frac{G^H}{K}}{6 + 2\frac{G^H}{K}}. \tag{9.65}$$

They found that Poisson's ratio increased with P up to a value of 0.45 for $P/K \simeq 0.23$. Thus the value of Poisson's ratio for the inner core found seismically is completely in accord with the hypothesis that the core is composed of a polycrystalline iron-like metal with central interatomic forces. By small adjustments in α_2 and β_2, they were able to fit the seismic PREM model's values of ν versus depth reasonably well.

9.9. Experiments for C_{44} versus P and v_s versus P for NaCl: a test for the central force ionic equations

9.9.1. *Introduction*

As shown above, the shear constants C_{44} and $C_S = 1/2(C_{11} - C_{12})$ and their pressure derivative are very sensitive to details of the crystallographic structure (the point group), whereas the bulk modulus K and its pressure derivative are not. We showed above that at least one of the shear constants has a pressure derivative that is small (or negative) compared to the bulk modulus pressure derivative. This means that the pressure derivative of the shear velocity of cubic solids where the coordination is low will be very small and possibly negative. A good candidate for a negative dv_s/dP is ZnO, where the coordination of the cation is 4. The value of dv_s/dP has been reported to be negative for ZnO, although dK/dP is normal (Soga and Anderson, 1967).

In the following we compare the NaCl values C_{11}, C_S, C_{44}, v_s, and v_p computed from three atomic potentials given in Chapter 8 against measurements of these quantities at high pressure.

Three atomic potentials and their associated equations of state will be considered. The first and simplest is the Born-Mie potential, (8.9). It has a simple coulombic attraction—a summing of the $1/r$ potential over all lattice sites leading to a Mädelung constant in the numerator. The repulsion is a power law term, b/r^n, where b and n are eliminated by using the equilibrium condition yielding K_0 and K_0'.

The second example is the Demarest potential (Demarest, 1974, 1975). The repulsion term in $\phi_v(r)$ is given by an exponential form representing the nearest neighbor (nn) interaction. Specific account is made of the next nearest neighbor (nnn) interactions. The nnn term is assumed to be exponential repulsion, similar to nn repulsion. Parameters in this potential are evaluated using elastic constants, including the shear constants, and their pressure derivatives.

The third example is the Decker (1971) potential. Decker tried to account for all significant forces on the ions, including very weak bonds, and to distinguish between various kinds of ion–ion interactions (Seitz, 1940). He took into account the energy due to dipole–dipole interactions of the ionic charges, and thus introduced the dielectric constant and the index of refraction into the potential, following Seitz (1940). There are seven parameters to be evaluated.

9.9.2. *Elastic constant equations at high pressure*

(a) The Born-Mie potential

Equations for C_{11}, C_{12}, and C_{44} can be evaluated in terms of K_{S_0}, K', and P from the potentials given in Chapter 7; for example, the Born-Mie potential (8.9) and the Born-Meyer potential (8.21). The various constants determined by experiment and used to describe the Born-Mie potential for NaCl are $r_0 = 2.82$ Å; $K_0 = 23.94$ GPa; $K_0' = 5.28$; and $n = 8.8$.

(b) Demarest's next nearest neighbor model

Demarest (1974) used a potential that involved three terms

$$\phi = -\frac{\mathbf{A}Z_1Z_2e^2}{r} + \sum_{1,j}^{nn} D_1 e^{-r_{ij}/r_1} + \sum_{1,j}^{nnn} D_2 e^{-r_{ij}/r_2}. \tag{9.66}$$

The first term is a pure Mädelung sum, and all other possible attractive forces are neglected. The constants $\mathbf{r_1}$ and $\mathbf{r_2}$ are chosen from the Born-Meyer repulsive parameter of Vukcevich (1970).

The pressure and elastic constants appropriate to (9.66) are listed in Table 9.7. To apply these equations, the energies must be separated into nn and nnn terms. Let $\phi_{nnn}(r)$ describe the nnn forces. The constants are shown at the bottom of Table 9.7. (The prime refers to differentiation with respect to r). For the simple potential given above, the parameters are chosen arbitrarily from zero pressure data and the equilibrium condition. Demarest (1975) also considered an nnn repulsive potential that separated the last term in (9.66) into two terms, one for the cation–cation interaction and one for the anion–anion interaction.

Table 9.7. Elastic constant expressions for the NaCl structure from the Demarest potential (9.66) with a Born-Meyer type potential for the next-nearest neighbor corrections ($P_{TH} = 0$)

$$P = \frac{1}{r}(-B_1 - B_2) - \frac{0.58252 Z_1 Z_2 e^2}{rV},$$

$$K = \tfrac{1}{r}\left(\tfrac{1}{3}A_1 - \tfrac{2}{3}B_1 - \tfrac{2}{3}A_2 - \tfrac{4}{3}B_2\right) - \frac{0.77669 Z_1 Z_2 e^2}{rV},$$

$$C_{11} = \tfrac{1}{3}(A_1 + A_2 + B_2) - \frac{2.55604 Z_1 Z_2 e^2}{rV},$$

$$C_{12} = \tfrac{1}{r}\left(\tfrac{1}{2}A_2 - B_1 - \tfrac{5}{2}B_2\right) + \frac{0.11298 Z_1 Z_2 e^2}{rV},$$

$$C_{44} = \tfrac{1}{r}\left(B_1 + \tfrac{1}{2}A_1 + \tfrac{3}{2}B_2\right) + \frac{1.27802 Z_1 Z_2 e^2}{rV}$$

$$A_1 = \phi''_{nn}(r) \qquad B_1 = \tfrac{1}{r}\phi'_{nn}(r),$$

$$A_2 = \phi''_{nnn}(r) \qquad B_2 = \tfrac{1}{r}\phi'_{nn}(r).$$

Values of A_1, A_2, B_1, and B_2 calculated using procedures from Peckham, 1967.

(c) The Decker potential

The Decker (1971) potential is given by

$$\phi = -\frac{\mathbf{A} Z_1 Z_2 e^2}{r} - \frac{c}{r^6} - \frac{d}{r^8} + \sum_{i=1}^{nn} \mathbf{b} e^{r_1/r} + \sum_{i=1}^{nnn}\left[\mathbf{b}_1 e^{-2^{1/2} r_1/r} + \mathbf{b}_2 e^{-2^{1/2} r_1/r}\right], \qquad (9.67)$$

where the attractive terms, coulombic terms, van der Waals interaction, and quadrapole–dipole interaction are summed over the entire lattice. The Decker equation has terms for nearest neighbors (nn) and next nearest neighbors (nnn). Decker evaluated many of the constants from theory. This potential was considered an ideal for an ionic solid (Seitz, 1940).

9.9.3. *Estimating the shear velocities at high P for NaCl*

Data now exist on the variation of the shear velocity with pressure at high pressure. Some of these data were taken for polycrystalline samples, so an averaging scheme must be used to convert the extrapolated values of the elastic constants to shear velocities.

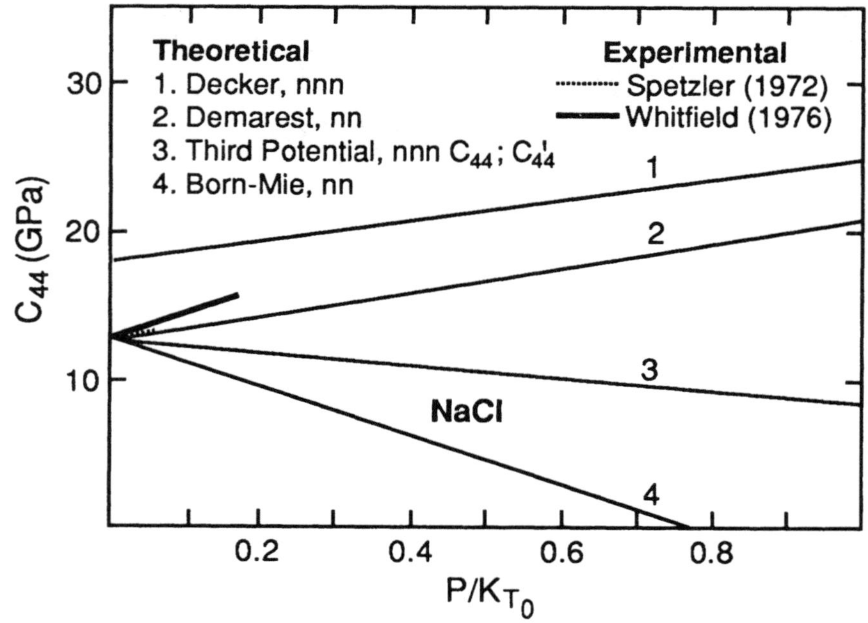

Fig. 9.6. C_{44} versus P/K_{T_0} from four potentials (8.9), (9.66), and (9.67). Comparison with experimental results of Spetzler et al. (1972) (dots) and Whitfield and Bassett (1976) for NaCl (modified from Anderson and Mammone, 1979).

Spetzler et al. (1972) measured C_{44} and C_S versus P of NaCl up to 0.3 GPa. Whitfield and Bassett (1976) measured C_S and C_{44} up to about 1 GPa. Morris et al. (1976) measured v_s of NaCl up to 23 GPa.

We have seen that simple potentials whose repulsive terms are purely nearest neighbors have a very low or zero value of dC_{44}/dP. At high pressure an instability is indicated where C_{44} becomes soft. When other attractive forces such as the *nnn* potential given by Decker and by Demarest are considered, then C_{44} may rise with pressure. This indicates the great sensitivity of C_{44} to details of the potential in comparison to the insensitivity of K_0'. A plot of C_{44} versus pressure is given for three computed potentials in Fig. 9.6. Shown with this are the experimental data of Whitfield and Bassett (1976) and Spetzler et al. (1972).

The curve illustrates the failure of the Decker equation of state to give the experimental value of C_{44} at zero pressure. The Demarest third potential has the best fit, which is not too surprising, as the *nnn* boundary conditions require it to do so. This is seen in Table 9.6; the constants B_1, A_1, and B_2 are chosen such that C_{44} is also the experimental C_{44}. Further manipulation can be made such that $(dC_{44}/dP)_0$ is also the measured dC_{44}/dP. This was done so as to conform to Spetzler's data.

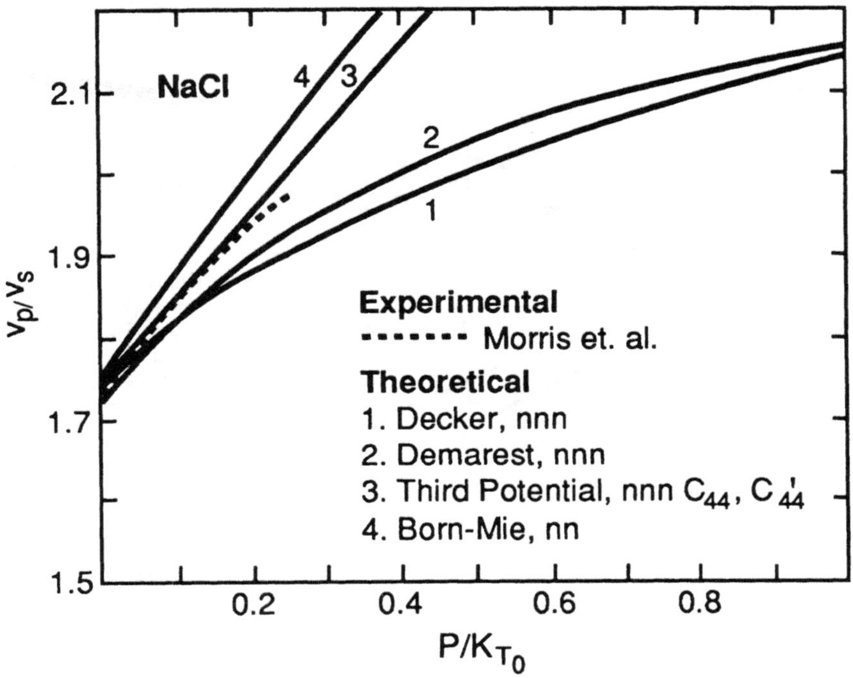

Fig. 9.7. Mean values of v_p/v_s as a function of P/K_{T_0} computed from four potentials. They are compared to the experimental results of Morris et al. (1976) for NaCl (modified from Anderson and Mammone, 1979).

The Demarest potential, based on *nnn* using the Leonard Jones potential, is fairly good. The Born-Mie, with no regard for *nnn*, does quite well for low P values of C_{44}, but large deviations occur at high pressure. The values of the ratio v_p/v_s at high pressure can be estimated for the various potentials. This is shown in Fig. 9.7 and compared against the measurements of Morris et al. (1976). At higher pressures the Demarest potential and the Demarest third potential give good agreement with the experiment, where for the latter, $(C_{44})'_0$ and $(C_{44})_0$ are boundary conditions.

In general, the predicted v_p/v_s curves give better agreement with experiment than the predicted C_{44} curves. This arises because v_p and v_s include information on C_{11}, C_S and C_{12} as well as C_{44}. There is good agreement on C_{11} for all the potentials. For C_{12} all but the Decker potential give good agreement. Therefore the average values of G are not off as much as the C_{44} values, and, of course, K is not affected at all by C_{44}.

The Decker potential, which is the most detailed potential, is not as good as the simple Born-Mie potential. The Demarest third potential stands out because it was set up so that the experimental values of C_{44} and dC/dP are used to evaluate two of the constants.

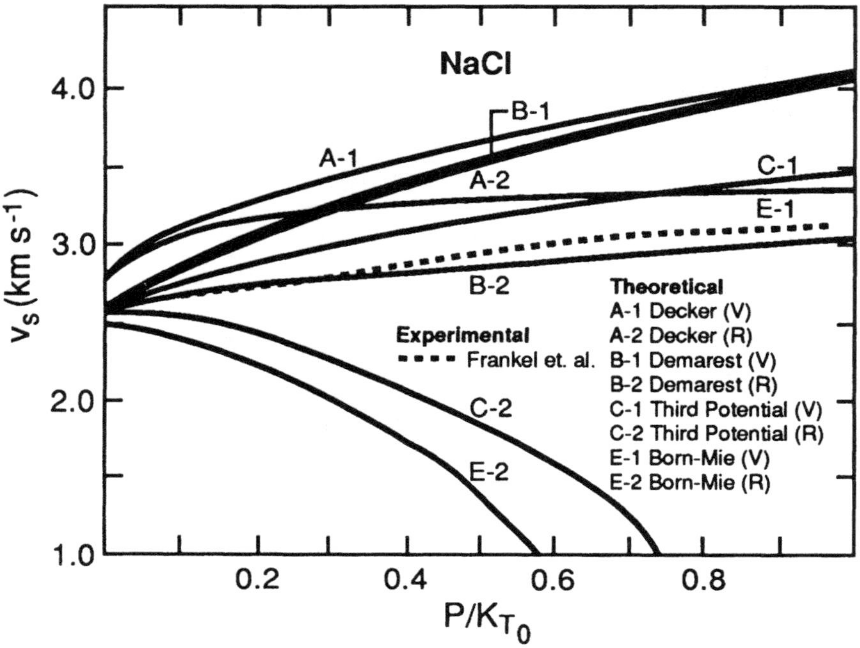

Fig. 9.8. Values of v_s as a function of P/K_{T_0} computed from four potentials. The upper and lower limits of v_s, rather than the mean values of v_s as in Fig. 9.7, are shown. Comparison is made with the experimental results of Frankel et al. (1976) for NaCl (modified from Anderson and Mammone, 1979).

Frankel et al. (1976) reported measurements of the shear velocity of polycrystalline NaCl against pressure up to $P/K_{T_0} = 0.95$. These data are compared with the predicted values of shear velocity in Fig. 9.8. We used the Reuss and Voigt calculations of G to obtain v_s. The comparison between the experimental v_s and the mean of the v_s^H and v_s^R must be made. Frankel et al.'s data fall considerably outside the limits of the Decker potential prediction, so the Decker potential fails to represent the shear velocity of NaCl. The data fall close to the lower limit of the Demarest potential and between the upper and lower Born-Mie potential.

In testing the Decker EoS against experimental data of the shear velocity at high pressure, it is found that other, simpler potentials give better agreement. The Demarest potential gives quite good agreement even considering that $(C_{44})_0$ and $(C_{44})'_0$ are used as boundary conditions. The Born-Mie equations give values of C_{44} too low and C_S too high at high P, but these values compensate to give reasonable values of v_s.

The calculations of Hemley and Gordon (1985) for NaCl give low values of C_{44} (Fig. 9.2) compared to experiment (Fig. 9.6), but nevertheless give reasonable values of C_S compared to experiment.

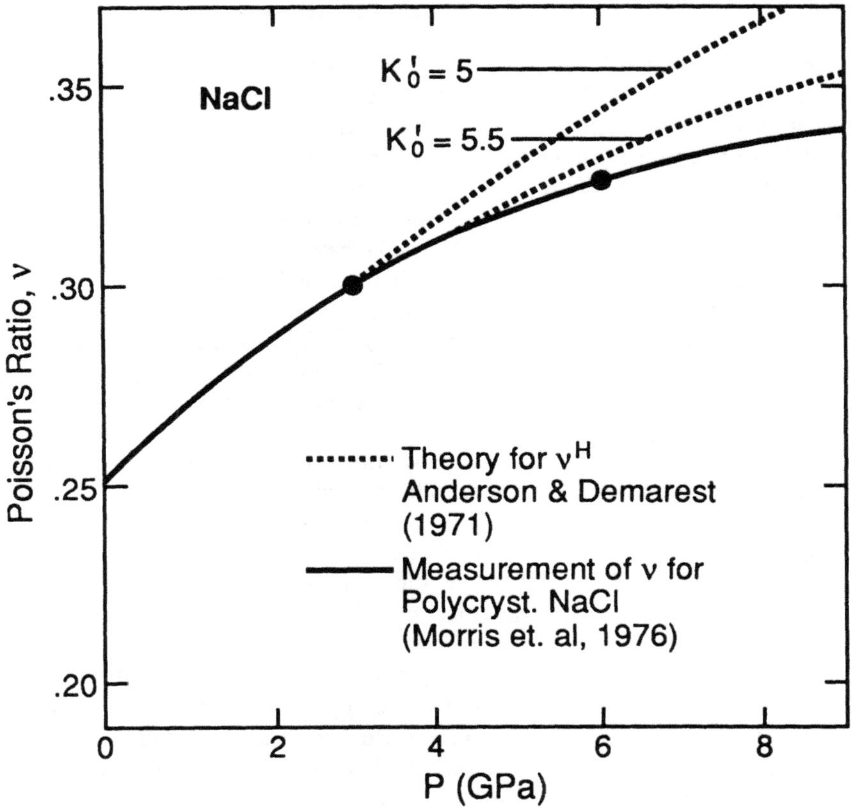

Fig. 9.9. The change of Poisson's ratio ν with P for polycrystalline NaCl up to 9 GPa. The experimental results of Morris et al. (1976) are given by the solid line. The dashed lines show how ν^H varies with P according to the theory of Anderson and Demarest (1971), where the superscript H means the Hill average.

9.9.4. *The variation of ν with P for polycrystalline NaCl*

Morris et al. (1976) measured the variation of ν with P for NaCl, as shown in Fig. 9.9. This demonstrates the typical way in which ν varies with P: a large slope, $d\nu/dP$, at low pressure, decreases as pressure increases. Their experimental data are compared with the polycrystalline ν^H for NaCl predicted by Anderson and Demarest (1971) based on use of the Born-Mie repulsion ϕ_v in the energy term. The theory is presented for two cases of K_0' (for the range of possible K', see Table 1.6). It is clear that the theoretical and experimental data are sufficiently close that a different repulsion potential to insure exact agreement could be found. From this agreement and that described in Section 9.8, it is a general result that ν increases with P along an isotherm.

9.10. Negative values of dv_s/dP for silicates and oxides

For simple cubic solids we have found that if the coordination number is low (such as 4 for ZnS), the derivative dv_s/dP is low or negative, and if it is high (such as 8 for CsCl), the derivative dv_s/dP is relatively high; NaCl (coordination number 6) is an intermediate case. This same principle holds for silicates. Values for the isotropic velocity pressure derivative (taken at $P = 0$) fall into four groups according to the value of the ratio $(dv_s/dP)/(dv_b/dP)$. MgO and Al_2O_3 have the highest value of this ratio (greater than 0.4). Minerals with intermediate values (0.2–0.4) include CaO, Mg_2SiO_4, olivine, and pyrope garnet. Those with low values (near zero) include BaO, Fe_2O_3, and TiO_2. Solids with negative values of the ratio include α-SiO_2, Ca-garnet, $CaCO_3$, beryl, $MgAl_2O_4$ (spinel), and ZnO. However, β-Mg_2SiO_4 (spinel structure) and γ-Mg_2SiO_4 have positive values of dv_s/dP (Rigden et al., 1992).

Thus for oxides with close packing and relatively high structural coordination, dv_s/dP is high. But for silicates with tetrahedral coordination and corner-linked tetrahedra, where compression can occur by the change of angles between the tetrahedra without significant change in cation–anion distance, the value of dv_s/dP is low and sometimes negative. Thus oxides and silicates classified by the magnitude of dv_s/dP fall into the same groups (C, B, A) as discussed in Chapter 5.

In deep planet interiors, where minerals will be stable in the most dense phases, the value of dv_s/dP will be typical of that found for Al_2O_3 and MgO. In the minerals of the earth's crust, on the other hand, the value of dv_s/dP can be low or even negative. This will be of importance in the seismic interpretation of surface rocks made of limestone and sandstone.

It has been noted that for many silicates the ratio G/K is a constant (solid to solid) (D.L. Anderson, 1987) at room temperature. This constancy of the ratio may not persist as pressure increases. Fig. 9.4 shows that in the CsCl structure v_s and v_b are rising with P in a proportional way, and we could expect G/K to be invariant with P. But on the other hand for the NaCl, ZnS, and CaF_2 structures, the spread between v_b and v_s increases very rapidly with P, so that G/K would decrease appreciably with pressure. For minerals in the deep planet, we expect that G/K will be invariant with P to a good approximation. But for the crust of the earth, the appropriate minerals may not follow the rule that G/K is invariant with P.

9.11. Calculating the velocity of sound near melting

If any two elastic constants are known in an isotropic solid, then all elastic constants are known. At $P = 0$, the behavior of the shear modulus and the bulk modulus versus temperature appears sufficiently well understood that extrapolations can be made to high temperatures near T_m.

Soga and Anderson (1966) found that G is linear with T up to high T for polycrystalline oxides. The computed isotropic G for single crystals at high T also appears linear with G (see, for example, the cases of NaCl and KCl ahead in Fig. 10.5; see also Fig. 15 in Anderson et al., 1992). Linearity persists at all measured temperatures above the Debye temperature (An exception is iron at seismic frequencies (Jackson, 1994)).

The adiabatic bulk modulus K_S is slightly curved with T, as shown in Fig. 2.4. However, the cross plot of K_S with the enthalpy $\mathcal{H}$ is linear. Because $\mathcal{H}$ is well known up to 3000 K, the bulk modulus can be extrapolated to near the melting temperature using the K_S–$\mathcal{H}$ plot.

Anderson (1989) found that the values of G near T_m are 54.65 GPa, 115.7 GPa, and 70.9 GPa for Mg_2SiO_4, Al_2O_3, and MgO, respectively. The values of K_S near T_m are 98 GPa, 210 GPa, and 110 GPa for Mg_2SiO_4, Al_2O_3, and MgO, respectively. This leads to values of v_s and v_p of 4.27 km/s and 7.46 km/s, 5.54 km/s and 9.80 km/s, and 4.78 km/s and 7.96 km/s for Mg_2SiO_4, Al_2O_3, and MgO, respectively. The values of T_m used are 2163 K, 2325 K, and 3098 K for Mg_2SiO_4, Al_2O_3, and MgO, respectively.

The method to extrapolate G with T recommended by D.L. Anderson (1988) is to assume that certain dimensionless parameters are independent of T; that is, $\delta_T = -(1/\alpha K_T)(\partial K_T/\partial T)_P =$ constant and $\Gamma = -(1/\alpha G)(\partial G/\partial T)_P =$ constant. Integrating $\gamma\, dT$ along the $P = 0$ isobar yields

$$\frac{G(T)}{G(T_0)} = -\exp\int_{T_0}^{T} \Gamma\alpha(T)\, dT. \tag{9.69}$$

Providing $\alpha(T)$ at high T is constrained sufficiently, $G(T)$ may be extrapolated to high T. A method of constraining $\alpha(T)$ into high temperature regions where it is not measured is to use (4.8).

In summary, the procedure I recommend for extrapolating the isotropic constants to high T at $P = 0$ is to use the linearity between K_S and $\mathcal{H}$ to find $K_S(T)$ and to use the linearity between G and T to find $G(T)$. Neither of these extrapolation methods depends on knowledge of α at high T.

9.12. The intrinsic $(\partial G/\partial T)_V$ for oxides and silicates

For oxides and silicates, $\{G\}_V = (1/\alpha K_T)(\partial G/\partial T)_V$ is about 1 to 2 less than $W = (1/\alpha K_T)(\partial K_T/\partial T)_V$, as shown by D.L. Anderson (1988) (Fig. 9.10). With a few exceptions, $\{G\}_V$ is negative. The large negative values of $\{G\}_V$ mean that the T effect on G is strong at constant V, and there is a T effect (the "intrinsic" effect) even without dilation. Expanding ΔG,

$$\Delta G = \left(\frac{\partial G}{\partial V}\right)_T \Delta V + \left(\frac{\partial G}{\partial T}\right)_V \Delta T. \tag{9.70}$$

It is ordinarily assumed that G arises from the stress deviator and is independent of volume. If ΔV has no effect on G, then the first term on the right of (9.70) vanishes, and (9.70) becomes

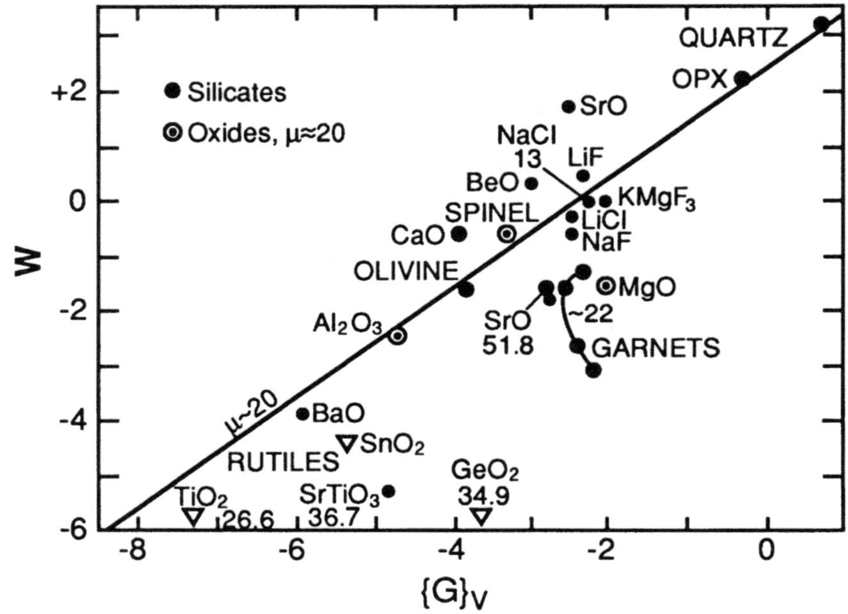

Fig. 9.10. Intrinsic temperature derivatives W and $\{G\}_V$ for oxides and silicates (after D.L. Anderson, 1988). MgO data point recalculated below.

$$\frac{\Delta G}{\Delta T} = \alpha K_T \{G\}_V. \tag{9.71}$$

I now compute $\{G\}_V$ and W of MgO, as shown in Fig. 9.10. Anderson et al. (1992) found $(\partial G/\partial T)_P = -2.48 \times 10^{-2}$ GPa/K at ambient conditions. αK_T is 5.04×10^{-3} GPa/K. Using the calculus equation (3.14), and taking $(\partial G/\partial P)_T = 3.01$, $(\partial G/\partial T)_V = -1.0 \times 10^{-2}$ GPa/K, and $\{G\}_V = -2.0$.

Alternately, $\Delta K_T/\Delta T = (\partial K_T/\partial V)_T \, \Delta V + (\partial K_T/\partial T)_V \, \Delta T$, or

$$\frac{\Delta K_T}{\Delta T} = \alpha K_T \, (K' + W). \tag{9.72}$$

The data from Anderson et al. (1992) show that at ambient conditions, $(\partial K_T/\partial T)_P = -2.7 \times 10^{-2}$ GPa/K, and $K' = 3.8$, from which we find $(\partial K_T/\partial T)_V = -0.78 \times 10^{-2}$ GPa/K and $W = -1.6$. (Using the *ab initio* calculations of Isaak et al. (1990), Anderson and Isaak (1993) found $\delta_T - K' = -W = 0.6$ at $T = 300$ K and $P = 0$ (see Fig. 3.3)). The comparison of the $\{G\}_V$ value, -2.0, with the $(K'+W)$ value, 2.2, shows that the effect of T on G is smaller than the effect of T on K_T. However, in the absence of a volume change, the effect of T is the comparison between the value of $\{G\}_V$, -2.0, and the value of W, -1.6. Thus the simple potentials used to establish the EoS in Chapter 10 are inadequate to explain the T effect on G. Unless T is built into the potentials, $(\partial G/\partial T)_V = 0$.

PART II REFERENCES

Al'tshuler, L.V., Sinakov, G.V., and Trunin, R.F., 1968, "On the composition of the earth's core," *Izv. Earth Phys.* **1**, 1–3.

Anderson, D.L., 1987, "A seismic equation of state. II. Shear properties and thermodynamics of the lower mantle," *Phys. Earth Planet. Inter.* **45**, 307–327.

Anderson, D.L., 1988, "Temperature and pressure derivatives of elastic constants with application to the mantle," *J. Geophys. Res.* **93**, 4688–4700.

Anderson, O.L., 1966, "Use of ultrasonic measurements at modest pressures to estimate high-pressure compression," *Phys. Chem. Solids* **27**, 547–565.

Anderson, O.L., 1970, "Elastic constants of the central force model for three cubic structures: Pressure derivatives and equations of state," *J. Geophys. Res.* **75**, 2719–2740.

Anderson, O.L., 1986, "Properties of iron at earth's core conditions," *Geophys. J. Roy. Astr. Soc.* **84**, 561–579.

Anderson, O.L., 1989, "The relationship between the adiabatic bulk modulus and enthalpy for mantle-related minerals," *Phys. Chem. Miner.* **16**, 559–569.

Anderson, O.L., 1992, "Francis Birch and the phase diagram of iron," (abstract), *EOS Trans. AGU* **73**, 67–68.

Anderson, O.L., 1993, "The phase diagram of iron and the temperature of the inner core," (abstract), *J. Geomag. Geoelectr.* **45**, 1235–1248.

Anderson, O.L., and Andreatch, P., 1966, "Pressure derivatives of the elastic constants of single crystal MgO, at 23°C and −195.8°C," *J. Am. Ceram. Soc.* **19**, 404–409.

Anderson, O.L., and Demarest, H., 1971, "Elastic constants of the central force model for cubic structures: Polycrystalline aggregates and instabilities," *J. Geophys. Res.* **76**, 1349–1369.

Anderson O.L., and Isaak, D., 1993, "The dependence of the Anderson-Grüneisen parameter δ_T upon compression at extreme conditions," *J. Phys. Chem. Solids* **54**, 221–227.

Anderson, O.L., and Liebermann, R.C., 1970, "Equations for the elastic constants and their pressure derivatives for three cubic lattices and some geophysical applications," *Phys. Earth Planet. Inter.* **3**, 61–85.

Anderson, O.L., and Mammone, J., 1979, "A test of the NaCl equation of state by the behavior of the shear velocities at high pressure," *High Pressure Science and Technology*, Vol. 2 (K.O. Timerhorn and M.S. Barber, Eds.). New York: Plenum Publishing Corp., pp. 181–194.

Anderson, O.L., and Sumino, Y., 1980, The thermodynamic properties of the earth's lower mantle," *Phys. Earth Planet. Inter.* **23**, 314–331.

Anderson O.L., Isaak, D.G., and Oda, H., 1992, "High-temperature elastic constant data on minerals relevant to geophysics," *Rev. Geophys.* **30**, 57–90.

Andrews, D.J., 1973, "Equation of state of the alpha and epsilon phases of iron," *J. Phys. Chem. Solids* **84**, 825–840.

Bardeen, J., 1938, "Compressibilities of the alkali metals," *J. Chem. Phys.* **6**, 372–378.

Barsch, G.R., 1968, "Relation between third-order elastic constants of single crystals and polycrystals," *J. Appl. Phys.* **39**, 3780–3793.

Barton, M.A., and Stacey, F.D., 1985, "The Grüneisen parameter at high pressure: A molecular dynamical study," *Phys. Earth Planet. Inter.* **39**, 167–177.

Barton, M.A., and Stacey, F.D., 1989, "Calculation of a high-pressure equation of state from low-pressure properties," *Phys. Earth Planet. Inter.* **56**, 311–315.

Birch, F., 1938, "The effect of pressure on the elastic properties of isotropic solids according to Murnaghan's theory of finite strain," *J. Appl. Phys.* **9**, 279–288.

Birch, F., 1952, "Elasticity and constitution of the Earth's interior," *J. Geophys. Res.* **57**, 227–286.

Birch, F., 1972, "The melting relations of iron, and temperatures in the earth's core," *Geophys. J. Roy. Astr. Soc.* **29**, 373–387.

Boehler, R., 1986, "The phase diagram of iron to 430 kbar," *Geophys. Res. Lett.* **13**, 1153–1156.

Boehler, R., 1993, "Temperatures in the Earth's core from melting-point measurements of iron at high static pressures," *Nature* **363**, 534–536.

Boehler, R., 1994, "The phase diagram of iron to 2 Mbar: New static measurements," *High-Pressure Science and Technology—1993* (S.C. Schmidt, J.W. Shaner, G.A. Samara, and M. Ross, Eds.). Woodbury, N.Y.: American Institute of Physics, pp. 919–922.

Born, M., 1923, "Atomtheorie des festes zustandes (dynamic der kristallzitter)," *Fortschritte der Mathematischen Wissenchaften*, Heft 4. Leipzig: B.G. Teubner, pp. 529–789.

Bottinga, Y., 1985, "On the isothermal compressibility of silicate liquids at high pressure," *Earth Planet. Sci. Lett.* **74**, 350–360.

Brennan, B.J., and Stacey, F.D., 1979, "A thermodynamically based equation of state for the lower mantle," *J. Geophys. Res.* **84**, 5535–5539.

Brown, J.M., and McQueen, R.G., 1980, "Melting of iron under shock conditions," *Geophys. Res. Lett.* **7**, 533–536.

Brown, J.M., and McQueen, R.G., 1982, "The equation of state for iron and the Earth's core," *High Pressure Research in Geophysics* (A. Akimoto and M. Manghnani, Eds.). Tokyo: Center for Academic Publications, pp. 611–625.

Brown, J.M., and McQueen, R.G., 1986, "Phase transitions, Grüneisen parameter, and elasticity for shocked iron between 77 GPa and 400 GPa," *J. Geophys. Res.* **91**, 7485–7494.

Bukowinski, M.S.T., 1976, "On the electronic structure of iron at core pressures," *Phys. Earth Planet. Inter.* **13**, 57–66.

Bukowinski, M.S.T., and Knopoff, L., 1976, "Electronic structure of iron and models of the Earth's core," *Geophys. Res. Lett.* **3**, 45–48.

Bukowinski, M.S.T., and Wolf, G.H., 1990, "Thermodynamically consistent decompression: Implications for lower mantle composition," *J. Geophys. Res.* **95**, 12583–12593.

Bullen, K.E., 1968, "Compression in the earth," *J. Roy. Astr. Soc.* **16**, 31–36.

Bullen, K.E., 1969, "Compressibility-pressure gradient and the constitution of the Earth's outer core," *Geophys. J. Roy. Astr. Soc.* **18**, 73–79.

Carter, W.S., Marsh, S.P., Fritz, J.N., and McQueen, R.G., 1971, "The equation of state of selected materials for high pressure references," *Accurate Characterization of the High Pressure Environment* (E.C. Lloyd, Ed.). NBS Spec. Publ. **326**, 147–158.

Cohen, R.E., 1987, "Calculation of elasticity and high pressure instabilities in corundum and stishovite with the potential induced breathing model," *Geophys. Res. Lett.* **14**, 37–40.

Davis, L.A., and Gordon, R.B., 1967, "Compression of mercury at high pressure," *J. Chem. Phys.* **46**, 2650–2660.

Decker, D.L., 1971, "High pressure equation of state for NaCl, KCl, and CsCl," *J. Appl. Phys.* **42**, 3239–3244.

Demarest, H., 1974, "Lattice model calculation of Hugoniot curves and the Grüneisen parameter at high pressure for the alkali halides," *J. Phys. Chem. Solids* **35**, 1394–1398.

Demarest, H., 1975, "Lattice model calculations of elastic and thermodynamic properties at high pressure and temperature," *Phys. Earth Planet. Inter.* **6**, 146–150.

Dziewonski, A.M., Hales, A.I., and Lapwood, E.R., 1975, "Parametrically simple Earth models consistent with geophysical data," *Phys. Earth Planet. Inter.* **10**, 12–48.

Falzone, A.J., and Stacey, F., 1980, "Second order elasticity theory: An improved formulation of the Grüneisen parameter at high pressure," *Phys. Earth Planet. Inter.* **24**, 284–290.

Fazio, D., Mulargia, F., and Boschi, E., 1979, "Equations of state of close-packed phases of iron and their implications for the Earth's core," *Phys. Earth Planet. Inter.* **18**, 13–15.

Filipov, S.I., Kazakov, N.B., and Pronin, L.A., 1966, "Velocities of ultrasonic waves, compressibility of liquid metals and their relation to various physical properties," *Izv. Vyssh. Uchebn. Zaved. Chern. Metall.* **9**, 8–14.

Frankel, F., Rich, F.J., and Homan, C.G., 1976, "Acoustic velocities in polycrystalline NaCl at 300°K measured at static pressures from 25 to 270 kbar," *J. Geophys. Res.* **81**, 6357–6363.

Fritz, J.N., Marsh, S.P., Carter, W.J., and McQueen, R.G., 1971, "The Hugoniot equation of state of sodium chloride in the sodium chloride structure," *Accurate Characterization of the High Pressure Environment* (E.C. Lloyd, Ed.). NBS Spec. Publ. **326**, 201–208.

Fürth, R., 1944, "On the equation of state for solids," *Proc. Roy. Soc. Lond.* **183**, 87–110.

Grover, R., Getting, J.C., and Kennedy, G.C., 1973, "Simple compressibility relation in solids," *Phys. Rev. B* **7**, 567–571.

Guinan, M.W., and Beshers, O.N., 1968, "Pressure derivatives of the elastic constants of α-iron to 10 kbar," *J. Phys. Chem. Solids* **29**, 541–549.

Gwanmesia, G.D., Rigden, S., Jackson, I., and Liebermann, R.C., 1990, "Pressure dependence of elastic wave velocity for β-Mg_2SiO_4 and the composition of the earth's mantle," *Science* **250**, 794–797.

Hashin, Z., and Shtrikman, S., 1962, "A variational approach to the theory of the elastic behavior of polycrystals," *J. Mech. Phys. Solids* **10**, 343–352.

Hazen, R.M., and Finger, L.W., 1979, "Bulk modulus-volume relationship for cation-anion polyhedra," *J. Geophys. Res.* **84**, 6723–6727.

Hazen, R.M., and Finger, L.W., 1982, "High-temperature and high-pressure crystal chemistry," *High Pressure Researches in Geoscience* (W. Schreyer, Ed.). New York: Academic Press, pp. 151–176.

Hemley, R.J., and Gordon, R.G., 1985, "Theoretical study of solid NaF and NaCl at high pressures and temperatures," *J. Geophys. Res.* **90**, 7803–7813.

Hill, R.W., 1952, "The elastic behavior of a crystalline aggregate," *Proc. Phys. Soc. London* **65 A**, 349–354.

Hofmeister, A.M., 1993, "Interatomic potentials calculated from equations of state: Limitations of finite strain to moderate K′," *Geophys. Res. Lett.* **20**, 635–638.

Isaak, D.G., and Masuda, K., 1993, "Can pure iron satisfy the inner core density constraint?" (abstract), *EOS Trans. AGU* **74**, 305.

Isaak, D.G., Cohen, R.E., and Mehl, M.J., 1990, "Calculated elastic and thermal properties of MgO at high pressures and temperatures," *J. Geophys. Res.* **95**, 7055–7067.

Jackson, I., 1994, "Viscoelastic relaxation in iron and the shear modulus of the inner core," *High-Pressure Science and Technology–1993* (S.C. Schmidt, J.W. Shaner, G.A. Samara, and M. Ross, Eds.). Woodbury, NY: American Institute of Physics, pp. 939–942.

Jeanloz, R., 1989, "Shock wave equation of state and finite strain theory," *J. Geophys. Res.* **94**, 5873–5886.

Jephcoat, A.P., Mao, H.K., and Bell, P.M., 1986, "The static compression of iron to 78 GPa with rare gas solids as pressure-transmitting media," *J. Geophys. Res.* **91**, 4677–4684.

Kalinin, V.A., 1960, "An equation of state for some metals at high pressure," *Problems of Theoretical Seismology and Physics of the Earth's Interior* (V. A. Magnitskii, Ed.). Jerusalem: Israel Program for Scientific Translations, National Science Foundation, pp. 56–59.

Keane, A., 1954, "An investigation of finite strain in anisotropic material subjected to hydrostatic pressure and its seismological applications," *Aust. J. Phys.* **7**, 323–333.

Knittle, E., and Jeanloz, R., 1987, "Synthesis and equation of state of (Mg,Fe)SiO_3 to over 100 GPa," *Science* **235**, 668–670.

Knittle, E., Jeanloz, R., and Smith, G.L., 1986, "Thermal expansion of silicate perovskite and stratification of the earth's mantle," *Nature* **319**, 214–215.

Knopoff, L., 1963, "The theory of finite strain and compressibility of solids," *J. Geophys. Res.* **68**, 2929–2932.

Knopoff, L., 1965, "Equations of state of solids at moderately high pressures," *High Pressure Physics and Chemistry*, Vol. 1 (R.S. Bradley, Ed.). New York: Academic Press, pp. 227–245.

Knopoff, L., and Shapiro, J.N., 1969, "Comments on the interrelationship between Grüneisen's parameter and shock and isothermal equations of state," *J. Geophys. Res.* **74**, 1439–1450.

Kumazawa, M., 1969, "The elastic constant of polycrystalline rocks and nonelastic behavior inherent to them," *J. Geophys. Res.* **74**, 5311–5320.

Landau, L.D., and Lifshitz, E.M., 1958, *Statistical Physics*. London: Pergamon Press, Ltd., pp. 236–240.

Libby, L., and Libby, W.F., 1972, "One parameter equation of state for metals and certain other solids," *Proc. Nat. Acad. Sci.* **69**, 3305–3306.

Liu, L.-G., 1975, "On the (γ, ϵ, ℓ,) triple point of iron and the earth's core," *Geophys. J. Roy. Astr. Soc.* **43**, 697–705.

Love, A.E.H., 1927, *A Treatise on the Mathematical Theory of Elasticity*, 4th ed. New York: Dora Press.

MacDonald, J.R., 1969, "Review of some experimental and analytical equations of state," *Rev. Mod. Phys.* **41**, 316–349.

Mammone, J.F., 1980, "Effects of ultra-high pressure on methanol, ethanol, calcium oxide, rutile, cassiterite, and germanium dioxide," Ph.D. thesis, University of California, Los Angeles.

Mao, N., 1970, "Empirical equation of state for high compression," *J. Geophys. Res.* **70**, 7508–7512.

Mao, H.K., and Bell, P.M., 1979, "Equations of state of MgO and ϵ Fe under static pressure conditions," *J. Geophys. Res.* **84**, 4533–4536.

McClellan, A.G., 1974, "Virial theorem generalized," *Am. J. Phys.* **42**, 239.

McMillan, W.G., and Latter, A.L., 1958, "On the compressibilities of simple metals," *J. Chem. Phys.* **29**, 15–17.

McQueen, R.G., Marsh, S.P., Taylor, J.W., Fritz, J.N., and Carter, W.J., 1970, "The equation of state of solids from shock wave studies," *High Velocity Impact Phenomena* (R. Kinslow, Ed.). New York: Academic Press, Inc., pp. 293–417.

Mehl, M.J., Hemley, R.J., and Boyer, L.L., 1986, "Potential-induced breathing model for elastic moduli and high pressure behavior of the cubic alkaline-earth oxides," *Phys. Rev. B* **33**, 8685–8696.

Monfort, C.E., and Swenson, C.A., 1965, "An experimental equation of state for potassium metal," *J. Phys. Chem. Solids* **26**, 291–301.

Morris, C.E., Jamieson, J.C., and Yarger, F.L., 1976, "Ultrasonic measurements at elevated pressures (9 GPa) to determine Poisson's ratio and other elastic constants," *J. Appl. Phys.* **47**, 3976–3986.

Morse, P.M., 1929, "Diatomic molecules according to wave mechanics. II: Vibrational levels," *Phys. Rev.* **34**, 57–64.

Murnaghan, F.D., 1937, "Finite deformation of an elastic solid," *Am. J. Math.* **59**, 235–239.

Murnaghan, F.D., 1944, "The compressibility of media under extreme pressures," *Proc. Nat. Acad. Sci. U.S.A.* **30**, 244–247.

Peckham, G., 1967, "The phonon dispersion relation for magnesium oxide," *Proc. Phys. Soc. Lond.* **90**, 657–670.

Rao, R.R., Chaplot, S.L., Choudhury, N., and Ghose, S., 1988, "A lattice dynamical calculation of phonon density of states and specific heat of forsterite, Mg_2SiO_4," *Phys. Chem. Miner.* **16**, 83–97.

Rice, M.H., 1965, "Pressure-volume relations for the alkali metals from short-wave measurements," *J. Phys. Chem. Solids* **26**, 483–492.

Rigden, S.M., Gwanmesia, G., Jackson, I., and Liebermann, R.C., 1992, "Progress in high-pressure ultrasonic interferometry, the pressure dependence of elasticity of Mg_2SiO_4 polymorphs and constraints on the composition of the transition zone of the earth's mantle," *High Pressure Research: Application to Earth and Planetary Sciences* (Y. Syono and M.H. Manghnani, Eds.). Tokyo: Terra Scientific Publishing Co., pp. 167–182.

Rose, J.H., Smith, J.R., and Ferrante, J., 1983, "Universal features of bonding in metals," *Phys. Rev. B* **28**, 1835–1845.

Ross, M., Young, D., and Grover, R., 1990, "Theory of the iron phase diagram at earth-core conditions," *J. Geophys. Res.* **95**, 21713–21717.

Rotter, C.A., and Smith, C.S., 1966, "Ultrasonic equations of state of iron: 1. low pressure, low temperature," *J. Phys. Chem. Solids* **27**, 267–276.

Sammis, C.G., 1970, "The pressure dependence of the elastic constants of cubic crystals in the NaCl and spinel structures from a lattice model," *Geophys. J. Roy. Astr. Soc.* **19**, 285–297.

Sammis, C.G., Anderson, D., and Jordan, T., 1970, "Application of finite strain theory to ultrasonic and seismological data," *J. Geophys. Res.* **75**, 4478–4480.

Saxena, S.K., Shen, G., and Lazor, P., 1993, "Experimental evidence for a new iron phase and implications for Earth's core," *Science* **260**, 1312–1313.

Seitz, F., 1940, *The Modern Theory of Solids*. New York: McGraw-Hill Book Co., p. 698.

Slater, J.C., 1939, *Introduction to Chemical Physics*, 1st edition. New York: McGraw-Hill Book Co., pp. 26, 239.

Slater, J.C., 1968, *Quantum Theory of Matter*. New York: McGraw-Hill.

Smith, C.A., and McLean, K.O., 1973, "A common equation of state?—LiF, NaF, NaCl, NaI," *J. Phys. Chem. Solids* **34**, 1143–1145.

Soga, N., and Anderson, O.L., 1966, "High temperature elastic properties of polycrystalline MgO and Al_2O_3," *J. Am. Ceram. Soc.* **49**, 355–359.

Soga, N., and Anderson, O.L., 1967, "Anomalous behavior of shear sound velocity under pressure for polycrystalline ZnO," *J. Appl. Phys.* **38**, 2985–2988.

Spetzler, H., Sammis, C.G., and O'Connell, R.J., 1972, "Equation of state of NaCl: Ultrasonic measurements to 8 kbar and 800°C and static lattice theory," *J. Phys. Chem. Solids* **33**, 1727–1750.

Stacey, F.D., Brennan, B.J., and Irvine, R.D., 1981, "Finite strain theories and comparison with seismological data," *Geophys. Surv.* **4**, 189–232 .

Stassis, C., 1994, "Inelastic neutron scattering of γ-iron, and the determination of the elastic constants by lattice dynamics," *High-Pressure Science and Technology—1993* (S.C. Schmidt, J.W. Shaner, G.A. Samara, and M. Ross, Eds.). Woodbury, N.Y.: American Institute of Physics, pp. 955–958.

Thomsen, L., 1970, "On the fourth-order anharmonic equation of state of solids," *J. Phys. Chem. Solids* **31**, 2003–2016.

Thomsen, L., 1977, "Theoretical foundation for equation of state for the terrestrial planets," *Ann. Res. Earth Planet. Sci.* **5**, 491–513.

Truesdell, C., 1966, *The Mechanical Foundation of Elasticity and Fluid Dynamics.* New York: Gordon and Breach.

Ullmann, W., and Pan'kov, V.L., 1976, "A new structure of the equation of state and its application in high-pressure physics and geophysics," Publ. No. 41, Potsdam: Central Institute of Physics of the Earth.

Vashchenko, V.Y., and Zubarev, V.N., 1963, "Concerning the Grüneisen constant," *Sov. Phys.-Solid State* (Engl. transl.) **5**, 653–655.

Vinet, P., Ferrante, J., Rose, J.H., and Smith, J.R., 1987, "Compressibility of solids," *J. Geophys. Res.* **92**, 9319–9325.

Vukcevich, M.R., 1970, "On the elastic properties of covalent crystals," *Phys. Status Solidi* **40**, 193–205.

Watt, J.P., Shankland, T.J., and Mao, N., 1975, "Uniformity of mantle composition," *Geology* **3**, 92–97.

Webb, S.L., and Jackson, I., 1990, "Polyhedral rationalization of variation among the single crystal elastic moduli for upper-mantle silicates, garnet, olivine and orthopyroxene," *Am. Miner.* **75**, 731–738.

Welch, D.O., Paskin, A., and Dienes, G.J., 1977, "A molecular dynamical study of the equation of state of solids at high temperature and pressure," *J. Phys. Chem. Solids* **39**, 489–603.

Whitfield, C.H., and Bassett, W.A., 1976, "Elastic moduli of NaCl by Brillouin scattering in a diamond anvil cell," *Rev. Sci. Instr.* **47**, 942–947.

Williams, Q., Jeanloz, R., Bass, J., Svendsen, B., and Ahrens, T.J., 1987, "The melting curve of iron to 250 Gigapascals: A constraint on the temperature at Earth's center," *Science* **236**, 181–182.

Williams, Q., Knittle, E., and Jeanloz, R., 1991, "The high-pressure melting curve of iron: A technical discussion," *J. Geophys. Res.* **96**, 2171–2184.

Yamamoto, S., Ohno, I., and Anderson, O.L., 1987, "High temperature elasticity of sodium chloride," *J. Phys. Chem. Solids* **48**, 143–151.

Young, D.A., and Grover, R., 1984, "Theory of the iron equation of state and melting curve to very high pressures," *Shock Waves in Condensed Matter–1983* (J. Asay, R.A. Graham, and G.K. Straub, Eds.). Amsterdam: Elsevier, pp. 67–87.

Zaretsky, J., and Stassis, C., 1987, "Lattice dynamics of γ-Fe," *Phys. Rev. B* **35**, 4500–4502.

Zharkov, V.N., and Kalinin, V.A., 1971, *Equations of State for Solids at High Pressures and Temperatures* (Engl. transl.). New York: Consultants Bureau, pp. 5–31.

Zharkov, V.N., Kalinin, V.A., and Pan'kov, V.L., 1972, "Equations of state and their geophysical applications," *Phys. Earth Planet. Inter.* **5**, 332–343.

Zharkov, V.N., and Trubitsyn, V.P., 1978, *Physics of Planetary Interiors* (Engl. transl. W.B. Hubbard). Tucson: Pachart Publishing House, pp. 148–152.

PART III
THERMAL PROPERTIES AT HIGH PRESSURE

The human mind has never invented a labor-saving machine equal to algebra.

—J. Willard Gibbs

At first these words may sound strange to many of us who know Gibbs's work and compare it with the ideas of algebra taught in high school. The foundation of physical chemistry created by Gibbs is based on more profound mathematics than found in most algebra primers. It should be remembered that Gibbs was appointed a professor of mathematical physics at Yale after he had spent 6 years in Paris, Heidelberg, and Berlin attending lectures from outstanding mathematics professors of the great universities in those cities. Thus he must have acquired considerable knowledge of abstract algebra. His algebra was more akin to the definition by Alfred North Whitehead (Kline, 1985), who wrote, "Algebra is the intellectual instrument which has been created for rendering clear the quantitative aspects of the world." Indeed, in the *Encyclopedic Dictionary of Mathematics*, 2nd edition, the discussion of algebra requires 83 pages in which there are listed as subclasses 10 mathematics topics, including words such as topology, rings, groups, fields, and number theory.

Thus the scope of mathematical methods found in Gibbs work is to him an aspect of algebra. The use of algebraic symbols and equations enables one to rewrite lengthy prose in a compact form in which the eye can quickly see what is being said. That is the labor-saving machine spoken of by Gibbs.

The four chapters in Part III encompass four complex subjects that are greatly simplified by the mathematics employed. Algebra, in the Gibbs context, is needed to help comprehend these subjects, and the sequence of the chapters reflects an increasing level of complexity in mathematical symbolism.

Many advanced subjects involving the EoS require knowledge of certain thermal properties at high pressure and conversely. Four such subjects, including melting, thermal pressure, shock waves, and thermodynamic function, are chapters in this part.

The chapter on thermal pressure P_{TH} shows how some of the properties in Part I can be added to the isothermal EoS of Part II to obtain a complete EoS. The underlying problems are: how does P_{TH} vary with temperature, and how does it vary with volume? It is shown that with rare exceptions, P_{TH} is independent of volume at high T. It is also shown that for $T > \theta$, P_{TH} varies linearly with T for all solids considered in the book. This is indeed a valuable approximation for the EoS at high T. Two important applications, the thermal pressure of silicate perovskite and that of iron, are discussed in detail.

The chapter on melting outlines the derivation of the Lindemann law of melting and shows some of the reasons for its longevity, in spite of its weak theoretical foundation. The Lindemann law of melting is strictly valid only for a monatomic solid that can be well represented by a single vibrational frequency. Thus the assumptions behind the Debye theory are woven into the fabric of the Lindemann law. It is no surprise that the Lindemann law has been used successfully for monatomic metals, including iron. Nevertheless, it works fairly well for a few important oxides and silicates, including that silicate especially important to geophysics, silicate perovskite. It also has been verified for MgO. These oxides and silicates have well packed structures and comprise a subset, called "Debye–like" solids, discussed in Chapter 5.

The chapter on shocked solids concerns the EoS as defined by the shock-wave Hugoniot and its relationship to the isothermal EoS. Included is the method of calculating the temperature along the Hugoniot. An application is finding the melting temperature in the phase diagram of iron.

The chapter on thermodynamic functions deals with calculations of the entropy $\mathcal{S}$, internal energy $\mathcal{U}$, and the free energy $\mathcal{F}$, based on the thermoelastic parameters discussed in Chapter 3. Provided the thermoelastic parameters δ_T and K' are known as a function of T and V, then $\mathcal{S}$, $\mathcal{U}$, and $\mathcal{F}$ can be computed. Further, if the equation of state is known, $P = f(V, T)$, then the free energy $\mathcal{G}$ and the enthalpy $\mathcal{H}$ can be determined from $\mathcal{S}$, $\mathcal{U}$, and F. The complete set of thermodynamic functions, over a wide range of temperatures and pressures is calculated and presented for MgO.

Of especial interest is the isothermal variation of $\mathcal{S}$ with V, which depends only on the integration of αK_T. Thus $\mathcal{S}(V, T)$ is defined without prior knowledge of C_V, $\mathcal{F}$, or the EoS, depending only on the measurements of α and of K_T at $P = 0$ and the volume variation of δ_T and K'. Once $\mathcal{S}(V, T)$ has been established, then the Grüneisen ratio is defined by taking a special derivative of $\mathcal{S}$ again without prior knowledge of C_V, $\mathcal{F}$ or the EoS.

10

THE THERMAL PRESSURE

10.1. Introduction

A useful analogy exists between the thermal pressure of a solid and the kinetic energy of a gas. The perfect gas law is an EoS for an ensemble of ideal gas particles,

$$PV = 3NkT, \tag{10.1}$$

where N is Avogadro's number; k is the Boltzmann constant; and 3 accounts for the three degrees of freedom of an individual particle. This equation, one of the introductory concepts of kinetic theory, is found by equating the mean kinetic energy of a particle to momentum transfer in a unit time over a unit area. The kinetic energy, $E_K = 3NkT$, is identical to the kinetic pressure. Thus (10.1) can be written

$$P_K = \frac{E_K}{V}. \tag{10.2}$$

The analog to the kinetic energy in gases is the thermal pressure in solids, where we take thermal pressure proportional to thermal energy, (1.34),

$$P_{TH} = \frac{\gamma_{mg}}{V} E_{TH}. \tag{10.3}$$

In solids the kinetic energy arises from the motion of the atoms as they vibrate back and forth, each constrained to oscillate around a particular lattice point. There are significant differences between (10.2) and (10.3). First, although $P_K = P$ for a gas, P_{TH} is only a component and often a small one of P for a solid. Second, γ_{mg}, the Mie-Grüneisen version of the Grüneisen parameter, so important for solids, is absent in the EoS of a gas. No such parameter arises because in ideal gases, $F = F_K = E_K - TS$, and $(\partial E_K/\partial V)_T = 0$.

Anharmonic terms are unimportant when the equipartition of energy assumption is appropriate. Assuming this is true, then for a solid,

$$E_{TH} = 3pRT, \tag{10.4}$$

where p is the number of atoms per molecule. Equation (10.3) becomes

$$P_{TH}V = 3p\gamma_{mg}NkT. \tag{10.5}$$

Contrast this to (10.1) for a perfect gas.

In general, γ_{mg} is a function of both T and V at constant P. However, in special cases γ_{mg} may only be a function of V; then $\gamma_{mg} \rightarrow \gamma$ (see (1.34)). One can make an assumption corresponding to the primitive assumption in (10.1) for gases—that γ_{mg} in (10.3) or (10.5) is independent of V and T. This turns out to be a very crude assumption for solids, for in that event, $P_{TH}V = (\text{const.})E_K$ at all T and V.

The next degree of sophistication in the EoS for gases, going beyond (10.1), is to account for the finite size of the molecules, giving them a volume b' in the contained volume V so that the V in (10.1) is replaced by $(V - b')$. Corrections are also made for the cohesion of the molecules when they are close. These two corrections on the free energy of a gas yield the van der Waals EoS. This improved EoS greatly extends the applications of gas thermodynamics and enables the prediction of physical phenomena, such as triple points, not possible with the perfect gas law, (10.1).

The second degree of sophistication for solids is to consider how each vibrational mode affects the thermal energy. A typical method is to take the statistical mechanical description of a solid to be E_{TH}, as was done in Chapter 2, and to assume a system of uncoupled Einstein oscillators. When defining E_{TH} to be the vibrational energy, it is assumed that each modal frequency is volume dependent but not temperature dependent. This is the quasiharmonic approximation (see Chapter 2),

$$E_{TH} = kt \sum^{3pN} \frac{\omega_j h}{kT} \left(\frac{1}{2} + \frac{1}{e^{k\omega_j/kT} - 1} \right) = \sum^{3pN} E_{TH_j}, \tag{10.6}$$

where p is the number of atoms per molecule, and E_{TH_j} is the thermal energy of the jth mode. At high temperature, (10.6) converges to

$$E_{TH} \rightarrow 3pNkT = 3pRT. \tag{10.7}$$

So the second degree of sophistication, the choice of (10.6) for E_{TH}, corresponds to imposing the Dulong and Petit limit upon the specific heat, C_V. The many, although finite, mode frequencies, ω_i, in (10.7) now are replaced by an equivalent single parameter, Θ, at high T. Then at high T just below the Dulong and Petit limit,

$$E_{TH} = \mathrm{f}\left(\frac{\Theta}{T}\right) \tag{10.8}$$

and

$$P_{TH} = \frac{1}{V}\left(\frac{\partial \, \ell\text{n}\, \Theta}{\partial \, \ell\text{n}\, V}\right)_T E_{TH} = \frac{1}{V}\gamma_D E_{TH}. \tag{10.9}$$

An important example of (10.8) is the Debye function, described in Chapters 1 and 5, but there could also be other functions of (Θ/T).

The second degree of sophistication for gases does not always work satisfactorily. Sometimes new corrections that arise from the electrostatic forces of electron and nuclear interactions have to be considered. These terms, called virial coefficients, require the $1/V$ in (10.2) to be replaced by a polynomial in $1/V$.

Similarly, the third degree of sophistication is sometimes required for solids. This is necessary when, at high T, (10.7) is not satisfactory, as evidenced by some solids where

$$\left(\frac{\partial C_V}{\partial T}\right)_P \neq 0. \tag{10.10}$$

Equation (10.10) defines an anharmonic condition. Thus the third degree of sophistication in thermal pressure of solids is to account for anharmonicity, which for high temperature is written as

$$E_{TH} = 3pRT + A_1(V)T^2 + A_0(V) + A_{-2}(V)T^{-2} \tag{10.11}$$

(see (2.54)) where the last three terms in (10.11) might correct for departure from the quasiharmonic model. To get P_{TH} in this case requires the determination of the so-called mode gammas. We have

$$P_{TH} = \frac{\sum \gamma_j E_{ij}}{V}, \tag{10.12}$$

where

$$\gamma_j = \frac{\partial \ln \omega_j}{\partial \ln V}. \tag{10.13}$$

We could determine all the mode gammas, at least in principle, but this would require the evaluation of $3pN$ terms. Although it is possible to determine a few of the mode gammas from experiment, most are unmeasurable. Fortunately for many high temperature problems, the uncertainty of this fundamental approach can be bypassed because of the high temperatures involved in geophysical and ceramics problems, and further, because geophysical problems often deal with closely packed minerals (see Chapter 5). Figure 2.1 indicates that (10.10) applies to forsterite and two garnets, whereas for MgO, Al_2O_3, and CaO, $(\partial C_V/\partial T)_P = 0$.

Just as in the general EoS for gases, where the third degree of sophistication reduced to the van der Waals EoS in a wide P, T range, so for solids the third degree of sophistication may reduce to a simplified form at high temperature over a wide temperature and pressure range. In cases where the product $\gamma(V,T)E_{TH}$ is simply proportional to T, the third degree of sophistication of solids reduces to the quasiharmonic approximation:

$$P_{TH} = \frac{\gamma(V)}{V} 3pRT, \tag{10.14}$$

as shown in (2.38). Equation (10.14) is called the quasiharmonic approximation at high temperature, because the mode frequencies in (10.6) are uncoupled, but ω is a function of V, so the spring constants are not independent of atomic spacing.

When C_V is independent of T at high T,

$$\left(\frac{\partial C_V}{\partial T}\right)_P = 0, \tag{10.15}$$

the anharmonicity is said to be nil.

10.2. Is there anharmonicity in thermal pressure?

In oxides and silicates, there is sometimes evidence of anharmonicity in specific heat as a result of a test of whether $(\partial C_V/\partial T)_P = 0$. However, any anharmonic terms placed in $\mathcal{F}$ will affect all properties that are determined by the derivatives of $\mathcal{F}$, as well as C_V. Here we consider the effect of anharmonicity on the thermal pressure, P_{TH}.

From calculus we have

$$P(V,T) - P(V,T_0) = \int_{T_0}^{T} \left(\frac{\partial P}{\partial T}\right)_V dT. \tag{10.16}$$

Substituting (1.22) in the above,

$$P(V,T) - P(V,T_0) = \int_{T_0}^{T} (\alpha K_T)\, dT. \tag{10.17}$$

Equation (10.17) is especially appropriate at high temperatures where $T > \Theta$ and $T_0 = \Theta$. The last term in (10.17) is independent of V to a first approximation. To show this, use (1.24) to define

$$\alpha K_T = \frac{\gamma}{V} C_V. \tag{10.18}$$

At high T, C_V is not very sensitive to V, and for minerals, γ is roughly proportional to V given by $\gamma = \gamma_0(V/V_0)$. In many cases, αK_T depends only slightly on V at high T (see Fig. 3.7). Thus (10.18) can be taken as independent of V. Replacing (10.17) by

$$P_{TH} = \int_0^T (\alpha K_T)\, dT, \tag{10.19}$$

we note the integrand. At high temperature, α increases with T and K_T decreases with T, so that to a good approximation, αK_T is often independent of T (Fig. 2.5). Equation (10.19) then becomes

$$\Delta P_{TH} = \alpha K_T (T - T_0) + P_{TH}(T_0), \tag{10.20}$$

where T_0 is the temperature at which linearity of αK_T begins. It is given by

$$P_{TH}(T_0) = \int_0^{T_0} (\alpha K_T)\, dT. \tag{10.21}$$

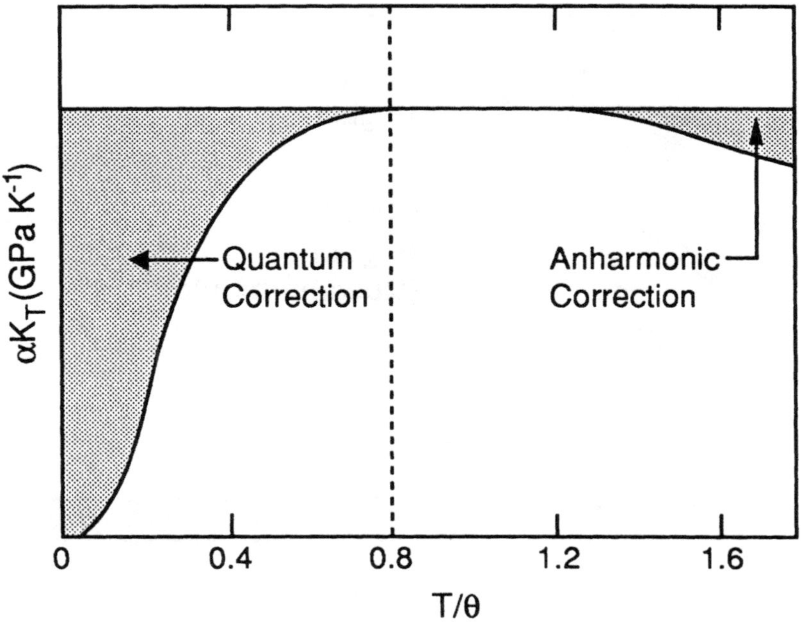

Fig. 10.1. A schematic diagram showing the corrections that must be made to the approximation $(\partial P_{TH}/\partial T)_V$ = constant for a crystalline solid at high temperature. The abscissa is T/Θ, where Θ is the Debye temperature.

Equation (10.21) is a constant, determined by integrating αK_T from absolute zero up to the onset of the high temperature regime, T_0, which starts somewhere between 0.8Θ and Θ, depending on the material. Below T_0 the behavior of αK_T resembles the low temperature part of the specific heat curve (see Fig. 2.6), and (10.21) therefore defines the quantum correction. Above T_0 the departure of αK_T from constancy is a measure of the anharmonicity of αK_T, as shown in Fig. 10.1. The data for $\int_{T_0}^{T} \alpha K_T \, dT$ for seven minerals and two alkali halides is shown in Fig. 2.3. Linearity in $P_{TH}(T) - P_{TH}(300)$ is demonstrated for $T > \Theta$. This shows that the average value of $\overline{\alpha K_T}$, above Θ, is independent of T (see (2.42)). It also demonstrates that $\overline{\alpha K_T}$ is large or small as $\rho_0/(M/p)$ is large or small.

10.3. Anharmonicity effect for thermal pressure at $V < V_0$

If T^* is the limit in linearity of the P_{TH} T curve found for the isochore $V = V_0$, the application of pressure extends T^* to higher values. This is proven by Hardy's (1980) theory, which shows that the application of pressure extends the boundary between the anharmonic and the classical regime to higher temperatures. Thus anharmonic effects die off with increasing P.

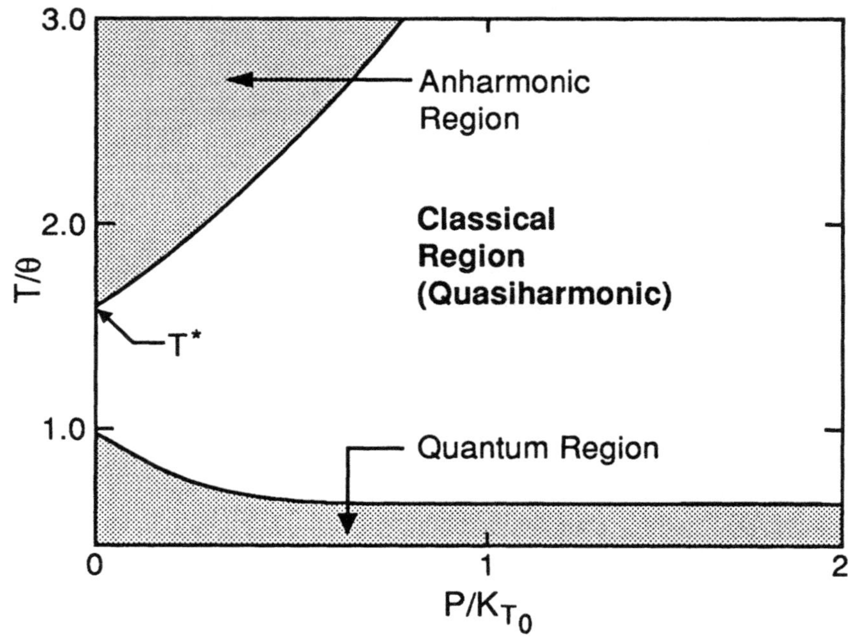

Fig. 10.2. A schematic diagram showing the pressure effect on the separation of the classical region (where $P_T \sim T$) from the anharmonic region for the vibrational contribution to the pressure equation. The actual lines of separation depend on the assumption of the repulsive potential function and the accuracy demanded in definition of the departure from linearity between P_{TH} and T (after Hardy, 1980).

Figure 10.2 is an interpretation of one of Hardy's (1980) published curves. The value of T^* marks the boundary between the classical harmonic and the (shaded) anharmonic region at $P = 0$. According to Hardy's results, the value of T^* may increase by as much as threefold as the pressure increases to $P \approx K_{T_0}$. In the lower mantle of the earth, where $P \approx K_{T_0}$, we might expect that T^* is several times greater than the value it would have in its uncompressed state (i.e., $T^* > 3000$ K at the core–mantle boundary). As a consequence, the linear curves in Fig. 2.3, measured to $T/\Theta_0 = 2.0$, are in the range of the lower mantle temperature ratio ($T/\Theta_0 = 1.9$; see Table 1.10) due to the pressure suppression of anharmonicity.

From the ΔP_{TH} data on mantle minerals given in Fig. 2.3, it seems reasonably safe to assume at this time that the whole of the lower mantle is approximately classical and that a linear law,

$$\Delta P_{TH} = a + bT, \tag{10.22}$$

where b is controlled by ρ_0 and M/p, applies throughout the mantle.

Linearity in P_{TH} and T should not extend all the way to the solidus; nonlinearity would probably occur in the vicinity of the melting boundary of the solid. But even near the core–mantle boundary, the T of the mantle may be far below the solidus.

10.4. Experimental results of the dependence of P_{TH} on V

Evidence is presented below showing that for some solids, P_{TH} is independent of V and proportional to T, and for other solids the dependence on V is marked, whereas P_{TH} is proportional to T.

10.4.1. *Evidence from alkali metals and alkali halides*

The work of Swenson and his collaborator (Swenson, 1968; Monfort and Swenson, 1965) on alkali metals is very illustrative because the bulk modulus is sufficiently low that experiments are possible up to $P/K_{T_0} > 1$. Further, the Debye temperature is sufficiently low that experiments are conducted well into the classical range.

The experiments on potassium where P versus V is measured at constant T are especially instructive. From this experiment $P(T) - P(0)$ versus T at constant V can be found. Monfort and Swenson's results go up to 2 GPa ($K_{T_0} = 3.35$ GPa, so $P/K_{T_0} = 6$); they range from -78 K to 300 K ($\Theta = 95$ K), so $0.82 < T/\Theta < 8.2$; and the volume goes from 33 cc/mole to 45.5 cc/mole, so $V/V_0 < 0.74$. Their basic results, illustrated by the case for potassium, are shown in Fig. 10.3 (top). From these results, the change in pressure with T at constant V can be calculated, which gives ΔP_{TH} with T at constant V, as shown in Fig. 10.3 (bottom). Above Θ the change in ΔP_{TH} is linear with T and independent of V.

It is seen that at constant V the change in pressure is virtually proportional to T, and that the higher the temperature, the larger the ΔP. This is seen in Fig. 10.3 (top), where it is demonstrated that $(\partial P/\partial V)_T = 0$, as the P-T curves for various isochores all overlap. Figure 10.3 (top) shows that the change in pressure versus T, as shown in Fig. 10.3 (bottom), is linear as well as being independent of V.

Thus for potassium

$$\left(\frac{\partial P}{\partial T}\right)_V = \text{constant}, \tag{10.23}$$

and

$$\left[\frac{\partial\left(\frac{\partial P}{\partial T}\right)_V}{\partial T}\right]_V = 0. \tag{10.24}$$

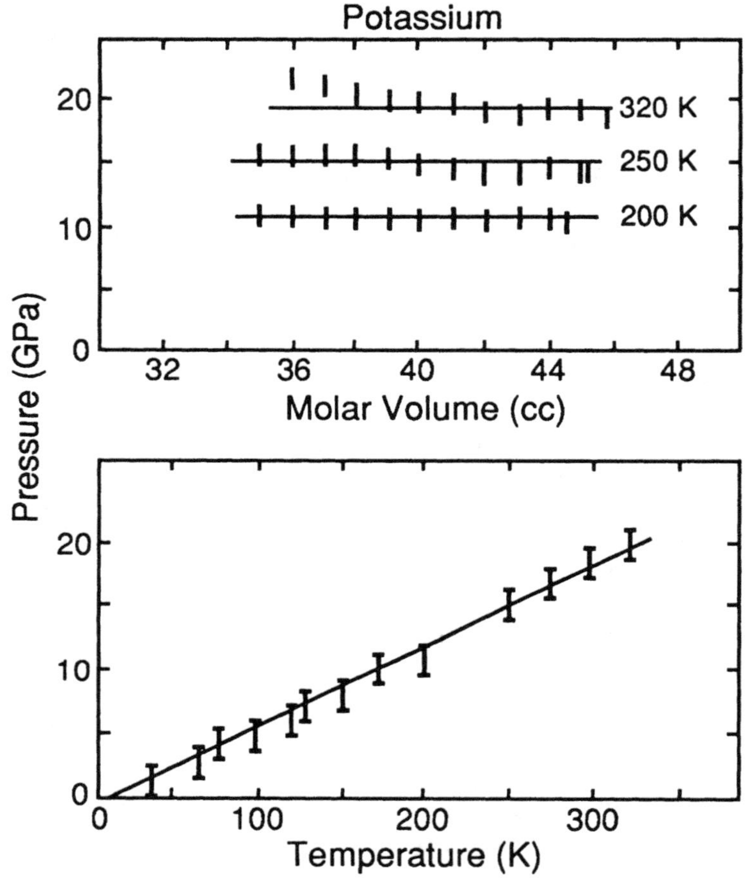

Fig. 10.3. (top): P versus V at three isotherms for potassium using data in Fig. 10.8. (bottom): P versus T for potassium. All isochores such as given by the top of this figure overlap, so P_{TH} for potassium is volume independent (modified from Monfort and Swenson, 1965).

But as $P = P_0(V) + P_{TH}$, the two equations above can be replaced by

$$\left(\frac{\partial P_{TH}}{\partial T}\right)_V = \text{constant} \tag{10.25}$$

and

$$\left[\frac{\partial\left(\frac{\partial P_{TH}}{\partial T}\right)_V}{\partial T}\right]_V = \left[\frac{\partial(\alpha K_T)}{\partial T}\right]_V = 0. \tag{10.26}$$

Thus (10.22) is valid in potassium, at least up to $T = 3.4\Theta$, and $P = 6K_{T_0}$.

Equation (10.26) means that the slopes of the isochores on a P-T plot are parallel at any given T, as shown in Fig. 10.3 (top). This means that the thermal pressure is independent of volume, as shown in Fig. 10.3 (bottom). We can demonstrate this mathematically by the principle that the order of differentiation can be reversed in (10.29), giving

$$\left[\frac{\partial\left(\frac{\partial P_{TH}}{\partial T}\right)_V}{\partial V}\right]_T = 0; \tag{10.27}$$

that is, $(\partial P_{TH}/\partial T)_V$ is constant and independent of V. It is in fact zero, from the upper plot in Fig. 10.3. This means that if the slopes of the isotherms on a P-V plot are independent of T (as in Fig. 10.3 (top)) for constant V, then all isotherms are parallel.

Now $(\partial P_{TH}/\partial T)_V = \alpha K_T$. Using the lower part of Fig. 10.3 combined with (10.27),

$$\left(\frac{\partial \alpha K_T}{\partial V}\right)_T = 0. \tag{10.28}$$

Using (3.2), (10.28) yields

$$\left(\frac{\partial K_T}{\partial T}\right)_V = 0. \tag{10.29}$$

Conversely, this means that a solid in which $(\partial K_T/\partial T)_V = 0$ has parallel isotherms on a P-V plot, but the isochores superimpose on a P-T plot.

M.S. Anderson and Swenson (1975) found that the thermal pressure isochore shifts slightly with V for solid xenon, solid krypton, and substantially more for solid argon, a different situation from that found for potassium metal (Fig. 10.4). This requires that $(\partial K_T/\partial T)_V$ be dependent on volume. Even though there is a small, detectable volume dependence, P_{TH} is quite linear with T (the total separation between isochores corresponds roughly to $10^{-3}V_0$). For the case of solid argon in Fig. 10.4, all measurements are below the Debye temperature ($\Theta = 100$ K), and quantum effects spread the distance between the isochores. For solid krypton and solid xenon, the isochores do not overlap but are close. For these solids, P_{TH} is strictly linear in T, but (10.29) is not true.

In Fig. 10.5, the plot of $K_T(V = V_0)$ versus T is made for NaCl and KCl, showing that $K_T(V = V_0)$ is parallel to the T axis for NaCl, but slopes downward perceptibly with T for KCl. This shows that for NaCl, $(\partial K_T/\partial T)_V = 0$, whereas for KCl, $(\partial K_T/\partial T)_V \neq 0$. The case for NaCl is similar to that of the alkali metals. Table 10.1 shows that for NaCl the thermal pressure is independent of volume along isotherms above Θ. On the other hand, KCl behaves more like the noble gas solids, because $(\partial K_T/\partial T)_V$ is nonzero, and P_{TH} depends on V.

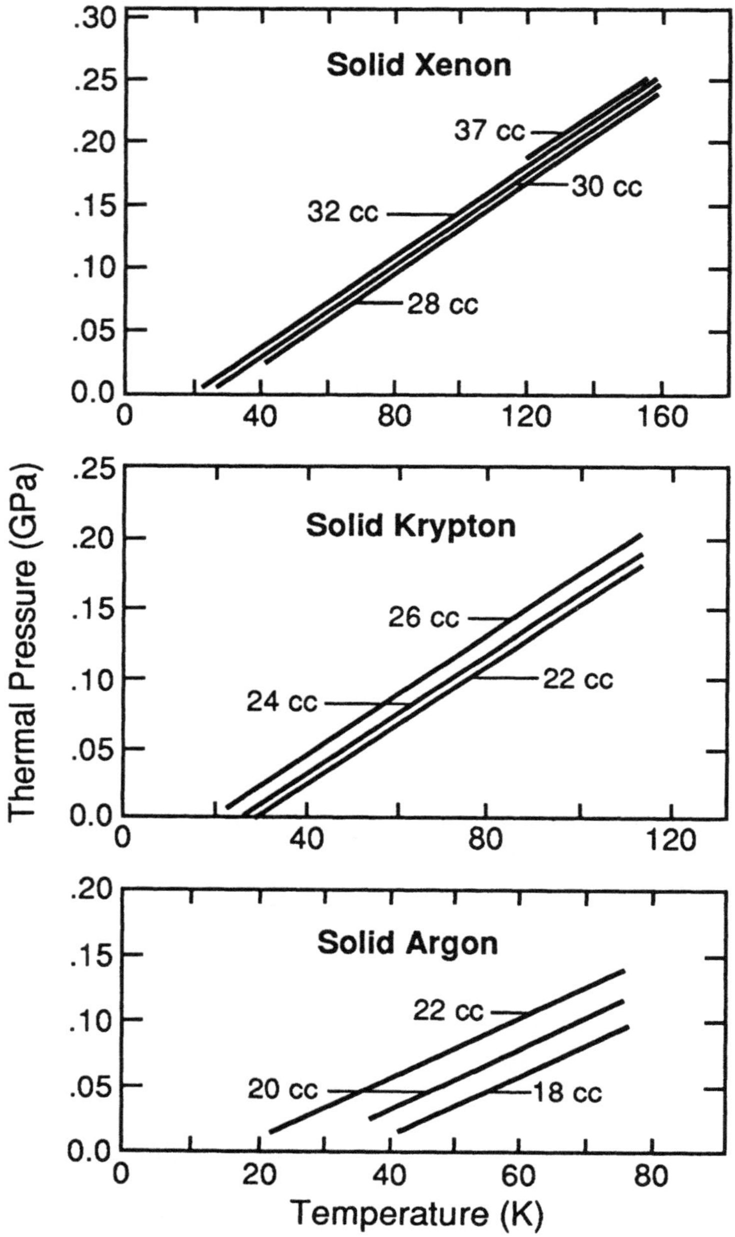

Fig. 10.4. P_{TH} versus T for solid xenon, solid krypton, and solid argon, showing that P_{TH} depends on V but is linear in T ($\Theta = 55°$ for xenon, 65° for krypton, and 100° for argon) (modified from M.S. Anderson and Swenson, 1975).

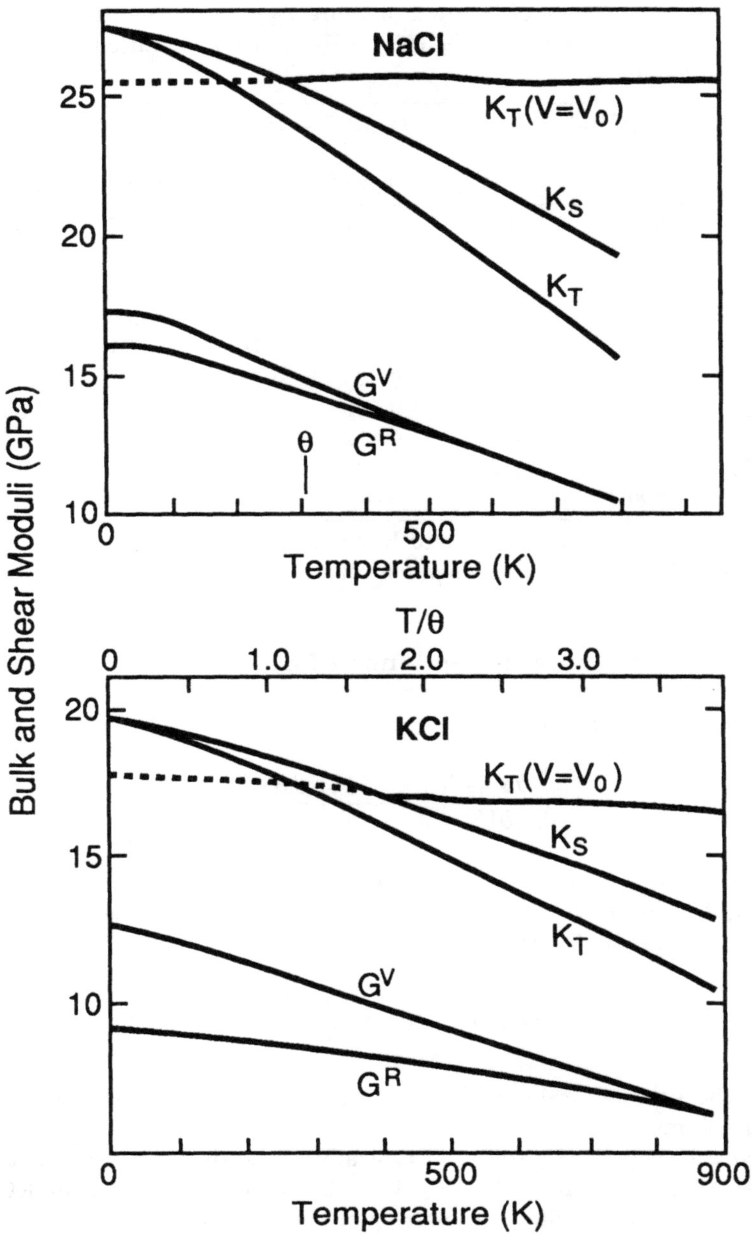

Fig. 10.5. The isochoric K_T versus T for NaCl and KCl. For NaCl, $K_T(V = V_0)$ is virtually independent of T, whereas for KCl, $K_T(V = V_0)$ decreases as T increases (after Yamamoto et al., 1987 and Yamamoto and Anderson, 1987).

Table 10.1. Change of thermal pressure P_{TH} from room temperature to reference temperature T, $P_{TH}(T) - P_{TH}(25°\text{C})$, at various compressions for NaCl.

V/V_0	P (GPa)	$P_{TH}(T) - P_{TH}(25°\text{C})$ (GPa)				
	25°C	100°C	200°C	300°C	400°C	500°C
1.00	0.000	0.216	0.501	0.785	1.067	1.349
0.99	0.247	0.22	0.500	0.784	1.067	1.348
0.98	0.509	0.223	0.510	0.783	1.068	1.348
0.97	0.788	0.215	0.499	0.783	1.068	1.348
0.96	1.84	0.214	0.499	0.784	1.071	1.348
0.95	1.389	0.214	0.499	0.786	1.071	1.349
0.94	1.732	0.214	0.502	0.789	1.078	1.351
0.93	2.087	0.213	0.502	0.792	1.084	--
0.92	2.464	0.213	0.50	--	--	--
0.91	2.864	0.213	--	--	--	--

Data computed from Boehler and Kennedy, 1980.
Note that P_{TH} is nearly independent of volume above $T = \Theta = 300$ K.

10.5. The volume dependence of αK_T

Take the identity given by (3.2)

$$\left(\frac{\partial \alpha K_T}{\partial V}\right)_T = -\frac{1}{V}\left(\frac{\partial K_T}{\partial T}\right)_V .$$

The right side is a small number compared with $-(1/V)(\partial K_T/\partial T)_P$. Assuming it to be small compared with αK_T at constant P, then by integrating the left side with respect to V

$$\alpha K_T(V, T) = \alpha K_T(V_a, T) + \left(\frac{\partial K_T}{\partial T}\right)_V \ell\text{n}\,(1-\eta), \tag{10.30}$$

where the volume strain $\Delta\eta$ is $(1 - V/V_a)$, and where V_a is the reference initial volume.

Now because $(\alpha K_T/\partial T)_V$ is typically a negative number, if not zero, as seen from the slope of the $K_T(V - V_0)$ curve in Fig. 10.5 for KCl, an increase in strain results in a lowering of αK_T at high T.

To see the effect of a nonzero value of $(\partial K_T/\partial T)_V$ on the thermal pressure, we apply the general formula for P_{TH} above $T = \Theta$,

$$P_{TH}(V,T) - P_{TH}(V,\Theta) = \int_0^T \alpha K_T \, dT.$$

Now αK_T can reasonably be taken to be independent of T above Θ, as discussed above in Section 10.2, so that by integrating (10.30) with respect to T,

$$P_{TH}(V,T) - P_{TH}(V,\Theta) = \left[\alpha K_T(V_a,T) + \left(\frac{\partial K_T}{\partial T}\right)_V \ell n\,(1-\eta)\right](T-\Theta). \tag{10.31}$$

The effect of strain on P_{TH} for the case of gold where $(\partial K_T/\partial T)_V$ is a large negative number is shown in Fig. 10.6. Here the slope of the $P_{TH} - T$ curve is reduced with increasing $\Delta\eta$. The large negative slope of $(\partial K_T/\partial T)_V$ with T for gold is shown in Fig. 10.7, illustrating the source of the excessive strain dependence of P_{TH}. However, Fig. 10.6 shows the slope of P_{TH} is straight, which indicates that $\left[\partial(\alpha K_T)/\partial T\right]_V = 0$ for gold, a normal behavior.

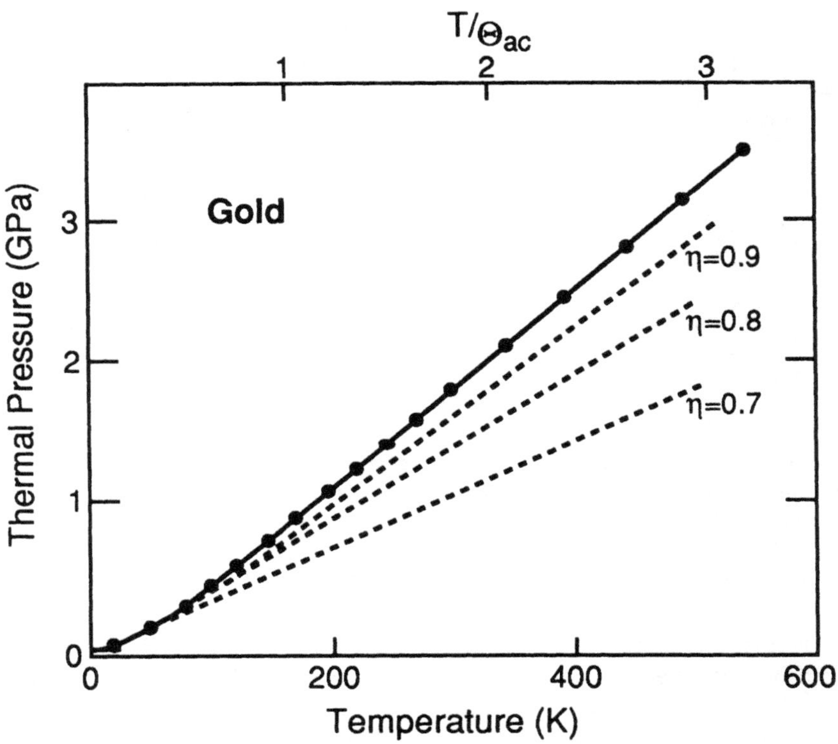

Fig. 10.6. Thermal pressure P_{TH} plotted against temperature for different values of strain $\Delta\eta$ for gold. Note the essentially linear characteristic of P_{TH} for T above Θ. The plotted data points are for zero compression, i.e., $\eta = 1$. The dashed lines show P_{TH} calculated for various compressions using the volume correction to αK_T (modified from Anderson et al., 1989).

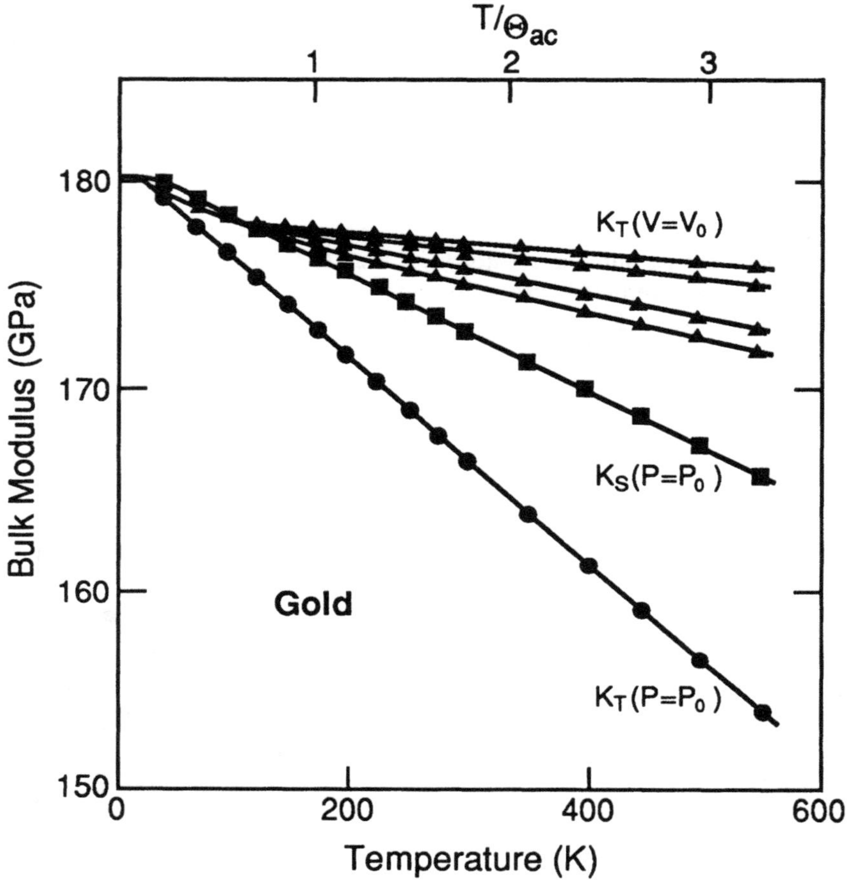

Fig. 10.7. Thermal data for gold. Note that $K_T(V = V_0)$ decreases with T for reported values of $(\partial K_{T_0}/\partial P)_T$, whereas $[\partial(\alpha K_T)/\partial T]_P$ is virtually independent of T at high T (see Fig. 10.6) (modified from Anderson et al., 1989).

10.6. $(\partial K_T/\partial T)_V$ for noble gas solids

In their work on the EoS's of the noble gas solids, M.S. Anderson and Swenson (1975) emphasized their result that $(\partial K_T/\partial T)_V$ is zero, provided T is above the Debye temperature. Figure 10.8 shows their work for solid xenon, solid krypton, and solid argon. For xenon and krypton below Θ, K_T at constant volume decreases perceptibly with T, but above Θ, K_T is independent of T. For solid argon, no data were taken above Θ, so the test cannot be made. We see these same features in Al_2O_3 (Fig. 10.9). Below Θ (1045 K) the slope of $(\partial K_T/\partial T)_V$ is negative. Above Θ, the slope of $K_T(V = V_0)$ is nearly parallel to the T axis.

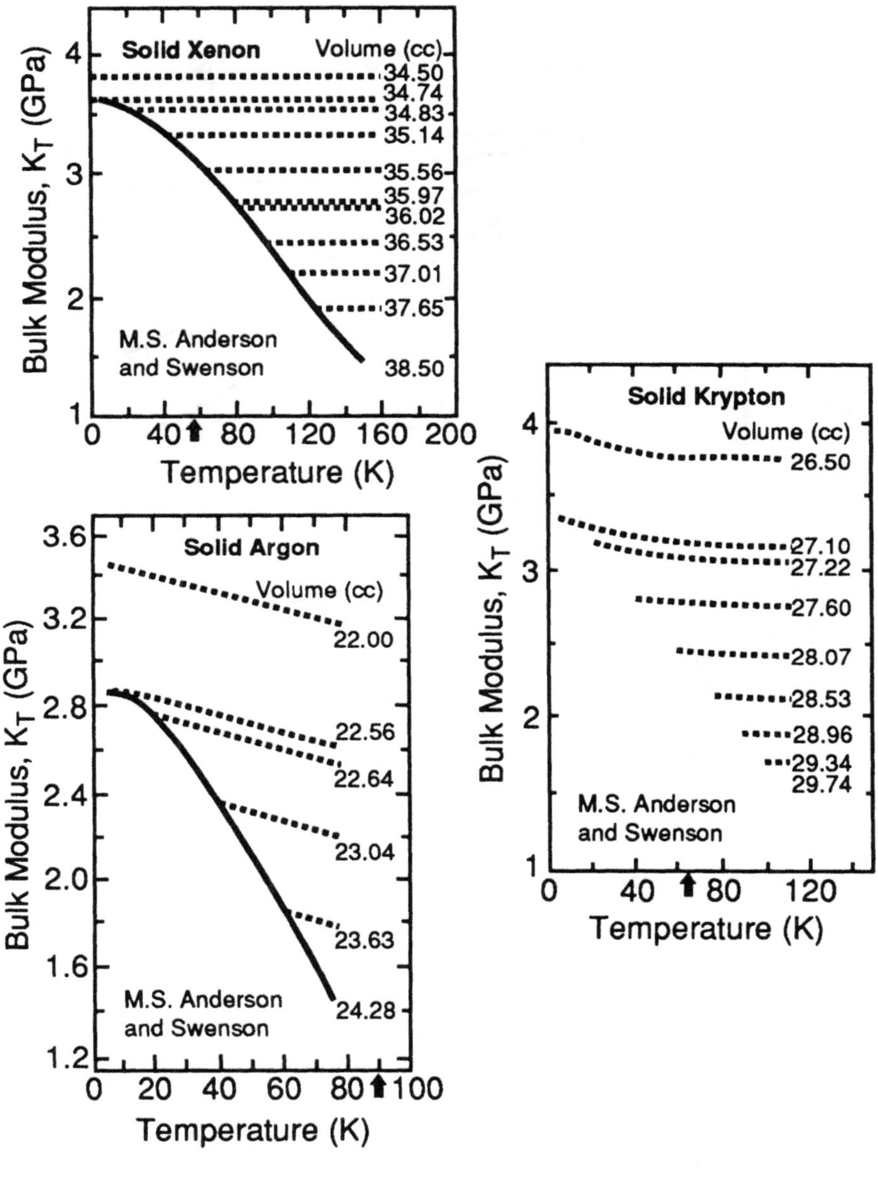

Fig. 10.8. The temperature and volume dependences of the isothermal bulk modulus for solid xenon, solid krypton, and solid argon, where isochores of K_T are plotted. The dashed lines terminate on the left, where $P = 0$. The arrows indicate the Debye temperature, Θ. Note that in each case the isochores are independent of T above Θ so that $(\partial K_T/\partial T)_V = 0$ above $T = \Theta$ (modified from M.S. Anderson and Swenson, 1975).

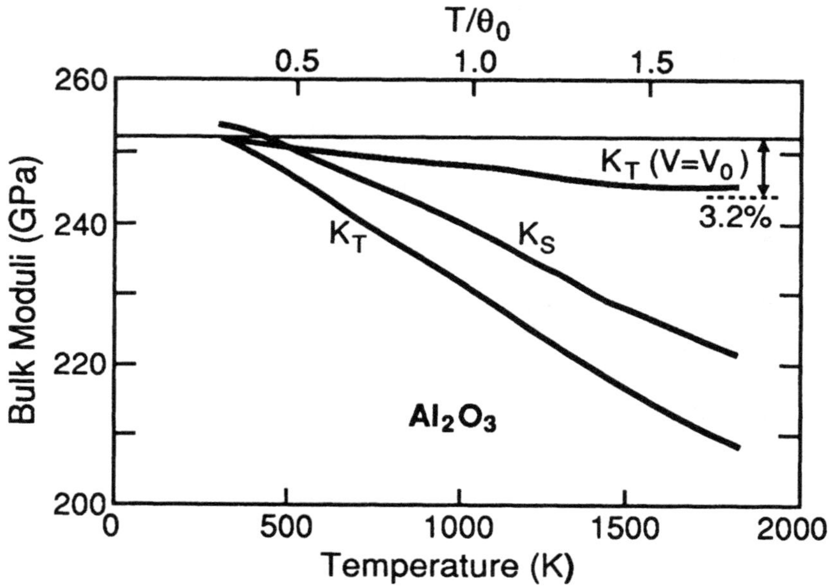

Fig. 10.9. The variation of K_S, K_T, and $K_T(V = V_0)$ with T for Al_2O_3. Note that the slope $(\partial K_T/\partial T)_V$ is negative at room temperature, but increases so that in the high temperature regime it approaches zero.

The data on the noble gas solids are instructive because as the volume decreases, the curves of K_T versus T (constant V) remain parallel. Thus we can see that the value of $(\partial K_T/\partial T)_V$ does not change for different isochores. This is in accord with the theory of Hardy (1980), who found that as the pressure increases, the anharmonicity must diminish. This information is quite valuable because it shows that the behavior we see for the few oxides and silicates we have measured is typical of that for a wide class of solids, even solid noble gases, giving confidence in the generality of the application to a wide class of minerals of direct interest in geophysics.

10.7. General comments on the behavior of P_{TH}

As we have seen, when integrated at constant V, the experimental values of αK_T, will give P_{TH} versus T, which above Θ is a straight line up to some quite high temperature T^*. At low temperature, P_{TH} will depart from linearity and approach absolute zero with zero slope, similar to the behavior of enthalpy. Near $T = 0$, the value of P_{TH} is in the so-called quantum region, where the effect of individual Einstein oscillators is distinguishable (Fig. 10.10). As the temperature is lowered, these oscillators one by one become quiescent, and finally near absolute zero P_{TH} versus T resembles the enthalpy, $\mathcal{H}$, versus T.

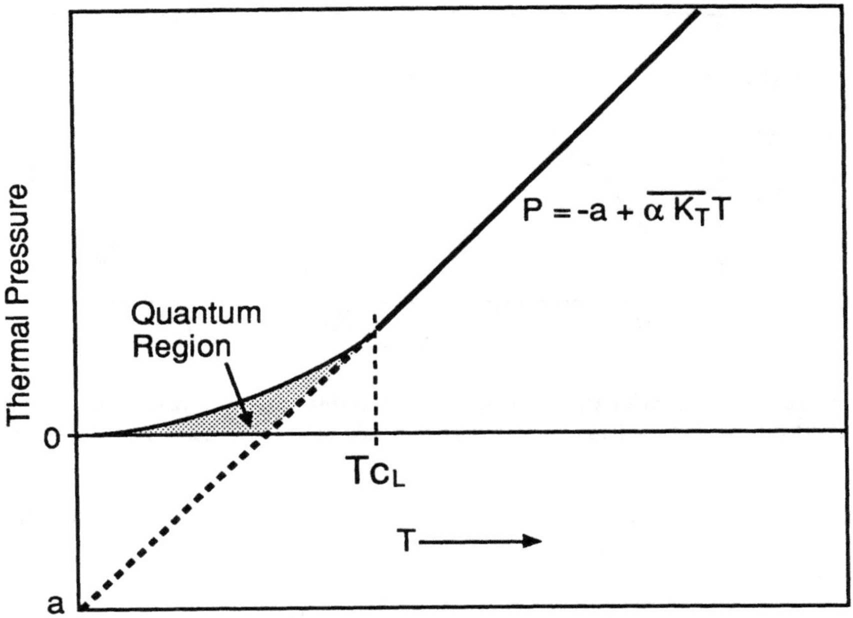

Fig. 10.10. Schematic variation of P_{TH} versus T, showing the relationship between the intercept a and T_{CL}, the onset of the classical region.

10.8. The thermal pressure of the lower mantle

10.8.1. *The Debye approximation of the earth's thermal pressure*

Some idea of the magnitude of the thermal pressure in the earth can be obtained by computing the thermal pressure from the Debye theory (Chapter 1). According to the Debye theory, the thermal part of the free energy at high T is

$$P_{TH} = RT\left[3\,\ell n\left(1 - e^{-e/T}\right) - D\left(\frac{\Theta}{T}\right)\right]. \qquad (10.32)$$

The thermal pressure is $P_{TH} = -\,(\partial F_{TH}/\partial V)_T$, so that

$$P_{TH} = 3RTp\left(\frac{\gamma_D}{V}\right) D\left(\frac{\Theta}{T}\right), \qquad (10.33)$$

where $D(\Theta/T)$ is the Debye function (Table A–6.2), and

$$\gamma_D = \frac{\partial\,\ell n\,\Theta}{\partial\,\ell n\,\rho}. \qquad (10.34)$$

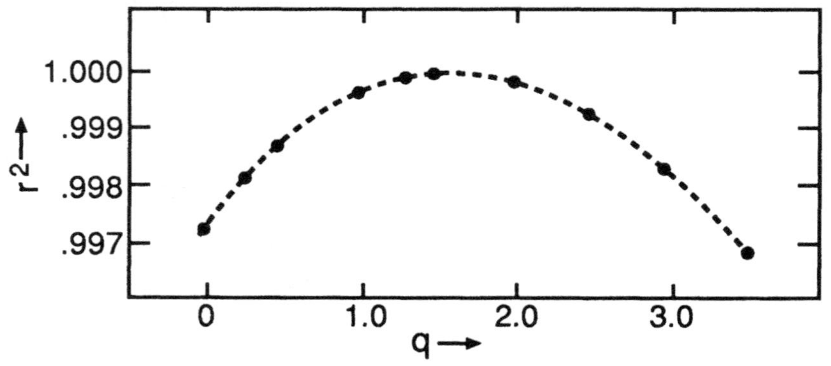

Fig. 10.11. A plot of the coefficient of determination, r^2 versus q, associated with the data of the seismic model of Dziewonski et al. 1975 using $\rho_0 = 4.2$ g/cc (modified from Anderson, 1979b).

Changing ρ/V to $\rho/(M/p)$, where p is the number of atoms in the molecule, and evaluating the numerical constants (taking $M/p = \mu 21$), (10.33) becomes

$$P_{TH} = 1.88\rho\gamma_D D\left(\frac{\Theta}{T}\right) T \text{ MPa}, \tag{10.35}$$

where ρ is in dimensions of g/ml. The Debye function will approach unity when $\Theta > T$, which is the case throughout the lower mantle, where $D(\Theta/T) = 0.99$. Thus αK_T becomes

$$\alpha K_T = \left(\frac{\partial P_{TH}}{\partial T}\right)_V = 1.186\rho\gamma_D \text{ MPa/deg}. \tag{10.36}$$

10.8.2. *The value of q in the lower mantle*

Equation (10.36) requires analysis of $\rho\gamma_D$ throughout the lower mantle, which is done from seismic data. Anderson (1979a) showed that the variation of γ_D with ρ found with (3.18) can be expressed as

$$\left(\frac{\gamma}{\gamma_0}\right) = \left(\frac{V}{V_0}\right)^q . \tag{10.37}$$

From (10.37) and (2.66) relating γ, Θ, and V, it follows that

$$\ell n\, \Theta = \ell n\, \Theta_0 + \frac{\gamma_0}{q}\left[1 - \left(\frac{\rho_0}{\rho}\right)^q\right], \tag{10.38}$$

where Θ_0 is the Debye temperature at zero pressure. Data for the acoustic version of Θ, (2.69), as a function of depth were found using seismic velocities of the PREM model (Dziewonski and D.L. Anderson, 1981) (Table A-5). Taking $\rho_0 = 4.2$ g/cc (the uncompressed but cold value of lower mantle density), the values of γ_0 and Θ_0 were determined for various assumed values of q according to the seismic model. The value of the coefficient of determination r^2 versus q was then computed and is shown in Fig. 10.11.

From Fig. 10.11 it is concluded that the seismic models of the lower mantle and especially the density distribution are not sufficiently precise to make a definitive choice of one value of q over any other in the band $0.5 < q < 2.2$. It is clear that $q = 1$ or $\rho\gamma_D = \rho_0\gamma_{D_0}$ is a quite reasonable solution and does not violate any of the primary data. Choosing $q = 1$ for evaluating (10.37) and (10.38) and assuming that the lower mantle is homogeneous, we evaluate γ_{D_0} and ρ_0. Thus (10.36) is replaced by the high T equation,

$$\alpha K_T = \left(\frac{\partial P_{TH}}{\partial T}\right)_V = 1.186\gamma_{D_0}\rho_0 \text{ MPa/deg.} \qquad (10.39)$$

The best fit of the lower mantle seismic model data, as reported by Anderson and Sumino (1980), yields ρ_0 (cold) $= 4.2$ g/ml, $\gamma_{D_0} = 1.3$ and $\Theta_0 = 920$. Thus, at high temperatures, (10.36) becomes

$$\alpha K_T = \left(\frac{\partial P_{TH}}{\partial T}\right)_V = 6.5 \text{ MPa/deg.} \qquad (10.40)$$

This value can be compared with the values found for the minerals. Reference to Fig. 2.5 shows that the high temperature value of αK_T for Al_2O_3 and MgO is very close to 6 MPa/deg, whereas the corresponding value for olivine (low pressure phase) and that for fayalite is close to 4 MPa/deg. Thus the uncompressed value of αK_T in the lower mantle is in reasonable accord with what is found for minerals, with ρ_0 and Θ close to their respective values for $MgSiO_3$ perovskite (see Fig. 1.3).

10.8.3. *P_{TH} in the lower mantle from thermal models*

In computing the thermal pressure in the lower mantle, there is the potential problem that αK_T may be a function of V at constant T, such as was found for gold (see Fig. 10.6). The recent values of αK_T versus P (Boehler et al., 1988; Chopelas, 1990) at room temperature indicate that this may be a problem, at least at modest pressure. But as shown in Fig. 3.7, for MgO at high T and high compression, αK_T is more or less independent of V.

From (3.7) we see that

$$\left(\frac{\partial \ln \alpha K_T}{\partial \ln V}\right)_T = K_T' - \delta_T. \qquad (10.41)$$

The values of $K_T' - \delta_T$ versus T and V have been solved for MgO and are shown in Table 3.4. At low compression, or η slightly less than 1, and at high temperature, $\delta_T > K_T'$ and $q > 1$, and αK_T varies with V. At high compression and high temperature, δ_T is close to or slightly less than K', so that αK_T varies very little with compression at high compression, as illustrated in Fig. 3.7. The evidence is that αK_T is indeed independent of depth in the P, T conditions of the mantle. The expression for the ΔP_{TH} using (10.40) is thus

$$\Delta P_{TH} = 6.5 \times 10^{-3}(T - 300) \quad \text{GPa}. \tag{10.42}$$

Anderson and Sumino (1980) found an empirical expression for the thermal pressure of the lower mantle by a self-consistent analysis of seismic and thermodynamic data on minerals. They discovered that all data examined were consistent, assuming a homogeneous lower mantle, if one also assumes

$$P_{TH} = -1.36 + .00634T \text{ GPa}. \tag{10.43}$$

Comparing (10.43) and (10.42), we have

$$P_{TH} = 0.53 + 6.5 \times 10^{-3}(T - 300) \quad \text{GPa}. \tag{10.44}$$

The variation of P_{TH} with depth in the mantle can be found from (10.42) if T is known at each depth. Most solutions for T with depth in the lower mantle arise from solving a thermal model of the earth, but such a model has built into it certain assumptions about γ, i.e., for each T distribution there is a certain assumed γ function. Table 10.2 shows the data on T for three thermal models.

Table 10.2. Temperature found from three thermal models of the lower mantle

	Temperature (K)		
Depth km	B&S†	A‡	Stacey*
771	1908	2096	2290
971	1973	2171	2372
1171	2034	2255	2445
1371	2092	2334	2518
1571	2147	2425	2585
1771	2199	2512	2650
1971	2249	2570	2713
2171	2296	2628	2776
2371	2341	2680	2840
2571	2384	2756	2905
2771	2426	2804	3005
2885	2449	2833	3157

†Brown and Shankland (1981); ‡Anderson (1981);
*Stacey (1977b)

Table 10.3. Thermal pressure referenced to 300 K from (10.3) using T from three thermal models

Depth km	B&S† ΔP_{TH} GPa	A‡ ΔP_{TH} GPa	Stacey* ΔP_{TH} GPa	Pressure P° GPa	B&S P_{TH}/P
771	10.40	11.68	12.94	28.29	0.39
971	10.86	12.16	13.47	37.29	0.31
1171	11.27	12.71	14.42	46.49	0.25
1371	11.65	13.22	14.85	55.90	0.22
1571	12.0	13.82	15.28	65.52	0.19
1771	12.34	14.38	15.68	75.36	0.17
1971	12.67	14.76	16.10	85.43	0.15
2296	12.97	15.13	16.51	95.76	0.14
2371	13.27	15.47	16.93	106.39	0.13
2571	13.55	15.97	17.58	117.35	0.12
2771	13.82	16.28	18.23	128.71	0.11
2885	13.96	16.46	18.57	135.75	0.10

†Brown and Shankland (1981); ‡Anderson (1981); *Stacey (1977b); °P from the PREM model.

Using the values of T from Table 10.2 and (10.44), P_{TH} can be computed for any of the thermal models. For example, using the Brown and Shankland model for T, P_{TH} versus depth of the mantle is computed by (10.44); the resulting values are listed in Table 10.3. The value of the total pressure is taken from PREM, and the ratio of P_{TH} to P is also given in Table 10.3. From this we see that P_{TH}/P is largest at the top of the mantle and smallest at the base of the mantle. According to Table 10.3 the change in P_{TH} across the lower mantle is only about a few GPa, whereas the change in P is 105 GPa.

10.8.4. *P_{TH} in the mantle from equations of state*

The thermal pressure can be found from the pressure determined by seismic methods and the pressure calculated by an equation of state for the 300° isotherm: $P_{TH} = P(T, \rho) - P_0(300, \rho)$, where P_0 represents the ambient isotherm. To compute P_0, we need the values of the equation of state constants: $\rho(300, 0)$, $K_T(300, 0)$, and $K'_T(300, 0)$. We need the uncompressed values of these constants from the PREM model.

Using the PREM values of pressure and density and following the third order Birch-Murnaghan EoS, Bukowinski and Wolf (1990) obtained the uncompressed (hot) values $\rho(T^*, 0) = 4.005$ and $K_S(T^*, 0) = 224$ GPa, where T^* represents the hot temperature of the solid on decompression.

Anderson and Sumino (1980) found almost identical results for their decompression calculation: $\rho(T^*, 0) = 4.00$, $K_S(T^*, 0) = 221$ GPa. Anderson and Sumino (1980) went further and reduced the temperature to 300 K and then changed from adiabatic to isothermal parameters, finding: $\rho(300, 0) = 4.22 \pm 0.02$; $K_T(300, 0) = 275 \pm 6$. In the Anderson and Sumino (1980) paper, they used assumed unmeasured values of dK_S/dT. There has been a great deal of recent experimental work done where measurements of $(dK_S/dT)_P$ are taken at very high T (Anderson et al., 1992). Consequently better temperature-reduced values now are $\rho(300, 0) = 4.2$ and $K_T(300, 0) = 270 \pm 5$.

The ambient value of $(\partial K_T/\partial P)_T$ is more elusive in the decompression analysis. In the PREM (D.L. Anderson, 1989) data of the lower mantle, dK/dP varies considerably with depth, having the following properties: 721–971 km, $dK/dP < 3.0$; 1071–1671 km, $dK/dP < 3.6$; 1771–2677 km, $dK/dP < 3.2$; 2471–2741 km, $dK/dP < 3.3$. For the Birch Murnaghan EoS, in the regions down to 1671 where the compression is < 1.17, it matters little whether dK/dP is 3.2 or 3.6. But in regions where the compression is > 1.2, the calculated pressure, $P_0(300, \rho)$, is sensitive to the actual value of dK/dP. Therefore there is little error made in the calculated value of P_0 by assuming $dK/dP = (\partial K_S/\partial P)_S$ is 3.25 for the whole lower mantle. In the decompression analyses, consideration needs to be given to values of $\partial^2 K_S/\partial P^2$ and $\partial^2 K_S/\partial P \partial T$. Isaak (1993) showed that these two high derivatives are of opposite signs and tend to cancel out over much of the mantle, so that we take $(\partial K_S/\partial T)_S$ (300, 0) to be 3.25. In changing $(\partial K_S/\partial T)_S$ to $(\partial K_S/\partial T)_T$, we need to account for the factor $(\partial K_S/\partial T)_P$ (found from δ_S) and the factor $(\partial T/\partial P)_S$ (found from the thermal gradient). The correction to $(\partial K_S/\partial P)_S$ is 0.15, so that we may expect for the lower mantle,

$$(\partial K_S/\partial P)_T\,(0, 300) = 3.4.$$

We note that these decompressed and ambient values are for the case of using a Birch-Murnaghan EoS in the decompression analysis. Bukowinski and Wolf (1990) stressed the point that the actual numerical values for K_S and K' depend on the equations used in the decompression analysis.

We now compute the 300 K isotherm of pressure from the third order Birch-Murnaghan EoS as given by (6.18), (6.15), and (6.25):

$$P_0(\rho, 300)/270 = -3\epsilon\,(1 - 2\epsilon)^{5/2}\,(1 + 2\xi\epsilon)\,, \tag{10.45}$$

where

$$\epsilon = (1/2)\left\{1 - (\rho/\rho_0)^{2/3}\right\}$$

$$\xi = (3/4)\,(4 - K') = 0.45$$

$$\rho = \text{density according to PREM.}$$

Results of (10.45) and PREM data for ρ are given in Table 10.4.

Table 10.4. Computations of $P_0(V)$ and P_{TH} for the lower mantle using the third order B-M EoS: $K_{T_0} = 270 \pm 5$ GPa; $K' = 3.4$

Depth km	$\rho^\dagger$ g/cc	ϵ	P_0/K_{T_0} (10.45)	P_0 GPa	P^* GPa	P_{TH} GPa	$\frac{\text{B\&S}^\ddagger}{\text{BM}}$
771	4.44	−.0187	0.061	16.5 $\pm$0.31	28.29	11.81	0.93
971	4.56	−.0282	0.095	25.5	37.29	11.79	0.97
1171	4.67	−.0366	0.127	34.2	46.49	12.28	0.96
1371	4.78	−.0450	0.161	43.5	55.90	12.40	0.98
1571	4.89	−.0534	0.196	53.0	65.52	12.51	1.00
1771	5.0	−.0616	0.232	62.5	75.36	12.82	1.00
1971	5.10	−.0690	0.268	72.4	85.43	13.00	1.02
2171	5.20	−.0765	0.305	82.3	95.76	13.42	1.00
2371	5.30	−.0840	0.343	92.5	106.39	13.83	1.00
2571	5.40	−.0912	0.382	103.1	117.35	14.27	0.99
2771	5.50	−.0985	0.422	114.0	128.71	14.71	0.98
2885	5.56	−.1028	0.467	120.4 $\pm$2.30	135.75	15.33	0.95

†PREM density and *PREM pressure (D.L. Anderson, 1989);
‡Brown and Shankland P_{TH} divided by Birch-Murnaghan EoS P_{TH}.

The last column of Table 10.4 shows that the value of P_{TH} from the equation of state calculation in which no mantle temperature is used agrees remarkably well with the value of P_{TH} from the Brown and Shankland (1981) temperature distribution (Table 10.2) used in the calculation of P_{TH} (ΔP_{TH} in Table 10.3, plus 0.53). From this we conclude: (1) there is no evidence of anharmonicity effects (a T^2 term) in P_{TH} of the lower mantle; (2) there is no evidence that P_{TH} is sensitive to volume change in the lower mantle; (3) the best agreement comes by assuming the lower temperature distribution found in Brown and Shankland (1981), in which $T = 1980$ K at 771 km and $T = 2500$ K at 2285 km; and (4) the adiabatic compression method of finding temperature is valid. Apparently there is no need to assume significant superadiabatic gradients, such as arise from modulus relaxation or inhomogeneity corrections (Shankland and Brown, 1985) because P_{TH} from the B-M EoS and the seismically determined P agree with the thermal calculation of P_{TH}.

We note that $K_T(300, 0) = 270 \pm 5$ GPa obtained from the decompressed mantle values obtained by Bukowinski and Wolf (1990) agrees with the experimental values obtained for orthorhombic silicate perovskite ($K_{T_0} = 266$ GPa, Knittle and Jeanloz, 1987; $K_{T_0} = 261$ GPa, Mao et al., 1991; $K_{T_0} = 243$ GPa, Yeganeh-Haeri et al., 1989).

It is also significant that the temperature distribution of the mantle obtained by Brown and Shankland and verified by Table 10.4 was obtained by using as a starting temperature 1873 K at 670 km. This value is close to that measured by Katsura and Ito (1989) for the transition from the γ spinel form of olivine to the perovskite form of olivine: they found 1873 K at 655 km.

The results of several equations for the thermal pressure versus temperature in the lower mantle are summarized in Fig. 10.12. The experimental results for P_{TH} versus T at $P = 0$ for three solids are also plotted.

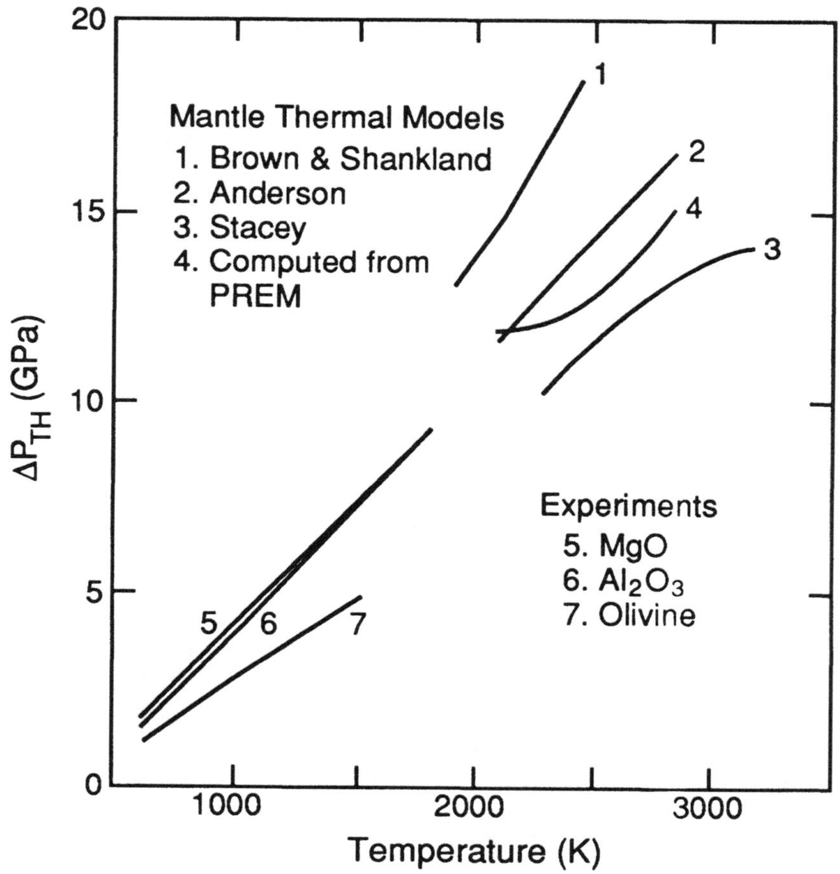

Fig. 10.12. Thermal pressure versus T for four thermal models of the lower mantle. Also, for comparison, the experimental results of thermal pressure versus T for three solids (MgO; Al_2O_3; Mg_2SiO_4; see Fig. 2.3). We note that the extrapolated values of P_{TH} at high T for minerals with high density values merge with the computed values of P_{TH} from PREM parameters at lower mantle conditions. Experimental data on P_{TH} from Anderson et al., 1992.

It is seen that the experimental results for MgO and Al_2O_3 extrapolate into the lower mantle solutions. However, the results for Mg_2SiO_4 do not extrapolate into the lower mantle solutions, nor should this be expected. Mg_2SiO_4 (olivine structure), reported in Fig. 10.12, is appropriate for the upper mantle. In the lower mantle, Mg_2SiO_4 would be in the perovskite structure, and having a denser packing in the high pressure phase would have a correspondingly larger value of αK_T, yielding thermal pressure slightly larger than the extrapolated Al_2O_3 thermal pressure.

The values of $P_0 = P_0(\rho, 300)$ and P from PREM taken from Table 10.4 are plotted in Fig. 10.13. The difference $P(\text{PREM}) - P_0$ is the thermal pressure, P_{TH}. The curves appear parallel, but they are not. The difference $P - P_0$ at 2885 km exceeds by 3.5 GPa the difference at 771 GPa. This shows that the thermal pressure is only a small fraction of P at the core–mantle boundary, whereas it is a much larger fraction (nearly 40%) at the upper–lower mantle boundary.

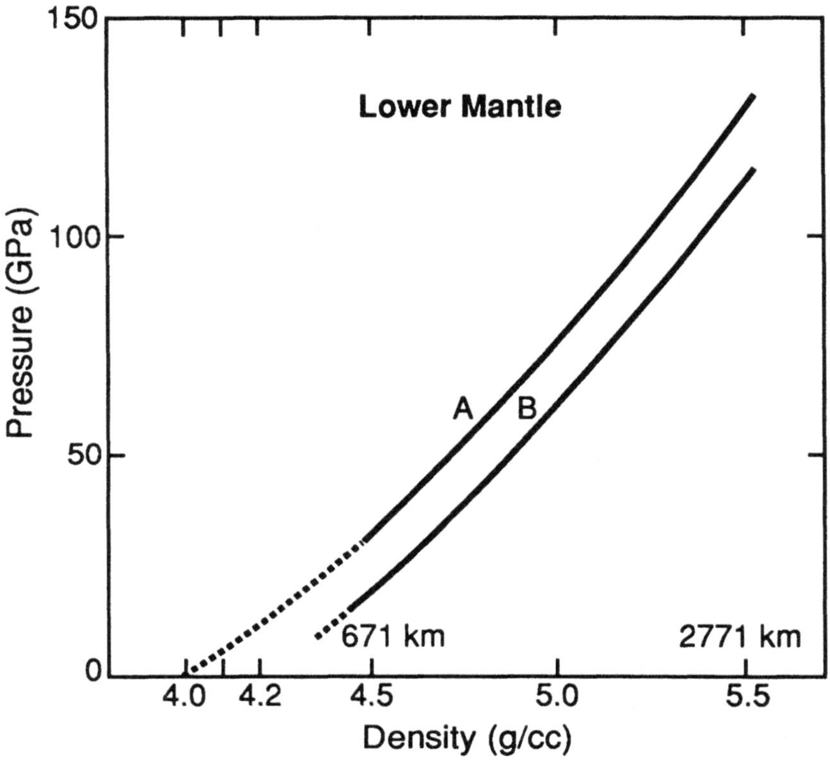

Fig. 10.13. The variation of pressure with density. Curve A is the density from PREM; curve B is the 300 K isobar computation of P using the third order Birch-Murnaghan EoS. ρ_0, K_{T_0}, and K'_{T_0} are obtained by decompressing PREM values, then reducing the temperature to 300 K (after Anderson and Sumino, 1980).

10.9. The thermal pressure of the inner core

10.9.1. αK_T *for the inner core*

The general form of thermal pressure, analogous to (10.33), is

$$\left(\frac{\partial P_{TH}}{\partial T}\right)_V = \alpha K_T = \frac{3R\gamma}{(M/p)}\rho. \tag{10.46}$$

M/p, the average atomic weight for iron, is 55.83.

Thus, for iron,

$$\left(\frac{\partial P_{TH}}{\partial T}\right)_V = 0.446\ \gamma\rho\ \text{MPa/deg.}, \tag{10.47}$$

where ρ is in g/cc.

An estimate of γ for the inner core is needed. The effect of γ_e on iron was discussed in Section 1.9. For the lattice contribution alone, γ was found by Jamieson et al. (1978) to be 1.5 ± 0.3 for the inner core. One does not simply add γ_e to γ to get the core γ, because they are of unequal weights. As shown by Jamieson et al. (1978) and Stacey (1977a,b), adding a small term to γ (they recommended 0.1) suffices for the electronic contribution to γ, in place of the detailed calculation, due to the relative smallness of the electronic contribution. Thus γ of iron at core conditions should be near 1.6 ± 0.3. From shock wave measurements, Jeanloz (1979) found $\gamma = 2.0$ at $P = 0$ for iron, becoming 1.5 at core conditions. Stacey (1977b) computed γ_{vz} from seismic density gradient data, finding 1.49 at 340 GPa. Anderson (1979a) computed γ_{ac} from seismic density gradient data, obtaining $\gamma = 1.6$ at 340 GPa. Brown and McQueen (1986) estimated γ at high P and T, $1.7 < \gamma_0 < 2.5$ from their shock wave measurements. I shall use $\gamma = 1.8$ for the inner core, and, using $\rho = 13$ g/cc, (10.47) is

$$\alpha K_T = \left(\frac{\partial P_{TH}}{\partial T}\right)_V = 10.5\ \text{MPa/deg;} \tag{10.48}$$

$$\Delta P_{TH} = P_{TH}(T) - P_{TH}(300) = 10.5 \times 10^{-3}(T - 300)\ \text{GPa}. \tag{10.49}$$

10.9.2. *Evaluation of* T *for the inner core*

To find P_{TH} from (10.49), we must assume the temperature of the inner core. There is some question of its exact value. A common procedure is to use the experimental information on the melting of pure iron at the inner core–outer core (IC-OC) boundary pressure and then estimate the melting point depression due to impurities in the outer core.

A great deal of effort has been expended in theories and experiments to attempt to find the value of T_m at the IC-OC boundary pressure. At this pressure the liquid and solid parts of the core are in thermal equilibrium. Therefore the experimentally or theoretically determined melting point of iron at this pressure gives insight into core temperatures. Some estimates of T_m at 330 GPa are shown in Table 10.5.

Table 10.5. Predicted melting temperatures T_m of pure iron at the pressure of the inner-outer core boundary (from Anderson, 1993).

	T_m °K
GILVARRY (1957)	6200
ZHARKOV (1962)	6200
BUNDY & STRONG (1962)*	6100–8100
HIGGINS & KENNEDY (1971)*	4250
BIRCH (1972)*	5100
BOSCHI (1975)	6000
LIU (1975)	5125
BUKOWINSKI (1977) (fcc iron)	5450
BOSCHI ET AL. (1979)	4500–7000
ABELSON (1981)‡	7800
STEVENSON (1981)	6300
BROWN & MCQUEEN (1982)§	6200 ± 500
ANDERSON (1982)§	5900 ± 700
SPILIOPOULOS & STACEY (1984)§	6140 ± 575
YOUNG & GROVER (1984)	6600
ANDERSON (1986)§	6210 ± 400
POIRIER (1986)	6000 ± 300
BROWN & MCQUEEN (1986)§	5800 ± 500
ROSS ET AL. (1990)	6400
WILLIAMS ET AL. (1987)	7700
YOO ET AL. (1993)	6830
POIRIER AND SHANKLAND (1994)	6160 ± 250 (α)
	6060 ± 260 (γ)
ANDERSON (1993)§	6200 ± 200 (ϵ)
BOEHLER (1994)	4850 ± 200

* Based on an extrapolation from experiments at 6 GPa.

‡ Based on the Monte Carlo theory.

§ Based on an extrapolation from experiments at 243 GPa.

The actual temperature of a pure iron core will be changed (probably lowered) by the presence of impurities. The temperature of the inner core will be influenced by the freezing point depression of the iron in the outer liquid core at the IC-OC boundary. Poirier and Shankland (1994) estimate this depression to be 500°C to 1000°C.

We use T_m from Anderson (1993) in Table 10.5 reduced by 700°K due to the estimated freezing point depression to yield $T_m = 5500 \pm 500$ K for the core. Taking this value in (10.49), we have

$$\Delta P_{TH} = 58 \pm 5.2 \text{ GPa}. \tag{10.50}$$

The error bars reflect uncertainty in T_m and in γ. This value will be used to find $\rho(z)$ by correcting the equation of state value of ρ computed along the 300 K isotherm, $\rho(300, T)$.

10.9.3. *The equation of state for the inner core*

Data for the parameters used in the equation of state are listed in Table 8.3. Calculations of the 300 K isotherm equations of state were made in Section 8.11 and were plotted in Fig. 8.5. To obtain these density trajectories at core temperatures, we keep ρ the same as in Fig. 8.5, while adding $\Delta P_{TH} = 60$ GPa to $P(\rho, 300)$, and the results are plotted in Fig. 10.14.

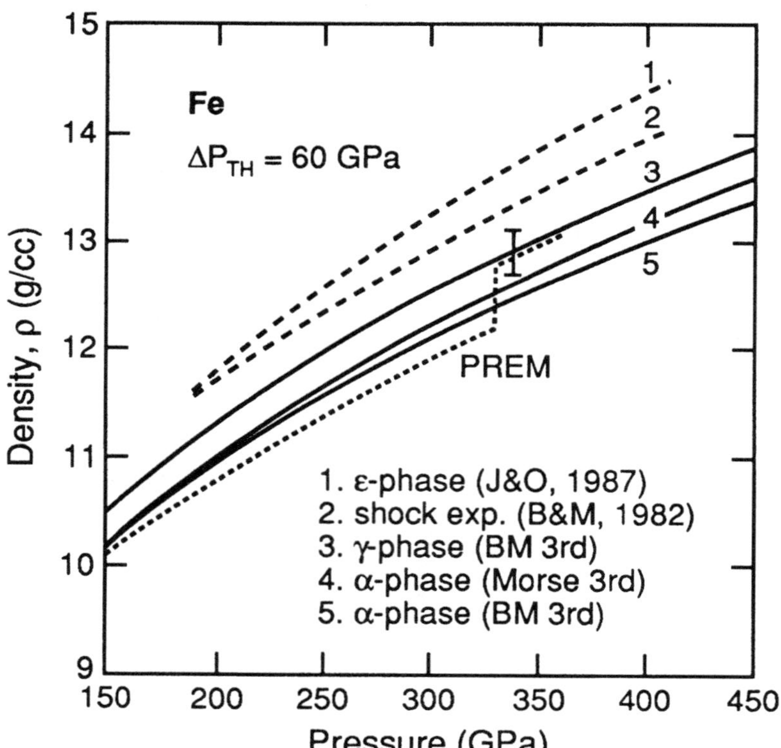

Fig. 10.14. The trajectories of ϵ-iron, γ-iron, and ferromagnetic α-iron at core pressures and temperatures found by shifting the trajectories in Fig. 8.5 to the right along the P axis ($\Delta P_{TH} = 60$ GPa). A comparison is made to PREM density. Two EoS's, the Birch-Murnaghan third order and the Morse potential, are used for two curves for α-iron. A pure ϵ phase has a density too high for PREM, but a pure fcc phase may have a density close to that of the inner core. Also shown is the experimental shock wave curve (Brown and McQueen, 1986).

10.9.4. *Impurities in the iron of the central core*

If the core is made of ϵ(hcp) iron, it is seen that the density at inner core conditions is too large for PREM, and consequently the inner core must be an alloy of iron and lighter impurities (Jephcoat and Olson, 1987). On the other hand, if the inner core is made up of fcc iron, the density of pure iron will satisfy the values found by PREM (see Fig. 10.14), providing the EoS parameters used for γ iron are accurate. In this case a model of the inner core's physical properties could be made of pure iron with negligible impurities. An fcc inner core implies that the crystallization of the inner core has resulted in the rejection to the outer core of most of its light element impurities in a process such as zone refining. The α phase curves are not physically meaningful, as the EoS parameters used are for ferromagnetic iron, and bcc iron at such high temperatures would not be ferromagnetic. Nonmagnetic iron is dynamically unstable at high temperatures (Stixrude and Cohen, 1994; Cohen et al., 1994; Sherman, 1994). No valid conclusion can be made about the composition of the inner core until the structure of iron phase at inner core conditions (see Fig. 8.2) is settled.

10.10. Thermal pressure and the EoS of silicate perovskite

I now demonstrate how the values of an equation of state can be generated with limited knowledge of the thermal pressure. As a good approximation, ΔP_{TH} is controlled by the value of αK_T at $T = \Theta$. The tacit assumption is that ΔP_{TH} is independent of V and proportional to T. The value of K_T at $T = 300$ measured by Mao et al. (1991) is 263 GPa. In Section 4.13.4 we found $\partial K_T/\partial T = 0.035$ GPa/K, so that $K_T(1000) = K_T(300) - .0035\,(1000-300) = 239$ GPa. Using Fig. 4.12 for $\gamma_0 = 1.5$, the value of α at 1000 K is 2.9×10^{-5}, so that $\alpha K_T = .00692$ GPa/K, and $\Delta P_{TH} = 6.92\,(T-300) \times 10^{-3}$ GPa. Figure 10.15 compares ΔP_{TH} for perovskite with that of MgO. The value of P_{TH} at $\eta = 1$ is shown as the top row in Table 10.6.

The EoS at $T = 300$ for perovskite was calculated from the data represented by Fig. 8 of Hemley and Cohen (1992) using a third order Birch Murnaghan EoS; the resulting $P(V)$ data are shown in Column 2 of Table 10.6. The remaining columns show the resulting $P(V,T)$, which includes ΔP_{TH}. These data are plotted in Fig. 10.16, except as isotherms in dimensions of $\rho(P)$.

The 2000 K isotherm of Fig. 10.16 is replotted in Fig. 10.17 and compared to the 2000 K isotherm calculated by Hemley et al. (1992), who derived their curve from experiments of Mao et al. (1991) taken up to 1100 K and 30 GPa. Considering the assumptions involved, the agreement is quite satisfactory. The Hemley et al. data show a shift of P_{TH} by an increase of 5 GPa over our calculation, so their ΔP_{TH} is about 20 GPa at 2000 K. The parallel curves in Fig. 10.17 confirm the assumption that ΔP_{TH} is independent of V, the result of which is listed in Table 10.6.

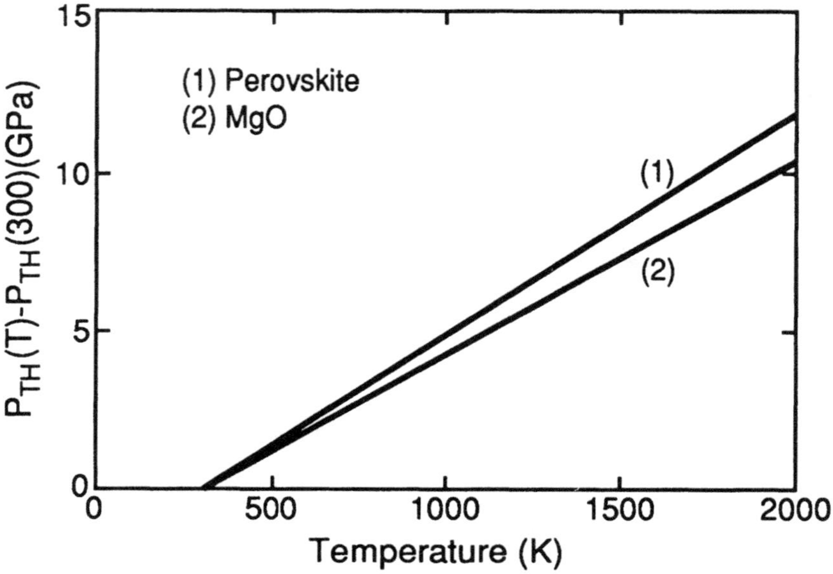

Fig. 10.15. Comparison of P_{TH} of silicate perovskite with that of periclase (modified from Anderson and Masuda, 1994).

Table 10.6. The EoS of silicate perovskite calculated from ΔP_{TH} using $\gamma_0 = 1.5$

$\eta = V/V_0$	P $T = 300$	P $T = 500$	P $T = 1000$	P $T = 2000$	P $T = 3000$
1.000	0.0	1.4	4.8	11.8	18.7
0.980	5.5	6.9	10.4	17.3	24.2
0.960	11.6	13.0	16.5	23.4	30.3
0.940	18.4	19.7	23.2	30.1	37.0
0.920	25.8	27.2	30.6	37.6	44.5
0.900	34.0	35.4	38.9	45.8	52.7
0.880	43.1	44.5	48.0	54.9	61.8
0.860	53.2	54.6	58.1	65.0	71.9
0.840	64.4	65.8	69.3	76.2	83.1
0.820	76.9	78.2	81.7	88.6	95.5
0.800	90.7	92.1	95.5	102.4	109.4
0.780	106.1	107.5	110.9	117.8	124.8
0.760	123.3	124.6	128.1	130.0	141.9
0.740	142.4	143.8	147.3	154.2	161.1
0.720	163.9	165.3	168.8	175.7	182.6

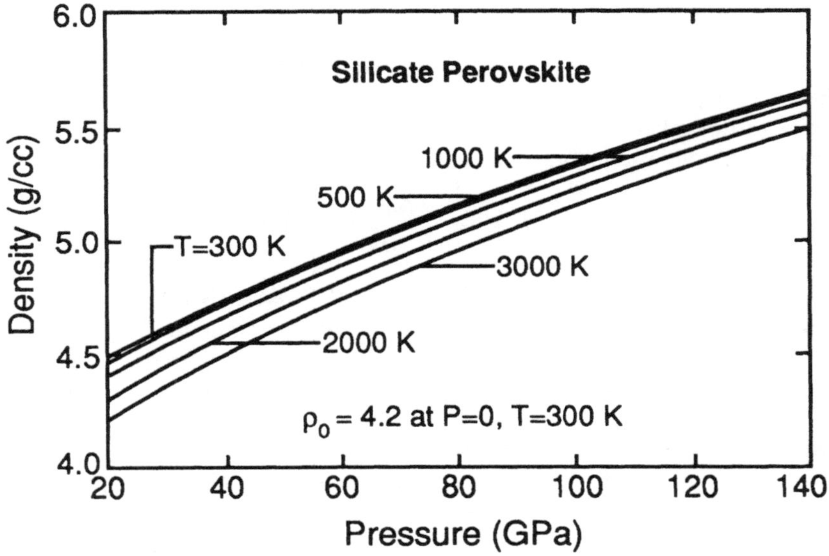

Fig. 10.16. The isothermal ρ versus P plots of silicate perovskite produced from the data in Table 10.6.

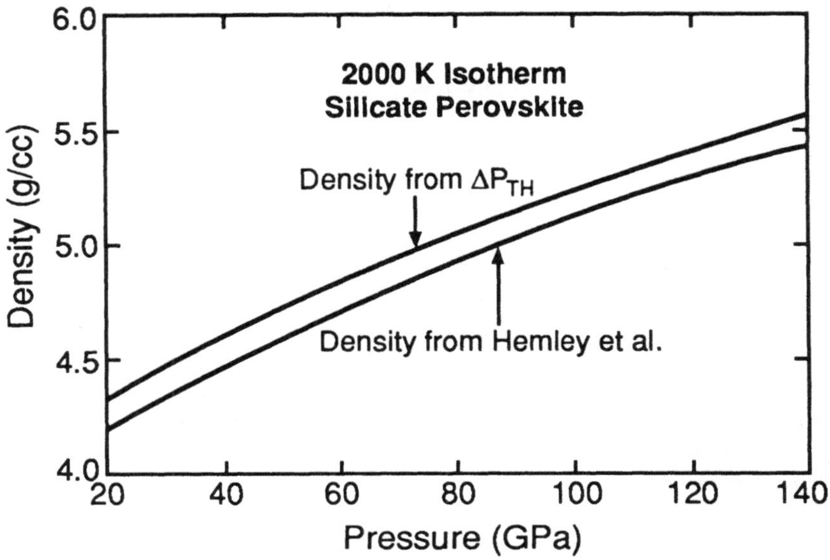

Fig. 10.17. Comparison of the 2000 K silicate perovskite isotherm in Fig. 10.16 with the density profile at 2000 K computed from Hemley et al. (1992) using the Mao et al. data measured up to 1100 K and 30 GPa.

10.11. Swenson's law

We have found that $(\partial K_T/\partial T)_V$ is close to zero for many classes of solids: alkali metals, noble gas solids, alkali halides, some metals, oxides, and ceramics (a notable exception is gold). Anderson (1979c) noted this common behavior in solids, and that using (3.2),

$$\left(\frac{\partial K_T}{\partial T}\right)_V \approx \left(\frac{\partial \alpha K_T}{\partial V}\right)_T \approx 0, \tag{10.51}$$

which he called "Swenson's law," because of Swenson's (1968) pioneering work. Many experiments described in this chapter confirm (10.51).

The approximation $(\partial \alpha K_T/\partial V)_T \approx 0$ may be regarded as a beginning point in defining P_{TH} for the EoS of solids, just as the ideal gas law, $PV = nRT$, is regarded as the beginning point for gases. (The ideal gas law is approximately true for gases over an extended V, T space, but requires modification as V gets quite small). Future theory for thermodynamics of solids may follow this lead and use (10.51) as a first order effect in defining P_{TH}. Following the analogy of the ideal gas law, we can expect (10.51) to be less and less true as V/V_0 gets smaller, as, for example, in the case of NaCl. On the other hand, as for MgO, $\mathbf{w} = (\partial K_T/\partial T)_V$, though small at $\eta = 1$, gets even smaller for lower η, as shown in Fig. 3.3. But these effects are minor and do not perturb (10.51) significantly.

The fact that Swenson's law works for so many different kinds of solids indicates that we are dealing with a correspondence principle that transcends solid type and is probably independent of chemical bond type and crystallographic class, as manifested in $(\partial P_{TH}/\partial T)_V \approx 0$.

10.12. Summary

As has been shown, the high temperature behavior of P_{TH} is quite similar for many classes of solids. The thermal pressure is linear in T, or very nearly so, over a large T range and depends only slightly, if at all, on V.

Of significance is P_{TH} of sodium metal measured up to near the solidus by Swenson (1968). His results show that P_{TH} is independent of volume and linear in T up to the melting point; no effects of anharmonicity are observed in P_{TH} even just below melting. In the case of oxides and silicates, such a high T measurement of P_{TH} is not possible. However, it has been shown that above $T = \Theta$ and at $P = 0$, the measured value of $\alpha K_T = (\partial P/\partial T)_V$ has some of the same characteristics as found for measured values of the other solids listed above. αK_T tends to be independent of T at high T, and departures from this independence disappear at high pressure.

For alkali metals, the data show that when the isochoric values of K_T are parallel to the T axis, (10.51) applies, and thus ΔP_{TH} is independent of volume. For the oxides and silicates measured so far, the data indicate also that αK_T is probably independent of V.

11

MELTING

11.1. Introduction

Over the last 50 years a sustained effort to quantify theoretically the solid–liquid phase transition has resulted in an enormous literature on the subject. This has not brought about uniform agreement to any satisfaction, and, indeed, as Cotterill (1980) pointed out, melting is still one of the least understood of the common physical phenomena known to man.

There are two opposing camps on the theory of melting. The first is that of the purists, who insist that melting requires imposing the thermodynamic condition that the Gibbs free energies of solids and liquids are equal, requiring that a detailed description of the atomic and electronic structures of the solid and of the liquid be quantified (Tallon, 1982). This approach is difficult because the details of the structure of the liquid are known with far less precision than those of the solid. The second camp insists that the nature of melting can often be understood by studying the stability properties of a solid near its melting point: melting is thought to result from a lattice instability. The earliest instability criterion was made by Lindemann (1910), who suggested melting occurs when the amplitude of the vibration of atoms reaches a value large enough that the solid lattice is shaken apart. Other early instability criteria were the ideas that the isothermal compressibility vanished at melting (Hertzfield and Goeppert-Meyer, 1934) or that a shear elastic constant vanished at melting (Born, 1939).

Sometimes the more fundamental approach is required. If a solid–liquid phase transition boundary is calculated from first principles, then the Gibbs free energy of the liquid phase must be specified. This approach is promising, and for a few cases of simple metals a lot of progress has been made. Of importance to geophysicists is the calculation by this method of the melting of iron at high pressures (Young and Grover, 1984; Cohen et al., 1994; Sherman, 1994; Stixrude and Cohen, 1994).

When extrapolation from existing experimental data on melting to extreme conditions is required, the second approach has proven useful, and much of this chapter will emphasize the instability approach to melting. Instability of the solid lattice is sometimes defined as the onset of the lack of long range order (as shown by x-rays) or the sudden loss of rigidity (as shown by the vanishing of the shear velocity). Some physicists have found these two phenomena to be equivalent (P.W. Anderson, 1984).

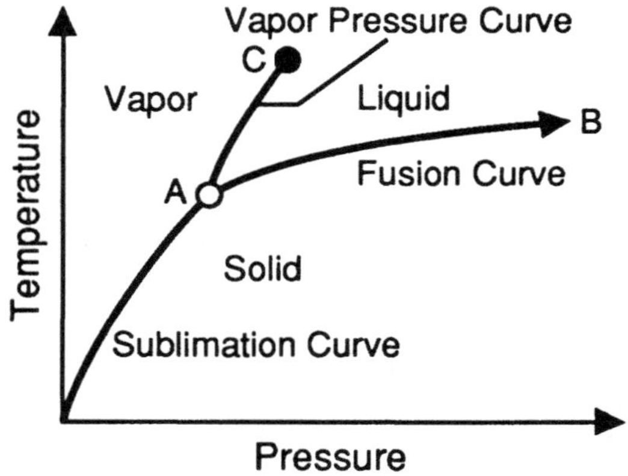

Fig. 11.1. Coexistence curves for a typical pure monatomic system in P, T space: A is the triple point, and C is the critical point; curve A–B is an example of a fusion curve with a positive slope.

Much of the work reviewed in this chapter is based on thermodynamics. The power of classical thermodynamics is available to melting theory because many applications do not require detailed knowledge of the structure of the solid treated. The consequence of ignoring structural considerations is that we cannot deal with theories of melting in which structural considerations are important, such as, for example, the structure of melts and the melting of molecular crystals, polymers, and rocks. Thus, this chapter emphasizes the phenomenological thermodynamics of melting. It is essentially restricted, therefore, to discussions of monatomic solids, such as metals, in which the vibrational density of states can be represented adequately by one frequency, the Debye frequency required in Lindemann's theory of melting. This approach can also be applied to oxides, silicates, and alkali halides that are Debye-like solids. For these solids, the vibrational density of states can be described adequately by a Debye frequency spectrum (see Chapter 5).

11.2. The Clausius-Clapeyron equation

For a monatomic substance, three phase boundaries coexist: the vapor–liquid, the solid–liquid, and the vapor–solid. All three phases can exist at a single point, the triple point (Fig. 11.1).

The Gibbs free energy is constant at and across a phase transition. Classification of a transition is often done through the derivatives of G. The n-order phase transition means that if the nth derivative of G is discontinuous, all $n-1$ are continuous. Because

$$\left(\frac{\partial \mathcal{G}}{\partial P}\right)_T = V, \tag{11.1}$$

and $V_{\text{SOLID}} \neq V_{\text{LIQUID}}$, and further, as

$$-\left(\frac{\partial \mathcal{G}}{\partial T}\right)_P = \mathcal{S}, \tag{11.2}$$

and $\mathcal{S}_{\text{SOLID}} < \mathcal{S}_{\text{LIQUID}}$, the melting is a first order (discontinuous) phase transition. We usually find

$$\Delta V_F > 0 \quad \text{and} \quad \Delta\mathcal{S}_f > 0.$$

A special case of critical melting might exist. Several theories treat the solid $\rightleftharpoons$ liquid transition as continuous. Such a transition might occur if $\Delta V_v \to 0$ and $\Delta\mathcal{S}_f \to 0$.

With two phases at equilibrium, $\mathcal{G}$ is formally the same for both phases. Let $\mathbf{L}$ be the latent heat going from phase 2 to phase 1, or

$$\mathbf{L} = \mathcal{H}_2 - \mathcal{H}_1 \tag{11.3}$$

at equilibrium. From the relationship between $\mathcal{H}$ and $T\mathcal{S}$, we have at melting $\mathbf{L} = T(\mathcal{S}_2 - \mathcal{S}_1)$, or $\mathcal{H}_2 - T\mathcal{S}_2 = \mathcal{H}_1 - T\mathcal{S}_1$, which is another statement that

$$\mathcal{G}_1 = \mathcal{G}_2. \tag{11.4}$$

This must be true at all points on the P-T curve. The phase boundary may then be described as

$$d(\mathcal{G}_2 - \mathcal{G}_1) = 0. \tag{11.5}$$

Expanding $d\Delta G$ out as a function of dT and dP, we have

$$\begin{aligned} d(\mathcal{G}_1 - \mathcal{G}_2) &= \left(\frac{\partial(\mathcal{G}_2 - \mathcal{G}_1)}{\partial T}\right)_P dT + \left(\frac{\partial(\mathcal{G}_2 - \mathcal{G}_1)}{\partial P}\right)_T dP \\ &= -(\mathcal{S}_2 - \mathcal{S}_1)\, dT + (V_2 - V_1)\, dP. \end{aligned} \tag{11.6}$$

The above is equal to

$$d(\mathcal{G}_1 - \mathcal{G}_2) = \mathbf{L}\,\frac{dT}{T} + (V_2 - V_1)\, dP. \tag{11.7}$$

From this we find the Clausius-Clapeyron law along the solid-liquid transition curve. Because $d(\mathcal{G}_2 - \mathcal{G}_1) = 0$ at $T = T_m$, (11.7) is

$$\frac{dP}{dT} = \frac{\mathbf{L}/T_m}{\Delta V_m}, \tag{11.8}$$

where ΔV_m is the volume change of fusion. Using $L/T_m = \Delta\mathcal{S}_m$, at $T = T_m$

$$\frac{dT_m}{dP} = \frac{\Delta V_m}{\Delta\mathcal{S}_m}. \tag{11.9}$$

Values of $\Delta V_f/V$, $\Delta\mathcal{S}_f$, and dT_m/dP for a number of elements are given in Table 11.1. Note that for some elements, like germanium, dT_m/dP is negative and accompanied by a negative volume change.

Table 11.1. Entropy and volume changes across a melting boundary with resulting dT_m/dP

Element	Crystal Structure	$\Delta V_f/V$	ΔS_f cal/mol°	dT_m/dP deg/GPa
Lithium	Body-centered cubic	0.0165	1.59	35
Sodium	Body-centered cubic	0.025	1.68	86
Potassium	Body-centered cubic	0.0255	1.65	169
Rubidium	Body-centered cubic	0.025	1.79	187
Cesium	Body-centered cubic	0.026	1.69	257
Aluminum	Face-centered cubic	0.060	2.74	55
Copper	Face-centered cubic	0.0415	2.30	33
Silver	Face-centered cubic	0.038	2.19	45
Gold	Face-centered cubic	0.051	2.21	59
Lead	Face-centered cubic	0.035	1.90	83
Magnesium	Hexag. close-packed	0.041	2.31	63
Zinc	Hexag. close-packed	0.042	2.55	37
Cadmium	Hexag. close-packed	0.040	2.44	53
Indium	Tetrag. face-centered	0.020	1.82	43
Tin	Tetrag. body-centered	0.028	3.41	32
Antimony	Rhombohedral	−0.0095	5.25	− 8
Bismuth	Rhombohedral	−0.0335	4.78	−38
Germanium	Cubic diamond	−0.05	6.28	−27
Argon	Face-centered cubic	−0.144	3.35	260
Krypton	Face-centered cubic	0.151	3.36	328
Xenon	Face-centered cubic	0.151	3.40	406

After Ubbelohde, 1965.

11.3. Development of the Lindemann law for melting

11.3.1. *Energy of a vibrating linear chain*

The statistical measure of amplitude motion is helpful in defining the melting temperature if one assumes that melting occurs with the onset of the loss of distant order. Consider a system of p coupled atoms. The degree of oscillatory motion in this one-dimensional chain is given by

$$<a^2> = \frac{1}{p}\overline{(a_1^2 + a_2^2 + \cdots a_p^2)}, \tag{11.10}$$

where a_1, a_2, ... are the amplitudes of the oscillations at a given instant of time. The bar signifies a time average.

The basic physical idea behind Gilvarry's (1956a) formulation is that when $< a^2 >$ is sufficiently large, a fraction of atoms in the system will have a very large amplitude and will be in a state more similar to that of a classical gas than to that of a solid. $<a^2>$ is a measurable quantity found from x-ray experiments representing the average condition of the lattice at high T (The critical amplitude a_c^2 appropriate to those atoms in the gaseous state is much larger than $<a^2>$).

We proceed to find the mean energy of a Planck oscillator by integrating over the elastic frequency spectrum of the solid. Brillouin (1946) derived the energy of a monatomic linear chain in oscillatory motion as follows:

Let the departure from equilibrium for an atom at site j be given by

$$y_j = a_j \sin(\omega t - \psi_j). \tag{11.11}$$

At high temperature the values of a_j are distributed over a spectrum at any given instant. The kinetic energy of the total chain is given by

$$\begin{aligned} E_{\text{kin}} &= \frac{1}{2}m\left(\dot{y}_1^2 + \dot{y}_2^2 + \cdots \dot{y}_p^2\right) \\ &= \frac{1}{2}m\sum_{j=1}^{p}\dot{y}_j^2 \\ &= \frac{1}{2}m\omega^2\sum a_j^2\cos^2(\omega t - \phi_j). \end{aligned} \tag{11.12}$$

Take the time-averaged value of the kinetic energy. We then have

$$\bar{E}_{\text{kin}} = \frac{1}{2}m\omega^2 <a^2> \sum \overline{\cos^2(\omega t - \phi_j)}.$$

Now

$$\overline{\cos^2(\omega t - \phi_j)} = \frac{1}{2}, \tag{11.13}$$

and

$$\bar{E}_{\text{kin}} = \bar{E}_{\text{potential}} = \frac{1}{2}E.$$

Thus the potential energy of the chain is

$$E = 2\bar{E}_{\text{kin}} = \frac{1}{2}m\omega^2 <a^2>, \tag{11.14}$$

where E is understood now to be the energy per atom.

Equation (11.14) is a greatly simplified representation of the energy in a solid at high T. For a monatomic isotropic crystal, the Debye model requires that ω be replaced by $\omega = k/(4\pi h)\Theta$, so that (11.14) is replaced by

$$E = m\left(\frac{k^2}{h^2 4\pi^2}\right)\Theta^2 <a^2> Q(x), \tag{11.15}$$

where $x = \Omega/T$ and $Q(x) = 1 + x^2/36 + x^4/360 + \ldots$ (Gilvarry, 1956a). The Debye theory is now mixed with Lindemann's theory.

For a polyatomic solid, m in (11.15) can be replaced by the reduced mass and $Q(x)$ by a more realistic value, by integration of the frequency spectrum. These refinements are in the numerical constants. The values of V, Θ, and $< a^2 >$ are the important variables. These numerical constants are removed by a subsequent differentiation.

11.3.2. *Melting conditions using the Gilvarry criterion*

Lindemann's (1910) theory was a relation between melting and the Einstein vibrational frequency. To find the Einstein frequency, melting was assumed to occur when the space between surfaces of atomic spheres on lattice points vanished. Gilvarry (1956a) made this a theory of melting by proposing that melting occurs when the root-mean-square amplitude is a critical fraction $\bar{e}$ (called the Gilvarry criterion) of the distance separating nearest neighbor atoms; that is, melting occurs when

$$< a^2 > = \bar{e}^2 r_0^2. \tag{11.16}$$

Taking the logarithm of (11.15) and using (11.16), we have

$$\ell\text{n } E = 2\ \ell\text{n } \Theta - (2/3)\ \ell\text{n } \rho + \text{constant}. \tag{11.17}$$

All numerical values in the parens of (11.15), and also $\bar{e}$, are in the constant of (11.17).

Under adiabatic conditions of compression, the Debye temperature is related to the density by (Grüneisen, 1926)

$$\left(\frac{\Theta}{\Theta_0}\right) = \left(\frac{\rho}{\rho_0}\right)^{\gamma}, \tag{11.18}$$

where γ is the Grüneisen ratio. Using (11.18) in (11.17), the Grüneisen theory is introduced into the Lindemann-Gilvarry formulation:

$$d\ \ell\text{n } E = 2\,(\gamma - (1/3) + f)\ d\ \ell\text{n } \rho, \tag{11.19}$$

where

$$f = \frac{d\ \ell\text{n } \bar{e}}{d\ \ell\text{n } \rho}. \tag{11.20}$$

The constants identifying the energy belonging to either a one-dimensional or three-dimensional lattice disappear in (11.19) when the derivative of a constant is taken. Equation (11.19) is appropriate for a solid at high T for which a single value Θ is a reasonable representation of the thermal properties, e.g., the solid is Debye-like. For a criterion on the identification of a Debye-like solid, see Chapter 5. Note that a test of (11.19) may be a test of the Debye model as much as of the Lindemann model. Equation (11.19) is for adiabatic compression. For isothermal compression the amplitude is higher for constant E, and (11.20) is replaced by

$$f = \left(\frac{d\,\ell\text{n}\,\bar{e}}{d\,\ell\text{n}\,\rho}\right)_S \left[1 + \left(\frac{d\,\ell\text{n}\,e}{d\,\ell\text{n}\,T}\right)_P\right]. \tag{11.21}$$

11.3.3. *The Lindemann law with the Gilvarry criterion*

We now have a criterion giving ΔE in terms of $\Delta\rho$, the Grüneisen constant, and $< a^2 >$. Equation (11.19) is applied to a solid at the fusion temperature.

We now use (11.19) at melting and replace E with the classical formula nRT_m, where n accounts for the degrees of freedom. Thus (11.19) becomes the Gilvarry criterion of melting,

$$d\,\ell\text{n}\,T_m = 2\left(\gamma - \left(\frac{1}{3}\right) + f\right) d\,\ell\text{n}\,\rho. \tag{11.22}$$

The above equation is usually written without the parameter f, but to eliminate f requires an additional assumption. To evaluate f we must know how the value of $\bar{e}$ changes with density. Gilvarry proposed that $\bar{e}$ remains constant as ρ increases. We now have a simple formula that shows how T_m changes with ρ, which is now widely represented as the Lindemann law for melting, though it should be called Gilvarry's law,

$$\frac{d\,\ell\text{n}\,T_m}{d\,\ell\text{n}\,\rho} = 2\left(\gamma - \frac{1}{3}\right). \tag{11.23}$$

By taking f to be zero, we assume that (11.23) is true everywhere on the fusion curve. Equation (11.23) is therefore a scaling law; that is, if we know $\gamma(\rho)$ and we know one value of T along the fusion curve, we can find T at every point along the fusion curve. Equation (11.23) is valid only for a Debye-like solid because of Gilvarry's inclusion of (11.18) in the derivation. A failure of (11.23) may be expected for a non Debye-like solid.

11.4. The Simon law: a special case of the Lindemann law

To evaluate the pressure variation of T_m, we need to use an equation of state. Using the definition of the bulk modulus, (11.22) becomes

$$\frac{d\,\ell\text{n}\,T_m}{dP} = \frac{2(\gamma - (1/3) + f)}{K_T}, \tag{11.24}$$

where $(1/K_T) = (1/\rho)(\partial\rho/\partial P)_T$. Let us assume K_T is linear in P. Then

$$\begin{aligned}\frac{T_m}{T_{m_0}} &= \left[1 + K_0'\left(\frac{P}{K_0}\right)\right]^{2(\gamma - 1/3 + f)} \\ &= \left[1 + K_0'\left(\frac{P}{K_0}\right)\right]^C.\end{aligned} \tag{11.25}$$

Equation (11.25) is a variation of the Simon melting law (Simon and Glatzel, 1929). Gilvarry showed (1956b) that the Simon melting law is really a combination of the Murnaghan equation of state and the Lindemann law.

11.5. The Kraut-Kennedy law based on the Lindemann law

11.5.1. *Kraut-Kennedy law description*

Consider (11.22). By replacing dT_m with $T_m - T_{m_0}$ and $d\rho/\rho$ with $-\Delta V/V_0$, (11.22) can be written as

$$T_m = T_{m_0}\left(1 + C\frac{\Delta V}{V}\right), \tag{11.26}$$

where

$$C = 2\left(\gamma - \frac{1}{3} + f\right), \tag{11.27}$$

and where T_m, γ, and $\Delta V/V$ are appropriate to the melting boundary.

Equation (11.26), known as the Kraut-Kennedy (1966) equation of melting, shows that T_m is linear with $\Delta V/V_0$. Alkali metals obey the Kraut-Kennedy equation to very large compressions (Luedemann and Kennedy, 1968), as do some metals (Fig. 11.2). But for many solids, (11.26) holds only for small compressions; Fig. 11.3 shows the case for some alkali halides, a silicate, and an oxide.

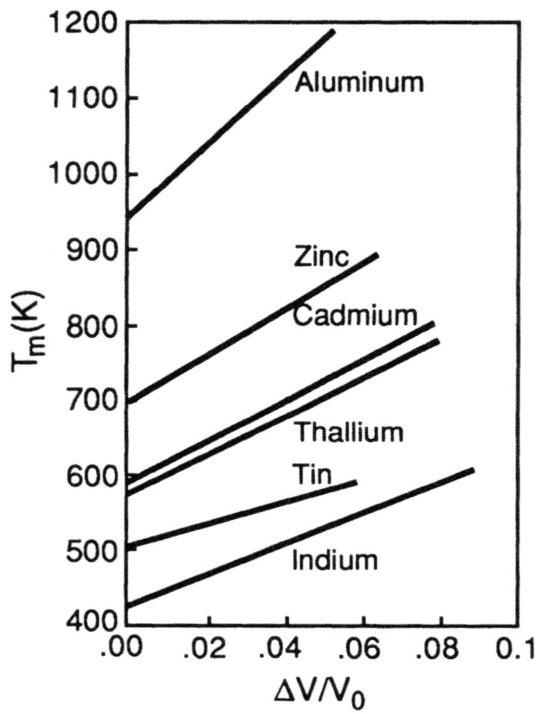

Fig. 11.2. Demonstration that the Kraut-Kennedy law (T_m linear in $\Delta V/V_0$) is true for many metals (modified from Kennedy and Vaidya, 1970).

The Kraut-Kennedy law is a simple restatement of the Lindemann law, but, as shown in Fig. 11.3, it may be valid only at low pressure. The linearity of T_m with $\Delta V/V$ was derived by Gilvarry (1956a) 10 years before Kennedy's paper, but Gilvarry's contribution was not generally recognized until Kennedy firmly pointed out the importance of relating the melting temperature to volume instead of to pressure, as was the customary practice before Kennedy's papers on the subject.

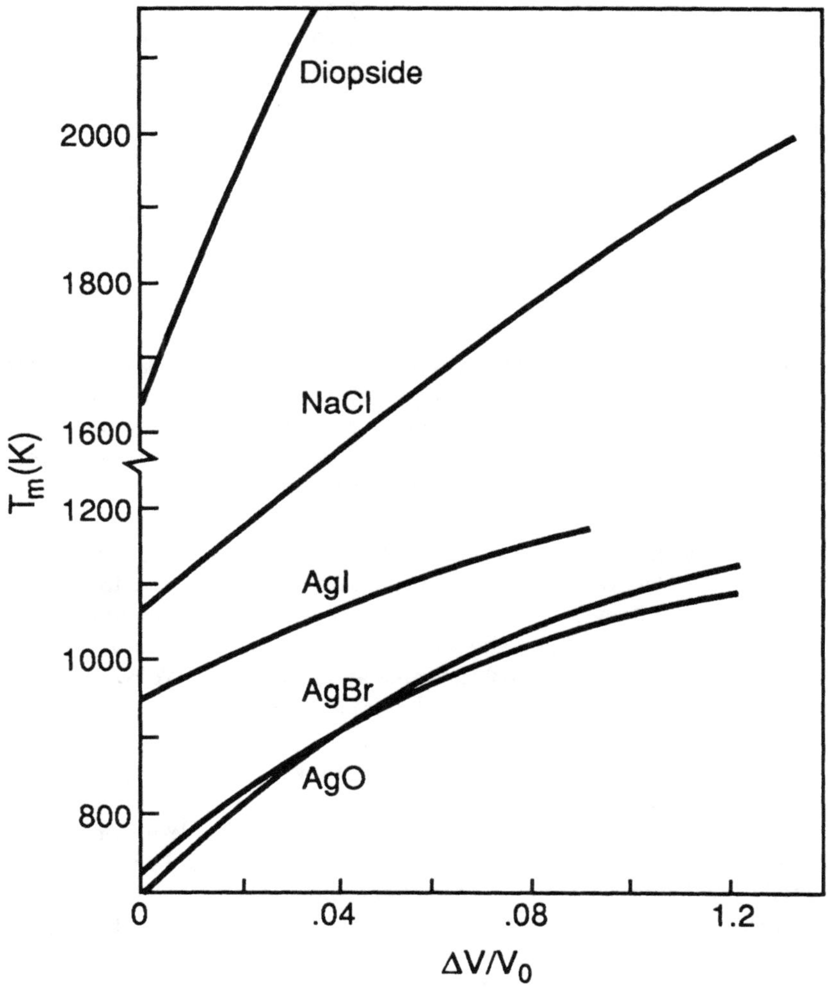

Fig. 11.3. Demonstration that the Kraut-Kennedy law (T_m linear in $\Delta V/V_0$) does not hold for a class of solids, except at low compressions (modified from Kennedy and Vaidya, 1970). This class of solids includes alkali halides, some silicates, and some oxides (diopside and AgO are examples of the latter two).

Table 11.2. Constants in (11.25) and (11.26)

Substance	C/deg measured	γ room temp.	f from (11.26)
Li	0.55	0.90	−0.18
Na	1.66	1.33	−0.17
K	1.60	1.37	−0.24
Rb	1.75	1.85	−0.65
Cs	1.68	1.49	−0.32
Cu	3.96	1.92	+0.39
Ag	5.26	2.46	+1.00
Au	8.12	3.09	+1.30
Al	4.70	2.17	−0.50
Te	4.45	0.92	+1.72
NaCl	4.88	2.0	+1.10
Ge	−2.67	0.8	−1.73
Sb	−0.28	0.84	−0.65

In most applications of the Kraut-Kennedy law, f is taken to be zero, and γ is the room temperature (low pressure) value. Using this value of f may create some error (as γ at melting is seldom known), but we can test the validity of (11.26) by using values of C derived from Kennedy's many experiments and the reported value of γ to find the value of f shown in Table 11.2.

Examination of Table 11.2 leads to the following: For the alkali metals (Li, Na, K, Rb), the value of f is small compared with that of γ. In this table the value of γ is taken from room or low temperature experiments, whereas the value of γ should be the value appropriate to the fusion curve. There is no reason to believe γ is exactly the same at T_m as at T_0. We conclude that (11.26) is reasonably satisfactory. The value of f may in fact be close to zero due to the use of an incorrect value of γ. We see that if γ is less than about 1/3, the gradient of T_m with density may be negative by (11.25), as in the case of germanium and antimony (Kraut and Kennedy, 1966) (see Table 11.1). Thus it is possible to account for a negative fusion gradient by the most elementary assumptions in the Lindemann law, the Kraut-Kennedy law being a restricted case of the Lindemann law.

In the framework of the Lindemann law, f should be zero because $\bar{e}$ is presumed to be independent of P. The values of f in Table 11.2 may be non-zero due to errors in γ or C. There have been few reports on how f varies with P. One theoretical paper for MgO shows that f is zero up to at least 300 GPa (see Section 11.9).

Recognizing that γ is a decreasing function of density, C is not invariant with density but should gradually also decrease with density for higher values of $\Delta V/V$. This changing slope behavior is shown experimentally in some of Kennedy's later works (Akella et al., 1969; Kennedy and Vaidya, 1970) (Fig. 11.3). Thus we expect that the Kraut-Kennedy law is only strictly applicable in the pressure range where γ is insensitive to pressure (Leudemann and Kennedy, 1968). In the lower mantle of the earth, γ decreases steadily with density increase, satisfying $\gamma\rho =$ constant (Anderson, 1979b). Similarly, a change of γ with ρ will be found when the pressure is sufficiently high.

11.5.2. *From the Clausius-Clapeyron equation*

Using the Clausius-Clapeyron equation (11.9), it is easy to show that

$$T_m - T_{m_0} = \left(\frac{\Delta V_f}{\Delta S_f}\right)\left(\frac{K_T \Delta\rho}{\rho_0}\right), \tag{11.28}$$

which is valid over the same range of density that $d\rho/\rho = (\rho - \rho_0)/\rho_0$ is considered valid. Equation (11.28) can be placed in the Kraut-Kennedy form given by (11.26) (Libby, 1966) if

$$C = \left(\frac{\Delta V_f}{\Delta S_f}\right)\left(\frac{K_T}{T_{m_0}}\right). \tag{11.29}$$

The value of C from thermodynamic data using (11.29) compares reasonably well with the value of C measured by Kennedy and his colleagues for a number of solids (Table 11.3). Thus the Kraut-Kennedy law has some support from the Clausius-Clapeyron equation.

Table 11.3. Computation of Kraut-Kennedy constant C from the Clausius-Clapeyron equation using (11.28)

Substance	V cc/g-atom	$\Delta V_f/V$	K_T kbar	T_m	ΔS_f entropy units	C cal	C obs
Li	13.02	0.016	132	454	1.58	0.94	0.55
Na	23.79	0.0755	66	390	1.68	1.49	1.66
K	45.61	0.025	40	335	1.65	2.01	1.60
Rb	56.07	0.025	32	331	1.80	1.83	1.75
Cs	69.1	0.026	20	302	1.68	1.73	1.68
Cu	7.12	0.051	1371	1356	2.30	3.82	3.96
Ag	10.27	0.033	1087	1234	2.25	3.23	5.26
Au	10.22	0.051	1803	1336	2.21	7.75	8.12
Al	10.00	0.060	764	932	2.74	4.37	4.70
Tl	17.22	0.072	366	575	1.77	10.86	4.45
NaCl	27.10	0.250	232	1093	6.7	5.19	4.88

11.6. The Lindemann law at high compression

At high compression the Lindemann law is more reliable if γ is a function of ρ, unlike the assumption γ = constant used in deriving the Simon law or the Kraut-Kennedy equation. In past chapters we have found that, over certain ranges of P and T, a reasonable approximation is

$$\frac{\gamma}{\gamma_0} = \left(\frac{V}{V_0}\right)^q = \left(\frac{\rho}{\rho_0}\right)^{-q} \tag{1.32}$$

(Figures 3.12 and 3.13 show the regions where (1.32) is not valid). Using (1.32) in (11.23) and integrating, the relationship between T_m and ρ is (Anderson, 1986)

$$\frac{T_m}{T_{m_0}} = \left(\frac{\rho_0}{\rho}\right)^{2/3} \exp\left\{\frac{2\gamma_0}{q}\left[1 - \left(\frac{\rho_0}{\rho}\right)^q\right]\right\}. \tag{11.30}$$

The Spiliopoulos and Stacey (1984) equation takes $q = 1$ or

$$\frac{T_m}{T_{m_0}} = \left(\frac{\rho_0}{\rho}\right)^{2/3} \exp\left\{2\gamma_0\left[1 - \left(\frac{\rho_0}{\rho}\right)\right]\right\}. \tag{11.31}$$

11.7. The Lindemann law at $P = 0$

11.7.1. *Critical fractional amplitude for melting*

Redefine (11.14) using (11.16) and assume a Debye-like solid by taking ω_D for ω:

$$3RT_m = \frac{8}{3}\pi^2 m\omega_D^2 \bar{e}^2 r^2. \tag{11.32}$$

Here $8/3\pi^2$ replaces $1/2$ as a more reasonable factor representing the integration of the elastic frequency spectra for a three dimensional solid (Grüneisen, 1926), and ω_D is the Debye frequency. We then have

$$\omega_D^2 = \frac{9RT_m}{8\pi^2\bar{e}^2 m V_0^{2/3}}. \tag{11.33}$$

The dimensionless value of $\bar{e}$ is of great interest. Solving for $\bar{e}$,

$$\bar{e} = \frac{1.19}{\Theta}\left(\frac{T_m}{mV_0^{2/3}}\right)^{1/2}, \tag{11.34}$$

where Θ is the Debye temperature ($\bar{e}$ should not vary within a crystal class). Failure of (11.34) when it is tested against experimental data may not necessarily imply failure of the Lindemann criteria. Such a failure may arise because the solid tested is not a Debye-like solid (see Chapter 5).

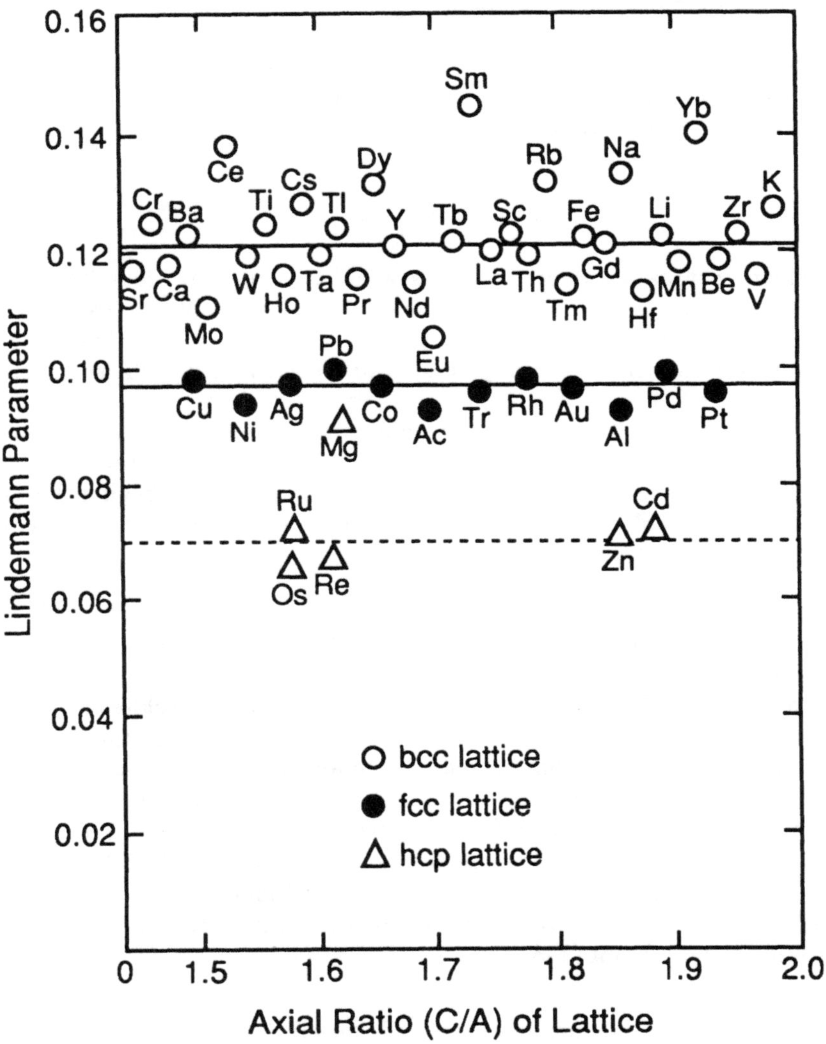

Fig. 11.4. Lindemann parameters calculated for monatomic metals by (11.34) (modified from Cho, 1982).

The Lindemann derivation used the simplest model for a vibrating solid, namely, a single characteristic vibrational frequency, here represented by Θ (Ubbelohde, 1965). Gilvarry (1956c) found that $\bar{e}$ in (11.16) is of the order of 0.1, but the value apparently depends on the crystal class. Cho (1982) did an extensive study of metals to determine $\bar{e}$ and found that the values of $\bar{e}$ did indeed depend on crystal class. He determined that $\bar{e} = 0.12$, 0.096, and 0.07 for bcc, fcc, and hcp, respectively (see Fig. 11.4).

11.7.2. *The Lindemann constant, $\mathcal{L}$*

Equation (11.34) can be viewed in another way. Rearranging the right side of (11.34) and letting $\bar{e}$ be a constant, we have another scaling law relating T_m, Θ, and m. By this rearrangement we define a constant, traditionally called the Lindemann constant, $\mathcal{L}$

$$\mathcal{L} = \Theta V_0^{1/3} \left(\frac{m}{T_m} \right)^{1/2}, \qquad (11.35)$$

in which all parameters are readily available. Cho (1982) found $\mathbf{L} = 126$ for bcc metals, 154 for fcc metals, and 218 for hcp metals. Values of $\mathbf{L}$ for alkali halides and diamond structured elements are given in Table 11.4. Cho showed that the uniformity of $\mathbf{L}$ is fairly good within each crystal class structure.

J.P. Poirier (1988) found that oxides and fluorides in the perovskite structure had a common value of $\bar{e}$ near 0.13, and he felt that this was sufficient evidence that perovskites should be classified as obeying the Lindemann law (Table 11.5). Poirier (1988) defined a different $\mathcal{L}$ parameter, essentially the square of (11.35) without T_m:

$$\mathcal{L}' = mV^{2/3}\,\Theta^2 \times 10^7. \qquad (11.36)$$

He applied (11.36) to oxides and fluorides of the perovskite structure, correlating $\mathcal{L}'$ and T_m, as shown in Table 11.5 and Fig. 11.5. The linearity between $\mathcal{L}'$ and T_m was additional evidence that the perovskites obey Lindemann's law at low pressures given by (11.35), in spite of the fact that silicate perovskite is a complicated silicate and certainly not a monatomic solid (a founding assumption of the Lindemann law). Silicate perovskite, however, is a Debye-like solid (see Chapter 5), and therefore can be treated as a monatomic solid. Poirier used the linearity in Fig. 11.5 to estimate T_{m_0} for orthorhombic perovskite, Mg_2SiO_3, and found $T_{m_0} = 2564$ K (see Fig. 11.5), close to the zero pressure measurements of Heinz and Jeanloz (1987) and Ohtani (1983).

Table 11.4. Lindemann parameter **L** for alkali halides and diamond-like (d.l.) structures

Struc. fcc	**L**	T_m °K	Struc. bcc	**L**	T_m °K	Struc. (d.l.)†	**L**	T_m °K
NaCl	210	1081	RbBr	206	955	Ge	209	1232
KCl	195	1049	AgBr	229	705	Si	195	1683
AgCl	239	728	LiF	200	1115	C	*	$\simeq 4000$
KI	180	959	CaF_2	211	1633	H_20	217	273
RbI	189	915						

*154–198; †diamond-like.

Table 11.5. Physical properties of the perovskites: M_{mol}, molar mass in g/mole; V, molar volume in cc/mole; v_m, average acoustic velocity in km/s; T_m, melting temperature in °K; Θ, acoustic Debye temperature in °K; $\mathcal{L}'$, Poirier ratio (Poirier, 1988); $\bar{e}$, Gilvarry ratio

ABX_3	M_{mol}	V	v_m	T_m	Θ	$\mathcal{L}'$	$\bar{e}$
$KMgF_3$	120.41	38.23	4.29	1413	315	2.75	0.13
$KMnF_3$	151.04	44.16	3.43	1308	240	2.21	0.14
$KZnF_3$	161.32	40.13	3.43	1143	248	2.36	0.12
$KNiF_3$	154.97	38.84	3.53	1403	258	2.40	0.13
$KCoF_3$	155.01	40.58	3.38	1305	243	2.19	0.14
$RbMnF_3$	197.40	45.69	3.13	1259	217	2.41	0.13
$RbCoF_3$	201.54	42.34	3.31	1148	235	2.74	0.11
$ScAlO_3$	120.00	28.04	6.19	2143	505	5.71	0.11
$GdAlO_3$	232.25	31.22	4.43	2303	348	5.64	0.11
$SmAlO_3$	225.35	31.39	4.57	2373	358	5.81	0.11
$EuAlO_3$	226.94	31.30	4.41	2213	347	5.49	0.11
$YAlO_3$	163.89	30.63	5.09	2223	403	5.27	0.11
$CaTiO_3$	135.98	33.55	5.63	2248	431	5.33	0.11
$BaTiO_3$	232.24	38.62	3.77	1898	276	4.11	0.12
$SrTiO_3$	138.52	35.84	5.29	2213	397	6.37	0.11
$MgSiO_3$	100.41	24.43	7.39		630	6.78	

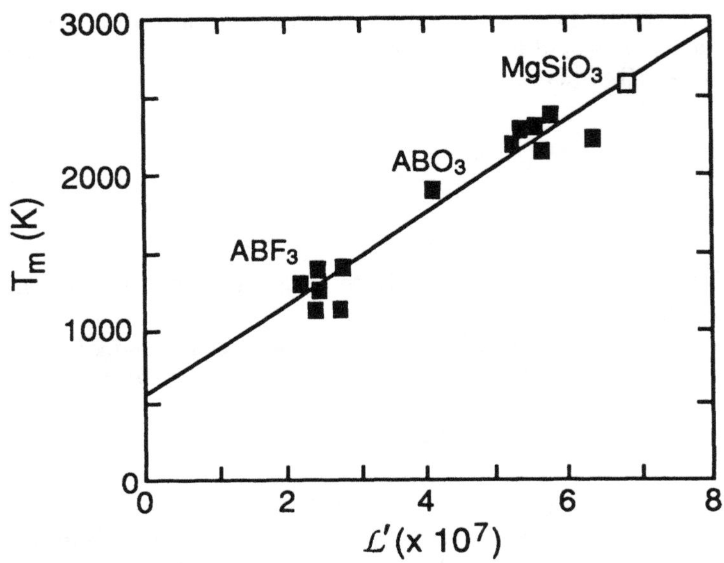

Fig. 11.5. Correlation of T_m with the quantity $\mathcal{L}' = mV^{2/3}\Theta \times 10^{-7}$ at $P = 0$ for 15 oxide (ABO_3) and fluoride (ABF_3) perovskites. The open square is for silicate perovskite (modified from Poirier, 1988).

11.8. Improvements on the Lindemann formulation

Many authors are dissatisfied with the assumptions behind the Lindemann law of melting, and there have been several attempts to derive it from more fundamental considerations. Lindemann's (1910) hypothesis was that melting occurred when the vibrations of the atoms became large enough to cause atomic collisions. This vague criterion was quantified by Gilvarry (1956a), who proposed that melting occurs when $< a^2 >^{1/2}$ exceeds a fraction of the interatomic distance, r_0.

Here we comment on four attempts to improve and/or justify (11.23). In all these theories, refinements of the assumptions lead to placing a very small correction term in (11.23).

(1) Gilvarry (1956a) started with the Debye-Waller formula for the Bragg angle of an isotropic monatomic crystal

$$M = \left(\frac{6h^2T}{mk\Theta^2}\right)\left(\sin\frac{\Theta}{\lambda}\right)^2 Q\left(\frac{h\nu}{kT}\right), \tag{11.37}$$

where M = constant; Θ = Bragg angle; λ = wavelength; and Q is defined in terms of the Debye function. Gilvarry's final equation is

$$\frac{\partial \ell\text{n } T_m}{\partial \ell\text{n } \rho} = 2\left(\gamma - \frac{1}{3}\right)\left(1 + \gamma C_V \frac{\Delta V}{V} M\right). \tag{11.38}$$

The last term in the second set of parentheses is much less than unity.

(2) Stevenson (1980) derived a melting equation from a theory of the liquid state that would be more appropriate for the liquid side of the transition curve. His equation is

$$\frac{\partial \ell\text{n } T_m}{\partial \ell\text{n } \rho} = \frac{2\left(\gamma - \dfrac{k}{C_V}\right)}{\left(2 - 3\dfrac{k}{C_V}\right)}, \tag{11.39}$$

where k is the Boltzmann constant. In a classical solid, $C_V = 3k$, so the above formula is close to (11.23).

(3) Stacey and Irvine (1977) derived their equation using the Clausius-Clapeyron equation, (11.9), replaced by the equivalent form

$$\frac{\partial \ell\text{n } T_m}{\partial \ell\text{n } \rho} = \left(\frac{K_T}{\mathbf{L}}\right)\Delta V_f. \tag{11.40}$$

They calculated the value of $\Delta V_f/\mathbf{L}$ by considering a thermodynamic cycle involving melting at constant pressure and at constant volume, a kind of Carnot cycle passing through the liquidus and solidus. They found

$$\frac{\Delta V_f}{\mathbf{L}} = \frac{2(\gamma - 2\gamma^2\alpha T_m)}{K_T}. \tag{11.41}$$

Because αT_m is a number near 0.07 for all solids, $2\gamma^2\alpha T_m = 0.23$, not substantially different from the 1/3 found in Lindemann's law (11.23). What is important is that their basic assumption involves the Clausius-Clapeyron equation, which is derived from the equilibrium of the Gibbs free energy of solid and liquid as shown in Section 11.2.

(4) Poirier (1986) based his derivation on dislocation theory. He noted that a concentration of dislocations introduces a dilation in the lattice ϵ_d and that melting occurs when the concentration of dislocations destroys the order locally; that is, when saturation occurs, or $\epsilon_d \to \epsilon_d^{\rm sat}$. His theory rests on the equality of the Gibbs free energy of the solid and molten states. Saturation occurs, as determined by molecular dynamics simulation, when about one-third of the atoms of a crystal are on a dislocation line. The final equation found by Poirier is

$$\frac{\partial \ \ell{\rm n}\ T_m}{\partial \ \ell{\rm n}\ \rho} = 2\left(\gamma - \frac{1}{3}\right)\left[1 - \epsilon_d^{\rm sat}\left(\gamma - \frac{1}{3}\right)\right]^{-1}. \tag{11.42}$$

The quantity in brackets is only slightly larger than unity, so that to a good approximation, the original Lindemann law is verified.

11.9. Verification of the Lindemann law for a dense oxide

Cohen and Gong (1994) performed large scale simulations of clusters of MgO using molecular dynamics for melting up to 300 GPa. They found that melting occurs when the r.m.s. displacement of the ions reaches about 18% of the near neighbor distance (Fig. 11.16), in agreement with the Lindemann-Gilvarry criterion (11.16). They concluded that for MgO, melting is indeed related to intrinsic instability of crystals, and then verified Gilvarry's basic assumption that $d\ \ell{\rm n}\ \bar{e}/d\ \ell{\rm n}\ \rho$ vanishes over a large pressure range. In their theory they made no reference to a critical frequency, and their result is not handicapped by the assumption of a single value of Θ. They did not assume a Debye-like solid.

Their potential is the nonempirical and many-body approach, called the potential induced breathing (PIB) model (Boyer et al., 1985), which was used to find thermoelastic properties of MgO (Isaak et al., 1990). The thermoelastic properties of MgO found by PIB are in Chapters 3 and 4.

Cohen and Gong show that although melting is a first-order thermodynamic phase transition, it is related in a fundamental way with an intrinsic instability of the solid. This results in ΔV_m changing from 5.4 cc/mol to 0.07 cc/mol K and ΔS_m changing from 27 J/mol K to 16 J/mol K over the equivalent pressure range. The liquid changes its coordination number M from 4.5 at $P = 0$ to about 5.8 at 300 GPa, and, as a result, its bulk modulus increases with P. At large P, the bulk modulus of the liquid approaches that of the solid. The large change in K_T with P results in ΔV changing faster than ΔS, and consequently, dT_m/dP decreases with P (200 K/GPa at $P = 0$; 4 K/GPa at $P = 300$ GPa).

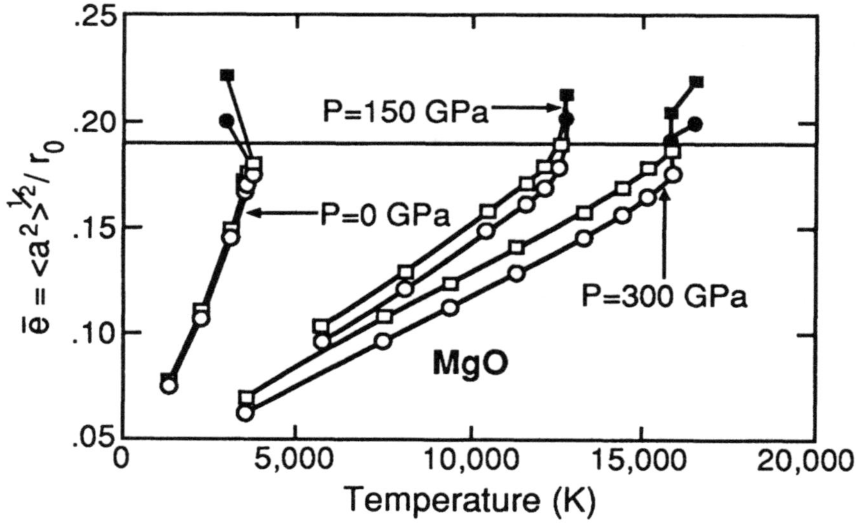

Fig. 11.6. The Gilvarry criterion $\bar{e} = < a^2 >^{1/2} / r_0$ versus T along isobars for MgO using MD calculation of 1000 atom clusters (modified from Cohen and Gong, 1994). Open symbols are for the crystalline state, closed symbols for the liquid state. We see that $\bar{e} = 0.18$, and $d\bar{e}/dT$ is infinite at the transition. Here is proof of Gilvarry's basic assumption, (11.20), because $\bar{e}$ is independent of density over a wide pressure range.

11.10. The Lindemann law for oxides and silicates

The most quoted objection to the Lindemann law is that it works for some classes of solids but not for other classes. But for oxides and silicates, the Lindemann law is not always valid, as shown by the evidence of Wolf and Jeanloz (1984), who proposed a modification of the Lindemann theory to include anharmonicity and tested four minerals. They found (11.33) invalid for diopside, fayalite, and pyrope (see Fig. 11.7 for diopside), but there was excellent agreement with the data for forsterite (see Fig. 11.8).

In Section 11.3, we saw that the Lindemann formula is based on the assumption of a monatomic solid, and, in particular, that a single frequency, ω_D, enters the formula for the r.m.s. amplitude. Obviously an important criterion is: how valid is the approximation that the frequency spectrum of the oxide or silicate can be represented by a single number?

In Chapter 5 we found that if the structure has an efficient packing (e.g., the packing fraction is large), then a single value of Θ will suffice to describe thermoelastic properties, but if the structure has a small packing fraction, a single value of Θ will not suffice.

Similarly we found in Section 5.5 that the moments of γ, a single number, can represent a property. The moment $\gamma(-2)$ is appropriate for the

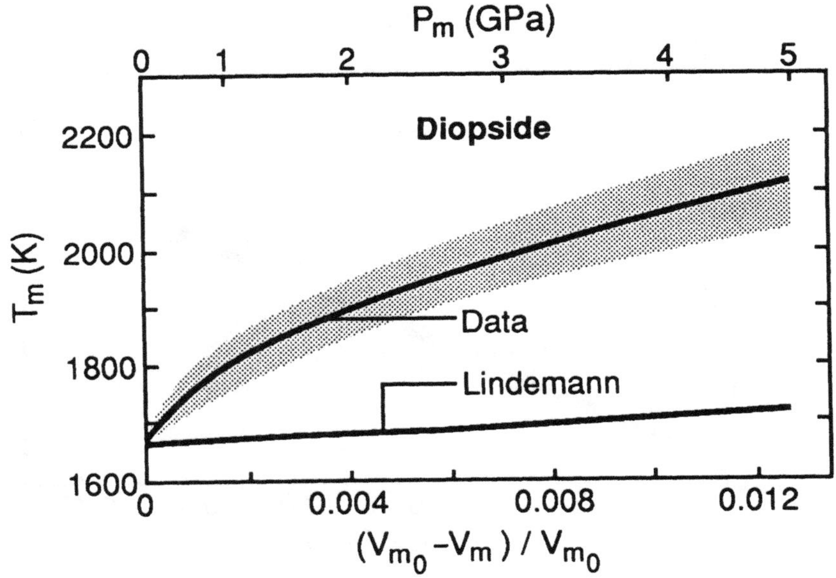

Fig. 11.7. Test of the Lindemann equation as a scaling law for diopside. The Lindemann law fails for diopside (after Wolf and Jeanloz, 1984).

Lindemann formula (Wolf and Jeanloz, 1984), but it is still a single number and inadequate for the class of solids described as Class A in Chapter 5.

Thus, according to the packing fraction (Table 5.2), diopside and fayalite are not Debye-like solids, and forsterite is a marginal case.

On the other hand, the dense silicates and oxides, especially MgO, Al_2O_3, and $MgSiO_3$ perovskite, have well packed structures, and are therefore Debye-like solids. Just as MgO was shown to follow Lindemann's law (see Fig. 11.6), silicate perovskite should also follow Lindemann's law. There are some theoretical and experimental data for silicate perovskite on whether T_m versus P follows a Lindemann pattern, but the results are conflicting.

Using Lindemann's law, Poirier (1986) computed T_m versus P for $MgSiO_3$ perovskite (shown in Fig. 11.9). Although T_m rose with P in the Lindemann fashion, Ohtani (1983), also using Lindemann's law, found T_m rose faster. Stixrude and Bukowinski (1990) found that T_m rose at an even smaller rate by computing the Gibbs energy of both solid and liquid.

The experimental results are also mixed. Several separate experiments of the Berkeley school (Heinz and Jeanloz, 1987; Knittle and Jeanloz, 1989; Sweeney and Heinz, 1993) all found that there is little tendency of T_m to rise with P in the Lindemann law fashion. On the other hand, Zerr and Boehler (1993) of the Mainz group found experimentally that T_m rises with P, close to the prediction of Ohtani (Fig. 11.9). From Table 11.5 and

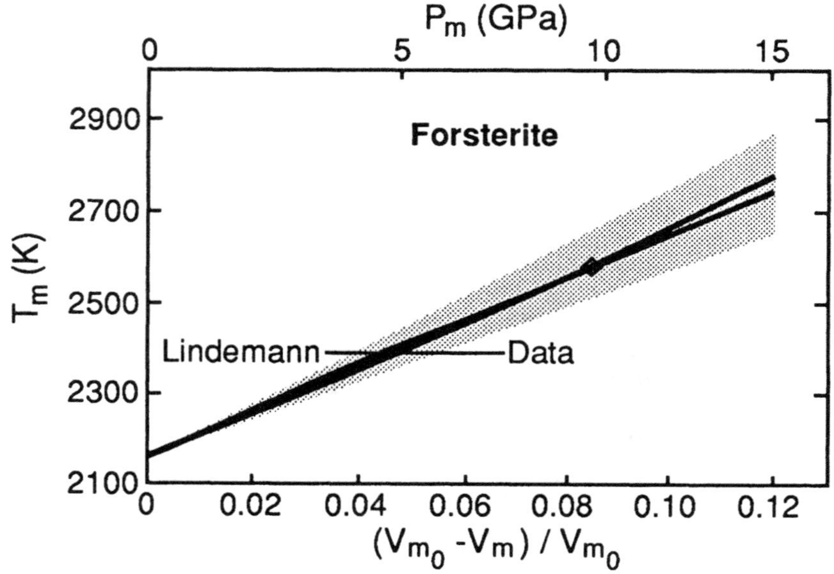

Fig. 11.8. Test of the Lindemann equation as a scaling law for forsterite. The Lindemann law succeeds for forsterite (modified from Wolf and Jeanloz, 1984).

Figs. 11.5 and 11.9, I conclude that a silicate perovskite probably obeys Lindemann's law. Marginal cases involve forsterite and pyrope garnet, according to the packing fraction criteria (Table 5.2). But, according to Fig. 11.7, it appears that forsterite does follow Lindemann's law. The criterion for Lindemann law application to oxides and silicates is whether Θ_D, a single number, adequately represents the thermal properties of that solid.

Lindemann's law, (11.22) and (11.32), is derived in terms of one thermal parameter. If the thermal properties were represented by two parameters instead of one, then a larger number of solids could be represented by a simple (two parameter) theory. Mulargia and Quareni (1988) elaborated upon the simple Lindemann theory by imposing separate Debye temperatures (or frequencies) for the longitudinal waves and for the shear waves, following the suggestion of Brillouin (1946). This was also the strategy of Brown and Shankland (1981) to calculate the entropy (see Section 5.10.). Mulargia and Quareni defined dT_m/dP in terms of the pressure derivatives of the two sound velocities. Using experimental data on sound velocities versus pressure, they calculated dT_m/dP and compared it with the measured dT_m/dP of many solids, including oxides and silicates. They concluded that the theory can be used for minerals and "quite confidently to calculate...the melting curve in the earth's interior from seismic data." They found that their interpretation of the Lindemann law worked for diopside. That presented by Wolf and Jeanloz (Fig. 11.7) did not.

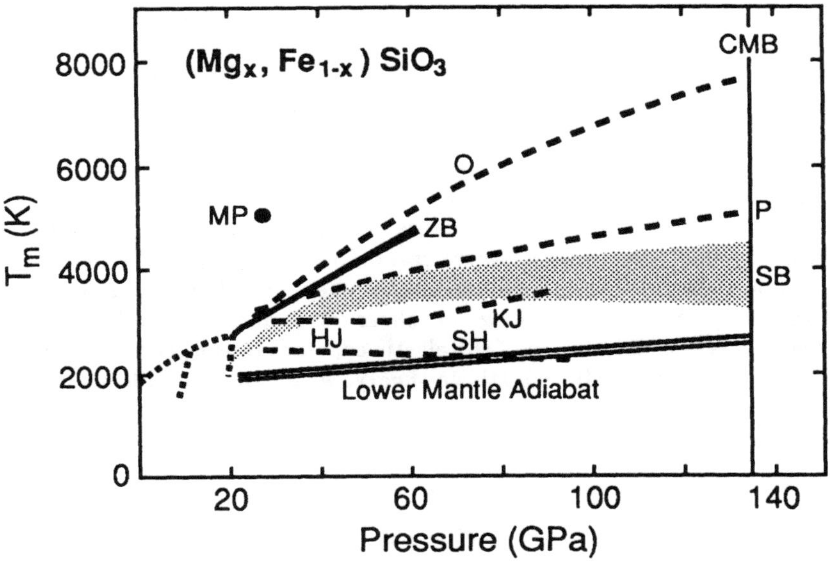

Fig. 11.9. Calculation of T_m for silicate perovskite according to: the Lindemann law in T, V space, Poirier, 1988 (P); the Lindemann law in T, P space, Ohtani, 1983 (O); equating solid and liquid Gibbs energies, Stixrude and Bukowinski, 1990 (SB); and molecular dynamics, Matsui and Price, 1991 (MP). Measurement of T_m of silicate perovskite according to: Heinz and Jeanloz, 1987 (HJ); Knittle and Jeanloz, 1989 (KJ); Sweeney and Heinz, 1993 (SH); and Zerr and Boehler, 1993 (ZB) (after Zerr and Boehler, 1993).

The reason Mulargia and Quareni got better results on diopside than did Wolf and Jeanloz is that the former authors effectively used two Debye temperatures, one for shear velocity waves and one for longitudinal velocity waves, resulting in a spread out Debye spectrum with two peaks. This spread out Debye spectrum represents the density of states of diopside better than a single parabolic curve. Wolf and Jeanloz (1984) concluded from their study that the Lindemann law could not be extended to minerals with complex interatomic forces and without close packing. This would eliminate minerals with low packing fractions, like quartz and feldspar.

Summarizing the existing data, it appears that the Lindemann melting law may not be valid for minerals and oxides with low coordination numbers, such as quartz and calcium carbonate, and, as a consequence, is not valid for minerals in the earth's crust. If the Mulargia-Quareni formulation is used, it is marginally valid for the crust. The Lindemann law appears to be valid for the lower mantle itself and for dense oxides and silicates such as those identified for the lower mantle. Because the Lindemann law requires a single value of Θ, upper mantle mineral structures may or may not obey the Lindemann law (forsterite does, but fayalite does not).

11.11. The elastic constant instability criterion for melting

From a physical point of view, the Gilvarry criterion of melting is somewhat unsatisfactory. Although the Lindemann law appears to work in many cases, it has not satisfied some readers who look for the physics principles involved in melting. That is probably why there is a continuous literature published on instability as a criterion for melting.

Born (1939) suggested that the loss of resistance to shear occurs at the melting point. "In actual fact there can be no ambiguity in the definition, or criterion, for melting. The difference between a solid and a liquid is that the solid has elastic resistance against shear stress and the liquid does not." If this idea is placed in terms of the elastic constant tensor, the cubic lattice is stable, according to Born's lattice theory, when three conditions hold:

$$C_{11} + 2C_{12} > 0;$$

$$C_{11} - C_{12} > 0;$$

$$C_{44} > 0.$$

These three conditions guarantee that the Helmholtz free energy, expanded as a function of T and six independent strain components, is positive definite; i.e., any increase in any strain increases the Helmholtz energy. The first condition requires the bulk modulus to be non-zero; the second requires the two shear moduli to be non-zero. Note that the criterion that the elastic shear constants be non-zero brings an associated criterion that the bulk modulus be non-zero. Born (1939) proposed that melting would be preceded by a continuous decrease of one of the shear elastic constants to zero at T_m, and so the vanishing of either shear constant is his criterion of melting.

Durand (1936) showed experimentally that melting of the rocksalt structure (type B1) occurred near a temperature T_c when

$$C_S = \frac{1}{2}(C_{11} - C_{12}) \rightarrow 0;$$

that is, he extrapolated to the condition that $C_{11} = C_{12}$ to find melting.

However, much later experiments showed that the solid melts at a slightly lower temperature than T_c, and this has led to the concept that mode softening (as distinct from mode disappearance) leads to melting. Calling the temperature of critical instability T_c, then $T_c < T_m$. Jackson and Liebermann (1974), for example, find a good correlation between T_c and T_m for alkali halides (Table 11.6). They also find good agreement between the pressure derivatives of T_c and T_m. The onset of shear instability also produces a divergent thermal expansion coefficient. An excellent review of the literature of shear instability criteria of melting is found in Boyer (1985).

Table 11.6. a) Comparison of critical temperatures T_c at which $C_S = 0$ and melting temperatures T_m for rocksalt halides ($C_S = 1/2(C_{11} - C_{12})$)

Material	C_S kbar	$(\partial C_S/\partial T)_P$ kbar/deg	T_c C	T_m C	T_c/T_m
LiF	330	−0.370	914	842	1.09
LiCl	133	−0.210	655	614	1.07
LiBr	104	−0.189	572	547	–
LiI	74	−0.155	499	450	–
NaF	364	−0.331	1122	988	1.14
NaCl	183	−0.201	932	801	1.16
NaBr	146	−0.174	861	755	–
NaI	108	−0.144	772	651	–
KF	255	−0.255	1022	846	–
KCl	169	−0.186	931	776	1.20
KBr	145	−0.167	890	730	–
KI	116	−0.145	822	686	–
RbF	207	−0.221	959	775	–
RbCl	151	−0.174	890	715	–
RbBr	133	−0.154	886	682	–
RbI	111	−0.136	838	642	1.31

b) Comparison of critical temperatures T_c at which $C_S = 0$ and melting temperatures T_m for rocksalt oxides

Material	C_S kbar	$(\partial C_S/\partial T)_P$ kbar/deg	T_c C	T_m C	T_c/T_m
MgO	1009	−0.340	2993	2800	1.07
CaO	820	−0.266	3108	2580	1.20
SrO	640	−0.211	3058	2430	1.26

c) Comparison of critical temperatures T_c at which $C_S = 0$ and melting temperatures T_m for B2 halides

Material	C_S kbar	$(\partial C_S/\partial T)_P$ kbar/deg	T_c C	T_m C	T_c/T_m
CsCl	80	−0.103	797	646	1.23
ÇsBr	75	−0.098	785	636	1.23
CsI	62	−0.081	785	621	1.26

The idea that the isothermal compressibility diverges; that is, $K_T \to 0$ as $T \to T_m$, goes back to Hertzfield and Goeppert-Meyer (1934). This idea preceded Born's (1939) suggestion, but it was disregarded by many during the late 1940's and 1950's as attempts to prove that a shear instability triggered melting dominated the scientific scene of melting theories as a result of Born's (1939) advances. The idea of a bulk modulus softening (often called compressibility divergence) has returned through the research of Boyer (1985). In fact, Boyer has found that compressibility divergence must occur simultaneously with the shear mode softening. For this reason Boyer coined the name "thermoelastic instability" in treating instability related to the criteria $C_{11} + 2C_{12} \leq 0$; it requires consideration of stability that is both thermodynamic and elastic in nature. Boyer also points out that T_c for thermoelastic instability occurs such that $T_m < T_c$.

The Lindemann law of melting is also an instability model with a special criterion. In the form given by (11.24), K_T occurs in the denominator. It is often overlooked that in the Lindemann formula K_T must have a value appropriate to the proximity of melting, and, as K_T must descend fast near T_m, it should be named K_{T_m}. Actually, K_{T_m} does not vanish because the shear mode has not vanished either, but just as the shear constant is small at T_m, so is K_{T_m}. As a consequence, the use of (11.24) will produce a much lower value of T_m than will (11.23) if the assumed value of K_T is too high. Great care is needed when using (11.24) to evaluate correctly the value of K_{T_m}, which is bound to be smaller than found by extrapolating K_T using measured high T values of $(\partial K_T/\partial T)_P$.

11.12. Compressibility divergence

The following derivation is parallel to the method used by Boyer (1985). He used a theory of potentials, whereas this approach uses thermodynamics. We break up the pressure according to (2.19) into its thermal pressure and its isothermal component

$$P(V,T) = P_0(V) + P_{TH}(V,T). \tag{11.43}$$

We evaluate $P_{TH}(V,T)$ using the formula

$$\left(\frac{\partial P_{TH}}{\partial T}\right) = \alpha K_T,$$

so that

$$\Delta P_{TH} = P_{TH}\,(T,V) - P_{TH}\,(300,V) = \alpha K_T(T - 300). \tag{11.44}$$

To get the pressure dependence of P_{TH}, we use the identity

$$\left(\frac{\partial \alpha K_T}{\partial V}\right)_T = -\frac{1}{V}\left(\frac{\partial K_T}{\partial T}\right)_V = \frac{w}{V}. \tag{11.45}$$

From the identity (3.7), we find

$$-\left(\frac{\partial K_T}{\partial T}\right)_V = w = \alpha K_T\left(\frac{\partial\ \ell\text{n}\ \alpha K_T}{\partial\ \ell\text{n}\ V}\right) = \alpha K_T(\delta_T - K'). \qquad (11.46)$$

Thus $w(\eta, T)$ is the product of αK_T and $(\delta_T - K')$ if solved separately. We solve for αK_T explicitly using identity (3.7), finding

$$\alpha K_T = (\alpha K_T)_0 \exp \int (\delta_T - K')\frac{\partial\eta}{\eta}. \qquad (11.47)$$

A plot of lines of constant w in the T, η plane is shown in Fig. 3.3.

Solving for ΔP_{TH} in terms of w, we have

$$\Delta P_{TH} = (T - 300)\frac{w}{(\delta_T - K')} = (T - 300)(\alpha K_T)_0 \exp \int (\delta_T - K')\frac{\partial\eta}{\eta}. \qquad (11.48)$$

The behavior of (11.48) is quite different for solutions where $\eta > 1$ than for solutions where $\eta < 1$. Solutions for ΔP_{TH} are quite asymmetric around the $\eta = 1$ axis, especially at high T.

This can be seen by solving for αK_T in terms of w, assuming that w is a small number and a constant along an isotherm. In this case

$$\alpha K_T - (\alpha K_T)_0 = w\ \ell\text{n}\ \eta,$$

$$\Delta P_{TH} = (T - 300)\left[2\ \ell\text{n}\ \eta + (\alpha K_T)_0\right].$$

We see that $\ell\text{n}\ \eta$ is positive for $\eta > 1$ and negative for $\eta < 1$, which produces asymmetry in the P_{TH} curves.

The effect of the upward curvature of the ΔP_{TH} curve in Fig. 11.10 is to shift the minimum in the P curve to lower values of compression. This is similar to the effect Boyer (1979) found by using first principles calculations on an assumed interatomic potential for alkali halides.

ΔP_{TH} for various isotherms is plotted and shown in Fig. 11.10. Also plotted is the isothermal ($T = 0$) value of pressure designated by $P_0(\eta)$. We note the rapid increase in slope of ΔP_{TH} with η for the isotherms with large T. Thus, the total pressure is asymmetric.

We plot the total pressure P in Fig. 11.11 resulting from the static and thermal pressure, as shown in Fig. 11.10. By taking account of the pressure dependence of the thermal pressure, we see how the total pressure varies with η. In Fig. 11.11, the bulk modulus K_T at $P = 0$ is the slope of the intercept of P with the zero pressure isobar. We see that for increasing temperature, the slope of the intercept for the isotherms becomes smaller as T rises, showing that K_T decreases with T such as shown, for example, in Fig. 2.4. However, there is a temperature isotherm T_c for which the isotherm is tangent to the zero pressure isobar; that is, the slope is zero. Thus $K_{T_c} \to 0$, the condition that Boyer (1985) called compressibility divergence, and that is his proposed criterion for melting.

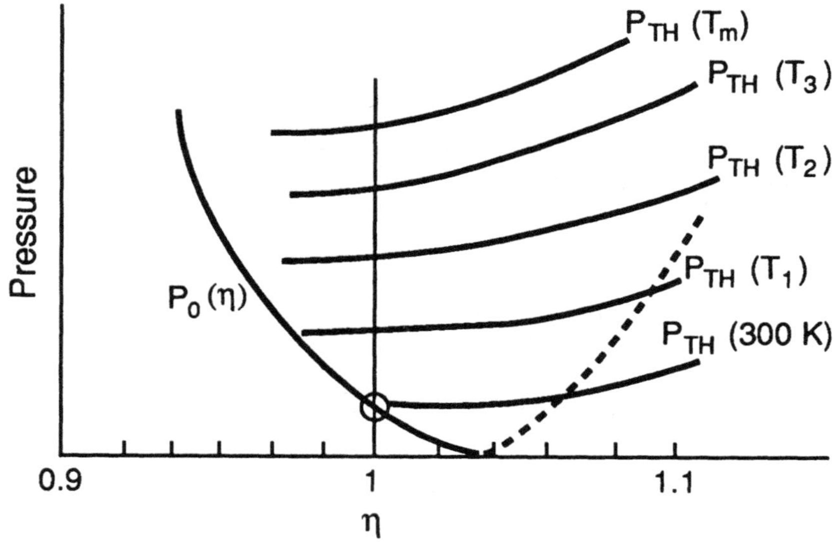

Fig. 11.10. The variation of thermal pressure with η for high T and η. Note that P_{TH} curves upwards for high T and $\eta > 1$.

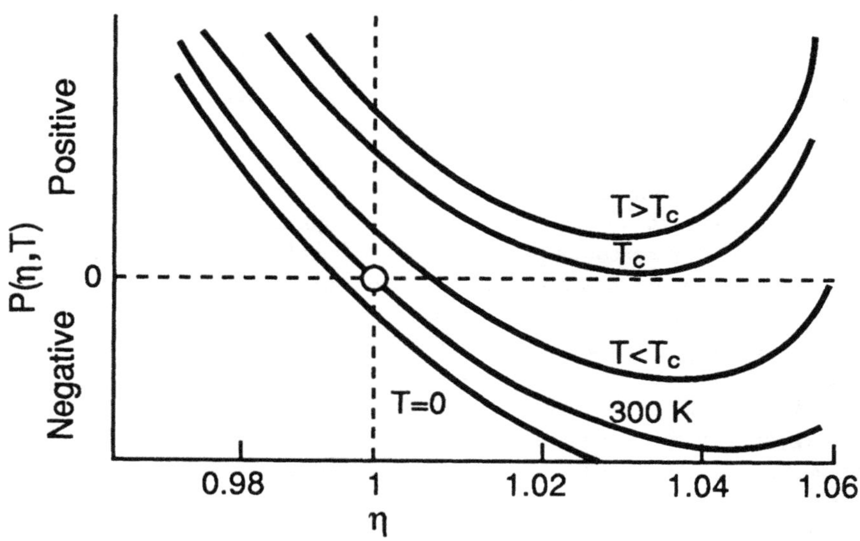

Fig. 11.11. Plot of external pressure, $P(\eta, T)$ versus η along isotherms. The centering locating ambient conditions is at $P = 0$ and $\eta = 1$. Values of P_{TH} are referenced to the centering position. The external pressure $P(\eta, T)$ is equal to $P_0(\eta)$ and P_{TH} given in Fig. 11.10.

For isotherms close to T_c, the bulk modulus is falling very rapidly with T. At an isotherm T_c the slope of the T versus η curve vanishes; this is related to the melting temperature T_m. Thus the bulk modulus is theoretically zero at the critical temperature T_c. Of course, this is never realized, just as the vanishing of the shear constant is never realized.

Starting from the 300 K isotherm, where $K_T = K_{T_0}$, K_T becomes progressively smaller as T increases, decreasing steadily with T at first, and then somewhat more precipitously near T_c. Thus, as Boyer (1985) showed, one can describe melting as resulting from a continuous thermoelastic transition, notwithstanding the first order nature of melting.

Just as the compressibility is divergent at T_c, so is the specific heat. The thermal diffusivity, $\mathbf{d} = C_P\rho/\mathbf{k}$, where $\mathbf{k}$ is the thermal conductivity, increases rapidly before T_c is reached, thus hastening the transition that appears discontinuous and leading to the name "thermoelastic transition" for T_c. The value of $\mathbf{d}$ is equivalent to an internal friction, preventing K_T from actually going to zero because T never reaches T_c. K_{T_m} may be smaller than obtained by simple extrapolation of measured K_T versus T data. In Fig. 11.12 we show the relationship of K_{T_m} to the extrapolated value of K_T using $\partial K_T/\partial T$ measured at lower T. This graphically illustrates that the value of K_T appropriate for the Lindemann's law transition may be much lower than K_T expected from experiment.

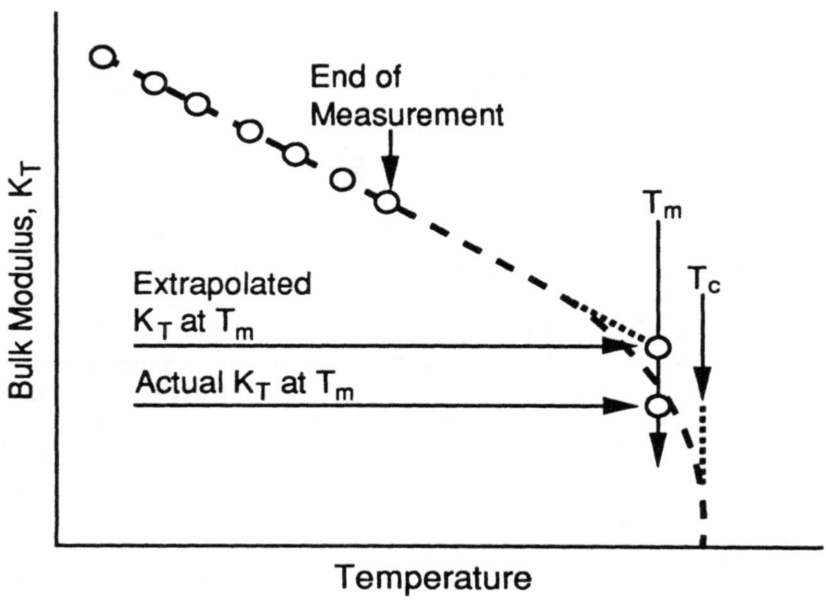

Fig. 11.12. A schematic representation of the variation of bulk modulus with temperature in the neighborhood of melting.

11.13. Compressibility divergence in the Lindemann law

In dealing with a rock composed of minerals in which the thermodynamic properties are known, the melting temperature of the rock is a complex consideration and requires a truly fundamental approach. Unlike the mineral under pressure, the rock has a variety of Lindemann numbers, one for each mineral. So the Lindemann law is of little help except as a description of a mineral in the rock. A further complication is that the iron is redistributed from mineral to mineral as P and T increase, so the composition is not invariant in a mineral as depth increases, as once was thought.

As an example, we describe how Ohtani estimated the melting temperature of silicate perovskite using many of the techniques discussed in this chapter. He estimated the initial slope dT_m/dP for silicate perovskite from the Clausius-Clapeyron equation (77 ± 8 K/GPa), but this required values of ΔV and ΔS. To obtain ΔV, he found the EoS of the melt and the EoS of the mineral, and these gave him the change in volume at the conditions of the beginning of the lower mantle. He obtained ΔS between the liquid and the melt from calorimetric data on the solid. He then used the Kraut-Kennedy equation (11.26) to get T_m as a function of $\Delta V/V_m$ by evaluating the constant C from (11.29).

The value of C in the Kraut-Kennedy law given in (11.26) is found from the derivative taken at low P

$$\left(\frac{dT_m}{dP}\right)_0 = T_{m_0}\frac{C}{K_T} \tag{11.49}$$

or

$$C = \left(\frac{dT_m}{dP}\right)\frac{K_T}{T_{m_0}}. \tag{11.50}$$

A measurement of dT_m/dP over a limited range of pressure, the measured value of T_{m_0}, and the value of K_T establish the value of C. K_T is not reliably known at melting. It is customarily estimated by extrapolation using knowledge of K_T and $(\partial K_T/\partial T)$ at temperatures substantially below melting. We note that if K_T is overestimated, then C will be overestimated. Finding that $dT_m/dP = 77$K/GPa, and $T_{m_0} = 2900$°C, Ohtani computed $C = 29.6$ from (11.50) using the extrapolated value of K_T. He obtained 7500°K for T_m at 130 GPa, an increase over T_{m_0} by 225%. This apparently very rapid increase of T_m must be tempered by the lack of knowledge of K_T right at melting, K_{T_m}. If K_{T_m} is smaller than the linear extrapolated K_T due to compressibility divergence, as shown in Fig. 11.10, then C will be correspondingly smaller, producing a smaller increase in T_m at the higher pressures.

Ohtani's curve for T_m for silicate perovskite, shown in Fig. 11.9, is 2800° larger at 125 GPa than found by Poirier (1988), who used the form of the Lindemann law not requiring knowledge of K_{T_m} and given by (11.31). It is to be emphasized that Poirier and Ohtani both used the same Lindemann theory and the same initial value of T_{m_0}, except one was the formulation in T_m, ρ space and the other was that in T_m, P space.

Following Poirier's method, we put $q = 1$ (Anderson, 1979a) and $\gamma_0 = 1.3$ (Anderson, 1979b) in (11.30), finding

$$\frac{T_m}{T_{m_0}} = 1.586$$

for the value of (ρ/ρ_0), where ρ corresponds to the core-mantle boundary, and ρ_0 corresponds to the 670 km depth condition ($4.38/5.5 = .796$). Thus if T_{m_0} is taken to be 2900 K (Ohtani, 1983) at 670 km, then we find $T_m = 4600$ K at 2880 km, in agreement with Poirier's calculations (see Fig. 11.9). But is 2900° less than T_m at the core-mantle boundary computed by Ohtani.

Thus K_T should be replaced by K_{T_m}, meaning the value of K_T at T_m that may be influenced by compressibility divergence. That is, if compressibility divergence occurs, (11.49) and (11.50) should be replaced by

$$\frac{dT_m}{dP} = T_{m_0}\frac{C}{K_{T_m}} \qquad (11.51)$$

or

$$C = \left(\frac{\partial T_m}{\partial P}\right)\frac{K_{T_m}}{T_{m_0}}, \qquad (11.52),$$

where $K_{T_m} <$ the extrapolated K_T at T_c from lower T measurements (see Fig. 11.10).

11.14. Melting of iron

The definition of melting of iron varies from author to author and experiment to experiment. A variety of definitions were used by Boehler et al. (1990) and Boehler (1993) in recent reports on the melting of iron up to 200 GPa in the diamond anvil cell.

Boehler used four different methods to find the onset of melting for T_m of iron, as shown in Fig. 11.13. He observed the onset of convective motion by microscopic methods. He also measured the transition from a solid to a liquid by observing the power–temperature response: melting occurred at a sharp discontinuity. Another method was to monitor the intensity of a laser beam reflected from the hot spot. At the onset of melting, the intensity of the reflected light dropped sharply. These three methods agreed quite well, as shown in Fig. 11.13. The fourth method is to observe a jump in the temperature-resistance curve when electrically insulated iron wire was heated along an isobar (Boehler, 1986).

Williams et al. (1987) measured melting of iron by quenching the sample and looking for visual evidence of melting on the cold solid sample.

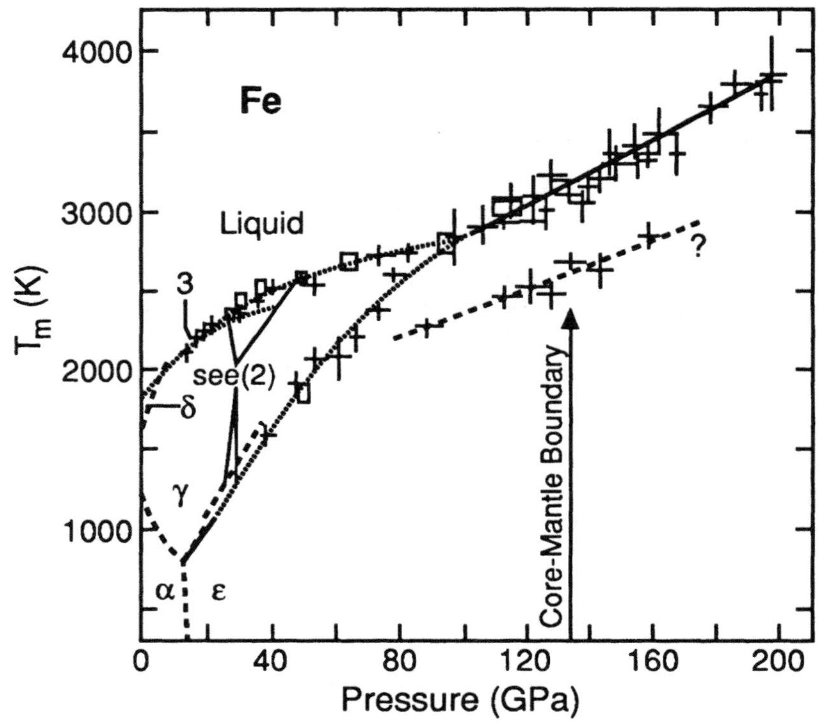

Fig. 11.13. Melting of Fe below 200 GPa (2 Mbar) (after Boehler, 1993). 3 refers to Boehler 1986; 2 refers to Boehler et al., 1990. The squares are taken from changes in optical reflectivity at T_m. The large + signs represent detection of the onset of convective motion. The ? refers to a new solid-solid phase boundary.

Their measurements of T_m above 30 GPa are consistently higher than those of Boehler, and they show much larger error bars. The larger error bars are typical of quenched samples found in petrology studies.

A different experimental approach to melting is to use the concept that the shear velocity of a solid vanishes at T_m. The presence of a solid–liquid transition in iron was detected in shock wave techniques by observing the pressure at which $v_p \rightarrow v_b$, because when $v_s = 0$, $v_p = v_b$ (Brown and McQueen, 1986; Fig. 8.3).

Also, at the onset of melting, the magnetic susceptibilities of some metals show large, abrupt changes. This is especially true for iron and iron-rich alloys in the bcc phase, where the change is about 20% (Ubbelohde, 1965); the effect is less pronounced for iron in the fcc structure.

From the theoretical point of view, the fundamental basis of melting is the cooperative appearance of disorder. In many theories of melting. disorder is caused by a critical concentration of lattice defects that break the crystalline order locally. These are called defect-mediated theories of melting (Poirier, 1991).

Poirier developed a dislocation-mediated theory of melting for iron, considering dislocations as a type of defect. He derived the equation of the elastic energy stored in a crystal, taking into account the interactions of dislocations (Poirier, 1986). It is defined in terms of the shear modulus and other dislocation parameters, as well as γ. He was able to find the total extra free energy due to the dislocations and obtain the expressions for $\Delta\mathcal{S}_m$ and ΔV_m due to a large concentration of dislocation. From the Clausius-Clapeyron equation and the integrated dT_m/dP, he found $\mathcal{F}$ as a function of dislocation concentrations C_d and T and found that $\mathcal{F}$ vanished at a particular value of the concentration of dislocations, called dislocation saturation, but only for a particular value of T, identified as T_m. For T less than T_m, $\mathcal{F}$ did not vanish. Using such a dislocation theory, Poirier and Shankland (1994) found T_m for iron to depend on the structure, yielding 6160 K, 6060 K, and 5610 K for hcp iron, fcc iron, and bcc iron, respectively.

11.15. The fundamental two-phase theory of phase transition

A quantitative theory of melting demands a reliable expression for the Gibbs energy of the liquid phase for a solid–liquid transition. This requires one to quantify the disorder of the liquid. The entropy of melting is

$$\Delta\mathcal{S}_m = \left(\frac{\Delta\mathcal{S}}{\Delta V}\right)_T \Delta V_m + \Delta\mathcal{S}_d, \qquad (11.53),$$

where $\Delta\mathcal{S}_d$ is the entropy of disorder arising from the entropy difference between the solid and the liquid. The liquid, always being more disordered, requires $\Delta\mathcal{S}_d$ to be positive upon melting. Equation (11.53) was verified for metals by Oriani (1951). Using (1.22), we have

$$\Delta\mathcal{S}_m = \alpha K_T \,\Delta V_m + \Delta\mathcal{S}_d. \qquad (11.54)$$

Stishov (1975) gave a review of the thermodynamics of melting that emphasized two-phase theories. Stishov's papers are important because they give us techniques for finding the value of $\Delta\mathcal{S}_m$. He showed, for example, that there is a very simple relationship between the relative volume discontinuity, $\Delta V/V_m$, and the entropy discontinuity, $\Delta\mathcal{S}/R$. He found that for many simple solids, $\Delta\mathcal{S}_m/R \to \ell\text{n}\, 2$ as $\Delta V/V_m \to 0$. Further,

$$\frac{\Delta\mathcal{S}_m}{R} = \ell\text{n}\, 2 + a\frac{\Delta V_m}{V_m}, \qquad (11.55)$$

This means that $\Delta\mathcal{S}_d = R\,\ell\text{n}\, 2$. By taking V_m as a constant in (11.55), (11.54) becomes

$$\Delta\mathcal{S}_m = R\,\ell\text{n}\, 2 + \alpha K_T \,\Delta V_m. \qquad (11.56)$$

At temperatures below melting, we have found that αK_T is independent of T, and, except for unusual cases like that of gold, virtually independent of V. Therefore, dT_m/dP may be a constant for a limited range:

$$\frac{dT_m}{dP} = \frac{\Delta V_m}{\Delta \mathcal{S}_m} \cong \frac{1}{\alpha K_T}. \tag{11.57}$$

Equation (11.57) indicates that dT_m/dP is constant for a pressure interval that is not too large. When the pressure range is large, we may expect ΔV_m to change more slowly than $\Delta \mathcal{S}_m$ at high P, due to an effect arising from a difference of compressibilities for the solid and the liquid. In this case the value of $V_\ell - V_S$ decreases with increasing pressure more than $\Delta \mathcal{S}_m$, and the melting line is curved downward (Bassett and Weathers, 1994). It is possible that dT_m/dP can be negative, as, for example, in the cases of Ga, Sb, and Bi, where ΔV_m is negative (Ubbelohde, 1965).

Equations equivalent to those above apply to solid–solid transitions as well. However, in most cases, the solid–solid curve is straight because both $\Delta \mathcal{S}$ and ΔV are insensitive to volume and temperature change along the T_m curve. However, $\Delta \mathcal{S}$ may have an abrupt change near P, T conditions in which phenomena such as a ferromagnetic–paramagnetic transition occurs because extra disorder entropy terms are added to the free energy. In such regions the slope dT/dP will be curved (Bassett and Weathers, 1994), even for a solid–solid transition.

12

SHOCKED SOLIDS

12.1. Introduction

The generation of high pressures by shock waves relies on the inertial response to rapid acceleration, instead of equilibrium of forces as in a fixed static experiment. Although the time interval is short, very large pressures ($\approx$ 300 GPa) can be achieved with simultaneous high temperatures. The shock wave can be visualized as a single stress discontinuity traveling in an undisturbed medium followed by an approximately uniform stress and subsequently by a relaxation of pressure. This is the shock-front profile. There is a connection between the physical properties of the solid and the structure of the shock front.

The study of the behavior of solids under extreme shock conditions is rooted in hydrodynamics, finite elasticity, and plasticity. A highly stressed solid is not a fluid, although, as a good approximation, much of the mathematics of fluid hydrodynamics is applied to a shocked solid. Experimental data on solids are usually analyzed as though the total stress is a pressure and the total strain is a volume change, without regard to shear stresses. But the success of the hydrodynamic approach applied to many shocked solids should not lull the student away from the basic fact that the shock is a one-dimensional plane wave, and further that shear stresses and yield points are involved in the complete description of the state of the material under shock.

The basic assumption is that there is thermodynamic equilibrium between the shocked and unshocked solid across the shock front. A further assumption important to equations of state is that the stress observed in the shock-front profile is hydrostatic. Thus the aniostropic stress distribution must be treated separately. In many cases the anisotropic correction is small enough that it can be corrected for by a small offset in the pressure ordinate, in the P, V plane between the hydrostatic curve and in the measured curve. The yield stress of the solid, typically a few kilobars, is a measure of the offset (although the shocked yield limit is ordinarily different from the yield limit in a static tensile test).

The shock propagating in a linear direction starts out with a sound velocity that is the longitudinal velocity of sound; that is, the stress is not hydrostatic and cannot be represented strictly by hydrostatic pressure P. Therefore, as this axial stress cannot be supported above the yield point of the material, the mode of deformation changes at the volume appropriate to the yield stress.

In a shocked fluid where the stress is hydrostatic, and there is no yield point, the wave velocity is controlled by the bulk velocity of sound. The bulk sound velocity is numerically less than the longitudinal velocity of sound (Fig. 12.1). This is expressed by the condition

$$-\left(\frac{\partial P}{\partial V}\right)_{\text{hydrostatic}} < -\left(\frac{\partial \sigma_{11}}{\partial V}\right)_{\text{uniaxial}} .$$

As the stress is raised above the yield point σ_{yp} in a real solid, the $\sigma_{11} - V$ curve passes through a cusp, and at higher stresses traverses a path closely parallel to the hydrostatic P-V curve (Fig. 12.2). If the offset is small; that is, if the yield stress is small compared with the stress level achieved at state B, one may safely use the hydrostatic solution to describe the shock.

At state B, however, there is a certain amount of elastic strain energy, and a certain amount of nonrecoverable energy dissipated through plastic strain is stored in the solid. The elastic energy arises from σ_{yp}, and dissipation due to plastic energy arises from the fact that a uniaxial shock front creates large shear stresses on planes at angles of about 45° to the shock normal. Plastic flow (or its alternates, fracture or twinning) will occur along these shear planes so that the maximum resolved stress along these planes is held within the failure limits. The relief of these stresses occurs at a certain rate, and the amount of dissipated plastic energy depends on the time of the shock and the laws of plastic flow.

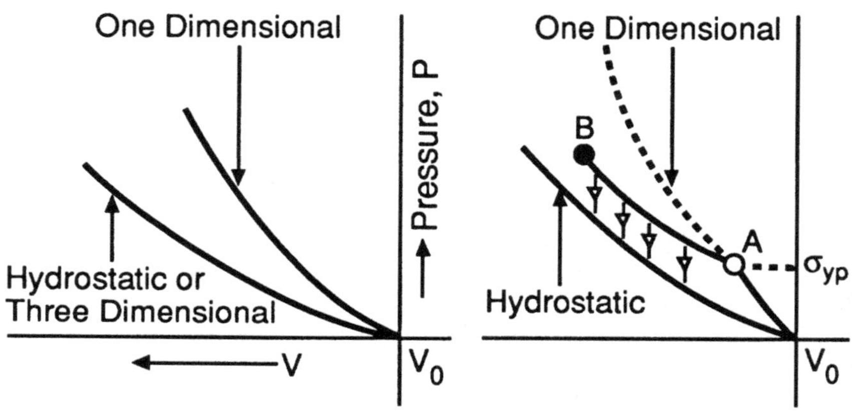

Fig. 12.1., left. The one-dimensional shock compared with the three-dimensional or hydrostatic shock.

Fig. 12.2., right. The one-dimensional shock (A—B) parallels the hydrostatic shock after the yield stress σ_{yp} has been exceeded.

Shocked solids are finite in size, and compressed waves are reflected from the surface so that, after the stress has reached the level appropriate to state B, it is eventually relieved (Fig. 12.3). If there were no further ability to sustain shear stresses in the solid, the path in the σ_{11}-V plane would be from B to C, to relieve the elastic energy. However, there is still the capacity of sustaining a negative shear stress. The path actually traversed is from B through C to D below the hydrostatic curve. The length CD depends on the relative time constants in the shock and the yield stress at state B. At D, the material yields and then traverses the path from D to E at zero stress, but often arrives at the volume V_0', which is different from V_0. (The curve BCD closely parallels OA, and DE closely parallels CO). Thus, when the fragments of a shocked solid are reassembled, the volume and shape of the reassembly is very often not equal to the shape or volume of the material before the shock.

To describe the action of the shocked material accurately would require that one describe all the ramifications implied by the curve OABCDE. This is occasionally done by experts in the field, who show that some shock-induced waves travel at longitudinal sound velocity, while others travel at hydrostatic sound velocity.

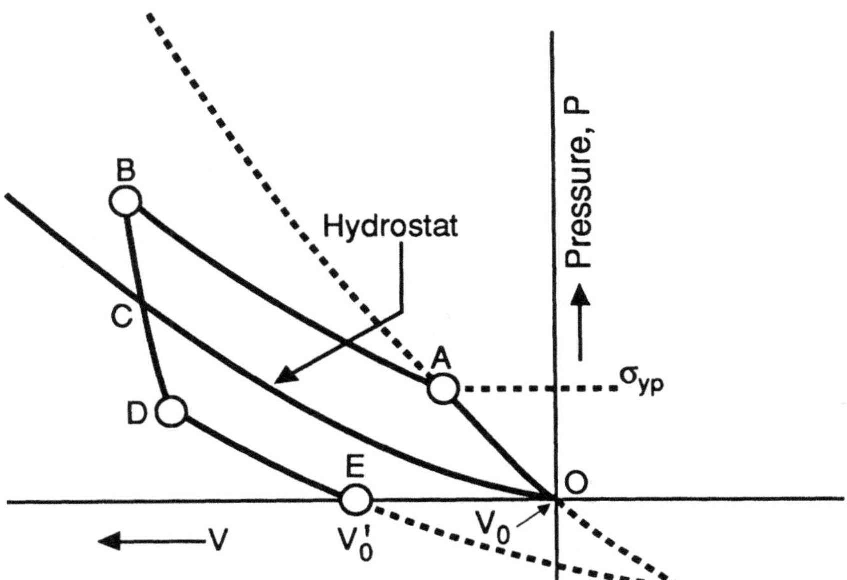

Fig. 12.3. The path traversed in a shock compression (O,A,B,C,D,E), including the decompression path DE.

For a great many solids shocked at high values of pressure, however, the curve OABCDE is a small departure from the curve OC. This permits one to use hydrostatic pressure in place of uniaxial stress and to treat the solid as if it were in hydrostatic condition, often a good approximation to the actual state of the material. It is pedagogically useful to begin a discussion of shock waves in the approximation that is called the hydrostatic Hugoniot, or simply the Hugoniot. Much useful information from a shock experiment can be determined by the assumption of hydrostaticity, but close observation shows that the compression phase of the shock has important, time-dependent details, often called precursors. In this chapter, we ignore these important physical details.

The next six sections rely heavily on the articles by McQueen et al. (1964, 1967, 1970) and Fritz et al. (1971), and on the book by Zharkov and Kalinin (1971). Some of the derivations that were sketched in the original articles are expanded here in more detail.

12.2. The hydrostatic Hugoniot

The passage of an intense shock wave through a solid permits one to determine the EoS in several of its forms. The EoS, as it is used in shock wave science, is often not the same set of thermodynamic variables as the set used in static thermodynamics. In static EoS work, one seeks the pressure as a function of volume and temperature, $P(V,T)$. In shock wave experiments, the EoS used in theory may be the caloric EoS, $\mathcal{S}(V,P)$, where $\mathcal{S}$ is entropy, or the mechanical EoS, $\mathcal{U}(V,P)$, where $\mathcal{U}$ is the internal energy. In fact, neither $\mathcal{S}(V,P)$ nor $\mathcal{U}(V,P)$ is the simplest EoS that arises from the shock wave experiments. One often uses the experimentally measured values, where $\mathcal{U}$ and P are replaced by experimental "Hugoniot" variables.

The experimentally observed quantities in shock wave work, measured optically by high speed photography or by time-of-flight apparatus, are the shock velocity U_S and the particle velocity U_P. An empirical relation U_S (U_P) that passes through the set of points (U_S, U_P) can be found, there being as many points as separate shock experiments on the solid. This relationship can be regarded as an EoS, as it contains all of the thermodynamic information determinable at high pressure from a single shock. The usual empirical relationship is found to be linear:

$$U_S = c_0 + \mathbf{s}U_P, \tag{12.1}$$

where $\mathbf{s}$ is the slope of the line, and c_0 is the shock velocity at vanishing pressure; it is identical to the bulk sound velocity, $c_0 = v_{b_0} = (K_{S_0}/\rho_0)^{1/2}$, where K_{S_0} is the adiabatic bulk modulus at zero pressure, and v_{b_0} is v_b at zero pressure. For a few solids, (12.1) is not adequate, and a quadratic relationship relating U_S and U_P must be used. A typical experimental result on the U_S-U_P plane is shown in Fig. 12.4, which is the shock data for iron (Marsh, 1980) and the resulting Hugoniot.

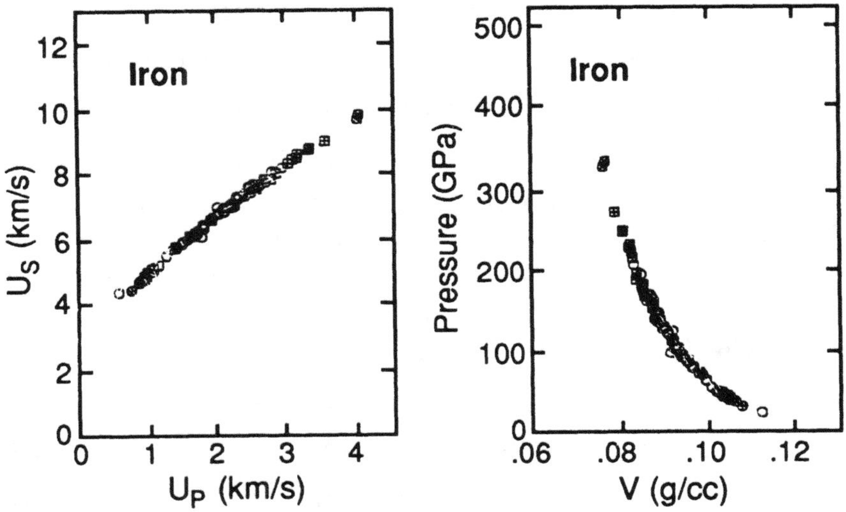

Fig. 12.4. The U_S versus U_P shock data for iron and the resulting Hugoniot pressure P_H versus V (after Marsh, 1980).

To transform the U_S-U_P empirical curve into a relationship between the customary thermodynamic variables, one applies the conditions of conservation to both sides of the shock. If one assumes that there is a steady state, there are the following relationships between any two points in the shocked and unshocked states. The variables are shown in Fig. 12.5.

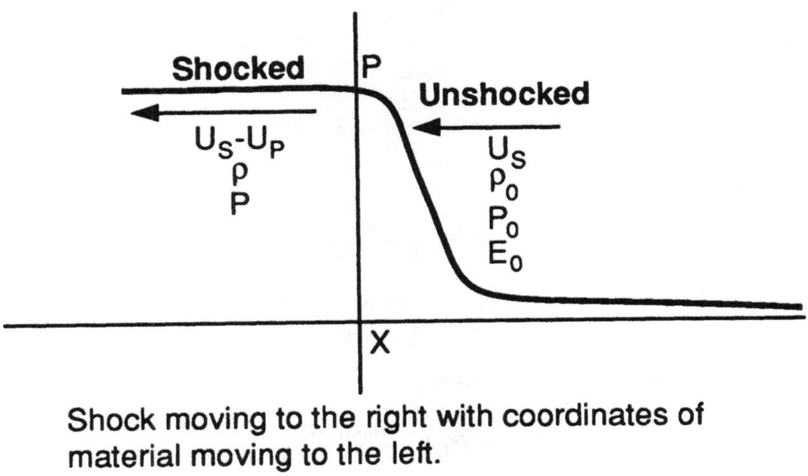

Fig. 12.5. The shock front is shown with the shocked variables on the left and the unshocked variables on the right.

$$\rho_0 U_S = \rho(U_S - U_P): \text{ mass conservation;}$$

$$\rho_0 U_S^2 + P_0 = \rho(U_S - U_P)^2 + P: \text{ momentum conservation;}$$

$$\rho_0(E_0 + 1/2U_S^2) + P_0 U_S = \rho\left[E_{\mathbf{H}} + 1/2(U_S - U_P)^2\right](U_S - U_P) + P(U_S - U_P): \text{ energy conservation.}$$

We are using the notation E here to represent the energy of the solid under deformation, following standard shock wave nomenclature. However, E is identical to the internal energy $\mathcal{U}$ described in Chapter 1, and $\mathcal{U}$ will be used subsequently.

12.3. The Hugoniot variables

The three conservation equations above can be manipulated to produce

$$\rho_0 U_S = \rho(U_S - U_P); \tag{12.2}$$

$$P_{\mathbf{H}} - P_0 = \rho_0 U_S U_P; \tag{12.3}$$

$$\mathcal{U}_{\mathbf{H}} - \mathcal{U}_0 = 1/2(P + P_0)(V_0 - V) = V_0 P_{\mathbf{H}}(V). \tag{12.4}$$

Equation (12.4) is the increase of internal energy in the solid due to the shock.

In (12.3) and (12.4), we have placed the subscript **H** on P and $\mathcal{U}$ because these values lie on the so-called Hugoniot. This identification is made because the locus of points (say, P and V), which satisfy (12.2), (12.3), and (12.4), and which are plotted on the P-V plane, is neither an isotherm nor an isentrope, as the temperature and entropy vary from point to point along the P-V locus.

However, $P_{\mathbf{H}}$ is only a function of V and $\mathcal{U}_{\mathbf{H}}(V)$; and $\mathcal{U}_{\mathbf{H}}$ is only a function of V and $P_{\mathbf{H}}(V)$. This can be demonstrated using (12.1) and a variable from the volume compression,

$$\Delta\eta = \frac{V_0 - V}{V_0} = 1 - \frac{V}{V_0} = 1 - \eta = 1 - \rho_0 V. \tag{12.5}$$

From (12.1),

$$U_S = \frac{c_0}{1 - \mathbf{s}\Delta\eta}; \tag{12.6}$$

$$U_P = \frac{c_0 \Delta\eta}{1 - \mathbf{s}\Delta\eta}. \tag{12.7}$$

Equations (12.3) and (12.4) become

$$P_{\mathbf{H}} = P_0 + \frac{\rho_0 c_0^2 \Delta\eta}{(1 - \mathbf{s}\Delta\eta)^2}; \tag{12.8}$$

$$\mathcal{U}_{\mathbf{H}} = \mathcal{U}_0 + \frac{P_0 \Delta\eta}{\rho_0} + \left(\frac{1}{2}\right) \frac{c_0^2 (\Delta\eta)^2}{(1 - \mathbf{s}\Delta\eta)^2}. \tag{12.9}$$

The Hugoniot variables $P_{\mathbf{H}}$ and $\mathcal{U}_{\mathbf{H}}$ are now functions of the experimental variables c_0 and $\mathbf{s}$. Equations (12.6) and (12.8) contain observed quantities on the right hand side. Equation (12.3) relates pressure with compression with various constants, all determinable from other experiments at vanishing pressure. The same can be said for $\mathcal{U}_{\mathbf{H}}$. Thus it is shown that $\mathcal{U}_{\mathbf{H}} = \mathcal{U}_{\mathbf{H}}(V)$, and $P_{\mathbf{H}} = P_{\mathbf{H}}(V)$. A consistent thermodynamics can be defined from the $U_S(U_P)$ empirical relationship without regard to the temperature. It is possible to define a "Hugoniot" bulk modulus by applying the operator $(-V\partial/\partial V)$ to (12.8).

$$K_{\mathbf{H}} = -V\left(\frac{dP_{\mathbf{H}}}{dV}\right) = \rho_0 c_0^2(1-\Delta\eta)\frac{1+\mathbf{s}\Delta\eta}{(1-\mathbf{s}\Delta\eta)^2}. \tag{12.10}$$

In this case $K_{\mathbf{H}}$ means $K_{\mathbf{H}}(V)$.

The Hugoniot gives P, V data without knowledge of the temperature. That is a great advantage in shock compression analyses. On the other hand, it is correspondingly difficult to get information on temperature from shock compression experiments. If one wishes to specify the internal energy, entropy, or temperature along the Hugoniot, or if one wishes to compute the isotherm or the isentrope from the above, some additional information besides $U_S(U_P)$ is needed. This additional information is provided by assumptions about the Grüneisen ratio γ, usually taking the form

$$\left(\frac{\partial P}{\partial \mathcal{U}}\right)_V = \gamma\rho, \tag{12.11}$$

or

$$\gamma = V\left(\frac{\partial P}{\partial \mathcal{U}}\right)_V. \tag{1.19}$$

We have shown in Chapter 1 that γ decreases with density, and that a good empirical approximation over a limited range of compression is (1.32)

$$\frac{\gamma}{\gamma_0} = \left(\frac{V}{V_0}\right)^q = \left(\frac{\rho}{\rho_0}\right)^{-q}. \tag{1.32}$$

Now if $q = 1$, which holds at high T (Fig. 3.12),

$$\gamma\rho = \text{constant} = \rho_0\gamma_0. \tag{3.58}$$

In this approximation we have from (1.24) an alternate form of (3.54),

$$\alpha K_T/C_V = \text{ constant}, \tag{12.12}$$

and if C_V is independent of P and T at high T, as it often is, then (12.11) reduces further to

$$\alpha K_T = \text{ constant}, \tag{12.13}$$

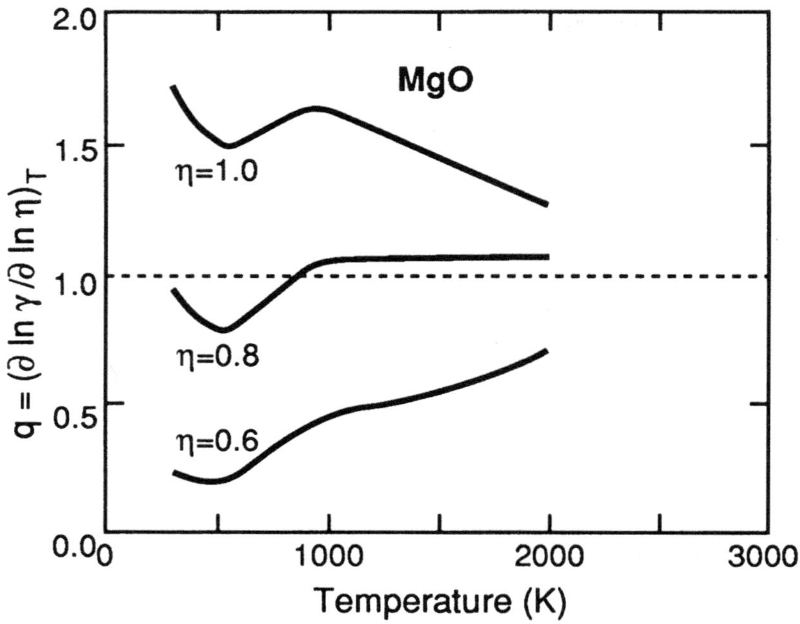

Fig. 12.6. The variation of q with T along various isochores for MgO. There is convergence to $q = 1.0$ at shock wave temperatures (modified from Anderson and Isaak, 1994).

an approximation recommended by Birch (1952) for use in the earth's deep interior.

There are many cases where $\gamma\rho =$ constant is used in shock wave data reduction. The main problem is the evaluation of γ_0. Thermodynamic identities show that $q \rightarrow 1$ as the temperatures becomes large, as shown in Fig. 12.6, which is related to Fig. 3.12 (Anderson et al., 1993). This situation is probably good for all high shocked conditions.

It has been suggested that the relationship between q and η over a wide range of T and η is given by

$$q = q_0 \eta^{\bar{q}}$$

(Bukowinski, 1988). A similar form was suggested by Jeanloz (1989). But this is not adequate to describe all the relationships plotted in Fig. 12.6, unless $\bar{q}$ is a strong function of η and T. At high T, we see that dq/dT is negative at $\eta = 1$ and positive at small η (see also Fig. 3.13, which shows the slope of $d\ \ell\text{n}\ q/d\ \ell\text{n}\ \eta$ to be curved and temperature sensitive). The above equation may be useful to describe the variation of q in limited T, η zones. But it is not sufficient to compare shock wave results (where $q = 1$) with static compression measurements at $T < \Theta$.

12.4. The isentropic bulk modulus

To use (12.11) in developing corrections to the "Hugoniot" EoS, we need to develop some additional thermodynamic relationships. These relationships will help us in comparing pressure and temperature derivatives of the prime variables in a Hugoniot. Let us replace (12.4) with the following thermodynamic function, called the Rankine-Hugoniot function,

$$\mathbf{H}(P, V, P_0, V_0) = \mathcal{U}(P, V) - \mathcal{U}_0 - \frac{1}{2}(P + P_0)(V_0 - V). \tag{12.14}$$

The Rankine-Hugoniot equations given above can be usefully applied to (12.4) and (12.10) by requiring that

$$d\mathbf{H} = 0 \text{ along a Hugoniot} \tag{12.15}$$

and where

$$\mathbf{H} = 0 \text{ defines the Hugoniot centered at } P_0, V_0.$$

(Note that the function changes for each Hugoniot. It is a two-point scalar in the P, V plane. One usually varies P, V, not P_0,V_0.).

Using $d\mathcal{U} = T\,d\mathcal{S} - P\,dV$, we expand the Hugoniot in P,V variables:

$$d\mathbf{H} = T\,d\mathcal{S} - \frac{1}{2}(V_0 - V)\,dP - \frac{1}{2}(P - P_0)\,dV. \tag{12.16}$$

Using

$$dP = \left(\frac{\partial P}{\partial \mathcal{S}}\right)_V d\mathcal{S} + \left(\frac{\partial P}{\partial V}\right)_\mathcal{S} dV \tag{12.17}$$

in (12.16), we find

$$\begin{aligned} d\mathbf{H} = & \left[1 - \frac{\gamma}{2V}(V_0 - V)\right] T\,d\mathcal{S} \\ & + \left(\frac{1}{2}\right)\left(\frac{V_0 - V}{V}\right)\left[K_S - V\left(\frac{P - P_0}{V_0 - V}\right)\right] dV. \end{aligned} \tag{12.18}$$

Expanding the Hugoniot out in general terms as a function of

$$d\mathbf{H} = \left(\frac{\partial \mathbf{H}}{\partial \mathcal{S}}\right)_V d\mathcal{S} + \left(\frac{\partial \mathbf{H}}{\partial V}\right)_\mathcal{S} dV. \tag{12.19}$$

Compare coefficients in (12.19) and (12.18), and we find the definitions of $(\partial\mathbf{H}/\partial\mathcal{S})_V$ and $(\partial\mathbf{H}/\partial V)_\mathcal{S}$. We need differentials at constant $\mathbf{H}$ ($d\mathbf{H} = 0$) that help us obtain information from the shock wave experiment.

From calculus, and using a dummy variable X,

$$\left(\frac{\partial X}{\partial V}\right)_{\mathbf{H}} = \left(\frac{\partial X}{\partial V}\right)_{S} - \left(\frac{\partial X}{\partial S}\right)_{V} \left[\frac{(\partial \mathbf{H}/\partial V)_S}{(\partial \mathbf{H}/\partial S)_V}\right]. \tag{12.20}$$

Using (12.18)

$$\left(\frac{\partial X}{\partial V}\right)_{\mathbf{H}} = \left(\frac{\partial X}{\partial V}\right)_{S} - \frac{\left(\frac{V_0 - V}{2V}\right)\left(K_S - V\frac{P - P_0}{V_0 - V}\right)}{1 - \gamma\frac{V_0 - V}{2V}} \frac{1}{T} \left(\frac{\partial X}{\partial S}\right)_{V}$$

$$= \left(\frac{\partial X}{\partial V}\right)_{S} - \frac{\left(\frac{V_0 - V}{2V}\right)\left(K_S - V\frac{\Delta P}{\Delta V}\right)}{1 - \gamma\frac{\Delta V}{2V}} \frac{1}{T} \left(\frac{\partial X}{\partial S}\right)_{V}. \tag{12.21}$$

Now for X, substitute P. From (3.34) and (3.38), $(\partial P/\partial S)_V = \gamma T/V$. Placing this in (12.21), we find relations between the bulk modulus and various constants:

$$K_{\mathbf{H}} = K_S + \frac{\left(\frac{V_0 - V}{2V}\right)\gamma\left(K_S - V\frac{P - P_0}{V_0 - V}\right)}{1 - \gamma\frac{V_0 - V}{2V}}$$

$$= K_S + \frac{\gamma\frac{\Delta V}{2V}\left(K_S - V\frac{\Delta P}{\Delta V}\right)}{1 - \gamma\frac{\Delta V}{2V}}. \tag{12.22}$$

Equation (12.22) shows the relationship between the Hugoniot bulk modulus and the adiabatic bulk modulus. It can be written as:

$$(K_{\mathbf{H}} - K_S)\left(1 - \gamma\frac{\Delta V}{2V}\right) = \gamma\frac{\Delta V}{2V}\left(K_S - V\frac{P - P_0}{V_0 - V}\right). \tag{12.23}$$

Because γ and $\Delta V = V_0 - V$ are positive and γ is near unity, $[1 - \gamma(\Delta V/2V)]$ is positive. The sign of $(K_{\mathbf{H}} - K_S)$ is determined by the sign on the right.

Define the chord bulk modulus,

$$K_C = -V\frac{P - P_0}{V_0 - V},$$

which is a straight line drawn from V_0 of the origin to a V on the Hugoniot. Then (12.23) becomes

$$(K_{\mathbf{H}} - K_S) = \left(\frac{1 - \gamma\frac{\Delta V}{2V}}{\gamma\frac{\Delta V}{2V}}\right)(K_S - K_C), \tag{12.24}$$

or

$$\left(\frac{\partial P}{\partial V}\right)_{\mathcal{S}} - \left(\frac{\partial P}{\partial V}\right)_{\mathbf{H}} = \left(\frac{1-\gamma\dfrac{\Delta V}{2V}}{\gamma\dfrac{\Delta V}{2V}}\right)\left[\frac{P-P_0}{V_0-V} - \left(\frac{\partial P}{\partial V}\right)_{\mathcal{S}}\right]. \quad (12.25)$$

In terms of the slopes near the origin of the P-V plot, we see that $(P - P_0)/(V_0 - V)$ is larger than $(\partial P/\partial V)_S$, the former being a straight line, or chord, and the latter curving upward. Therefore the right hand side of (12.25) is positive. As a consequence, $(\partial P/\partial V)_{\mathcal{S}}$ is less than the chord. But the left hand side is also positive, and therefore the adiabatic slope is less than the Hugoniot slope.

$$\left|\frac{dP}{dV}\right|_{\text{chord}} > \left|\left(\frac{\partial P}{\partial V}\right)_{\mathbf{H}}\right| > \left|\left(\frac{\partial P}{\partial V}\right)_{\mathcal{S}}\right|.$$

But beyond a certain compression, the isentropic slope is greater than the chord slope; the signs reverse; and

$$\left(\frac{\partial P}{\partial V}\right)_{\mathbf{H}} > \left(\frac{\partial P}{\partial V}\right)_{\mathcal{S}} > \left(\frac{\partial P}{\partial V}\right)_{\text{chord}}. \quad (12.26)$$

12.5. Differentials along the Hugoniot

We now wish to show how entropy varies along the Hugoniot. The expansion of $\mathcal{S}$ in terms of V and T is

$$\partial\mathcal{S} = \left(\frac{\partial\mathcal{S}}{\partial V}\right)_T dV + \left(\frac{\partial\mathcal{S}}{\partial T}\right)_V dT. \quad (12.27)$$

Recalling that $(\partial\mathcal{S}/\partial V)_T = \alpha K_T$ and $(\partial\mathcal{S}/\partial T)_V = C_V/T$, and further from (12.4) that $\alpha K_T/C_V = (\gamma/V)$, we have

$$\frac{d\mathcal{S}}{C_V} = \frac{\gamma}{V}dV + \frac{dT}{T}. \quad (12.28)$$

Use of (12.28) in (12.18) yields

$$d\mathbf{H} = \left(1-\gamma\frac{V_0-V}{2V}\right)C_V\,dT$$
$$+\left[\left(1-\gamma\frac{V_0-V}{2V}\right)C_V\,T\frac{\gamma}{V} + \left(\frac{V_0-V}{2V}\right)(K_{\mathbf{H}}-K_{\mathcal{S}})\right]dV. \quad (12.29)$$

Using (12.23) in the above,

$$\frac{d\mathbf{H}}{1-\gamma\dfrac{V_0-V}{2V}C_V} = dT + \left(\gamma\frac{T}{V} + \frac{K_{\mathbf{H}}-K_{\mathcal{S}}}{\gamma C_V}\right)dV. \quad (12.30)$$

Comparing (11.28) to the expansion of **H** in terms of T and V,

$$d\mathbf{H} = \left(\frac{\partial \mathbf{H}}{\partial T}\right)_V dT + \left(\frac{\partial \mathbf{H}}{\partial V}\right)_T dV,$$

we can define $(\partial \mathbf{H}/\partial T)_V$ and $(\partial \mathbf{H}/\partial V)_T$. Thus

$$\left(\frac{\partial X}{\partial V}\right)_{\mathbf{H}} = \left(\frac{\partial X}{\partial V}\right)_T - \left(\gamma\frac{T}{V} + \frac{K_{\mathbf{H}} - K_S}{\gamma C_V}\right)\left(\frac{\partial X}{\partial T}\right)_V. \tag{12.31}$$

Immediately we see that

$$-\left(\frac{\partial T}{\partial V}\right)_{\mathbf{H}} = \gamma\frac{T}{V} + \frac{K_{\mathbf{H}} - K_S}{\gamma C_V}, \tag{12.32}$$

$$-\left(\frac{dS}{dV}\right)_{\mathbf{H}} = \frac{K_{\mathbf{H}} - K_S}{\gamma T}. \tag{12.33}$$

Proceeding along the same line, we find

$$\frac{d\mathbf{H}}{a} = \frac{VdP}{\gamma} + \frac{K_{\mathbf{H}}}{\gamma} dV, \tag{12.34}$$

where $a = \Big\{1 - \gamma\big[(V_0 - V)/2V\big]\Big\}$. Solving for dP from (12.34)

$$dP = \left(\gamma^2\frac{C_V}{V}T - K_S\right)\frac{dV}{V} + \gamma\frac{C_V}{V}dT. \tag{12.35}$$

These differentials along the Hugoniot demonstrate the importance of knowing γ at every value of V, P. Equation (12.29) shows that as $K_{\mathbf{H}} - K_S$ is positive, and γT is positive, $(\partial S/\partial V)_{\mathbf{H}}$ is negative; therefore, $(dS/dP)_{\mathbf{H}}$ is positive. As the pressure increases on the Hugoniot, the entropy increases. From Fig. 12.6, we see that at high temperature γ/V can be replaced by γ_0/V_0 or $\gamma_0\rho_0$, as a good approximation.

12.6. Changing from EoS parameters to shock parameters

Start with the expansion of S with the variables P and V:

$$dS = \left(\frac{\partial S}{\partial P}\right)_V dP + \left(\frac{\partial S}{\partial V}\right)_P dV.$$

The partial derivatives above are given by Maxwell's relations, (3.38) and (3.39), so that

$$dS = -\left(\frac{V}{\gamma T}\right)dP + \left(\frac{K_S}{\gamma T}\right)dV. \tag{12.36}$$

In an isentropic process, $dS = 0$, so that

$$\frac{dP}{K_S} = -\frac{dV}{V}. \tag{12.37}$$

In isentropic compression K_S is approximately represented by a linear expansion in P,

$$K_S = K_{S_0} + K'P. \tag{12.38}$$

This equation is closely obeyed over most of the lower mantle, for example, as shown in Fig. 7.3. (This is not to imply, however, that such a linear relationship exists between the isothermal bulk modulus K_T and P. It does not, even when (12.38) is true).

Integrating (11.38) for P between 0 and P, and V between V_0 and V, we have the Murnaghan EoS (Section 7.4):

$$P_S = \frac{K_{S_0}}{K'}\left[\left(\frac{V_0}{V}\right)^{1/K'} - 1\right]. \tag{12.39}$$

This gives us the value of P_S at V when the centering point is at $P = 0$. Comparison of (12.39) to the P_{TH} is best made when the variables above are changed to the shock wave variables c_0 and **s**.

The variable **s** found in the U_S-U_P experimental relationship is defined in terms of the derivatives of P_{H} with respect to V. Ruoff (1967) showed that by combining (12.1), (12.2), (12.3), and (12.39) and by assuming $U_P \to 0$ at $\eta \to 1$,

$$c_0 = \sqrt{\frac{K_{S_0}}{\rho_0}}. \tag{12.40}$$

Using the definitions of K_S and $(\partial K_S/\partial P)_S$ (see Section 1.5), he found that

$$\mathbf{s} = \frac{1}{4}\left[1 + \left(\frac{\partial K_S}{\partial P}\right)_S\right]_{P=0} = \frac{1}{4}(1 + K'). \tag{12.41}$$

Equations (12.40) and (12.41) are the main connections between shock wave variables and static equations of state. There are two shock wave variables (s and c_0) and three static EoS variables (K_{S_0}, K' and ρ_0). Thus the linear U_S-U_P relationship corresponds to the third order EoS. In the derivation above, a Murnaghan-type EoS was used in isentropic compression. Pastine and Piascesi (1966) showed that a quadratic U_S-U_P curve gave the above equations, (12.40) and (12.41), but that additional equations involving K'' arise. They based their derivation on repulsive and attractive potentials, such as found in Chapter 8. Jeanloz (1989) showed that (12.40) and (12.41) could also be derived by using an isothermal Birch-Murnaghan EoS of third degree in the derivation. Details about additional terms arising when a quadratic U_S-U_P curve is used will be found in Ruoff (1967) and Jeanloz (1989). Thus (12.40) and (12.41) appear to represent any reasonable third order EoS, and an additional equation involving K_0'' appears for any reasonable fourth order EoS.

In the course of their derivation, Pastine and Piascesci (1966) showed that near the origin, $P = 0$,

$$\left(\frac{\partial P}{\partial \eta}\right)_{\mathbf{H}} = \left(\frac{\partial P}{\partial \eta}\right)_S \tag{12.42a}$$

and

$$\left(\frac{\partial^2 P}{\partial \eta^2}\right)_{\mathbf{H}} = \left(\frac{\partial^2 P}{\partial \eta^2}\right)_S, \tag{12.42b}$$

which is helpful in changing from Hugoniot variables to isentropic variables.

We evaluate P_S, (12.39), from shock wave variables. From (12.41)

$$\frac{1}{K'} = \frac{1}{4s - 1}.$$

Using (12.40) and (12.42), we can re-evaluate (12.39), finding

$$P_S = \frac{c_0^2 \rho_0}{4s - 1}\left[\left(\frac{1}{\eta}\right)^{1/(4s-1)} - 1\right]. \tag{12.43}$$

This is to be compared with the definition of $P_{\mathbf{H}}$, (11.8), with $P_0 = 0$:

$$P_{\mathbf{H}} = \frac{\rho_0 c_0^2 \Delta\eta^2}{(1 - s\Delta\eta)^2}. \tag{12.44}$$

12.7. Computing the temperature along the Hugoniot

We need T along the isentrope centered at V_0. From (1.29),

$$\gamma = -\left(\frac{\partial\, \ell\mathrm{n}\, T}{\partial\, \ell\mathrm{n}\, V}\right)_S. \tag{1.29}$$

The temperature at V along the isentrope is thus

$$T_0 = T_0 \exp\left(-\int_{V_0}^{V} \gamma \frac{dV}{V}\right). \tag{12.45}$$

At high temperatures $\gamma\rho = $ constant, as justified by Fig. 12.6, so

$$T_S = T_0 \exp(\gamma_0 \Delta\eta). \tag{12.46}$$

Using (12.16) and taking $d\mathbf{H} = 0$, dS along the Hugoniot is:

$$dS = \frac{(V_0 - V)\, dP + (P - P_0)\, dV}{C_V}. \tag{12.47}$$

dS expanded as a function of T and V is:

$$dS = \left(\frac{\partial S}{\partial T}\right)_V dT + \left(\frac{\partial S}{\partial V}\right)_T dV = \left(\frac{C_V}{T}\right) dT + \left(\frac{\gamma C_V}{V}\right) dV.$$

Thus

$$\frac{dT}{T} = -\gamma \frac{dV}{V} + \frac{dS}{C_V}. \qquad (12.48)$$

Placing (12.47) in (12.48), T along the Hugoniot $T_{\mathbf{H}}$ is

$$dT_{\mathbf{H}} = -\gamma \frac{T_S}{V}\, dV + \frac{(V_0 - V)\, dP + (P - P_0)\, dV}{2C_V}. \qquad (12.49)$$

The first term on the right side of (12.49) is the increment of temperature due to isentropic compression by dV, dT_S, arising from (12.46), and the second term is due to the change of temperature arising from $(P_{\mathbf{H}} - P_S)$ at compression dV in order to shift from the isentrope to the Hugoniot. Thus at constant V, the terms on the right are positive, and $T_{\mathbf{H}} > T_S$.

Equations (12.47) and (12.48) are differential equations that must be solved to get the temperature and entropy along the Hugoniot. To evaluate $T_{\mathbf{H}}$, we note that, although the shock wave experiments give $P_{\mathbf{H}}(V)$, one also needs the functions $\gamma(V)$ and $C_V(V)$ along the Hugoniot. For many purposes it is adequate to assume C_V = constant and $\gamma\rho$ = constant.

12.8. The temperature of shocked iron along the Hugoniot

Brown and McQueen (1982) presented data for the longitudinal velocity of sound along the Hugoniot of iron. They determined the velocity of a pressure release wave that propagates behind and overtakes the shock front. This is the longitudinal velocity v_p, which exceeds the value of the bulk wave v_b traveling in the liquid. From their experiment, Brown and McQueen (1980) found two phase transitions in the solid: one at 200 GPa and one at 243 GPa (see Fig. 8.3) along the Hugoniot. The 243 GPa transition was found (Brown and McQueen, 1982) to be the transition from a solid to a liquid. The iron Hugoniot was reported by Brown and McQueen (1986) to be

$$U_S = 3.955(0.023) + 1.580(0.011) U_P.$$

The location of the pressure and volume of the solid–liquid and solid–solid transition is known with considerable accuracy. The determination of the temperature at these two phase transitions involves knowing the temperature of the Hugoniot at the two pressures. This involves using (12.49) where we assume $\rho\gamma = \rho_0\gamma_0$ = constant, written in the form

$$dT_{\mathbf{H}} = (\rho_0\gamma_0)\, T_S\, dV + \frac{(V_0 - V)\; dP + (P - P_0)\; dV}{2C_V}. \qquad (12.50)$$

It has been shown (see Fig. 3.12) that $\gamma\rho = \gamma_0\rho_0 =$ constant holds for typical shock wave temperatures. So the first problem is to evaluate $\gamma_0\rho_0 =$ constant.

Because ρ_0 of iron is known providing the phase is specified, the focus is on the determination of γ_0. Brown and McQueen (1986) find a reasonable value to be $\gamma_0 = 1.7$, by using ρ_0 for the liquid state (7.0 Mg/m^3). Thus $\gamma\rho = 12$ Mg/m^3 was used in the integration of (12.50). But if it is assumed that the properties of the solid side of the solid–liquid transition are needed, then ρ_0 is higher and $\gamma_0\rho_0$ may be higher than 12 Mg/m^3.

The next problem is to determine C_V. It is composed of the high temperature lattice specific heat and also the contribution for electronic states. The Brown and McQueen (1986) model of specific heat is

$$C_V = D\left(\frac{T}{\Theta}\right) + \beta_e\left(\frac{\rho_0}{\rho}\right)^{\gamma_e} T, \tag{12.51}$$

where γ_e is the electronic γ; β_e is proportional to the density of electrons at the Fermi level; and $D(T/\Theta)$ is the Debye function for specific heat. Boness et al. (1986) evaluated β_e and γ^e with band structure calculations and found $\beta_e = 90.8$ J Mg^{-1}K^{-2} and $\gamma^e = 1.34$.

The effect of the calculated electronic term is to add a small increment to the lattice specific heat $C_V = 3R$ from Debye theory at $T > 0$. Jamieson et al. (1978) found this electronic term to be equivalent to adding 1 to 2 R to the C_V of the lattice, $3R$. Other calculations appear to give a lower value for C_V^e. Thus C_V is expected to be $4R$ to $5R$, which might include a contribution to C_V from anharmonicity.

Using their own determination of C_V^e, Brown and McQueen computed the values of T along the Hugoniot and presented the data shown in Table 12.1. For transitions shown in Fig. 8.3, this computation yields the following T_m:

s-ℓ transition (243 GPa) $\quad T_m = 5800 \pm 500$ K

s-s transition (200 GPa) $\quad T_m = 4400 \pm 500$ K.

Both transitions are along the Hugoniot and are connected. To change parameters so as to raise (or lower) the first transition T in (12.50) would correspondingly raise (or lower) the second transition T.

Equation (12.50) can be thought of as composed of an isentrope term plus a correction. The correction is relatively small unless the reduced pressure P/K_{T_0} is high. Because P/K_{T_0} is 1.2 to 1.4 in the 200 to 243 GPa range, the correction term will be of the order of 20% of the total $T_{\mathbf{H}}$. Therefore for the calculated $T_{\mathbf{H}}$ to shift upward or downward, there has to be a rather large change in the value of γ_0.

Scientists who find T_m by measuring the radiance of shocked iron get values much higher than listed above. Yoo et al. (1993) report that at 255 GPa, $T_m = 6784 \pm 500$ K, giving a T_m at 243 GPa some 800° higher than above, and Bass et al. (1990) report yet even larger T_m values.

So there is the question of whether the Brown and McQueen value of T_m (at 243 GPa) is too low. On the other hand, there is an opposite conflict with Boehler's *s*-ℓ transition T_m at 200 GPa, which is 4250 ± 200 K, lower by about 150° than the *s*-*s* transition T found by Brown and McQueen (4400 ± 500 K). The problem is, how can an *s*-ℓ transition have a lower value of T_m than the T of an *s*-*s* transition, both at the same P? That would make the solid-solid transition at a higher temperature than the solid-liquid transition. This raises the question of whether the Brown and McQueen temperatures reported for 200 GPa are too high. On the other hand, if one focuses on the radiance-type shock wave results, one might argue that the solid-liquid value of T_m at 243 GPa reported by Brown and McQueen is too low. It is clear from (12.50) that sets of values of γ_0 and C_V^{ht} could be found to either raise or lower the Hugoniot temperature.

Taking the current values of C_V, the consequence of lowering the calculated T_m substantially to satisfy the DAC data of Boehler is that γ_0 must be much lower than 1.5. The consequence of raising the calculated T_m substantially to satisfy the shock wave radiance measurement of Yoo et al. (1993) would be that $\gamma_0 > 2.0$.

Table 12.1. Calculated properties for iron on the Hugoniot

Pressure GPa	Temperature K	Density Mg/m^3	Bulk Sound Velocity km/s
40	655	9.49	5.67
60	969	9.99	6.21
80	1355	10.40	6.67
100	1795	10.74	7.08
120	2274	11.04	7.45
140	2783	11.30	7.89
160	3311	11.54	8.19
180	3854	11.75	8.41
200	4407	11.94	8.69
220	4967	12.12	9.96
240	5531	12.28	9.22
260	6096	12.44	9.47
280	6662	12.58	9.71
300	7228	12.71	9.94
320	7792	12.84	10.17
340	8354	12.96	10.39
360	8914	13.07	10.60
380	9471	13.18	10.81
400	10024	13.28	11.02

From Brown and McQueen, 1986.

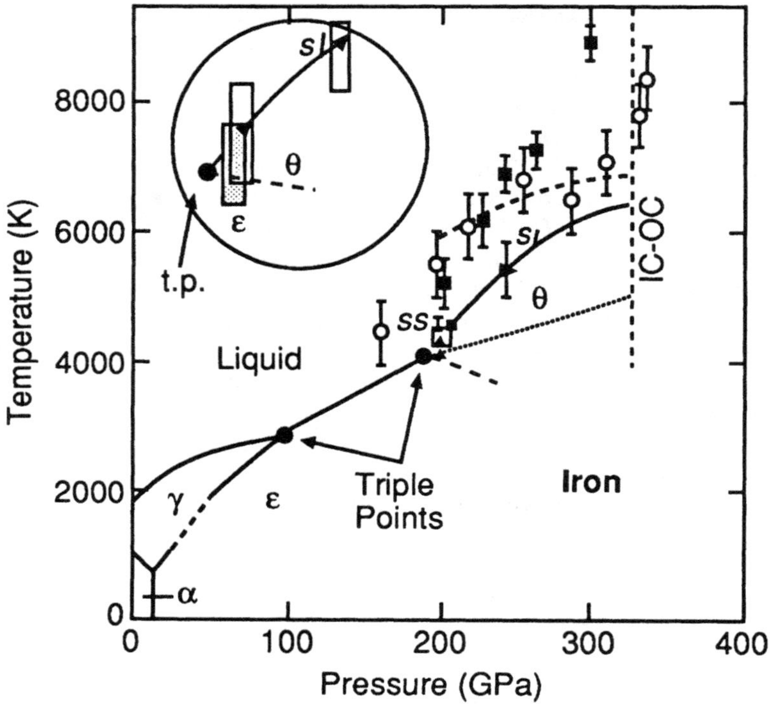

Fig. 12.7. The radiance shock wave data and the solid–liquid melting boundary. The s-s transition at 200 GPa and the s-ℓ transition at 243 GPa from Brown and McQueen (1986) (Fig. 8.3 and Table 12.1) are shown as triangles. The large open square is the upper point of the s-ℓ transition at 200 GPa found by Boehler (1993): below 200 GPa, Boehler's data are represented by a solid line. The dashed-dotted line between 200 GPa and 243 GPa is proposed by Anderson (1993) to separate the θ phase from the ϵ phase, but this requires the addition of a triple point at 190 GPa (see inset). The inset shows how the branches of the assumed triple point go through the data (Anderson, 1994); Anderson finds $T_m(330) = 6000$ K. The open circles are the shock data of Yoo et al. (1993), and the closed squares are the shock data of Bass et al. (1987). The dashed line is the solution proposed by Yoo et al. (1993), yielding $T_m(330) = 6830$ K. The dotted line is the solution proposed by Boehler (1994), yielding $T_m(330) = 4250$ K.

12.9. T_m of iron at 330 GPa: the inner–outer core interface

Data for T_m of iron for $P \geq 200$ GPa are shown in Fig. 12.7, where all the radiance shock wave data are plotted as circles and solid squares, along with Boehler's (1993) data on T_m from his diamond-anvil-cell measurements plotted as solid lines, and the Brown and McQueen (1986) two shock-wave data points plotted as triangles.

As of May 1994, there is disagreement on T_m of iron at core conditions, especially at 330 GPa. Boehler (1994) proposed that the Brown and McQueen s-s transition at 200 GPa and the s-ℓ transition at 243 GPa should have been reported at lower temperatures. By assuming the lowering of the B&M T_m at 243 GPa by 1100 K, he extrapolated his T_m curve from 200 to 330 GPa using the Kraut-Kennedy law (Fig. 12.8), obtaining thereby $T_m = 4250 \pm 200$ K at 330 GPa, the lowest value on record.

Yoo et al. (1994), ignoring the Boehler data and the B&M data, find from their shock wave data $T_m = 6830 \pm 500$ K at 330 GPa. The Williams et al. (1987) solution for T_m at $P = 330$ GPa is 7600 ± 500 K, the highest value on record, the high points having been determined by shock waves (Bass et al., 1987). These two shock wave determinations of T_m are found from a measurement of radiance from the shock, which arises from a two-body system (iron foil against corundum block). The calculation of the iron radiance alone requires one to compute the thermal diffusivity of the shocked two-body system (Ahrens et al., 1990; Tan and Ahrens, 1990).

By setting a new triple point (t.p.) at 190 GPa, Anderson (1994) reconciled the Boehler datum at 200 GPa and the Brown and McQueen data (see insert, Fig. 12.7). This t.p. increased dT_m/dP, and at 330 GPa, $T_m = 6000 \pm 300$ K. Although this solution passes within one of the error bars of the shock wave data of Yoo et al. (1993), the radiance-type shock wave data are ignored in Anderson's proposed solution.

Saxena et al. (1994) proposed that the β-ϵ boundary passes through the s-s boundary of B&M at 200 GPa and 4000 K (Fig. 12.8), and that it intercepts the ϵ-liquid boundary, establishing a triple point at 216 GPa and 4500 K. This places his T_m of pure iron near 6000 K for 330 GPa.

Neglecting error bars, the temperature spread for $T_m(330)$ from all extrapolations (Williams et al., 1991, to Boehler, 1994) is 4250–7600 K, a large difference. If one disregards the uppermost point of the Bass et al. data (Fig. 12.7) in view of the new, unpublished shock wave work of Ahrens, who now estimates $T_m(330)$ to be near 6930 K, the temperature spread is about 2600 K. Poirier and Shankland (1994) calculated $T_m(330)$ from dislocation theory (Section 11.8) and found it to be 5600 K to 6160 K, depending on the crystallographic structure (Section 11.14). This agrees with the Saxena et al. (1994) T_m extrapolation and the Bukowinski (1977) calculation. Until the conflicting data on T_m at high P from diamond cell and shock wave experiments are reconciled, I recommend that a T_m value of 6000 K for iron at 330 GPa be used for geomagnetism calculations.

12.10. What is the dominant iron phase in the inner core?

There is little agreement on the structure of pure iron at inner core conditions (as of March, 1994). The hcp phase (ϵ) has been justified for the inner core by Saxena et al. (1994) by including the B&M s-s 200 GPa point but ignoring the θ phase proposed by Boehler (1986) and by ignoring the radiance-type shock wave results shown in Fig. 12.7. Williams et

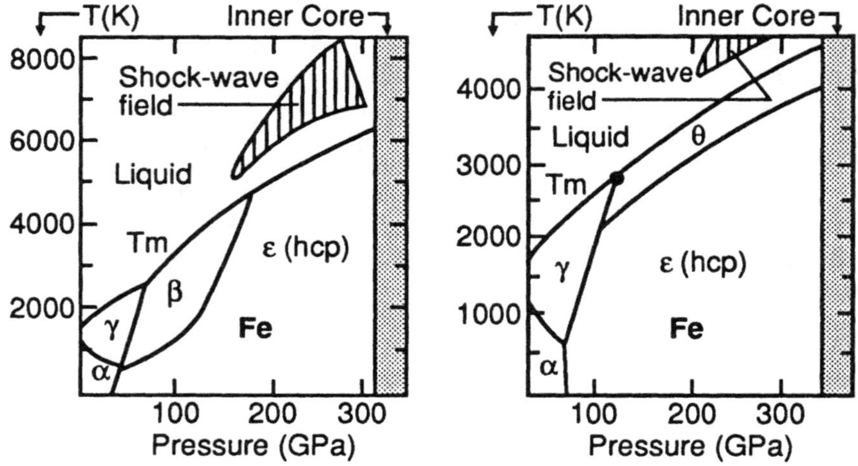

Fig. 12.8. Two cases for the phases of the inner core: left) Saxena et al.'s (1994) proposal for ϵ iron; and right) Boehler's (1994) proposal for the θ phase of iron. Anderson's (1994) proposal (Fig. 12.7) is similar to Boehler's except that T_m is higher; θ is suggested to have the fcc structure; and θ is cut off from β by ϵ.

al.'s (1991) phase diagram shows ϵ as the inner core phase, but they ignore the B&M s-ℓ 243 GPa point and all Boehler's diamond cell work. Boehler's interpretation is that the B&M data are to be lowered and the radiance-type shock wave work is to be ignored; the θ phase (Boehler, 1986) is the inner core, disallowing hcp. Boehler (1994) extended θ to include the new β phase. The structure of the inner core phase of iron has not yet been established. Theoreticians (Cohen et al., 1994; Stixrude and Cohen, 1994) say this phase cannot be bcc. Could the inner core iron phase be fcc (γ)? This possibility harkens back to the theory of fcc iron at inner core conditions by Bukowinski (1977). Two current proposals for the phase diagram of iron, showing the phase appropriate to the inner core, are shown in Fig. 12.8.

The amount of impurities in the iron of the inner core needed to match PREM density depends on the density of the iron phase dominating the inner core. An ϵ inner core requires a large amount of light impurities (Jephcoat and Olson (1987)). We do not confidently know the density of fcc iron at inner core conditions. The neutron scattering results of Stassis (1994) indicate that K_T of fcc iron is 143 GPa at 1428 K (the ambient value is unknown. Extrapolation gives 170 GPa at ambient, higher than the value in Table 8.3). Considering that ρ_0 is high (8.00 g/cc), this high value of K_{T_0} would make the density of fcc iron lower at inner core conditions than indicated by the EoS calculations shown in Fig. 10.14. If so, an fcc inner core phase may require some light impurities. But an fcc inner core certainly requires fewer impurities than an hcp inner core. Thus it is important to find the ambient values of K_{T_0} and K' of fcc iron quite precisely.

13

THERMODYNAMIC FUNCTIONS

13.1. Introduction

It is well known that the computations of thermodynamic functions (entropy, for example) require knowledge of the thermal expansivity α and compressibility κ, as well as the specific heat. The pressure (or volume) and temperature variation of α and κ contribute to the pressure (or volume) and temperature variation of the thermodynamic functions. The tabulation of these thermodynamic functions in the extreme ranges of temperature and pressure requires a good understanding of the EoS, elasticity at very high temperature, anharmonicity of the solid, and the thermal expansivity at high temperature.

Recent experimental progress in the thermal EoS has clarified the volume and high-temperature corrections to the thermal pressure P_{TH} and to the isothermal bulk modulus K_T (the reciprocal of the compressibility, κ) for MgO, Mg_2SiO_4, and Al_2O_3 (Goto et al, 1989; Isaak et al., 1989a,b). This progress is represented by the ability to measure elastic constants at high temperatures (in fact, as high as 1825 K) using new resonance measurement techniques (Goto and Anderson, 1988).

The product αK_T is an important correction needed for evaluating entropy at high compression and volume. It has become apparent in the last 10 years that it is best to treat αK_T as a single parameter, as the product varies little with pressure or temperature, whereas both α and K_T each varies significantly with pressure and with temperature.

Birch (1952) first proposed that for silicates and oxides, to a good approximation, αK_T should be independent of pressure. He based this idea on experiments made by Bridgman (Birch, 1952) on alkali metals up to 3 GPa. Anderson (1980) proposed that at high temperatures, αK_T is independent of T for oxides and silicates, and he followed up with a review paper that confirmed this for many oxides and silicates (Anderson, 1984). The research cited above, among others, has pinned down and supported Birch's proposal, by detailed measurements on individual minerals. Although it is of great value to geophysical theories to know that αK_T is approximately independent of pressure and temperature, for analysis of thermodynamic functions, it is still important to precisely quantify the amount that αK_T varies with pressure and with temperature. In this text we shall deal with αK_T as a function of V and T, although its variation is small, at high pressure and high temperature.

In this chapter, I present the equations for the thermodynamic functions appropriate to oxides and silicates. The equations are presented in such a way that αK_T is a major variable. I illustrate the procedure by presenting data on the thermodynamics for MgO, for the temperature range 300–2000 K and in the pressure range up to 150 GPa. The complete data set on all thermodynamic functions of MgO will be found in Anderson and Zou (1989). We also discuss the experiments needed to improve accuracy at extreme compression and temperature.

13.2. Basic equations for entropy

The entropy as a function of V and T can be found from

$$d\mathcal{S} = \left(\frac{\partial \mathcal{S}}{\partial V}\right)_T dV + \left(\frac{\partial \mathcal{S}}{\partial T}\right)_V dT. \tag{13.1}$$

Consider isotherms $dT = 0$. Equation (13.1) becomes

$$d\mathcal{S}\Big|_T = \left(\frac{\partial \mathcal{S}}{\partial V}\right)_T dV. \tag{13.2}$$

From calculus

$$\left(\frac{\partial \mathcal{S}}{\partial V}\right)_T = \alpha K_T = \left(\frac{\partial P}{\partial T}\right)_V,$$

so

$$d\mathcal{S}\Big|_T = \alpha K_T \, dV. \tag{13.3}$$

Integrating (13.3),

$$\Big[\mathcal{S}(V) - \mathcal{S}(V_0)\Big]_{T*} = \int \alpha K_T \, dV, \tag{13.4}$$

where T^* represents the isotherm.

We need the dependence of αK_T with volume at constant temperature. This is usually not available as direct experimental information. Using the thermodynamic identity

$$\left(\frac{\partial \alpha K_T}{\partial V}\right)_T = -\frac{1}{V}\left(\frac{\partial K_T}{\partial T}\right)_V, \tag{13.5}$$

because the properties on the right are known from high temperature experiments. The term on the right is usually a small number, often zero. We see this from the calculus equation

$$\left(\frac{\partial K_T}{\partial T}\right)_V = \left(\frac{\partial K_T}{\partial T}\right)_P + \left(\frac{\partial P}{\partial T}\right)_V \left(\frac{\partial K_V}{\partial P}\right)_T = \left(\frac{\partial K_T}{\partial T}\right)_P + \alpha K_T K'. \tag{13.6}$$

The first term on the right is negative; the second term on the right, involving αK_T from a high temperature experiment and $K_T' = (\partial K_T / \partial P)_T$ from a pressure experiment, is positive. These two terms are about equal in value. Dividing the right side by αK_T, we have $(\partial K_T/\partial T)_V = -\alpha K_T\,(\delta_T - K')$.

13.2.1. *Entropy when $(\partial K_T/\partial T)_V$ is independent of V*

Now consider that $(\partial K_T/\partial T)_V$ is a small number, often sufficiently small that its volume dependence, if any, can be ignored. In Chapter 3, we found

$$\left(\frac{\partial K_T}{\partial T}\right)_V = \mathbf{w}. \tag{13.7}$$

$\mathbf{w}$ is zero for NaCl at ambient conditions, but non zero although small for most oxides and silicates. The variation of $\delta_T - K' = -\mathbf{w}/\alpha K_T$ with V and T is shown in Fig. 3.3. Equation (13.5) becomes

$$\left(\frac{\partial \alpha K_T}{\partial V}\right)_T = -\frac{\mathbf{w}}{V}, \tag{13.8}$$

and upon integration

$$\alpha K_T = \mathbf{w}\ \ell\text{n}\, V + C. \tag{13.9}$$

To evaluate the integrating constant, take $\alpha K_T = (\alpha K_T)_0$ when $V = V_0$. Thus

$$\alpha K_T - (\alpha K_T)_0 = \mathbf{w}\ \ell\text{n}\, \eta. \tag{13.10}$$

Placing (13.9) into (13.4),

$$\Big[\mathcal{S}(V) - \mathcal{S}(V_0)\Big]_{T^*} = (\alpha K_T)_0\,(V - V_0) - \mathbf{w} V_0 \int_1^\eta \ell\text{n}\,\eta\, d\eta. \tag{13.11}$$

Equation (13.11) is then

$$\Big[\mathcal{S}(\eta) - \mathcal{S}(1)\Big]_{T^*} = -\,(\alpha K_T)_{T^*}\, V_0\,(1-\eta) - \mathbf{w} V_0\,(\eta\ \ell\text{n}\,\eta + \eta - 1)\,. \tag{13.12}$$

In the special case $\mathbf{w} = 0$ (or $(\partial K_T/\partial T)_V = 0$), we have along the isotherm T^*

$$\Big[\mathcal{S}(\eta) - \mathcal{S}(1)\Big]_{T^*} = -\,(\alpha K_T)_{T^*}\, V_0\,(1-\eta)\,. \tag{13.13}$$

The entropy of MgO computed using (13.12) is found in Anderson and Zou (1989). Because (αK_T) is nearly independent of η and T, [(10.39) and (10.40)], we see that for the simple case of (13.13) entropy varies as $(1-\eta)$.

Consider the isochore (constant volume) case, $\Delta V = 0$. Using (13.1) along with the identity $(\partial \mathcal{S}/\partial T)_V = C_V/T$,

$$\mathcal{S}(T) - \mathcal{S}(T_0) = \int_{T_0}^{T} C_V \left(\frac{dT}{T}\right). \tag{13.14}$$

At high temperature when C_V can be assumed to be independent of T, as found for many oxides and minerals (Fig. 2.1),

$$\Big[\mathcal{S}(T) - \mathcal{S}(T_0)\Big]_V = (C_V)T_0 \ \ell\text{n} \left(\frac{T}{T_0}\right). \qquad (13.15)$$

The isochore represents the volume at $P = 0$ and $T = T_0$ (standard room temperature, 300 K).

13.2.2. *Entropy when $(\partial K_T/\partial T)_V$ is a function of V*

Equation (13.6) is of the form

$$\ell\text{n}\ (\alpha K_T) = (\delta_T - K')\ \ell\text{n}\ V \qquad (13.16)$$

or

$$\alpha K_T = (\alpha K_T)_0 \exp \int (\delta_T - K')\frac{d\eta}{\eta}. \qquad (11.47)$$

Using (13.3),

$$d\mathcal{S} = \alpha K_T\, dV = V_0 \alpha K_T\, d\eta. \qquad (13.17)$$

Along isotherms for different values of η, we have from (13.17)

$$\mathcal{S}(\eta)_{T^*} - \mathcal{S}(\eta = 1) = V_0 \int_\eta \alpha K_T d\eta. \qquad (13.18)$$

From (13.17)

$$\mathcal{S}(\eta)_{T^*} - \mathcal{S}(\eta = 1) = V_0 \int_\eta \left[(\alpha K_T)_0 \exp \int (\delta_T - K')\frac{d\eta}{\eta}\right] d\eta. \qquad (13.19)$$

The change of entropy with volume going from V_0 to V, or from 1 to η along an isotherm T^* as given in (13.19), involves a double integral. But one proceeds numerically by finding $(\delta_T - K')$ as a function of T and ρ first.

Values for $\delta_T - K'$ as a function of η along isotherms are shown in Fig. 13.1 for the case of MgO. The value of $(\delta_T - K')$ is not zero, but it is small. It represents a measure of the intrinsic (isochoric) change of bulk modulus with T that can be seen from the identity given by (3.12), where

$$(K' - \delta_T) = \left(\frac{1}{\alpha K_T}\right)\left(\frac{\partial K_T}{\partial T}\right)_V = \frac{\mathbf{w}}{\alpha K_T}.$$

The solution for (11.47), αK_T versus η along isotherms, has already been plotted in Chapter 3 (Fig. 3.7) for MgO.

Units of $\alpha K_T V_0 = \alpha K_T/\rho_0$ are J/g-K. Typically ρ_0 is in g/cc, α is in units of 10^{-6}/K, and K_T is in J/cm^2 (K_T = 160 GPa = 1.60×10^5 J/cc; $\alpha = 3.12 \times 10^{-6}$; $\rho = 3.587$ g/cc, so $\alpha K_T/\rho_0 = 1.395$ J/gK).

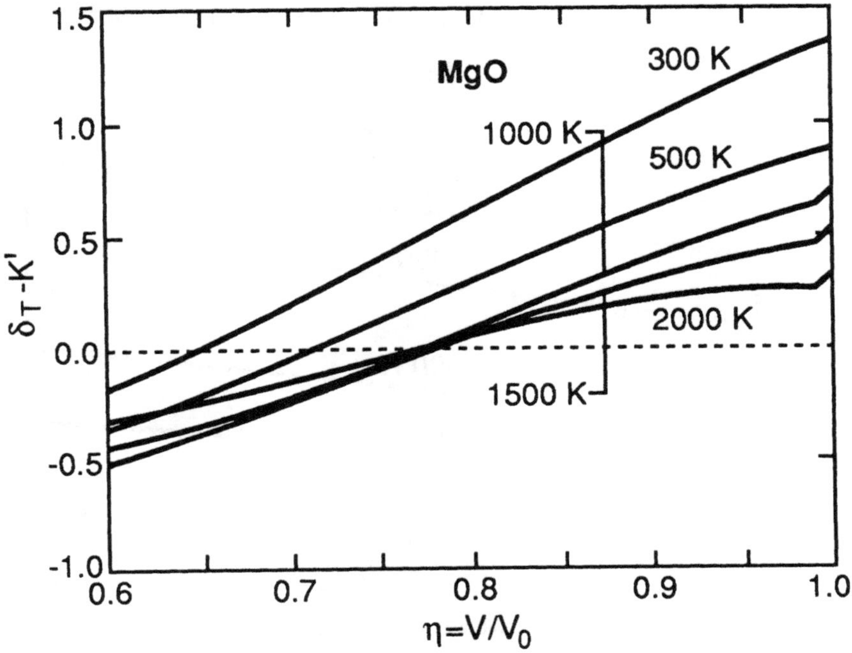

Fig. 13.1. A plot of $\delta_T - K'$ versus η at various isotherms for MgO. Above the Debye temperature $\delta_T - K'$ gets increasingly close to zero as the temperature rises. Data taken from Figs. 3.4 and 3.5 (modified from Anderson et al., 1993).

When we integrate (13.19) using data shown in Fig. 13.1, we need the constant of integration, which is the value of $\mathcal{S}$ at $T_0 = 300$ K. We are constructing our thermodynamics in the $\mathcal{S}(\eta, T)$ field, but measurements are taken at constant pressure, not constant η, so we must find the relationship between $\mathcal{S}$ at constant η (called here $\mathcal{S}_{\eta=1}$) and $\mathcal{S}$ at constant P (called here $\mathcal{S}_p$). The variation of entropy with constant η (or V/V_0), from T_0 to T^* is given by

$$\mathcal{S}_{\eta=1}(T^*) - \mathcal{S}_{\eta=1}(T_0) = \int_{T_0}^{T^*} C_V \frac{dT}{T},$$

where $C_V = C_P/(1+\alpha\gamma T)$. Standard conditions were taken as $T_0 = 300$ K and $P = 1$ bar. Thus at absolute zero $\eta = 1.08$. The entropy of $\mathcal{S}_{\eta=1}$ at 300 K is found by using the tabled values of $\mathcal{S}_P$ for MgO and correcting by calculating (Stull and Prophet, 1971)

$$\mathcal{S}_P(300) - \mathcal{S}_{\eta=1}(300) = \int_0^{300} (C_P - C_V) \frac{dT}{T}.$$

We use the published value of $\mathcal{S}_P(300)$ (Barron et al., 1959). We find $\mathcal{S}_{\eta=1}(300) = 0.5705$ J/g-K from the measurement $\mathcal{S}_P(300) = 0.6743$ J/g-K. The data for $\mathcal{S}_\eta$ are tabled in Table 13.1 and plotted in Figs. 13.2 and 13.3.

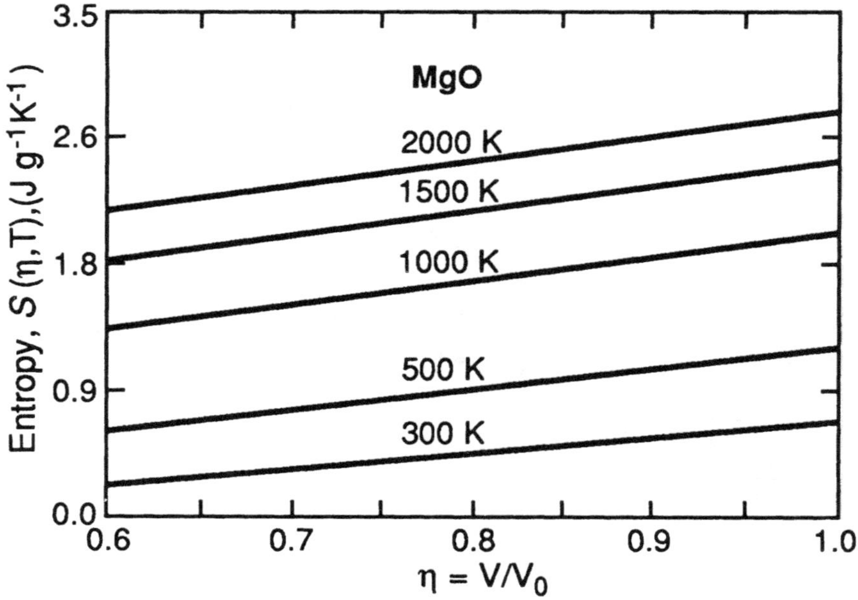

Fig. 13.2. $\mathcal{S}$ versus η along isotherms for MgO.

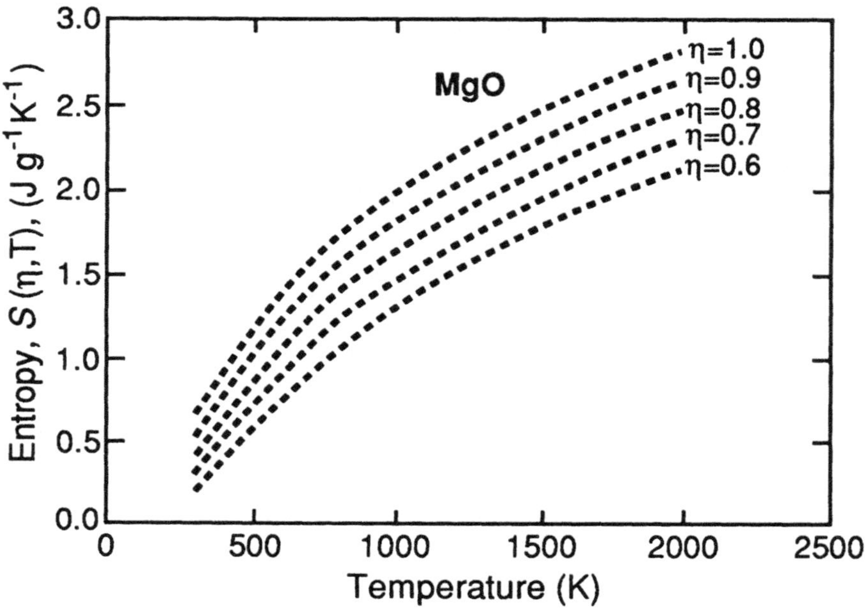

Fig. 13.3. $\mathcal{S}$ versus T along isochores for MgO (modified from Anderson and Zou, 1989).

Table 13.1. Entropy versus η at constant T, $\mathcal{S}(\eta, T)$, for MgO

$\eta = V/V_0$	$\mathcal{S}(\eta, 300)$	$\mathcal{S}(\eta, 500)$	$\mathcal{S}(\eta, 1000)$	$\mathcal{S}(\eta, 1500)$	$\mathcal{S}(\eta, 2000)$
0.60	0.203	0.58	1.30	1.77	2.12
0.62	0.225	0.68	1.34	1.81	2.15
0.64	0.246	0.64	1.37	1.85	2.19
0.66	0.267	0.67	1.41	1.88	2.22
0.68	0.289	0.71	1.44	1.92	2.26
0.70	0.310	0.73	1.48	1.95	2.29
0.72	0.332	0.76	1.51	1.99	2.33
0.74	0.353	0.79	1.54	2.02	2.36
0.76	0.376	0.82	1.58	2.05	2.40
0.78	0.380	0.85	1.61	2.09	2.43
0.80	0.420	0.88	1.64	2.12	2.46
0.82	0.443	0.90	1.68	2.16	2.50
0.84	0.466	0.94	1.71	2.19	2.53
0.86	0.489	0.97	1.74	2.22	2.57
0.88	0.513	1.00	1.78	2.26	2.60
0.90	0.538	1.03	1.81	2.29	2.64
0.92	0.564	1.06	1.85	2.33	2.67
0.94	0.589	1.09	1.88	2.36	2.71
0.96	0.616	1.13	1.92	2.40	2.74
0.98	0.642	1.16	1.95	2.43	2.78
1.00	0.6705	1.19	1.99	2.47	2.81
1.02	0.699	1.23	2.02	2.51	2.85
1.04	0.727	1.26	2.06	2.54	2.88
1.06	0.755	1.29	2.10	2.58	2.92
1.08	0.784	1.33	2.13	2.61	2.96
1.10	0.812	1.36	2.17	2.65	2.99
1.12	0.841	1.39	2.21	2.69	3.03
1.14	0.870	1.43	2.24	2.72	3.06
1.16	0.898	1.46	2.28	2.76	3.10
1.18	0.927	1.50	2.32	2.80	3.14

The variations of C_P with T at constant pressure ($P = 1$ bar) and the computed C_V taken from the entropy ($\eta = 1$), as well as the variation of entropy with $\mathcal{S}_P$ and $\mathcal{S}_V$ ($P = 1$ bar, $\eta = 1$), are shown in Fig. 13.4. We find C_V to be classical with little evidence of anharmonicity in our temperature range. This is $(\partial C_V/\partial T) = 0$, indicating that in the density of states, the frequencies are a function of V but not of T up to at least 2000 K.

$\mathcal{S}_V$, or $\mathcal{S}(V,T)$, is important for the Helmholtz energy, $\mathcal{F} = \mathcal{U} - T\mathcal{S}_V$. On the other hand, $\mathcal{S}_P$, or $\mathcal{S}(P,T)$, is important for the Gibbs energy because $\mathcal{G} = \mathcal{H} - T\mathcal{S}_P$. Here we use $\mathcal{S}(\eta, T)$ in place of $\mathcal{S}(V,T)$.

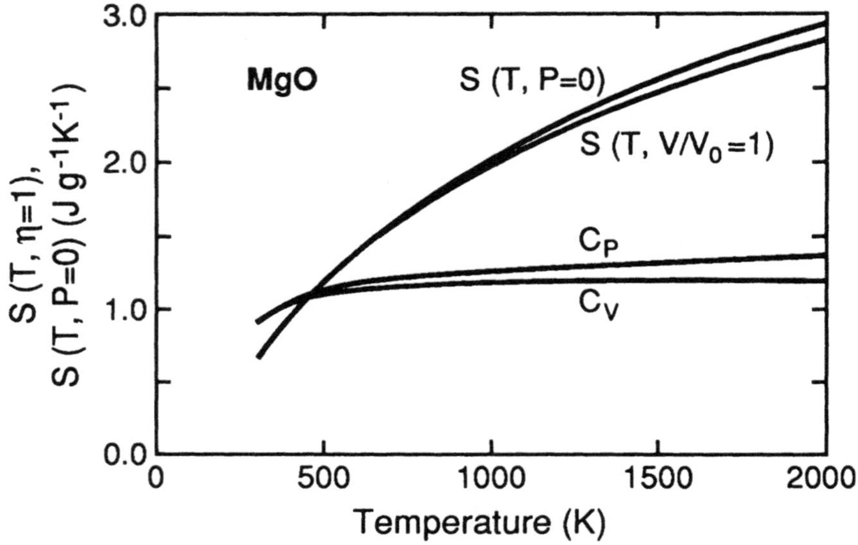

Fig. 13.4. Entropy for MgO at constant P and constant η versus T and C_P and C_V versus T (modified from Anderson and Zou, 1989).

13.3. The internal energy as a function of V and T for MgO

The variation of internal energy $\mathcal{U}$ with T at $\eta = 1$ above $T = 300$ is given by

$$\mathcal{U}(T,^* 1) - \mathcal{U}(300, 1) = \int_{300}^{T^*} C_V \, dT.$$

We find the value of $\mathcal{U}(300, 1)$ by integrating $C_P - C_V$, as tabled by Barron et al. (1959). This gives us the difference between the enthalpy and the internal energy at 300 K. At standard conditions we find

$$\mathcal{H}(300, P = 0) - \mathcal{U}(300, \eta = 1) = 2.4 \text{ J/g}$$

for MgO where the value $\mathcal{H}(300, P = 0)$ is 129.7 J/g for MgO, according to Garvin et al. (1988). To obtain the volume dependence on $\mathcal{U}$, we expand the internal energy out versus dT and dV,

$$d\mathcal{U} = \left(\frac{\partial \mathcal{U}}{\partial T}\right)_V dT + \left(\frac{\partial \mathcal{U}}{\partial V}\right)_T dV. \tag{13.20}$$

Consider the isotherm $dT = 0 (T = T^*)$ and the Maxwell relationship

$$\left(\frac{\partial \mathcal{U}}{\partial V}\right)_{T^*} = (\alpha K_T T^* - P). \tag{13.21}$$

Replace (13.11)

$$\Delta \mathcal{U}\Big|_{T^*} = T^* \int \alpha K_T \, dV - \int P \, dV. \tag{13.22}$$

From (13.3)

$$\Delta \mathcal{S}\Big|_{T^*} = \int \alpha K_T \, dV. \tag{13.23}$$

Integrating (13.21) we have

$$\mathcal{U}(T^*, \eta) = \mathcal{U}(T^*, 1) + T^* \left[\mathcal{S}(T^*, \eta) - \mathcal{S}(T^*, 1)\right] - \int P \, dV. \tag{13.24}$$

Separating P into its two parts $P = P_{(T=0)} + P_{TH} = P_0 + P_{TH}$, where P_0 is the isothermal EoS, and P_{TH} is the thermal pressure, we have

$$\int P \, dV = \int P_0 \, dV + \int P_{TH} \, dV. \tag{13.25}$$

Now P_{TH} is sometimes slightly dependent on V, as αK_T is sometimes slightly dependent on V and $P_{TH} = \int_0^T \alpha K_T \, dT$. The volume effect is observable if $\mathbf{w}$ is not zero; nevertheless, $\mathbf{w} = (\partial K_T / \partial T)_V$ is always small. The pressure effect arising from $\mathbf{w}$ is a small contribution to P_{TH}, which itself is small compared to P_0. Treating P_{TH} as independent of V creates an error in $\int P_{TH} \, dV$ that is negligible compared to $\int P_0 \, dV$.

Assume P_{TH} to be independent of V; then (13.35) becomes

$$\int P \, dV = \int P_0 \, dV + P_{TH} \, \Delta V. \tag{13.26}$$

The last term is

$$P_{TH} \, \Delta V = P_{TH}(V - V_0) = -P_{TH} V_0 \, (1 - \eta) \, . \tag{13.27}$$

From (13.24), (13.25), and (13.26), we see that the change in internal energy with η along the T^* isotherm is

$$\mathcal{U}(T^*, \eta) - \mathcal{U}(T^*, 1) = (T^* \Delta \mathcal{S}_x)_{T^*} - P_{TH} \left(\frac{1}{\rho_0}\right) (1 - \eta) - \int P_0 \, dV, \tag{13.28}$$

where ρ_0 refers to the zero pressure value appropriate to the isotherm, T^*. For P along the isotherm $T = 0$, we use the Birch-Murnaghan EoS of third degree (where the subscript 0 on P means P at absolute zero),

$$P(\eta, 0) = P_0(\eta) = \frac{3}{2} K_0 \left(w^{7/3} - w^{5/3}\right) \left[1 + \frac{3(K_0' - 4)(w^{2/3} - 1)}{4}\right]. \tag{13.29}$$

Here $w = V_0/V = 1/\eta$. V_0 is absolute zero volume so that for ρ_0 in (13.29), we used 3.603 g/cc for MgO.

Accordingly,

$$\int_{V_0}^{\eta} P_{T=0}\,dV = \frac{3}{4}K_0V_0\left[\frac{3}{2}\left(\frac{3}{2}K_0' - 7\right)\left(w^{4/3} - 1\right) + 3\left(4 - \frac{3}{4}K_0'\right)\left(w^{2/3} - 1\right) + \left(3 - \frac{3}{4}K_0'\right)\left(w^2 - 1\right)\right]. \quad (13.30)$$

The term K_0V_0 is also $K_0/\rho_0 = 166.2/3.603$ GPa/(g/cm) $= 4.61 \times 10^4$ J/g for MgO. The value of $\int_1^{\eta} P_{T=0}\,dV$ is useful and is tabled below as $\int_1^{\eta} P_0\,d\eta$. We need the value of $\mathcal{U}(300, 1)$, obtained by integrating $C_P - C_V$, such as tabled in Barron et al. (1959), giving us $\mathcal{H}(300, P = 0) - \mathcal{U}(300, 1) = 0.9$ J/g. Stull and Prophet (1971) list $\mathcal{H}(300, P = 0)$ as 182 J/g, so that $\mathcal{U}(300, \eta = 1)$ is 127.3 J/g.

Table 13.2. Pressure, $P(\eta, 0) = P_0(\eta)$, and $\int_1^{\eta} P_0 d\eta$ versus η for MgO (GPa)

$\eta = V/V_0$	$P(\eta,0)$	$\int_1^{\eta} P_0\,d\eta$	$\eta = V/V_0$	$P(\eta,0)$	$\int_1^{\eta} P_0\,d\eta$
1.10	−12.9894	0.7047	0.890	24.7070	1.2293
1.09	−11.9697	0.5799	0.880	27.7522	1.4914
1.08	−10.8962	0.4655	0.870	30.9665	1.7848
1.07	− 9.7663	0.3622	0.860	34.3603	2.1111
1.06	− 8.5767	0.2704	0.850	37.9453	2.4723
1.05	− 7.3245	0.1909	0.840	41.7326	2.8704
1.04	− 6.0063	0.1242	0.830	45.7357	3.3074
1.03	− 4.6186	0.0710	0.820	49.9680	3.7856
1.02	− 3.1577	0.0321	0.810	54.4444	4.3074
1.01	− 1.6195	0.0081	0.800	59.1812	4.8754
1.00	− 0.0000	0.0000	0.790	64.1949	5.4921
0.99	1.7054	0.0084	0.780	69.5049	6.1605
0.98	3.5019	0.0343	0.770	75.1304	6.8836
0.97	5.3939	0.0788	0.760	81.0934	7.6646
0.96	7.3871	0.1427	0.750	87.4168	8.5070
0.95	9.4873	0.2271	0.740	94.1260	9.4144
0.94	11.7004	0.3331	0.730	101.2482	10.3909
0.93	14.0334	0.4619	0.720	108.8132	11.4407
0.92	16.4925	0.6145	0.710	116.8524	12.5685
0.91	19.0861	0.7924	0.700	125.4013	13.7791
0.90	21.8212	0.9968	0.690	134.4976	15.0780

From Anderson and Zou, 1990.

Table 13.3. Internal energy versus T at constant η, $\mathcal{U}(\eta, T)$

T (K)	$\mathcal{U}(1.0,T)$	$\mathcal{U}(0.90,T)$	$\mathcal{U}(0.85,T)$ (J/g)	$\mathcal{U}(0.80,T)$	$\mathcal{U}(0.75,T)$	$\mathcal{U}(0.70,T)$
300	127.3	380	790	1440	2450	3910
350	175.0	440	840	1500	2490	3960
400	225.7	480	880	1540	2540	4000
450	278.5	530	930	1590	2590	4050
500	332.8	580	980	1640	2640	4110
550	388.3	640	1040	1700	2700	4170
600	444.5	690	1090	1750	2760	4230
650	501.4	750	1150	1810	2820	4290
700	558.8	810	1210	1870	2880	4360
750	616.6	860	1260	1930	2940	4420
800	674.7	920	1320	1990	3000	4490
850	733.1	980	1380	2050	3070	4550
900	791.8	1040	1440	2110	3130	4620
950	850.7	1100	1500	2170	3190	4690
1000	909.7	1160	1560	2230	3260	4750
1050	968.8	1220	1620	2300	3330	4830
1100	1028.1	1280	1690	2360	3390	4890
1150	1087.4	1340	1750	2430	3460	4970
1200	1146.7	1400	1810	2490	3520	5030
1300	1265.6	1520	1930	2620	3660	5180
1400	1384.5	1640	2060	2750	3800	5320
1500	1503.6	1770	2180	2880	3930	5470
1600	1622.8	1890	2310	3010	4070	5610
1700	1742.0	2010	2440	3140	4210	5760
1800	1861.2	2130	2560	3270	4340	5910
1900	1980.5	2260	2690	3400	4480	6050
2000	2099.9	2380	2810	3530	4620	6200

V_0 is standard volume at $T = 300$ K and $P = 1$ bar.

$P(\eta, 0)$ and the correction term $\int_1^\eta P(\eta)\, d\eta$ versus η for MgO are shown in Table 13.2. $\int_1^\eta P(\eta)\, d\eta = \int_1^\eta P_0\, d\eta$ is found in (13.28), the equation for internal energy $\mathcal{U}$. $\mathcal{U}$ versus T along selected isochores is shown in Table 13.3. $\mathcal{U}(T, \eta = 1.1)$ is not shown, but is close in value to $\mathcal{U}(T, 0.9)$.

The isochoric plot of $\mathcal{U}$ versus T shows a steady increase with T (almost a straight line) following the law $(\partial \mathcal{U}/\partial T)_V = C_V$. From T slightly below Θ to higher T, C_V does not change with T for MgO (see Fig. 2.1). The isochores of $\mathcal{U}(T, \eta)$ are symmetric about $\eta = 1$. $\mathcal{U}$ $(T, 1.1)$ has almost the same set of values as $\mathcal{U}(T, 0.9)$. $\mathcal{U}$ has a minimum at $\eta = 1$ for 300 K.

13.4. The entropy versus P and T

Consider $\mathcal{S}_P(T, P)$ expanded in dT and dP. Then

$$d\mathcal{S}_P = \left(\frac{\partial \mathcal{S}_P}{\partial T}\right)_P dT + \left(\frac{\partial \mathcal{S}_P}{\partial P}\right)_{T^*} dP. \tag{13.31}$$

Consider the isotherm $dT = 0$; then

$$\left(\frac{\partial \mathcal{S}_P}{\partial P}\right)_{T^*} = -V\alpha, \tag{13.32}$$

or

$$\mathcal{S}(T^*, P) - \mathcal{S}(T^*, 0) = -\int_0^P V\alpha\, dP. \tag{13.33}$$

The functional relationship between α and P is not helpful, so we change the differential parameter to dV using the definition of K_T:

$$-V\, dP = K_T\, dV. \tag{13.34}$$

Thus

$$-\int_0^P V\alpha\, dP = \int_{V_0}^{V} \alpha K_T\, dV = V_0 \int_i^{\eta} \alpha K_T\, d\eta. \tag{13.35}$$

Because η is a function of T, we have

$$\mathcal{S}(T^*, P) - \mathcal{S}(T^*, 0) = V_0 \int_{\eta(T^*,0)}^{\eta(T^*,P)} \alpha K_T\, d\eta. \tag{13.36}$$

We see that the right side of (13.36) is the same as (13.19) except that the limits of integration in η are determined by both P and T^*.

Assume now that $\mathbf{w}$ is independent of V as a good approximation. Then $\int \alpha K_T$ changes with η according to (13.11), and we have

$$\int \alpha K_T\, d\eta = \alpha K_T V_0(1-\eta) - \mathbf{w} V_0(\eta\ \ell\text{n}\ \eta + 1 - \eta). \tag{13.37}$$

In this approximation (Anderson and Zou, 1989),

$$\mathcal{S}_P(T^*, P) - \mathcal{S}_P(T^*, 0) = -(\alpha K_T)_{T^*}\left(\frac{1}{\rho_0}\right)(1-\eta_0)$$
$$- \mathbf{w}(T^*, 0)\left(\frac{1}{\rho_0}\right)(\eta_0\ \ell\text{n}\ \eta_0 + 1 - \eta_0), \tag{13.38}$$

where $\rho_0 = \rho(T^*, P = 0)$.

Along the isobar at $P = 0$

$$\mathcal{S}_P = \int_0^T C_P\, dT, \tag{13.39}$$

which gives the integrating base.

The values of $\mathcal{S}_P(T, P)$ versus T for MgO are plotted along isobars in Fig. 13.5. There is a convergence of $\mathcal{S}_P(T, P)$ isobars at high P.

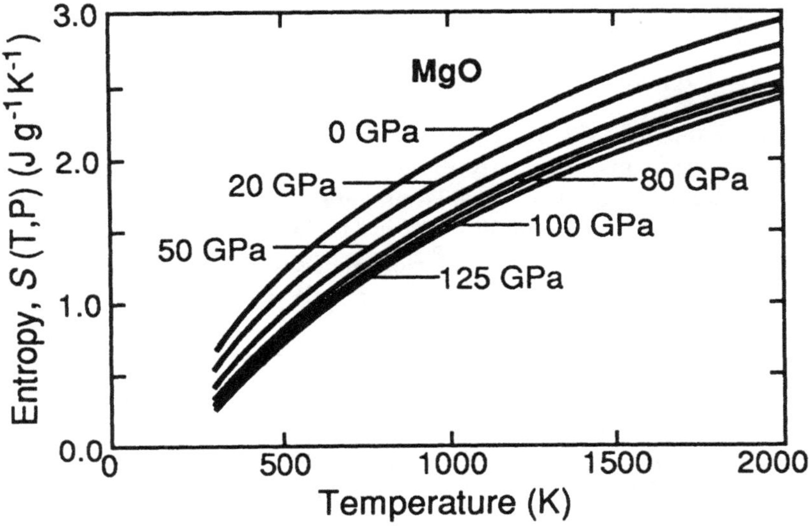

Fig. 13.5. The isobaric plots of entropy $\mathcal{S}_P$ at constant P versus T for MgO.

13.5. The enthalpy versus *P* and *T*

Enthalpy $\mathcal{H}$ is related to $\mathcal{U}$. Noting that $C_P = [\partial(\mathcal{U} + PV)/\partial T]_P$, the quantity $(\mathcal{U} + PV)$, heat content, occurs often enough to be given a symbol and a name, enthalpy. Start with the identity,

$$\left(\frac{\partial \mathcal{H}}{\partial P}\right)_T = V - T\left(\frac{\partial V}{\partial T}\right)_P, \tag{13.40}$$

or

$$\mathcal{H}(T^*, P) - \mathcal{H}(T^*, 0) = \int_0^P V\, dP - T^* \int_0^P V\alpha\, dP. \tag{13.41}$$

The computation of $\int V\alpha\, dP$ is given by (13.35). Thus we see that

$$\mathcal{H}(T^*, P) - \mathcal{H}(T^*, 0) = \int_0^P V\, dP_{T^*} + \left[\mathcal{S}(T^*, P) - \mathcal{S}(T^*, 0)\right]. \tag{13.42}$$

$\int_0^P V\, dP$ can be solved along an isotherm T^* by the inverse equation of state $V = f(0, P)$. A more straightforward way is to change variables

$$V_0 \eta\, dP = -K_T\, dV,$$

so that

$$V_0 \int_0^P \eta\, dP = -V_0 \int_{V(T^*,0)}^{V(T^*,P)} K_T\, d\eta. \tag{13.43}$$

The third degree Birch-Murnaghan equation for K_T as a function of density is

$$K_T(0,\eta) = K_{T_0} w^{5/3}\Big[1 + \tfrac{1}{2}(3K_0' - 5)(w^{2/3} - 1)\Big].$$

The temperature dependence of K_T is neglected; the thermal bulk modulus correction is sufficiently small compared to K_T at absolute zero to be safely ignored. However, there is temperature dependence in the limits of integration: $V(T^*, P)$ and $V(T^*, 0)$ in (13.43). We give the value of the integral $\int K_T\, d\eta$, (13.43), as $V_0 \int_0^P \eta\, dP$ in Table 13.4. The value of the integral $\int P\, d\eta$ and the ambient value of η for each P are also listed.

Table 13.4. The compression, $\eta = V/V_0$, and the functions $V_0 \int \eta\, dP$ and $V_0 \int P\, d\eta$ versus P for MgO

P (GPa)	$\eta = V/V_0$	$V_0 \int_0^P \eta\, dP$ (GPa)	$V_0 \int_1^\eta P\, d\eta$ (GPa)
0	1.000	0.0000	0.0000
5	0.972	4.9284	0.0684
10	0.948	9.9261	0.2466
15	0.926	14.4090	0.5200
20	0.907	18.9896	0.8509
25	0.889	23.4779	1.2542
30	0.873	27.8822	1.6934
35	0.858	32.2095	2.1805
40	0.845	36.4658	2.6666
45	0.832	40.6561	3.2168
50	0.820	44.7850	3.7856
55	0.809	48.8565	4.3621
60	0.798	52.8739	4.9948
65	0.788	56.8405	5.6216
70	0.779	60.7591	6.2303
75	0.770	64.6321	6.8836
80	0.762	68.4619	7.5036
85	0.754	72.2505	8.1624
90	0.746	75.9999	8.8619
95	0.739	79.7118	9.5089
100	0.732	83.3877	10.1899
105	0.725	87.0292	10.9064
110	0.719	90.6377	11.5500
115	0.712	94.2145	12.3365
120	0.706	97.7606	13.0425
125	0.701	101.2773	13.6542
130	0.695	104.7655	14.4172

From Anderson and Zou, 1990.

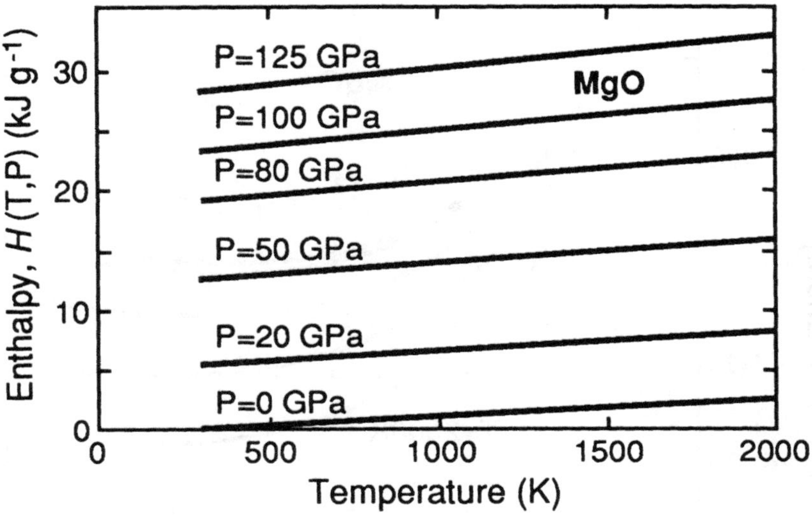

Fig. 13.6. Enthalpy versus T along isobars for MgO.

Using values of $V_0 \int_0^P \eta \, dP$ in Table 13.4 and the value $[\mathcal{S}(T^*, P) - \mathcal{S}(T^*, 0)]$ found in Section 13.4, the enthalpy is evaluated and plotted in Fig. 13.6.

As is well known from experiments, the enthalpy approaches 0 at $T = 0$ and $P = 0$ and also has zero slope at $T = 0$. Further, the enthalpy is proportional to the adiabatic bulk modulus, at least at $P = 0$ (Anderson, 1989). This can be seen from the definition of the Anderson-Grüneisen parameter given by (1.48), which can be written in the form

$$\left(\frac{\partial K_S}{\partial T}\right)_P = \left(-\frac{\delta_S \gamma C_P}{V}\right).$$

The evidence that δ_S and γ are independent of T at constant P is reviewed in Chapter 2. Taking $\delta_S \gamma$ to be independent of T and integrating the above

$$\begin{aligned} K_S(T) - K_S(T_0) &= -\,\delta_S \gamma \int \frac{C_P}{V} \, dT \\ &\approx \frac{\delta_S \gamma}{V_0} \left[\mathcal{H}(T) - \mathcal{H}(T_0)\right] + O\left[\alpha^2 (T^2 - T_0^2)\right]. \end{aligned} \qquad (13.44)$$

The last term in (13.44) is insignificant compared to the first and can be neglected. This demonstrates that K_S is proportional to $\mathcal{H}$ which is proven by demonstrating that a cross plot of $\mathcal{H}(T)$ and $K_S(T)$ is a straight line (Anderson, 1989). Because $\mathcal{H}(T)$ is measured at higher temperatures than is $K_S(T)$, the above correlation is useful for extrapolating K_S to high temperatures beyond the measurement limit (Anderson and Isaak, 1994).

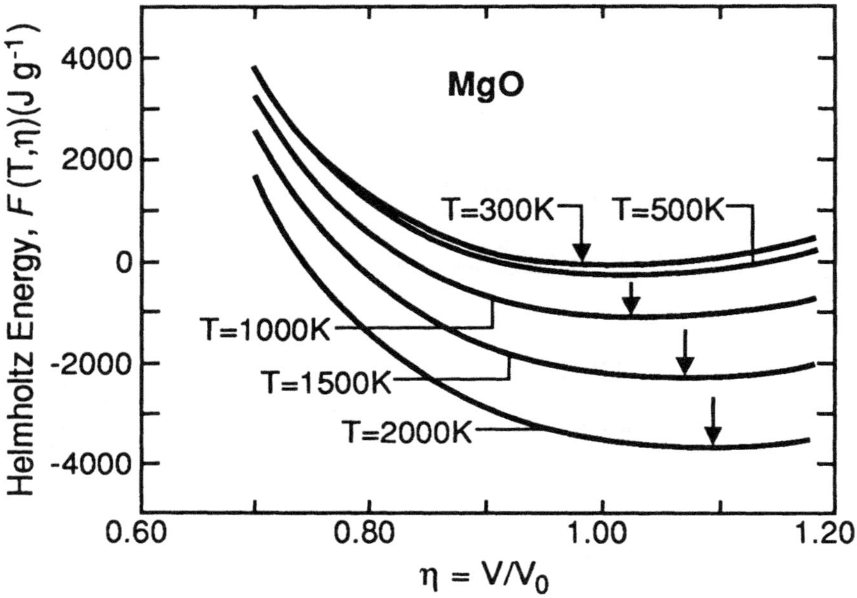

Fig. 13.7. $\mathcal{F}(\eta)$ along isotherms for MgO. The minimum is indicated by an arrow that sets the $P = 0$ value of volume for the isotherm. P_0 refers to ambient volume ($T = 300$, $P = 0$). The shift in the minimum with T represents the effect of α on V at $P = 0$ (modified from Anderson and Zou, 1989).

13.6. The Helmholtz free energy

The Helmholtz free energy is given by

$$\mathcal{F}(T, V) = \mathcal{U}(T, V) - T\mathcal{S}(T, V). \tag{13.45}$$

$\mathcal{U}(T, V)$ is in Table 13.3 and $\mathcal{S}(T, V)$ is in Table 13.1, so that the Helmholtz free energy, $\mathcal{F}(T, V)$, found from (13.45), is plotted in Fig. 13.7.

Figure 13.7 shows that the Helmholtz energy can be determined from experiments and used to test the theoretical energies found in first principles or *ab initio* calculations. First principles methods are often divided into two groups. There are self-consistent methods, such as linearized augment plane wave (LAPW) calculations. In these methods the crystalline energy is usually determined by solving Schrödinger's equation using no assumptions other than the local density approximation to account for exchange and correlation effects. There are also *ab initio* models in which a crystalline charge density is assumed. Electron gas models, in which the crystalline charge density consists of overlapping spherical atomic charges, are a type of model. In both cases, the dependence of $\mathcal{F}$ on strain, V, and T is used to calculate the thermoelastic properties of the solid. We emphasize that $\mathcal{F}$ found from experiments, as in this chapter, can test the theoretical $\mathcal{F}$.

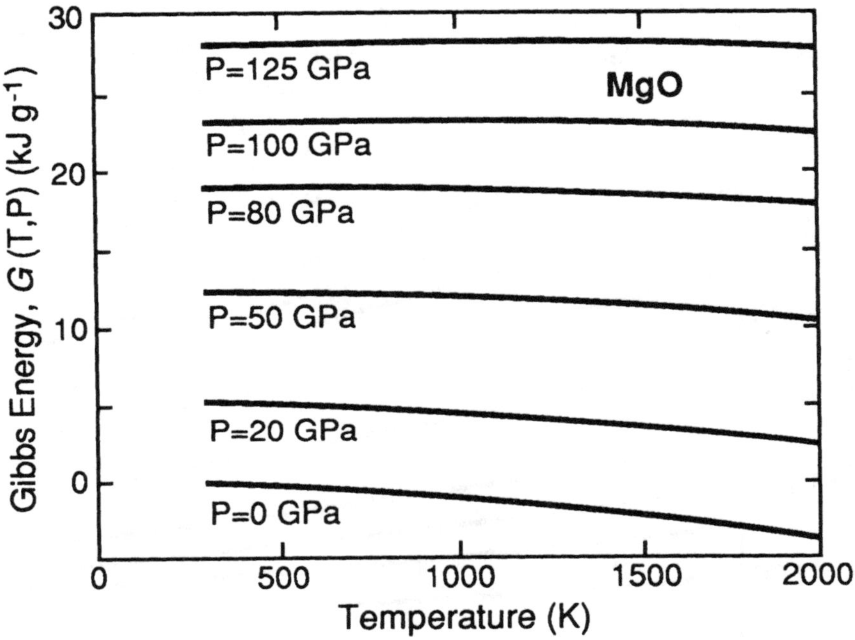

Fig. 13.8. Isobars of $\mathcal{G}(T)$ (modified from Anderson and Zou, 1989).

13.7. The Gibbs free energy

The Gibbs free energy is given by

$$\mathcal{G}(T,P) = \mathcal{H}(T,P) - T\mathcal{S}_P(T,P). \tag{13.46}$$

Taking $\mathcal{H}(T,P)$ from data plotted in Fig. 13.6 and $\mathcal{S}_P(P,T)$ from data plotted in Fig. 13.5, the Gibbs free energy is found and plotted in Fig. 13.8.

The behavior of $\mathcal{G}(T,P)$ can be related to entropy by the derivative

$$\left(\frac{\partial \mathcal{G}}{\partial T}\right)_P = -\mathcal{S}_P.$$

Thus the slopes of the curves in Fig. 13.8 are equal to the values of the respective ordinates in Fig. 13.5. Thus if $\mathcal{G}$ is known, $\mathcal{S}$ as a function of P and T can be determined. Similarly, as

$$\left(\frac{\partial \mathcal{G}}{\partial P}\right)_T = V,$$

the volume can be determined as a function of P and T, provided $\mathcal{G}$ (versus T) is known. The Gibbs free energy is particularly important because of

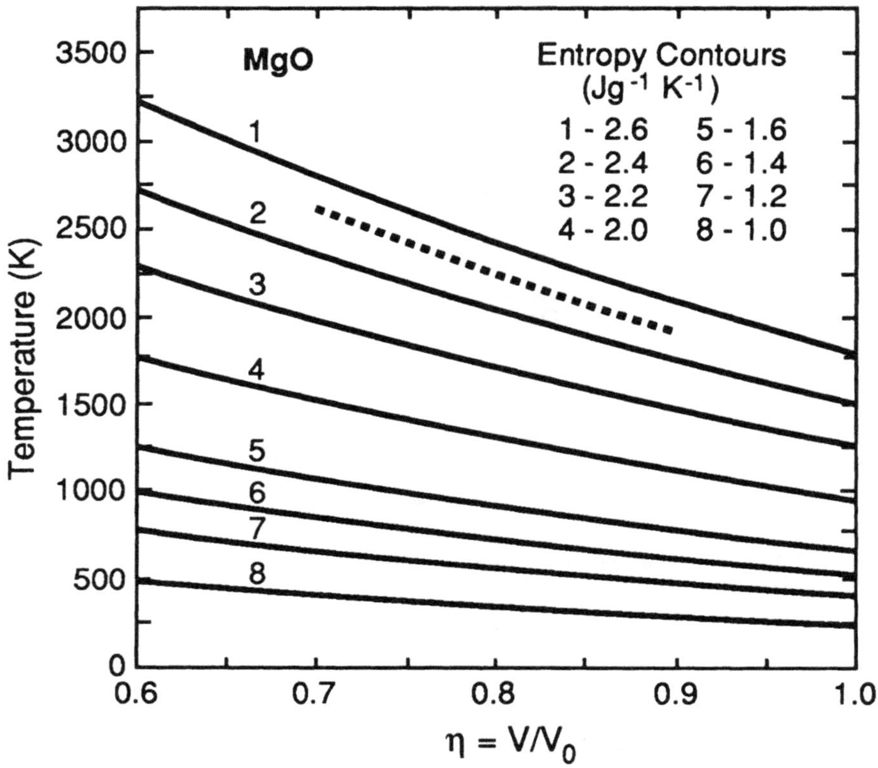

Fig. 13.9. T versus η along isentropes for MgO found from data represented by Table 13.1 plotted as solid lines. The isentrope found by Brown and Shankland (1981) for the earth's lower mantle is plotted as a dashed line for comparison.

the physical processes that occur at constant P and constant T, such as melting or change in phase.

13.8. Isentropes for MgO and the lower mantle

Using the data behind Table 13.1, contours of constant entropy for MgO were found; these are plotted in Fig. 13.9. Isentropes are of value when analyzing properties that vary with η at constant $\mathcal{S}$. It has been known since Birch's (1952) paper that the earth's lower mantle is compressed under nearly perfect isentropic conditions. Using the seismic velocities of the lower mantle and Debye theory, Brown and Shankland (1981) specified the value of entropy throughout the lower mantle. They found the temperature profile by assuming an adiabatic mantle and, by fixing the lower T to experiment, established the value of the lower mantle entropy. The isentrope of the lower mantle is compared with the isentropes of MgO in Fig. 13.9.

It is seen from Fig. 13.9 that the mantle isentrope exactly parallels the isentropes of MgO. This means that the average of the mixture of minerals comprising the mantle has isentropes paralleling the MgO isentropes. Thus the dominant mineral of the lower mantle, silicate perovskite, must have a set of parallel isotropes similar to those of MgO.

13.9. Finding γ from the entropy

The temperature of an oxide or silicate increases with density according to the adiabatic law (see Section 1.4.2.)

$$\left(\frac{\partial \ell n\, T}{\partial \ell n\, \eta}\right)_{\mathcal{S}} = -\gamma. \tag{1.29}$$

Ordinarily (1.29) is used to find T versus η by specifying the volume and temperature dependence of γ. However, if the values of T and η are known for constant $\mathcal{S}$, as in Table 13.1 and Fig. 13.9, then (1.29) can be used to find the isotherms of γ versus η or the isochores of γ versus T.

Using the contours of constant $\mathcal{S}$ as found in Fig. 13.9, γ is found by differentiation using (1.29) in the form

$$\gamma = -\frac{\eta}{T}\left(\frac{\partial T}{\partial \eta}\right)_{\mathcal{S}}. \tag{13.47}$$

The calculation of $\gamma(T, \eta)$ from (13.47) relies on the integration of αK_T with respect to η along isotherms to find the entropy, followed by the differentiation of the entropy contour. The method of calculating γ used in Chapter 3 relied on an entirely different path of integration of αK_T with respect to η. The result for γ according to (13.47) is shown in Fig. 13.10.

13.10. Extrapolations to absolute zero limits for K_T

Values of $K_T(0)$ and K'_T appropriate to absolute zero were needed, and computed as follows:

$$\frac{K_{\mathcal{S}}(300)}{K_{\mathcal{S}}(0)} = \left(\frac{\rho(0)}{\rho(300)}\right)^{K_S}. \tag{13.48}$$

For MgO, Jackson and Niesler's (1982) results show $K_{\mathcal{S}}(300) = 162.5$ GPa, and $K'_{\mathcal{S}} = 4.13$. These lead to the following values:

$$K_T(0) = 166.2 \text{ GPa}, K_T(300) = 160.5 \text{ GPa}, K'_T = 4.17.$$

13.11. Important sources of uncertainty

Of importance to the calculation of the thermodynamic functions is the parameter $\mathbf{w}$ given by (13.7); $\mathbf{w} = (\partial K_T/\partial T)_V$. In general this is a small

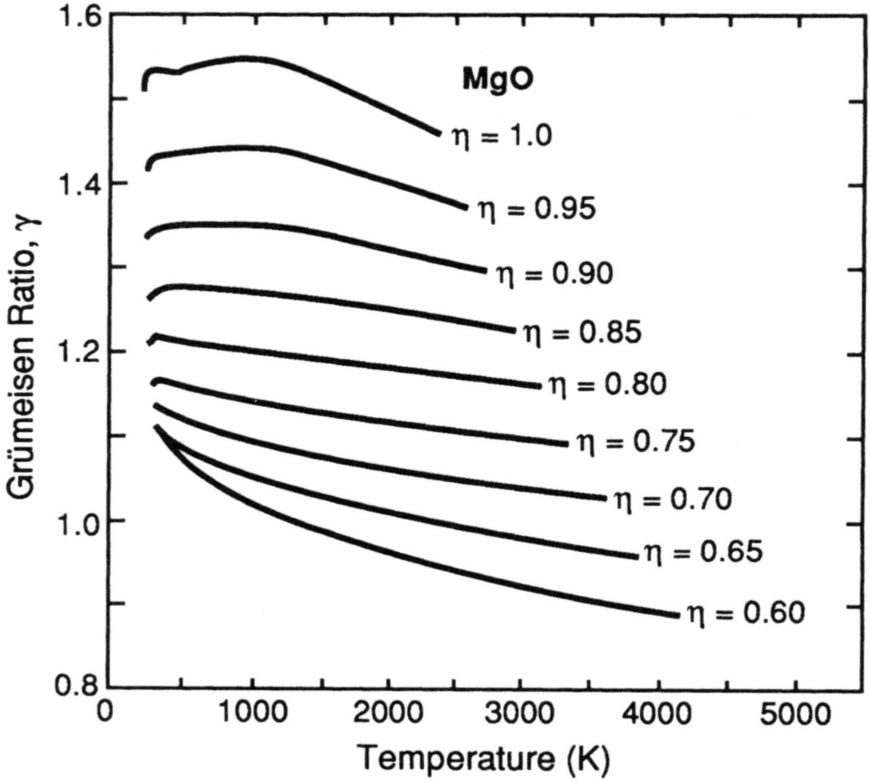

Fig. 13.10. The variation of γ versus T along isochores for MgO found from (13.47). Compare this with the γ plotted in Fig. 3.10. Above the Debye temperature and at high compression, there is very little difference.

number compared to $(\partial K_T/\partial T)_P$, but temperature and volume dependent, as shown in Fig. 3.3. Its $P = 0$ value is computed from the measurements given by the balance between $(\partial K_T/\partial T)_P$ and $\alpha K_T\,(\partial K_T/\partial P)_T$, the first term being negative and the second positive (see (13.7)). If (13.7) is close to zero, then we know from a thermodynamic identity that $(\partial \alpha K_T/\partial V)_T$ is zero. But in general (13.7) is not zero, although it is often a small quantity. In case it is not zero, we must measure or compute **w** through $\delta_T - K'$. To treat **w** as if it were independent of compression when it is not, leads to a small error in entropy at high T and large compression. Also important is the value of the temperature dependence of $(\partial K'/\partial T)_P$ or $d^2K_T/dPdT$. For MgO, Isaak (1993) found $d^2K_T/dTdP = 0.39 \times 10^{-3}\mathrm{K}^{-1}$ up to 800 K.

PART III REFERENCES

Abelson, R.S., 1981, "Monte Carlo calculations of the melting point of iron at core pressures," Ph.D. thesis, University of California at Los Angeles, 176 pp.

Ahrens, T.J., Tan, H., and Bass, J.D., 1990, "Analysis of shock temperature data for iron," *High Press. Res.* **2**, 145–157.

Akella, J., Vaidya, S.N., and Kennedy, G.S., 1969, "Melting of sodium chloride at pressures to 65 kbar," *Phys. Rev.* **185**, 1135–1140.

Anderson, D.L., 1989, *Theory of the Earth*. Boston: Blackwell Scientific Publications, Table AI.

Anderson, M.S., and Swenson, C.A., 1975, "Experimental equations of state for the rare gas solids," *J. Chem. Phys.* **36**, 145–162.

Anderson, O.L., 1979a, "The high temperature acoustic Grüneisen parameter in the Earth's interior," *Phys. Earth Planet. Inter.* **18**, 221–231.

Anderson, O.L., 1979b, "Evidence supporting the approximation $\gamma\rho$=constant for the Grüneisen parameter of the earth's lower mantle," *J. Geophys. Res.* **84**, 3537–3542.

Anderson, O.L., 1979c, "The Hildebrand equation of state applied to minerals relevant to geophysics," *Phys. Chem. Miner.* **5**, 33–51.

Anderson, O.L., 1980, "An experimental high-temperature thermal equation of state bypassing the Grüneisen parameter, " *Phys. Earth Planet. Inter.* **22**, 173–193.

Anderson O.L., 1981, "Temperature profiles of the Earth," *Evolution of the Earth* (R.J. O'Connell and W.S. Fyfe, Eds.). Washington, D.C.: American Geophysical Union, pp. 19–27.

Anderson O.L., 1982, "The earth's core and the phase diagram of iron," *Phil. Trans. Roy. Soc. London* **A**, 21–35.

Anderson, O.L., 1984, "A universal thermal equation of state," *J. Geodyn.* **1**, 185–214.

Anderson, O.L., 1986, "Properties of iron at earth's core conditions," *Geophys. J. Roy. Astr. Soc.* **84**, 561–579.

Anderson, O.L., 1989, "The relationship between the adiabatic bulk modulus and enthalpy for mantle-related minerals," *Phys. Chem. Miner.* **16**, 559–569.

Anderson, O.L., 1993, "The phase diagram of iron and the temperature of the inner core," *J. Geomag. Geoelectr.* **45**, 1235–1248.

Anderson, O.L., 1994, "Imperfections of the 1993 phase diagram of iron," *High-Pressure Science and Technology—1993* (S.C. Schmidt, J.W. Shaner, G.A. Samara, and M. Ross, Eds.). Woodbury, N.Y.: American Institute of Physics, pp. 907–910.

Anderson, O.L., and Isaak, D., 1994, "Elastic constants of mantle minerals," *AGU Handbook of Physical Constants*, in press.

Anderson, O.L., and Sumino, Y., 1980, "The thermodynamic properties of the earth's lower mantle," *Phys. Earth Planet. Inter.* **23**, 314–331.

Anderson, O.L., and Zou, K., 1989, "Formulation of the thermodynamic function for mantle minerals: MgO as an example," *Phys. Chem. Miner.* **16**, 642–648.

Anderson, O.L., and Zou, K., 1990, "Thermodynamic functions and properties of MgO and Mg_2SiO_4 at high compression and high temperature," *J. Phys. Chem. Ref. Data* **19**, 69–83.

Anderson, O.L., Isaak, D.G., and Yamamoto, S., 1989, "Anharmonicity and the equation of state for gold," *J. Appl. Phys.* **65**, 1534–1543.

Anderson O.L., Isaak, D.G., and Oda, H., 1992, "High-temperature elastic constant data on minerals relevant to geophysics," *Rev. Geophys.* **30**, 57–90.

Anderson, O.L., Oda, H., Chopelas, A., and Isaak, D.G., 1993, "A thermodynamic theory of the Grüneisen ratio at extreme conditions: MgO as an example," *Phys. Chem. Miner.* **19**, 369–380.

Anderson, P.W., 1984, *Basic Notions of Condensed Matter Physics*. London: Benjamin, pp. 49–69.

Barron, T.H.K., Berg, W.T., and Morrison, J.A., 1959, "On the heat flow of crystalline magnesium oxide," *Proc. Roy. Soc. London A* **250**, 70–83.

Bass, J.D., Svendson, B., and Ahrens, T.J., 1987, "The temperature of shock compressed iron," *High Pressure Research in Geophysics and Geochemistry* (M.H. Manghnani and Y. Syono, Eds.), Washington, D.C.: American Geophysical Union, pp. 393–402.

Bass, J.D., Ahrens, T.J., Abelson, J.R., and Hua, T., 1990, "Shock temperatures in metals: A new result for an Fe alloy," *J. Geophys. Res.* **95**, 21767–21776.

Bassett, W.A., and Weathers, M.S., 1994, "Thermodynamic significance of a high pressure/temperature bcc iron phase," *High-Pressure Science and Technology—1993* (S.C. Schmidt, J.W. Shaner, G.A. Samara, and M. Ross, Eds.). Woodbury, N.Y.: American Institute of Physics, pp. 915–918.

Birch, F., 1952, "Elasticity and constitution of the Earth's interior," *J. Geophys. Res.* **57**, 227–286.

Birch, F., 1972, "The melting relations of iron, and temperatures in the earth's core," *Geophys. J. Roy. Astr. Soc.* **29**, 373–387.

Boehler, R., 1986, "The phase diagram of iron to 430 kbar," *Geophys. Res. Lett.* **13**, 1153–1156.

Boehler, R., 1993, "Temperatures in the earth's core based on new melting measurements of iron to 2 Mbar," *Nature* **363**, 534–536.

Boehler, R., 1994, "The phase diagram of iron to 2 Mbar: New static measurements," *High-Pressure Science and Technology—1993* (S.C. Schmidt, J.W. Shaner, G.A. Samara, and M. Ross, Eds.). Woodbury, N.Y.: American Institute of Physics, pp. 919–922.

Boehler, R., and Kennedy, G.C., 1980, "Equation of state of sodium chloride up to 32 kbar and 500°C," *J. Phys. Chem. Solids* **41**, 517–523.

Boehler, R., von Bargen, N., Hoffbauer, W., and Huang, E., 1988, "$d\alpha/dT$ at high P and T from synchrotron measurements," *Geophys. J. Roy. Astr. Soc.* **66**, 579–596.

Boehler, R., von Bargen, N., and Chopelas, A., 1990, "Melting, thermal expansion and phase transitions of iron at high pressures," *J. Geophys. Res.* **95**, 21731–21736.

Boness, D.A., Brown, J.M., McMahon, A.K., 1986, "The electronic thermodynamics of iron under earth core conditions," *Phys. Earth Planet. Inter.* **42**, 227–240.

Born, M., 1939, "Thermodynamics of melting," *J. Chem. Phys.* **7**, 591–599.

Boschi, E., 1975, "The melting relations of iron and temperatures in the Earth's core," *Riv. Nuovo Cim.* **5**, 501–531.

Boschi, E., Mulargia, F., and Bonafede, M., 1979, "The dependence of the melting temperature on the choice of the interatomic potential," *Geophys. J. Roy. Astr. Soc.* **58**, 201–208.

Boyer, L.L., 1979, "Calculation of thermal expansion, compressibility and melting in alkali halides: NaCl and KCl," *Phys. Rev. Lett.* **42**, 584–587.

Boyer, L.L., 1985, "Theory of melting based upon lattice instability," *Phase Trans.* **5**, 1–48.

Boyer, L.L., Mehl, M.J., Feldman, J.L., Hardy, J.R., Flocken, J.W., and Fong, C.Y., 1985, "Beyond the rigid ion approximation with spherically symmetric ions," *Phys. Rev. Lett.* **54**, 1940–1943.

Brillouin, L., 1946, *Wave Propagation in Periodic Structures*. New York: McGraw-Hill Book Co., Inc., p. 246.

Brown, J.M., and McQueen, R.G., 1980, "Melting of iron under shock conditions," *Geophys. Res. Lett.* **7**, 533–536.

Brown, J.M., and McQueen, R.G., 1982, "The equation of state for iron and the Earth's core," *High Pressure Research in Geophysics* (A. Akimoto and M. Manghnani, Eds.). Tokyo: Center for Academic Publications, pp. 611–625.

Brown, J.M., and McQueen, R.G., 1986, "Phase transitions, Grüneisen parameter, and elasticity for shocked iron between 77 GPa and 400 GPa," *J. Geophys. Res.* **91**, 7485–7494.

Brown, J.M., and Shankland, T.J., 1981,"Thermodynamic parameters in the earth as obtained from seismic profiles," *Geophys. J. Roy. Astr. Soc.* **66**, 579–596.

Bukowinski, M.S.T., 1977, "A theoretical equation of state for the inner core," *Phys. Earth Planet. Inter.* **14**, 333–347.

Bukowinski, M.S.T., 1988, private communication.

Bukowinski, M.S.T., and Wolf, G.H., 1990, "Thermodynamically consistent decompression: Implications for lower mantle composition," *J. Geophys. Res.* **95**, 12583–12593.

Bundy, F.P., and Strong, H.M., 1962, "Behavior of metals at high temperature and pressure," *Solid State Physics*, vol. 13, (F. Seitz and D. Turnbull, Eds.). New York: Academic Press, pp. 81–143.

Cho, S.-A., 1982, "Role of lattice structure on the Lindemann fusion theory of metals," *J. Phys. F* **12**, 1069–1083.

Chopelas, A., 1990, "Thermal expansion, heat capacity, and entropy of MgO at mantle pressures," *Phys. Chem. Miner.* **17**, 142–148.

Cohen, R.E., and Gong, Z., 1994, "Melting and melt structures at high pressure," *High-Pressure Science and Technology—1993* (S.C. Schmidt, J.W. Shaner, G.A. Samara, and M. Ross, Eds.). Woodbury, N.Y.: American Institute of Physics, pp. 379–382.

Cohen, R.E., Stixrude, L., Papaconstanopoulos, D.A., 1994, "An accurate tight-binding model for high pressures, towards high temperature simulations of the earth's core," *High-Pressure Science and Technology—1993* (S.C. Schmidt, J.W. Shaner, G.A. Samara, and M. Ross, Eds.). Woodbury, N.Y.: American Institute of Physics, 891–894.

Cotterill, R.M.J., 1980, "The physics of melting," *J. Cryst. Growth* **48**, 582–588.

Durand, M.A., 1936, "The temperature variation of NaCl, KCl, and MgO," *Phys. Rev.* **50**, 449–455.

Dziewonski, A.M., and Anderson, D.L., 1981, "Preliminary reference earth model," *Phys. Earth Planet. Inter.* **25**, 297–356.

Dziewonski, A.M., Hales, A.I., and Lapwood, E.R., 1975, "Parametrically simple Earth models consistent with geophysical data," *Phys. Earth Planet. Inter.* **10**, 12–48.

Fritz, J.N., Marsh, S.P., Carter, W.J., and McQueen, R.G., 1971, "The Hugoniot equation of state of sodium chloride in the sodium chloride structure," *Accurate Characterization of the High Pressure Environment* (E.C. Lloyd, Ed.). NBS Spec. Publ. 326, 201–208.

Garvin, D., Parker, V.B., and White H.J. (Eds.), 1988, *CODATA Thermodynamic Tables*. Washington, DC: Hemisphere Publishing. Corp.

Gilvarry, J.J., 1956a, "The Lindemann and Grüneisen laws," *Phys. Rev.* **102**, 307–316.

Gilvarry, J.J., 1956b, "Equation of the fusion curve," *Phys. Rev.* **102**, 325–331.

Gilvarry, J.J., 1956c, "Variation of the amplitude of thermal vibration on the fusion curve," *Phys. Rev.* **104**, 908–913.

Gilvarry, J.J., 1957, "Temperatures of the Earth's interior," *J. Atmos. Terr. Phys.* **10**, 84–95.

Goto, T., and Anderson, O.L., 1988, "Apparatus for measuring elastic constants of single crystals by a resonance technique up to 1825 K," *Rev. Sci. Instrum.* **59**, 1405–1408.

Goto, T., Anderson, O.L., Ohno, I., and Yamamoto, S., 1989, "Elastic constants of corundum up to 1825 K," *J. Geophys. Res.* **94**, 7588–7602.

Grüneisen, E., 1926, "The state of a solid body," *Handbuch der Physik*, vol. 1. Berlin: Springer-Verlag, pp. 1–52; Engl. transl. NASA RE 2-18-59W, 1959.

Hardy, R.J., 1980, "Temperature and pressure dependence of intrinsic anharmonicity and quantum corrections to the equation of state," *J. Geophys. Res.* **85**, 7011–7015.

Heinz, D.L., and Jeanloz, R., 1987, "Measurement of the melting curve of $Mg_{0.9}Fe_{0.1}SiO_3$ perovskite at lower mantle conditions and its geophysical implications," *J. Geophys. Res.* **92**, 11437–11494.

Hemley, R.J., and Cohen, R.E., 1992, "Silicate perovskite," *Ann. Rev. Earth Planet. Sci.* **20**, 553–600.

Hemley, R.J., Stixrude, L., Fei, Y., and Mao, H.K., 1992, "Constraints on lower mantle composition from P, V, T measurements of (Fe,Mg)SiO_3-perovskite and (Fe,Mg)O," *High Pressure Research: Application to Earth and Planetary Sciences*, American Geophysical Union Monographs, no. 67. Tokyo: Terra Scientific Publishing Co., pp. 183–189.

Hertzfield, K.F., and Goeppert-Meyer, M., 1934, "On the theory of fusion," *Phys. Rev.* **46**, 995–1001.

Higgins, G., and Kennedy, G.C., 1971, "The adiabatic gradient and the melting point gradient in the Earth's core," *J. Geophys. Res.* **76**, 1870–1878.

Isaak, D.G., 1993, "The mixed P, T derivatives of elastic moduli and implications on extrapolating throughout Earth's mantle," *Phys. Earth Planet. Inter.* **80**, 37–48.

Isaak, D.G., Anderson, O.L., and Goto, T., 1989a, "Elasticity of single-crystal forsterite measured to 1700 K," *J. Geophys. Res.* **94**, 5895–5906.

Isaak, D.G., Anderson, O.L., and Goto, T., 1989b, "Measured elastic modulus of single-crystal MgO up to 1800 K," *Phys. Chem. Miner.* **16**, 703–704.

Isaak, D.G, Cohen, R.E., and Mehl, M.J., 1990, "Calculated elastic and thermal properties of MgO at high pressures and temperatures," *J. Geophys. Res.* **95**, 7055–7067.

Ito, Kiyosi (Ed.), 1993, *Encyclopedic Dictionary of Mathematics*, 2nd edition, vol. 1,2. Tokyo: Mathematical Society of Japan.

Jackson, I.N.S., and Liebermann, R.C., 1974, "Melting and elastic shear instability of alkali halides," *J. Phys. Chem. Solids* **35**, 1115–1119.

Jackson, I., and Niesler H., 1982, "The elasticity of periclase to 3 GPa and some geophysical implications," *High Pressure Research in Geophysics* (S. Akimoto and M. Manghnani, Eds.). Tokyo: Center for Academic Publications, pp. 93–113.

Jamieson, J.C., Demarest, H.H., and Schiferl, D., 1978, "A re-evaluation of the Grüneisen parameter for the Earth's core," *J. Geophys. Res.* **83**, 5929–5935.

Jeanloz R., 1979, "Properties of iron at high pressures and the state of the core," *J. Geophys. Res.*, **84**, 6059–6069.

Jeanloz, R., 1989, "Shock wave equation of state and finite strain theory," *J. Geophys. Res.* **94**, 5873–5886.

Jephcoat, A., and Olson, P., 1987, "Is the inner core of the Earth pure iron?" (letter), *Nature* **325**, 332.

Katsura, T., and Ito, E., 1989, "The system $Mg_2\,SiO_4$-$Fe_2\,SiO_4$ at high pressures and temperatures: Precise determination of stabilities of olivine, modified-spinel and spinel," *J. Geophys. Res.* **94**, 15663–15670.

Kennedy, G.C., and Vaidya, S.N., 1970, "The effect of pressure on the melting temperature of solids," *J. Geophys. Res.* **75**, 1019–1022.

Kline, M., 1985, *Mathematics for the Non-mathematician.* Mineola, NY: Dover Publications, p. 282.

Knittle, E., and Jeanloz, R., 1987, "Synthesis and equation of state of (Mg,Fe)SiO_3 to over 100 GPa," *Science* **235**, 668–670.

Knittle, E., and Jeanloz, R., 1989, "Melting curve (Mg,Fe)SiO_3 perovskite to 96 GPa: Evidence for a structural transition in the lower mantle," *Geophys. Res. Lett.* **16**, 421–424.

Kraut, E., and Kennedy, G., 1966, "New melting law at high pressures," *Phys. Rev.* **151**, 668–675.

Libby, W.F., 1966, "Melting points at high compression from zero-compression properties through the Kennedy relation," *Phys. Rev. Lett.* **17**, 423–424.

Lindemann, F.A., 1910, "Uber die Berechnung molecularer Eigenfrequenzen," *Phys. Z.* **11**, 609–612.

Liu, L.-G., 1975, "On the (γ, ϵ, ℓ) triple point of iron and the earth's core," *Geophys. J. Roy. Astr. Soc.* **43**, 697–705.

Luedemann, H.D., and Kennedy, G.C., 1968, "Melting curves of lithium, sodium, potassium and rubidium to 80 kilobars," *J. Geophys. Res.* **73**, 2795–2805.

Mao, H.K., Hemley, R.J., Fei, Y., Shu, J.F., Chen, L.C., Jepthcoat, A.P., and Wu, Y., 1991, "Effect of pressure, temperature and composition on lattice parameters and density of (Fe,Mg)SiO_3-perovskite to 30 GPa," *J. Geophys. Res.* **96**, 8069–8080.

Marsh, S.P., 1980, *LASL Shock Hugoniot Data*. Los Angeles: University of California Press, p. 89.

Matsui, M., and Price, G.D., 1991, "Simulation of pre-melting behavior of $MgSiO_3$ perovskite at high pressure and temperature," *Nature* **351**, 735–737.

McQueen, R.G., Marsh, S.P., and Fritz, J.N., 1964, "On the composition of the Earth's interior," *J. Geophys. Res.* **69**, 2947–2965.

McQueen, R.G., Marsh, S.P., and Fritz, J.N., 1967, "Hugoniot equation of state of twelve rocks," *J. Geophys. Res.* **72**, 4999–5036.

McQueen, R.G., Marsh, S.P., Taylor, J.W., Fritz, J.N., and Carter, W.C., 1970, "The equation of state of solids from shock wave studies," *High Velocity Impact Phenomena* (R. Kinslow, Ed.). New York: Academic Press, pp. 293–417; 515–568.

Monfort, C.E., and Swenson, C.A., 1965, "An experimental equation of state for potassium metal," *J. Phys. Chem. Solids* **26**, 291–301.

Mulargia, F., and Quareni, F., 1988, "Validity of the Sutherland-Lindemann law and melting temperatures in the Earth's interior," *Geophys. J.* **92**, 269–282.

Ohtani, E., 1983, "Melting temperature distribution and fractionation in the mantle," *Phys. Earth Planet. Inter.* **33**, 12–25.

Oriani, R.A., 1951, "Entropies of melting of metals," *J. Chem. Phys.* **19**, 93–97.

Pastine, D.J., and Piascesi, D., 1966, "The existence and implications of curvature in the relations between shock and particle velocities in metals," *J. Phys. Chem. Solids* **27**, 1783–1792.

Poirier, J.P., 1986, "Dislocation mediated melting of iron and temperature of the earth's core," *Geophys. J. R. Astr. Soc.* **85**, 315–328.

Poirier, J.P., 1988, "Lindemann law and the melting temperature of perovskites," *Phys. Rev. Lett.* **54**, 364–369.

Poirier, J.P., 1991, *Introduction to the Physics of the Earth's Interior*. New York: Cambridge University Press.

Poirier, J.P., and Shankland, T., 1994, "Dislocation theory of melting for iron, revisited," *High-Pressure Science and Technology-1993* (S.C. Schmidt, J.W. Shaner, G.A. Samara, and M. Ross, Eds.). Woodbury, NY: American Institute of Physics, 927–930.

Price, G.D., Parker, S.E., and Leslie, M., 1987, "The lattice dynamics and thermodynamics of polymorphs," *Phys. Chem. Miner.* **15**, 181–190.

Ross, M., Young, D., and Grover, R., 1990, "Theory of the iron phase diagram at earth-core conditions," *J. Geophys. Res.* **95**, 21713–21717.

Ruoff, A.L., 1967, "Linear shock velocity-particle velocity relationship," *J. Appl. Phys.* **38**, 4976–4980.

Saxena, S.K., Shen, G., and Lazor, P., 1994, "Temperatures in Earth's core based on melting and phase transformation experiments on iron," *Science*, **264**, 405–407.

Shankland, T.J., and Brown, J.M., 1985, "Homogeneity and temperatures in the lower mantle," *Phys. Earth Planet. Inter.* **38**, 51–58.

Sherman, D.M., 1994, "Electronic structure, entropy and the high-pressure stability of bcc iron," *High-Pressure Science and Technology—1993* (S.C. Schmidt, J.W. Shaner, G.A. Samara, and M. Ross, Eds.). Woodbury, N.Y.: American Institute of Physics, pp. 895–898.

Simon, F., and Glatzel, G., 1929, "Bemerkungen zur Schmelzruckkurve," *Z. Anorg. Allg. Chem.* **178**, 309–312.

Spiliopoulos, S., and Stacey, F.D., 1984, "The earth's thermal profile: Is there a mid-mantle thermal boundary layer?," *J. Geodynam.* **1**, 66–77.

Stacey, F.D., 1977a, "Applications of thermodynamics to fundamental earth physics," *Geophys. Surv.* **3**, 175–204.

Stacey, F.D., 1977b, "A thermal model of the Earth," *Phys. Earth Planet. Inter.* **15**, 341–348.

Stacey, F.D., and Irvine, R.D., 1977, "Theory of melting: Thermodynamic basis of Lindemann's law," *Aust. J. Phys.* **30**, 631–640.

Stassis, C., 1994, "Inelastic neutron scattering of γ-iron, and the determination of the elastic constants by lattice dynamics," *High-Pressure Science and Technology—1993* (S.C. Schmidt, J.W. Shaner, G.A. Samara, and M. Ross, Eds.). Woodbury, N.Y.: American Institute of Physics, pp. 955–958.

Stevenson, D.J., 1980, "Applications of liquid state physics to the Earth's core," *Phys. Earth Planet. Inter.* **22**, 42–52.

Stevenson, D.J., 1981, "Models of the Earth's core," *Science* **214**, 611–618.

Stishov, S.M., 1975, "The thermodynamics of melting of simple substances," *Sov. Phys. Usp* **17**, 625–634.

Stixrude, L., and Bukowinski, M.S.T., 1990, "Fundamental thermodynamic relations and silicate melting with implications for the constitution of D″," *J. Geophys. Res.* **95**, 19311–19325.

Stixrude, L., and Cohen, R., 1994, "First principles investigation of bcc, bct, fcc, and hcp phases of iron," *High-Pressure Science and Technology—1993* (S.C. Schmidt, J.W. Shaner, G.A. Samara, and M. Ross, Eds.). Woodbury, N.Y.: American Institute of Physics, 911–914.

Stull, D.R., and Prophet, H., (Eds.), 1971, *JANAF Thermochemical Tables*, 2nd edition, Washington, DC: Office of Standard Reference Data, National Bureau of Standards.

Sweeney, J.S., and Heinz, D.L., 1993, "Melting of iron-magnesium silicate perovskite," *Geophys. Res. Lett.* **20**, 855–858.

Swenson, C.A., 1968, "Equation of state of cubic solids: Some generalizations," *J. Phys. Chem. Solids* **29**, 1337–1348.

Tallon, J.L., 1982, "Crystal instability and melting," *Nature (London)* **299**, 188–190.

Tan, H., and Ahrens, T.J., 1990, "Shock temperature measurements for metals," *High Press. Res.* **2**, 159–182.

Ubbelohde, A.R., 1965, *Melting and Crystal Structure*. Oxford: Clarendon Press.

Williams, Q., Jeanloz, R., Bass, J., Svendson, B., and Ahrens, T.J., 1987, "Melting curve of iron to 250 GPa: A constraint on the temperature of the earth's center," *Science* **236**, 181–183.

Williams, Q., Knittle, E., and Jeanloz, R., 1991, "The high-pressure melting curve of iron: A technical discussion," *J. Geophys. Res.* **96**, 2171–2184.

Wolf, G.H., and Jeanloz, R., 1984, "Lindemann melting law: Anharmonic correction and test of its validity for minerals," *J. Geophys. Res.* **89**, 7821–7835.

Yamamoto, S., and Anderson, O.L., 1987, "Elasticity and anharmonicity of potassium chloride at high temperature," *Phys. Chem. Miner.* **14**, 332–340.

Yamamoto, S., Ohno, I., and Anderson, O. L., 1987, "High temperature elasticity of sodium chloride," *J. Phys. Chem. Solids* **48**, 143–151.

Yeganeh-Haeri, A., Weidner, D. J., and Ito, E., 1989, "Single crystal elastic moduli of magnesium metasilicate perovskite," *Perovskite* (A. Navrotsky and D.J. Weidner, Eds.). Washington, D.C.: American Geophysical Union, pp. 13–26.

Yoo, C.S., Holmes, N.S., Ross, M., Wolf, D.J., and Pike, C., 1993, "Shock temperatures, melting and phase diagram of iron at earth core conditions," *Phys. Rev. Lett.* **70**, 3931–3934.

Yoo, C.S., Holmes, N.C., Ross, M., and See, E., 1994, " Shock temperature measurements of iron to 350 GPa," *High-Pressure Science and Technology—1993* (S.C. Schmidt, J.W. Shaner, G.A. Samara, and M. Ross, Eds.). Woodbury, N.Y.: American Institute of Physics, pp. 959–962.

Young, D.A., and Grover, R., 1984, "Theory of the iron equation of state and melting curve to very high pressures," *Shock Waves in Condensed Matter–1983* (J. Asay, R.A. Graham, and G.K. Straub, Eds.). Amsterdam: Elsevier, pp. 67–87.

Zerr, A., and Boehler, R., 1993, "Melting of (Mg,Fe)SiO_3-perovskite to 625 kilobars: Indication of a high melting temperature in the lower mantle," *Science* **262**, 553–555.

Zharkov, V.N., 1962, "Physics of the Earth's core (trans.) (Trudy), *Inst. Earth Phys., Acad. Sci. USSR* **20**, 187.

Zharkov, V. N., and Kalinin, V. A., 1971, *Equations of State for Solids at High Pressures and Temperatures* (Engl. transl.). New York: Consultants Bureau, pp. 5–31.

APPENDIXES

What a sense of security
in an old book which Time
has criticized for us!

—James Russell Lowell

Table A–1. Physical constants and unit conversions

Avogadro number N	$6.02214 \times 10^{-23} mol^{-1}$
Planck constant h	6.62608×10^{-34} J s
Boltzmann constant k	1.38065×10^{-23} J K^{-1}
	1.38065×10^{-16} erg K^{-1}
Gas constant **R**	8.31447 $JK^{-1}mol^{-1}$
	1.98717 cal $K^{-1}mol^{-1}$
Electronic charge e	1.60218×10^{-19}C
Speed of light c	2.997925×10^{8}m s^{-1}
Bohr radius a_0	5.291×10^{-11}m
1 eV electron volt	1.6022×10^{-19}J
1 eV	23.05 kcal/mol
1000 cm^{-1} (spectroscopy)	1438.2 K Debye temp.
1000 K (Debye temp.)	695.3 cm^{-1} (spectroscopy)
1 J	10^7 erg
1 cal	4.1868×10^7erg = 4.1868 J
1 atm	1.01325×10^5**N** m^{-2}
1 **N**	1 Jm^{-1} = 10^5 dyn
1 Nm^{-2}	1 Pa

Table A–2. Prefix used for fractions and multiples

Fraction	Prefix	Symbol
10^{-2}	centi	c
10^{-3}	milli	*m*
10^{-6}	micro	m
10^{-9}	nano	n
10^{-12}	pico	p
Multiple	**Prefix**	**Symbol**
10^3	kilo	k
10^6	mega	M
10^9	giga	G

Table A–3. International system of units (SI units)

Table 3.1. Base units

Quantity	Unit	Symbol
Length	meter	m
Mass	kilogram	kg
Time	second	s
Thermodynamic temperature	kelvin	K
Amount of substance	mole	mol

Table 3.2. Derived units

Quantity	Unit	Symbol	Definition
Frequency	hertz	Hz	s^{-1}
Energy	joule	J	$kg\ m^2\ s^{-2}$
Force	newton	**N**	$kg\ m\ s^{-2} = J\ m^{-1}$
Power	watt	**W**	$kg\ m^2\ s^{-3} = J\ s^{-1}$
Pressure	pascal	Pa	$kg\ m^{-1}\ s^{-2} = \mathbf{N}\ m^{-2} = J\ m^{-3}$

Table 3.3. Cgs units and their relation to SI units

Quantity	Classical Units	Symbol	Definition	Relation to SI Units
Energy	erg	erg	$g\ cm^2 s^{-1}$	10^{-7}J
Energy	electron volt	eV		1.6022x10^{19}J
Energy	calorie	cal		4.18 J
Length	Angstrom	Å	10^{-8}cm	10^{-10}m
Length	micron	μ	10^{-4}cm	10^{-6}m
Mass	atomic mass unit	amu	1.6605 x10^{-24}g	1.6605 x 10^{-27}kg
Molecular mass	g/cc	**M**	g	10^{-3}kg
Volume	liter	l	$10^3 cm^3$	$10^{-3} m^3$
Concentration	molarity	M	mol/l	mol/d m^{-3}
Density	gm/cc	ρ		$10^3 kg/m^3$
Force	dyne	dyn	$g\ cm\ s^{-2}$	10^{-5}**N**
Pressure	bar	bar	10^6 dyn cm^{-2}	10^5 Pa
Pressure	GPa		10 kbar	0.01 Mbar
Pressure	MPa		.01 kbar	0.001 Mbar
Pressure	kbar		10^9 dyn	100 MPa

Table A–4. Atomic mass m of selected elements

z	Element	m	z	Element	m
1	Hydrogen (H)	1.0079	34	Selenium (Se)	78.96
2	Helium (He)	4.0026	35	Bromine (Br)	79.904
3	Lithium (Li)	6.941	36	Krypton (Kr)	83.80
4	Beryllium (Be)	9.0122	37	Rubidium (Rb)	85.468
5	Boron (B)	10.81	38	Strontium (Sr)	87.63
6	Carbon (C)	12.011	39	Yttrium (Y)	88.906
7	Nitrogen (N)	14.0067	40	Zirconium (Zr)	91.22
8	Oxygen (O)	15.9994	41	Niobium (Nb)	92.906
9	Fluorine (F)	18.9984	42	Molybdenum (Mo)	95.94
10	Neon (Ne)	20.179	43	Technecium	
11	Sodium (Na)	22.9898	44	Ruthenium (Ru)	101.07
12	Magnesium (Mg)	24.305	45	Rhodium (Rh)	102.906
13	Aluminum (Al)	26.9815	46	Palladium (Pd)	106.4
14	Silicon (Si)	28.086	47	Silver (Ag)	107.868
15	Phosphorus (P)	30.9738	48	Cadmium (Cd)	112.40
16	Sulphur (S)	32.06	49	Indium (In)	114.82
17	Chlorine (Cl)	35.453	50	Tin (Sn)	118.69
18	Argon (Ar)	39.948	51	Antimony (Sb)	121.75
19	Potassium (K)	39.098	52	Tellurium (Te)	127.60
20	Calcium (Ca)	40.08	53	Iodine (I)	126.905
21	Scandium (Sc)	44.9559	54	Xenon (Xe)	131.30
22	Titanium (Ti)	47.90	55	Cesium (Cs)	132.905
23	Vanadium (V)	50.9414	56	Barium (Ba)	137.34
24	Chromium (Cr)	51.996	73	Tantalum (Ta)	180.948
25	Manganese (Mn)	54.9380	74	Tungsten (W)	183.85
26	Iron (Fe)	55.847	75	Rhenium (Re)	186.2
27	Cobalt (Co)	58.9332	76	Osmium (Os)	190.2
28	Nickel (Ni)	58.71	77	Iridium (Ir)	192.2
29	Copper (Cu)	63.545	78	Platinum (Pt)	195.09
30	Zinc (Zn)	65.38	79	Gold (Au)	196.967
31	Gallium (Ga)	69.72	80	Mercury (Hg)	200.61
32	Germanium (Ge)	72.59	81	Thallium (Tl)	204.37
33	Arsenic (As)	74.9216	82	Lead (Pb)	207.2*

*variable

For each element, atomic number z, symbol and atomic mass are listed.

Table A–5. Values of the Earth model PREM condensed from Table A-1 in D.L. Anderson (1989)

Radius (km)	Depth (km)	Density (g/cc)	v_p (km/s)	v_s (km/s)	ϕ (km^2/sec^2)	K_S (GPa)	G (GPa)	σ	Pressure (GPa)	dK/dP
0.	6371.0	13.08	11.26	3.66	108.90	142	176	0.440	363.9	2.33
200.0	6171.0	13.07	11.25	3.66	108.80	142	176	0.440	362.9	2.33
400.0	5971.0	13.05	11.23	3.65	108.51	142	174	0.441	360.0	2.33
600.0	5771.0	13.01	11.20	3.62	108.02	141	171	0.441	355.3	2.33
800.0	5571.0	12.94	11.16	3.59	107.33	139	168	0.442	348.7	2.34
1000.0	5371.0	12.87	11.10	3.55	106.45	137	163	0.442	340.2	2.34
1200.0	5171.0	12.77	11.03	3.51	105.38	135	157	0.443	330.0	2.34
1221.5	5149.5	12.16	10.35	0	107.24	130	0	0.500	328.9	3.75
1400.0	4971.0	12.06	10.24	0	105.05	127	0	0.500	318.7	3.54
1600.0	4771.0	11.94	10.12	0	102.47	122	0	0.500	306.1	3.40
1800.0	4571.0	11.80	9.98	0	99.71	118	0	0.500	292.2	3.32
2000.0	4371.0	11.65	9.83	0	96.73	113	0	0.500	277.0	3.29
2200.0	4171.0	11.48	9.66	0	93.48	107	0	0.500	260.7	3.29
2400.0	3971.0	11.29	9.48	0	89.95	102	0	0.500	243.2	3.32
2600.0	3771.0	11.08	9.27	0	86.10	95.4	0	0.500	224.8	3.36
2800.0	3571.0	10.85	9.05	0	81.91	88.9	0	0.500	205.6	3.41
3000.0	3371.0	10.60	8.79	0	77.36	82.0	0	0.500	185.6	3.47
3200.0	3171.0	10.32	8.51	0	72.47	74.8	0	0.500	165.1	3.52
3400.0	2971.0	10.02	8.19	0	67.23	67.4	0	0.500	144.2	3.56
3480.0	2891.0	5.56	13.71	7.26	117.78	65.6	294	0.305	135.8	1.64
3600.0	2771.0	5.50	13.68	7.26	116.96	64.4	291	0.303	128.7	1.64
3630.0	2741.0	5.49	13.68	7.26	116.76	64.1	290	0.303	127.0	3.33
3800.0	2571.0	5.40	13.47	7.18	112.73	61.0	279	0.301	117.3	3.24
4000.0	2371.0	5.30	13.24	7.09	108.23	57.4	268	0.298	106.4	3.17
4200.0	2171.0	5.20	13.01	7.01	103.88	54.1	256	0.295	95.8	3.13
4400.0	1971.0	5.10	12.78	6.91	99.59	50.9	245	0.292	85.4	3.14
4600.0	1771.0	5.00	12.54	6.82	95.26	47.7	233	0.289	75.4	3.19
4800.0	1571.0	4.89	12.29	6.72	90.81	44.5	222	0.286	65.5	3.27

Table A–5. Values of the Earth model PREM condensed from Table A-1 in D.L. Anderson (1989) (continued)

Radius (km)	Depth (km)	Density (g/cc)	v_p (km/s)	v_s (km/s)	ϕ (km^2/sec^2)	K_S (GPa)	G (GPa)	σ	Pressure (GPa)	dK/dP
5000.0	1371.0	4.78	12.02	6.61	86.17	41.3	210	0.282	55.9	3.38
5200.0	1171.0	4.67	11.73	6.50	81.28	38.0	198	0.278	46.5	3.52
5400.0	971.0	4.56	11.41	6.37	76.08	34.7	186	0.273	37.3	3.67
5600.0	771.0	4.44	11.06	6.24	70.52	31.3	173	0.266	28.3	3.84
5650.0	721.0	4.41	10.91	6.09	69.51	30.7	164	0.273	26.1	3.00
5701.0	670.0	3.99	10.26	5.57	64.03	25.6	124	0.291	23.8	2.40
5771.0	600.0	3.97	10.15	5.51	62.61	24.9	121	0.290	21.0	2.37
5821.0	550.0	3.91	9.90	5.37	59.60	23.3	113	0.291	19.1	7.88
5921.0	450.0	3.78	9.38	5.07	53.78	20.4	97.7	0.293	15.2	7.46
5971.0	400.0	3.54	8.90	4.76	48.97	17.4	80.6	0.298	13.4	3.37
6061.0	310.0	3.48	8.73	4.70	46.71	16.3	77.3	0.295	10.2	3.30
6151.0	220.0	3.43	8.55	4.64	44.50	15.3	74.1	0.291	7.1	3.23

Table A–6. Debye functions for high temperatures at $P = 0$ (condensed from Tables 265Ac to 265 Ae from Landolt-Börnstein, *Physical Chemical Tables*, fifth enlarged printing. First supplementary volume, edited by Walther A. Roth and Karl Scheel, Verlag von Julius Springer, **Berlin, 1927)**

Table A–6.1. C_V (Specific Heat) Θ/T Between 0 and 11

Θ/T	0.0	0.1	0.2	0.3	0.4	0.5	0.6	0.7	0.8	0.9
0	5.957	5.954	5.945	5.930	5.909	5.883	5.851	5.813	5.770	5.722
1	5.669	5.611	5.549	5.482	5.412	5.337	5.259	5.178	5.094	5.007
2	4.918	4.827	4.737	4.638	4.543	4.444	4.345	4.246	4.147	4.047
3	3.947	3.849	3.750	3.653	3.556	3.459	3.364	3.270	3.178	3.086
4	2.996	2.909	2.822	2.737	2.653	2.573	2.497	2.416	2.343	2.270
5	2.197	2.127	2.059	1.992	1.928	1.867	1.806	1.747	1.690	1.634
6	1.582	1.532	1.481	1.433	1.386	1.341	1.297	1.255	1.214	1.175
7	1.137	1.100	1.065	1.031	0.9985	0.9964	0.9354	0.9062	0.8775	0.8495
8	0.8233	0.7977	0.7735	0.7501	0.7266	0.7042	0.6827	0.6619	0.6422	0.6231
9	0.6041	0.5876	0.5695	0.5523	0.5368	0.5213	0.5064	0.4921	0.4780	0.4647
10	0.4518	0.4392	0.4268	0.4153	0.4041	0.3942	0.3827	0.3725	0.3626	0.3532
11	0.3446	0.3351	0.3264	0.3188	0.3100	0.3022	0.2946	0.2873	0.2802	0.2733

Table A–6.2. $(\mathcal{U} - \mathcal{U}_0)/T$ (Internal Energy) Θ/T Between 0 and 9

Θ/T	0.0	0.1	0.2	0.3	0.4	0.5	0.6	0.7	0.8	0.9
0	5.957	5.7354	5.5221	5.3131	5.1110	4.9141	4.7234	4.5381	4.3590	4.1854
1	4.0175	3.8551	3.6983	3.5465	3.4002	3.2590	3.1229	2.9920	2.8660	2.7446
2	2.6280	2.5159	2.4090	2.3046	2.2054	2.1097	2.0182	1.9327	1.8468	1.7662
3	1.6891	1.6154	1.5450	1.4778	1.4134	1.3516	1.2903	1.2365	1.1830	1.1314
4	1.0824	1.0359	0.9912	0.9486	0.9077	0.8692	0.8330	0.7968	0.7635	0.7317
5	0.7007	0.6714	0.6435	0.6167	0.5914	0.5676	0.5443	0.5222	0.5012	0.4809
6	0.4610	0.4440	0.4266	0.4101	0.3942	0.3790	0.3644	0.3507	0.3374	0.3249
7	0.3129	0.3013	0.2903	0.2798	0.2698	0.2601	0.2508	0.2421	0.2337	0.2255
8	0.2179	0.2105	0.2037	0.1969	0.1900	0.1838	0.1778	0.1720	0.1665	0.1612

Table A–6.3. $(\mathcal{F} - \mathcal{F}_0)/T$ (Helmholtz Energy) Θ/T Between 0 and 9

Θ/T	0.0	0.1	0.2	0.3	0.4	0.5	0.6	0.7	0.8	0.9
0	∞	15.925	12.014	9.815	8.314	7.195	6.316	5.601	5.009	4.505
1	4.079	3.696	3.366	3.077	2.820	2.591	2.385	2.199	2.032	1.879
2	1.742	1.6168	1.5025	1.3975	1.3017	1.2134	0.1325	1.0587	0.9893	0.9259
3	0.8673	0.8132	0.7629	0.7165	0.6733	0.6332	0.5952	0.5613	0.5291	0.4990
4	0.4709	0.4449	0.4204	0.3976	0.3762	0.3563	0.3379	0.3200	0.3037	0.2884
5	0.2739	0.2602	0.2475	0.2353	0.2241	0.2136	0.2030	0.1940	0.1851	0.1766
6	0.1685	0.1614	0.1543	0.1477	0.1413	0.1353	0.1295	0.1242	0.1191	0.1143
7	0.1097	0.1054	0.1012	0.09730	0.09357	0.09002	0.08660	0.08342	0.08033	0.07737
8	0.07464	0.07199	0.06951	0.06706	0.06467	0.06247	0.06035	0.05831	0.05640	0.05455

Table A–6.4. $\mathcal{S}$ (Entropy) Θ/T Between 0 and 11

Θ/T	0.0	0.1	0.2	0.3	0.4	0.5	0.6	0.7	0.8	0.9
0	∞	21.650	17.536	15.128	13.425	12.109	11.039	10.139	9.368	8.690
1	8.097	7.551	7.064	6.624	6.220	5.850	5.508	5.191	4.898	4.624
2	4.3702	4.1327	3.9115	3.7021	3.5071	3.3231	3.1507	2.9914	2.8361	2.6921
3	2.5564	2.4286	2.3079	2.1943	2.0867	1.9848	1.8855	1.7978	1.7121	1.6304
4	1.5533	1.4808	1.4116	1.3462	1.2839	1.2255	1.1709	1.1168	1.0672	1.0201
5	0.9746	0.9316	0.8910	0.8520	0.8155	0.7812	0.7473	0.7162	0.6863	0.6575
6	0.6295	0.6054	0.5809	0.5578	0.5355	0.5143	0.4939	0.4749	0.4565	0.4392
7	0.4226	0.4067	0.3915	0.3771	0.3634	0.3501	0.3374	0.3255	0.3140	0.3029
8	0.2925	0.2825	0.2732	0.2640	0.2547	0.2463	0.2382	0.2303	0.2229	0.2158
9	0.2087	0.2025	0.1959	0.1897	0.1840	0.1784	0.1731	0.1680	0.1629	0.1582
10	0.1535	0.1491	0.1447	0.1405	0.1367	0.1333	0.1293	0.1257	0.1223	0.1190

Tables A–7.1–7.16. Physical properties and thermoelastic parameters of oxides and silicates at high T (Anderson and Isaak, 1994, Pt III ref.)

Table A–7.1. Al_2O_3: Thermal expansivity, specific heat, isotropic moduli

T K	ρ g/cc	α 10^{-5}/K	K_S GPa	G GPa	C_P J/(gK)	γ	C_V	K_T GPa
300	3.982	1.62	253.6	163.0	0.779	1.32	0.774	252.0
400	3.975	1.99	252.6	161.1	0.943	1.34	0.933	249.9
500	3.966	2.23	250.9	158.8	1.040	1.36	1.024	247.1
600	3.957	2.40	248.6	156.6	1.103	1.37	1.082	243.8
700	3.947	2.51	246.6	154.2	1.148	1.36	1.121	240.8
800	3.937	2.59	244.4	151.9	1.180	1.36	1.148	237.7
900	3.927	2.66	242.4	149.5	1.205	1.36	1.167	234.8
1000	3.916	2.73	240.0	147.1	1.223	1.37	1.179	231.4
1100	3.906	2.80	237.8	144.6	1.244	1.37	1.194	228.2
1200	3.894	2.88	235.2	142.2	1.257	1.38	1.199	224.5
1300	3.883	2.96	232.6	139.7	1.267	1.40	1.203	220.8
1400	3.872	3.03	230.0	137.2	1.277	1.41	1.205	217.1
1500	3.860	3.09	228.1	134.8	1.286	1.42	1.207	214.0
1600	3.848	3.15	225.9	133.5	1.296	1.43	1.209	210.7
1700	3.835	3.20	224.8	131.2	1.306	1.43	1.212	207.5
1800	3.823	3.25	221.8	127.5	1.318	1.43	1.216	204.7

Table A–7.2. Al_2O_3: Thermoelastic parameters and Debye temperature

T K	Θ K	δ_S	δ_T	Γ	$\frac{(\delta_T-\delta_S)}{\gamma}$	φ	αK_T MPa/K	ΔP_{TH} GPa
300	1034	3.30	5.71	5.71	1.82	1.60	4.08	0.00
400	1029	3.16	5.16	5.16	1.49	1.52	4.98	0.45
500	1022	3.20	5.03	6.27	1.35	1.46	5.53	0.98
600	1015	3.31	5.08	6.09	1.29	1.42	5.85	1.55
700	1008	3.43	5.17	6.05	1.28	1.40	6.03	2.15
800	1001	3.55	5.29	6.06	1.28	1.38	6.15	2.76
900	994	3.62	5.37	6.08	1.28	1.36	6.24	3.43
1000	986	3.66	5.42	6.09	1.29	1.36	6.30	4.01
1100	979	3.65	5.42	6.07	1.29	1.36	6.40	4.64
1200	971	3.60	5.39	6.03	1.30	1.37	6.45	5.93
1300	963	3.51	5.32	5.98	1.29	1.38	6.52	5.93
1400	955	3.39	5.22	5.93	1.30	1.40	6.57	6.59
1500	947	3.24	5.08	5.87	1.30	1.43	6.62	7.24
1600	939	3.06	4.92	5.80	1.30	1.47	6.64	7.91
1700	932	2.85	4.73	5.74	1.32	1.52	6.64	8.57
1800	924	2.60	4.50	5.66	1.32	1.58	6.66	9.24

Table A–7.3. MgO (units the same as in Table A–7.1)

T	ρ	α	K_S	G	C_P	γ	C_V	K_T
100	3.602	0.63	165.7	132.0	0.194	1.59	0.194	165.6
200	3.597	2.24	164.6	130.3	0.662	1.55	0.658	163.5
300	3.585	3.12	163.9	131.8	0.928	1.54	0.915	161.6
400	3.573	3.57	162.3	129.4	1.061	1.53	1.048	158.9
500	3.559	3.84	160.7	126.9	1.130	1.53	1.098	156.1
600	3.545	4.02	158.9	124.4	1.173	1.54	1.131	153.2
700	3.531	4.14	157.1	121.8	1.204	1.53	1.153	150.4
800	3.516	4.26	155.1	119.2	1.227	1.53	1.166	147.4
900	3.501	4.38	153.1	116.7	1.246	1.54	1.175	144.3
1000	3.486	4.47	151.1	114.1	1.262	1.54	1.181	141.4
1100	3.470	4.56	148.9	111.5	1.276	1.53	1.185	138.3
1200	3.454	4.65	146.7	109.0	1.289	1.53	1.188	135.1
1300	3.438	4.71	144.4	106.4	1.301	1.52	1.190	132.1
1400	3.422	4.80	142.0	103.8	1.312	1.52	1.191	128.1
1500	3.405	4.89	139.7	101.3	1.323	1.52	1.191	125.7
1600	3.388	4.98	137.3	99.0	1.334	1.51	1.191	122.5
1700	3.371	5.04	134.9	96.7	1.346	1.50	1.193	119.6
1800	3.354	5.13	132.7	94.5	1.358	1.50	1.193	116.6

Table A–7.4. MgO (units the same as in Table A–7.2)

T	Θ	δ_S	δ_T	Γ	$\frac{(\delta_T-\delta_S)}{\gamma}$	φ	αK_T	ΔP_{TH}
300	945	2.83	5.26	5.73	1.57	1.40	5.04	0.00
400	937	2.79	4.83	5.34	1.33	1.40	5.67	0.54
500	928	2.81	4.69	5.17	1.23	1.38	6.00	1.12
600	920	2.86	4.67	5.08	1.18	1.37	6.16	1.73
700	911	2.92	4.70	5.05	1.16	1.35	6.23	2.35
800	902	2.98	4.74	5.03	1.15	1.34	6.28	2.98
900	894	3.04	4.78	5.02	1.13	1.32	6.32	3.61
1000	885	3.12	4.84	5.05	1.12	1.31	6.32	4.24
1100	875	3.21	4.92	5.08	1.12	1.30	6.31	4.87
1200	806	3.30	4.99	5.09	1.11	1.28	6.28	5.50
1300	857	3.41	5.08	5.10	1.10	1.26	6.22	6.12
1400	847	3.47	5.12	5.04	1.09	1.24	6.19	6.74
1500	838	3.50	5.13	4.92	1.07	1.22	6.16	7.36
1600	828	3.46	5.07	4.75	1.07	1.21	6.13	7.97
1700	820	3.36	4.95	4.56	1.06	1.20	6.03	8.58
1800	811	3.12	4.66	4.34	1.03	1.23	6.00	9.20

Table A–7.5. Mg_2SiO_4 (units the same as in Table A–7.1)

T	ρ	α	K_S	G	C_P	γ	C_V	K_T
300	3.222	2.72	128.6	81.6	0.840	1.29	0.831	127.3
400	3.213	3.03	127.1	80.3	0.990	1.21	0.976	125.2
500	3.203	3.22	125.4	78.9	1.068	1.18	1.048	123.1
600	3.192	3.36	123.7	77.4	1.119	1.17	1.093	120.8
700	3.181	3.48	121.9	76.0	1.156	1.16	1.124	118.6
800	3.170	3.59	120.2	74.5	1.186	1.15	1.148	116.3
900	3.159	3.70	118.3	73.1	1.211	1.15	1.167	114.0
1000	3.147	3.81	116.6	71.6	1.235	1.14	1.183	111.7
1100	3.135	3.92	114.8	70.1	1.256	1.14	1.197	109.4
1200	3.122	4.05	112.9	68.6	1.277	1.15	1.210	106.9
1300	3.109	4.16	111.1	67.1	1.296	1.15	1.220	104.6
1400	3.096	4.27	109.2	65.6	1.315	1.15	1.231	102.2
1500	3.083	4.39	107.5	64.1	1.334	1.15	1.240	99.9
1600	3.069	4.50	105.6	62.7	1.352	1.15	1.249	97.6
1700	3.055	4.62	103.7	61.2	1.370	1.14	1.257	95.2

Table A–7.6. Mg_2SiO_4 (units the same as in Table A–7.2)

T	Θ	δ_S	δ_T	Γ	$\frac{(\delta_T-\delta_S)}{\gamma}$	φ	αK_T	ΔP_{TH}
300	763	4.45	5.94	6.07	1.16	1.20	3.46	0.00
400	757	4.20	5.58	5.66	1.14	1.21	3.80	0.36
500	751	4.15	5.49	5.54	1.14	1.20	3.97	0.75
600	744	4.15	5.48	5.50	1.14	1.20	4.07	1.16
700	738	4.16	5.49	5.46	1.15	1.20	4.13	1.57
800	731	4.13	5.47	5.45	1.17	1.18	4.18	1.98
900	724	4.08	5.46	5.44	1.20	1.20	4.22	2.40
1000	718	4.05	5.47	5.45	1.25	1.22	4.26	2.83
1100	711	4.00	5.46	5.43	1.28	1.20	4.31	3.25
1200	704	4.02	5.49	5.38	1.28	1.21	4.33	3.69
1300	697	3.97	5.44	5.32	1.28	1.20	4.36	4.13
1400	689	3.90	5.37	5.24	1.28	1.21	4.37	4.50
1500	682	3.92	5.38	5.22	1.27	1.23	4.39	5.07
1600	674	3.93	5.40	5.19	1.28	1.19	4.40	5.43
1700	668	3.96	5.42	5.16	1.28	1.20	4.39	5.87

Table A–7.7. Olivine $Fo_{90}Fa_{10}$ (units the same as in Table A–7.1)

T	ρ	α	K_S	G	C_P	γ	C_V	K_T
300	3.353	2.66	129.3	78.1	0.816	1.25	0.808	128.0
400	3.343	2.99	127.7	76.8	0.957	1.19	0.944	125.9
500	3.333	3.21	125.9	75.3	1.032	1.17	1.013	123.6
600	3.322	3.35	124.1	73.9	1.080	1.16	1.055	121.2
700	3.311	3.46	122.2	72.5	1.112	1.14	1.086	118.2
800	3.299	3.55	120.3	71.2	1.145	1.13	1.109	116.6
900	3.287	3.64	118.9	69.8	1.171	1.12	1.129	114.7
1000	3.275	3.71	117.0	68.5	1.194	1.11	1.147	112.1
1100	3.263	3.79	115.1	67.1	1.216	1.10	1.163	110.0
1200	3.251	3.86	113.2	65.8	1.236	1.09	1.177	107.8
1300	3.238	3.93	111.4	64.4	1.256	1.08	1.191	105.6
1400	3.225	4.00	109.6	63.1	1.275	1.07	1.203	103.4
1500	3.212	4.07	107.8	61.7	1.294	1.06	1.216	101.3

Table A–7.8. Olivine $Fo_{90}Fa_{10}$ (units the same as in Table A–7.2)

T	Θ	δ_S	δ_T	Γ	$\frac{(\delta_T-\delta_S)}{\gamma}$	φ	αK_T	ΔP_{TH}
300	731	5.24	6.59	6.56	1.07	1.17	3.37	0.00
400	725	4.70	5.95	5.92	1.03	1.17	3.76	0.36
500	719	4.46	5.65	5.63	1.02	1.17	3.97	0.75
600	713	4.33	5.51	5.50	1.02	1.18	4.05	1.15
700	706	4.25	5.44	5.42	1.04	1.18	4.11	1.56
800	700	4.21	5.40	5.38	1.06	1.18	4.14	1.97
900	699	4.16	5.36	5.36	1.07	1.18	4.18	2.38
1000	688	4.14	5.36	5.35	1.10	1.18	4.17	2.80
1100	681	4.13	5.37	5.35	1.13	1.18	4.17	3.22
1200	675	4.12	5.38	5.36	1.16	1.18	4.16	3.63
1300	669	4.07	5.35	5.32	1.18	1.18	4.15	4.05
1400	662	4.10	5.41	5.39	1.23	1.19	4.14	4.46
1500	665	4.10	5.43	5.41	1.26	1.19	4.13	4.88

Table A–7.9. MnO: (units the same as in Table A–7.1)

T	ρ	α	K_S	G	C_P	γ	C_V	K_T
300	5.378	3.46	149.0	68.3	0.632	1.51	0.623	146.7
350	5.369	3.58	148.0	67.6	0.653	1.51	0.641	145.2
400	5.359	3.68	146.9	66.7	0.669	1.51	0.655	143.7
450	5.349	3.77	145.8	65.6	0.682	1.51	0.665	142.2
500	5.339	3.85	144.8	64.4	0.692	1.51	0.673	140.7

Table A–7.10. CaO (units the same as in Table A–7.1).

T	ρ	α	K_S	G	C_P	γ	C_V	K_T
300	3.349	3.04	112.0	80.59	0.752	1.35	0.743	110.6
400	3.338	3.47	110.5	79.15	0.834	1.36	0.819	108.5
500	3.327	3.67	109.1	77.71	0.880	1.37	0.858	106.4
600	3.314	3.81	107.6	76.22	0.904	1.37	0.877	104.3
700	3.301	3.92	106.2	74.76	0.921	1.37	0.888	102.3
800	3.288	4.01	104.7	73.33	0.933	1.37	0.894	100.3
900	3.275	4.08	103.3	71.90	0.943	1.36	0.898	98.4
1000	3.262	4.14	101.7	70.40	0.952	1.36	0.901	96.3
1100	3.248	4.20	100.2	68.99	0.959	1.35	0.903	94.3
1200	3.234	4.26	98.7	67.56	0.965	1.35	0.903	92.3

Table A–7.11. NaCl (units the same as in Table A–7.1).

T	ρ	α	K_S	G	C_P	γ	C_V	K_T
300	2.159	11.8	25.3	14.71	0.868	1.59	0.822	24.0
350	2.146	12.2	24.8	14.27	0.883	1.60	0.826	23.2
400	2.132	12.7	24.2	13.81	0.897	1.61	0.829	22.4
450	2.118	13.2	23.7	13.39	0.910	1.62	0.830	21.6
500	2.104	13.7	23.2	12.96	0.923	1.64	0.830	20.8
550	2.089	14.3	22.5	12.53	0.937	1.64	0.830	19.9
600	2.074	14.8	21.7	12.11	0.950	1.63	0.830	19.0
650	2.059	15.4	21.1	11.68	0.964	1.63	0.829	18.1
700	2.043	16.0	20.5	11.25	0.979	1.63	0.828	17.3
750	2.026	16.6	19.8	10.80	0.997	1.63	0.829	16.5

Table A–7.12. KCl (units the same as in Table A–7.1).

T	ρ	α	K_S	G	C_P	γ	C_V	K_T
300	1.982	11.0	17.8	9.47	0.689	1.44	0.657	17.0
350	1.971	11.3	17.3	9.18	0.701	1.42	0.664	16.4
400	1.959	11.7	17.0	8.91	0.713	1.42	0.669	15.9
450	1.948	12.1	16.6	8.64	0.724	1.42	0.672	15.4
500	1.935	12.6	16.1	8.39	0.735	1.43	0.674	14.7
550	1.923	13.2	15.7	8.13	0.745	1.44	0.675	14.2
600	1.910	13.7	15.4	7.85	0.756	1.45	0.676	13.7
650	1.897	14.2	15.0	7.57	0.767	1.46	0.676	13.2
700	1.883	14.7	14.5	7.29	0.778	1.46	0.677	12.6
750	1.869	15.2	14.0	6.98	0.791	1.44	0.679	12.0
800	1.855	15.7	13.6	6.67	0.806	1.43	0.683	11.5
850	1.840	16.2	12.0	6.41	0.823	1.39	0.691	10.9

Table A–7.13. CaO (units the same as in Table A–7.2).

T	Θ	δ_S	δ_T	Γ	$\frac{(\delta_T-\delta_S)}{\gamma}$	φ	αK_T	ΔP_{TH}
300	671	4.15	6.19	6.00	1.51	1.24	3.36	0.00
400	666	3.75	5.54	5.38	1.31	1.24	3.73	0.36
500	660	3.60	5.27	5.13	1.22	1.23	3.90	0.74
600	654	3.54	5.14	5.01	1.17	1.23	3.98	1.13
700	649	3.52	5.07	4.99	1.13	1.23	4.01	1.53
800	643	3.53	5.03	4.95	1.10	1.23	4.02	2.34
900	637	3.55	5.01	4.93	1.07	1.22	4.01	2.34
1000	631	3.58	5.00	4.94	1.05	1.22	3.99	2.54
1100	625	3.62	5.01	4.96	1.03	1.22	3.96	3.13
1200	619	3.65	5.01	4.99	1.01	1.22	3.93	3.53

Table A–7.14. NaCl (units the same as in Table 7.2).

T	Θ	δ_S	δ_T	Γ	$\frac{(\delta_T-\delta_S)}{\gamma}$	φ	αK_T	ΔP_{TH}
300	304	3.47	5.56	5.05	1.32	1.29	2.82	0.00
350	300	3.56	5.62	5.00	1.29	1.26	2.83	0.14
400	296	3.65	5.69	4.95	1.27	1.23	2.84	0.28
450	291	3.72	5.74	4.90	1.25	1.20	2.86	0.43
500	287	3.80	5.82	4.86	1.24	1.18	2.86	0.57
550	283	3.91	5.95	4.83	1.25	1.16	2.84	0.71
600	278	4.03	6.10	4.79	1.27	1.13	2.81	0.85
650	274	4.14	6.24	4.77	1.29	1.11	2.78	0.99
700	269	4.23	6.37	4.76	1.31	1.10	2.77	1.13
750	264	4.34	6.53	4.76	1.35	1.08	2.73	1.27

Table A–7.15. KCl (units the same as in Table 7.2).

T	Θ	δ_S	δ_T	Γ	$\frac{(\delta_T-\delta_S)}{\gamma}$	φ	αK_T	ΔP_{TH}
300	230	3.77	5.84	4.66	1.34	1.17	1.87	0.00
350	227	3.86	5.88	4.77	1.34	1.17	1.86	0.09
400	224	3.92	5.88	4.86	1.32	1.17	1.86	0.19
450	221	3.97	5.88	4.93	1.30	1.17	1.86	0.28
500	218	4.02	5.88	4.97	1.28	1.17	1.86	0.37
550	214	4.05	5.87	5.02	1.26	1.17	1.87	0.47
600	211	4.06	5.84	5.10	1.23	1.18	1.88	0.56
650	208	4.09	5.83	5.19	1.21	1.18	1.88	0.65
700	204	4.18	5.90	5.30	1.23	1.18	1.86	0.75
750	200	4.27	5.98	5.44	1.25	1.18	1.83	0.84
800	196	4.34	6.04	5.61	1.27	1.19	1.81	0.93
850	192	4.50	6.19	5.76	1.33	1.19	1.77	1.02

Table A–7.16. $MgAl_2O_4$ (units the same as in Table A–7.1).

T	ρ	α	K_S	G	C_P	γ	C_V	K_T
300	3.576	2.11	209.9	108.2	0.819	1.51	0.811	207.9
350	3.572	2.18	208.2	107.7	0.899	1.41	0.889	205.9
400	3.568	2.25	207.1	107.2	0.963	1.36	0.952	204.6
450	3.564	2.32	205.3	106.6	1.014	1.32	1.001	202.5
500	3.560	2.38	203.9	106.0	1.055	1.30	1.039	200.8
550	3.555	2.45	202.8	105.5	1.088	1.28	1.069	199.4
600	3.551	2.51	201.6	104.9	1.115	1.28	1.094	197.8
650	3.547	2.57	200.3	104.3	1.139	1.28	1.115	196.1
700	3.542	2.63	199.0	103.6	1.160	1.27	1.133	194.4
750	3.537	2.69	197.7	103.0	1.179	1.27	1.149	192.7
800	3.532	2.74	196.2	102.4	1.180	1.27	1.164	190.9
850	3.528	2.80	194.7	101.8	1.213	1.27	1.178	189.0
900	3.523	2.85	193.4	101.1	1.229	1.27	1.190	187.3
950	3.518	2.90	192.0	100.5	1.243	1.27	1.201	185.5
1000	3.512	2.94	191.3	99.8	1.253	1.28	1.208	184.4

Table A–7.17. $MgAl_2O_4$ (units the same as in Table A–7.2).

T	Θ	δ_S	δ_T	Γ	$\frac{(\delta_T-\delta_S)}{\gamma}$	φ	αK_T	ΔP_{TH}
300	862	6.03	7.73	5.30	1.12	0.90	4.38	0.00
400	858	5.72	7.36	5.01	1.20	0.90	4.60	0.45
500	854	5.49	7.07	4.78	1.22	0.90	4.79	0.92
600	850	5.27	6.82	4.59	1.21	0.90	4.97	1.41
700	845	5.10	6.62	4.43	1.20	0.90	5.11	1.91
800	840	4.96	6.47	4.30	1.19	0.90	5.24	2.43
900	835	4.85	6.35	4.20	1.18	0.90	5.33	2.96
1000	830	4.74	6.24	4.11	1.17	0.90	5.43	3.49

Table A–7.18. MnO (Units the same as in Table A–7.2)

T	Θ	δ_S	δ_T	Γ	$\frac{(\delta_T-\delta_S)}{\gamma}$	φ	αK_T	ΔP_{TH}
300	534	4.14	5.96	8.33	1.20	1.56	5.07	0.00
350	531	4.03	5.82	8.14	1.18	1.57	5.20	0.26
400	527	3.94	5.71	8.01	1.17	1.57	5.29	0.52
450	523	3.88	5.64	7.95	1.16	1.57	5.36	0.79
500	519	3.83	5.58	7.94	1.16	1.58	5.41	1.05

Table A–7.19. Pyrope-rich garnet (units the same as in Table A–7.1)

T	ρ	α	K_S	G	C_P	γ	C_V	K_T
300	3.705	2.36	171.2	92.6	0.726	1.50	0.718	169.4
400	3.696	2.64	168.9	91.6	0.902	1.34	0.889	166.5
500	3.686	2.80	167.0	90.6	0.981	1.29	0.964	164.0
600	3.675	2.90	164.9	89.7	1.032	1.26	1.010	161.4
700	3.664	2.97	163.2	88.7	1.067	1.24	1.040	159.1
800	3.653	3.03	161.3	87.6	1.088	1.23	1.057	156.6
900	3.642	3.07	159.3	86.5	1.104	1.22	1.068	154.1
1000	3.631	3.11	157.3	85.5	1.116	1.21	1.076	151.6

Table A–7.20. Pyrope-rich garnet (units the same as in Table A–7.2)

T	Θ	δ_S	δ_T	Γ	$\frac{(\delta_T-\delta_S)}{\gamma}$	φ	αK_T	ΔP_{TH}
300	779	4.81	6.27	4.29	0.97	0.88	4.00	0.00
350	777	4.52	5.90	4.07	1.00	0.89	4.25	0.21
400	775	4.36	5.70	3.96	1.00	0.90	4.40	0.42
450	773	4.24	5.55	3.90	1.00	0.92	4.51	0.65
500	771	4.16	5.46	3.86	1.00	0.93	4.59	0.87
550	769	4.11	5.41	3.86	1.02	0.94	4.64	1.10
600	767	4.07	5.35	3.86	1.04	0.96	4.68	1.34
650	765	4.05	5.34	3.88	1.04	0.97	4.69	1.57
700	764	4.01	5.30	3.89	1.04	0.98	4.72	1.81
750	761	4.00	5.29	3.92	1.05	1.00	4.74	2.04
800	759	3.98	5.28	3.94	1.06	1.01	4.74	2.28
850	757	3.98	5.29	3.98	1.07	1.02	4.74	2.52
900	755	3.98	5.30	4.02	1.09	1.04	4.73	2.75
950	753	3.97	5.30	4.06	1.10	1.05	4.73	2.99
1000	751	3.97	5.32	4.10	1.11	1.06	4.71	3.23

Table A–7.21. Fe_2SiO_4 (units the same as in Table A–7.1)

T	ρ	α	K_S	G	C_P	γ	C_V	K_T
300	4.400	2.61	138.0	51.0	0.673	1.21	0.667	136.7
400	4.388	2.74	135.9	49.7	0.746	1.18	0.736	134.1
500	4.375	3.00	134.0	48.8	0.793	1.16	0.779	131.7
600	4.362	3.12	131.8	48.0	0.830	1.13	0.813	129.0
700	4.348	3.22	129.3	47.4	0.863	1.11	0.842	126.1

Table A–7.22. Grossular garnet (units the same as in Table A–7.1)

T	ρ	α	K_S	G	C_P	γ	C_V	K_T
300	3.597	1.92	167.8	106.9	0.736	1.22	0.730	166.6
400	3.589	2.28	166.2	105.7	0.865	1.22	0.855	164.4
500	3.581	2.49	164.9	104.5	0.945	1.21	0.931	162.5
600	3.571	2.61	163.3	103.1	0.995	1.20	0.977	160.3
700	3.562	2.71	161.6	101.8	1.028	1.19	1.006	158.1
800	3.552	2.78	160.3	100.5	1.052	1.19	1.025	156.2
900	3.542	2.83	158.9	99.1	1.072	1.19	1.041	154.3
1000	3.532	2.88	157.5	97.7	1.092	1.18	1.056	152.3
1100	3.522	2.92	156.2	96.4	1.113	1.16	1.073	150.6
1200	3.512	2.97	154.4	94.9	1.139	1.14	1.095	148.3
1300	3.501	3.00	152.6	93.4	1.170	1.12	1.121	146.2

Table A–7.23 Grossular garnet (units the same as in Table A–7.2)

T	Θ	δ_S	δ_T	Γ	$\frac{(\delta_T-\delta_S)}{\gamma}$	φ	αK_T	ΔP_{TH}
300	824	4.64	6.30	6.09	1.36	1.22	3.21	0.00
400	820	3.93	5.36	5.27	1.17	1.23	3.74	0.36
500	816	3.64	4.98	4.97	1.11	1.24	4.03	0.75
600	811	3.49	4.80	4.87	1.08	1.25	4.18	1.16
700	806	3.41	4.70	4.84	1.08	1.26	4.28	1.57
800	801	3.35	4.64	4.86	1.08	1.27	4.34	1.98
900	796	3.31	4.60	4.90	1.08	1.28	4.36	2.40
1000	791	3.29	4.58	4.96	1.09	1.29	4.38	2.83
1100	786	3.27	4.57	5.03	1.11	1.30	4.41	3.25
1200	780	3.26	4.57	5.11	1.15	1.31	4.40	3.69
1300	715	3.25	4.58	5.20	1.18	1.32	4.38	5.40

Table A–7.24. Fe_2SiO_4 (units the same as in Table A–7.2)

T	Θ	δ_S	δ_T	Γ	$\frac{(\delta_T-\delta_S)}{\gamma}$	φ	αK_T	ΔP_{TH}
300	511	5.99	7.34	9.34	1.12	1.54	3.56	0.00
400	506	5.56	6.85	7.49	1.09	1.33	3.82	0.37
500	501	5.35	6.62	6.02	1.09	1.09	3.95	0.76
600	497	5.24	6.50	4.69	1.11	0.85	4.02	1.16
700	494	5.18	6.45	3.43	1.14	0.60	4.06	1.56

GLOSSARY OF SYMBOLS

This grand book...the universe...
is written in the language of
mathematics and its characters...

—Galileo Galilei

α	coefficient of thermal expansion, constant V
α_a	α at ambient conditions
α_{T^*}	α at reference temperature, T^*
α_0	α at zero pressure
α_2	ratio of elastic constants for fcc lattice
α-iron	body centered cubic phase of iron
$\hat{\alpha}$	dimensionless constant involving α
αK_T	T derivative of P_{TH}
$\overline{\alpha K_T}$	average value of αK_T over T range
$(\alpha K_T)^{ht}$	high T limit in quasiharmonic approximation of αK_T
β	high pressure phase of Mg_2SiO_4
β_e	constant in definition of electronic specific heat
β_2	ratio of elastic moduli fcc lattice
β-iron	phase of iron at about 100 GPa
γ	Grüneisen ratio; $\gamma = (\alpha K_T)/\rho C_V$
γ-iron	face-centered cubic (fcc) phase of iron
$\overline{\gamma}$	approximation to γ from invoking quasiharmonic approximation to Helmholtz energy. Average over $3pN$ mode gammas
γ_{ac}	acoustic gamma, the average of p wave mode and two s wave modes
γ_{bj}	special case of γ_{mg} involving accounting of vibration perpendicular to interatomic bonds
γ_D	Debye representation of γ
γ_{Ds}	γ for Debye model at constant entropy
γ_e	γ mode for electronic density of states
γ_E	γ approximation arising from Einstein solid
γ_{ht}	high temperature limit of γ
γ_i	mode gamma of index i
γ_j	mode gamma of index j
γ_{mg}	Mie-Grüneisen EoS γ
γ_{oi}	the ith optical mode
γ_{opt}	γ approximation arising from optic modes and their pressure variations
γ_p	p wave mode; term in γ_{ac}

γ_s	s wave mode; term in γ_{ac}
γ_{sl}	Slater gamma
γ_{th}	Older literature symbol for γ
γ_{vz}	special case of γ_{mg} (vz refers to derivation by Vaschenko and Zubarev (1963))
γ_0	γ at $P=0$ or $\eta=1$
γ_{0_j}	$P=0$ value of jth mode
$\gamma_{T=0}$	γ at absolute zero
Γ	dimensionless derivative involving $(\partial G/\partial T)_P$
δ_S	Anderson-Grüneisen constant (adiabatic)
δ_T	Anderson-Grüneisen constant (isothermal)
δ_{T_0}	γ_T at $P=0$
Δ	$\Delta = V - V_0$, change in volume due to temperature change
ΔE	change in E
$\Delta \mathcal{G}$	change in $\mathcal{G}$
ΔP	change in P
ΔP_{TH}	change in P_{TH}
Δr	change in interatomic distance due to P
$\Delta \mathcal{S}$	change in entropy
$\Delta \mathcal{S}_m$	change in $\mathcal{S}$ at melting
ΔT	change in T
ΔV	change in V
ΔV_m	change in V at melting
$\dfrac{\Delta V_m}{V_m}$	relative ΔV of V at T_m
$\dfrac{\Delta V}{V_0}$	relative ΔV due to P
ϵ	strain variable
ϵ-iron	iron phase that is hexagonal close-packed
ϵ', ϵ'', ϵ'''	volume derivatives of ϵ
ϵ_r	departure from Cauchy condition
ϵ_s	shear strain in deformation of cation
ϵ'_0	derivative of ϵ with respect to at $V=V_0$
Υ	stress tensor
ε	dielectric constant
ζ	$(3K'_0 - 8)/9K_0$
η	compression, V/V_0
η_c	the value of η at which $\delta_T = K'$
θ-iron	iron phase found above 200 GPa (structure is unknown)
Θ	Debye temperature
Θ_{ac}	acoustically determined Θ
Θ_{cal}	Θ determined from specific heat data
Θ_p	Θ calculated from v_p
Θ_s	Θ calculated from v_s
Θ_0	Θ at $P=0$ or $\eta=1$
ϑ	symmetric finite strain tensor
κ	$\partial \ln \delta_T / \partial \ln V$
λ	length of sound wave
μ	mean atomic mass, $\mathbf{M}/p$
ν	Poisson ratio
ν^H	Hill average for finding ν
ν^R	Reuss average for finding ν
ν^V	Voigt average for finding ν
ν_0	Poisson ratio at zero pressur
ν_0^V	ν^V at $P=0$
ξ	$(3/4)(4-K'_0)$
π	numerical constant
ρ	density
σ	general symbol for stress component
σ_{11}	longitudinal stress
σ_{yp}	σ_{11} at yield point
Σ	summation sign
ϕ	interatomic potential

ϕ_a	interatomic potential of attractive neighbors
ϕ_{a_i}	ith component of ϕ_a
ϕ_v	repulsion potential due to close atoms
ϕ_{v_i}	i^{th} term of ϕ_v
$\phi_{nnn}(r)$	next nearest neighbor repulsion
$\phi'_{nn}(r)$	derivative of near neighbor repulsion
$\phi''_{nn}(r)$	second derivative, near neighbor repulsion
$\phi'_{nnn}(r)$	derivative of next nearest neighbor repulsion
$\phi''_{nnn}(r)$	2nd derivative, $\phi_{nnn}(r)$
ϕ_0	binding energy at static equilibrium
φ	$\partial \ln v_s/\partial \ln v_p$
$\mathbf{\Phi}$	$K_S/\rho = v_b^2$
ψ_j	phase lag of atom at site j
ω	modal frequency
$\overline{\omega}$	average frequency
ω_D	Debye frequency cutoff
ω_i	mode frequency, index i
ω_{io}	reflected spectral ω term at $P = 0$
ω_j	mode frequency, index j
$\omega_{k_{max}}$	frequency at maximum wave number
Ω	dummy variable
∂	partial derivative sign
a	numerical constant
$\mathbf{a}$	collection of numerical constants
$\bar{a}$	numerical constant
a_V	ratio of V_0 to V at $T = 0$
a_0	Bohr radius
A	arbitrary coefficient
$\mathbf{A}$	Madelung constant
$A_0(V)$	coefficient in anharmonic $\mathcal{F}_{anh}$
$A_1(V)$	coefficient in anharmonic $\mathcal{F}_{anh}$
$A_{-2}(V)$	coefficient in anharmonic $\mathcal{F}_{anh}$
b	numerical constant
$\mathbf{b}$	integrating constant
$\bar{b}$	numerical constant
b^-	numerical constant
b^+	numerical constant
c	centi, symbol for 10^{-2} when used as prefix
c	numerical constant
$\boldsymbol{c}$	speed of light
c_0	shock velocity at vanishing P; also v_0
c'	crystal structural factor
C	constant in T_m equation: Kraut-Kennedy law
$\mathcal{C}$	operator for C_{ij} (non-centrosymmetric)
C_i	specific heat capacity for Einstein mode in terms of ω_i
C_{ij}	generalized elastic constant of single crystal
C_n	constant of nth degree in $E^n(\epsilon)$
C_V^{ht}	high temperature limit of specific heat at constant volume
C_V	specific heat at constant volume
$C_{V_{anh}}$	contribution to C_V from anharmonic terms
C_{V_e}	electronic specific heat
C_{V_j}	that part of specific heat associated with frequency ω_j
C_2	constant arising in 2nd degree EoS; $E'(\epsilon)$
C_{11}	C_{ij} arising from longitudinal motion in single crystal
C_{11}^R	repulsive force contribution to C_{11}

C_{12}	elastic constant arising from torsion of single crystal
C_{44}	elastic constant arising from pure shear motion in single crystal
C_S	$1/2(C_{11} - C_{12})$
$\mathcal{C}_{44}$	C_{44} plus correction term in non-centrosymmetric lattices
$\mathcal{C}^R$	repulsive force contribution to $\mathcal{C}$
d	amplitude in sinusoidal oscillation
d	total derivative sign
$\bar{d}$	coefficient in Decker potential, or in lattice version, ratio of next nearest neighbor to nearest neighbor separation
$<\mathrm{d}^2>$	time-averaged value of d
$\mathcal{D}$	non-centrosymmetric operator for C_{ij}
$D(x)$	Debye function of parameter x, where $x = \Theta/T$
e	strain vector
e	electronic charge
$\bar{e}$	critical fraction of lattice spacing
e_d	dilation of lattice due to dislocation
e_d^{sat}	e_d at saturation
$\mathbf{e}_0$	e at $P = 0$
e_i	eigenvalues of energy in $\mathcal{Z}$
E	oscillator strength; becomes $\mathrm{E}_{\mathrm{TH_j}}$ when multiplied by kT
E	strain energy
E	superscript referring to Einstein
E_{el}	additive potential for $\mathcal{F}$ arising from electronic density of states
E_f	Fermi energy
E_K	kinetic energy
E_{ST}	potential of static lattice at absolute zero
E_{TH}	thermal energy
E_{TH_j}	j^{th} component of E_{TH}
E_{VIB}	vibrational energy
E_{ZV}	zero temperature vibrational energy
$E_{T=0}$	potential of static lattice at absolute zero and zero vibrational energy
$E(\epsilon)$	$\mathcal{F} - E_{ST}$, strain energy
E'	derivative of E with respect to ϵ
E''	second derivative of E with respect to ϵ
E'''	3rd derivative of E with respect to ϵ
E_0'	$(\partial E_{ST}/\partial V)_T$
EoS	equation of state
f	representing function as in $\mathrm{f}(\Theta/T)$
f	coefficient in Morse potenti
f	finite strain deformation constant
$f(w)$	dimensionless EoS factor
$\mathbf{F}_i$	force on ith particle of statistical system
$\mathcal{F}$	Helmholtz energy
$\mathcal{F}_{anh}$	anharmonic term added to Helmholtz energy
$\mathcal{F}^{ht}$	high temperature limit of $\mathcal{F}$ in quasiharmonic approximation
$\mathbf{F}_j^{\mathrm{ext}}$	external force of system acting on nuclei
$\mathcal{F}_{TH}$	$\mathcal{F}$ thermal component of energy arising from T excitation
$\mathcal{F}_{TH}^{ht}$	high temperature limit of $\mathcal{F}_{TH}$

$\mathcal{F}_{TH_j}$	j^{th} term in sum determining $\mathcal{F}_{TH}$
$\mathcal{F}_{VIB}$	vibrational energy due to motion of atoms as each is constrained to vibrate around a lattice point: $\mathcal{F}_{VIB} = \mathcal{F}_{TH} + E_{ZV}$
$\mathcal{F}_{VIB_i}$	i^{th} term in sum determining $\mathcal{F}_{VIB}$
$\mathcal{F}_{VIB}^{Cl}$	classic vibrational term in $\mathcal{F}$ for insulator
$\mathcal{F}_{VIB}^{dg}$	diatomic gas vibrational energy
g	gram
g	representing function, as in $g(\omega)$
g	spring constant in simple vibrating system
$g(U_e)$	density of states, electronic
$g(\nu)$	function of Poisson ratio ν
$g(\sigma)$	function of Poisson ratio
$g(\omega)$	density of states, vibrational
G	Giga
G	isotropic shear modulus
$\overline{G}$	geometric mean of G^V and G^R
$\mathcal{G}$	Gibbs energy
G'	pressure derivative of G
G^R	Reuss upper bound for isotropic G
G^V	Voigt upper bound for isotropic G
$\{G\}_V$	$(1/\alpha K_T)(\partial G/\partial T)_V$
GPa	GigaPascals
h	hecto: 10^2 when used as prefix
h	presumed coefficient in (anharmonic) P_{TH}
h	Planck's constant
$\hbar$	Planck's constant over 2π
H	Hugoniot symbol
$\mathcal{H}$	enthalpy
i	subscript & summation index
j	subscript & summation index
J	Joule: unit of energy
k	$k = (1/2)(K_0' - 1)$
k	Boltzmann constant
k	wave number
kg	kilogram
kbar	kilobar
kHz	kiloHertz
kT	thermal energy related to vibrating atom
K	(general) bulk modulus. In Chapter 9, K means K_S
K'	pressure derivative of K (generalized)
K_C	"chord" bulk modulus
$K_{\mathbf{H}}$	Hugoniot bulk modulus
K_S	adiabatic bulk modulus; in Chapter 9, K_S represented by K
$K_S' = K'$	K_S pressure derivative
$K_{S_0}' = K_0'$	K_{S_0} pressure derivative ($P = 0$)
K_{SS}'	$(\partial K_S/\partial T)_S$
K_{SS_0}'	K_{SS}' at $P = 0$
K_T	isothermal K
K_T^{ht}	isothermal bulk modulus in high temperature limit
K_{TH}	thermal correction: K_T
K_T'	K_T pressure derivative
K_{T_0}	bulk modulus at absolute zero
K_{T_0}'	K_{T_0} pressure derivative (at $P = 0$)

K_0	general K at $P = 0$; also short for K_{S_0} in Chapter 9
K_0'	pressure derivative K at $P = 0$
l	liter
ℓn	natural logarithm
L	number of optic modes
L	latent heat
$\mathcal{L}$	Lindemann constant
$\mathcal{L}'$	Poirier version of Lindemann constant
m	mass of atom
m	micro
$\boldsymbol{m}$	milli
m_1, m_2	third order elastic constants for isotropic symmetry
M	molarity
M	coordination number for nearest neighbors
M	molecular mass
$\mathcal{M}$	generalized modulus of elasticity
M'	coordination number for anion-anion
M''	coordination number for cation-cation
$\mathcal{M}_i$	elastic modulus used in v_i equation
n	index of refraction
n	repulsion parameter
n	number of atoms in unit cell
$\boldsymbol{n}$	nano: symbol for 10^{-9}
nn	subscript on v for nearest neighbor repulsion
nnn	next nearest neighbor in repulsion
$\mathbf{n}_i$	number of atoms of type i in unit cell
N	Avogadro's number
N	Newton: unit of force
p	pico: symbol for 10^{-12} when used as a prefix
p	no. of atoms in vibrational cell (2 for NaCl)
p	running index in series of equations
P	pressure
P^R	repulsive force contribution to pressure
$\mathcal{P}$	operator for lattice sum
$\mathcal{P}^R$	operator for lattice sum for repulsion
Pa	Pressure in Pascals
P_{B-M}	Birch-Murnaghan EoS for $P(V)$
P_{B-S}	Brennan-Stacey EoS for $P(V)$
$P_{e\ell}$	$P(V)$ from free electrons
P_{H}	P along Hugoniot path
P^{ht}	pressure in high T limit: quasiharmonic theory
P_K	kinetic pressure
P_{TH}	thermal pressure
$P_{TH_{anh}}$	anharmonic component of thermal pressure
$P_{T=0}$	$= P_0$; P at $T = 0$
P_{ZV}	zero vibrational component of pressure
P_0	$= P_{T=0}$; P at $T = 0$
q	$(3K_0' - 8)/9K_0$
q	$(\partial \ln \gamma / \partial \ln V)_T$
$\bar{q}$	$(\partial\ \ell\mathrm{n}\ q / \partial\ \ell\mathrm{n}\ V)_T$
$\boldsymbol{q}$	number of atoms in cell
q^{ht}	q in high T limit: quasiharmonic theory
q_0	q at $\eta = 1$
Q	$K_{T_0}' V_0 / \gamma$
$\mathcal{Q}$	operator for lattice sum
$\mathcal{Q}^R$	operator for lattice sum for repulsion
r	interatomic distance
r_a	average of cation-anion bond distance

R	equilibrium distance for
	interatomic separation
$\mathbf{R}$	residual after truncating
s	series
$\mathcal{S}$	gas constant, kN
$\mathcal{S}^{ht}$	slope of U_S, U_P curve.
	entropy
t	$\mathcal{S}$ in high T limit: quasi-
T	harmonic theory
T_f	time
$T = 0$	temperature
T^*	Fermi temperature
	absolute zero
$T_{\mathbf{H}}$	limit in T where anhar-
T_S	monic effects begin
	T along Hugoniot path
u	T along isentropic
$\mathcal{U}$	path
$\mathcal{U}_{anh}$	$(1/3)(2 - K_0')$
	internal energy
$\mathcal{U}_e$	anharmonic terms of in-
	ternal energy
	electronic energy
$\mathcal{U}_{\mathbf{H}}$	internal energy along Hugo-
	niot path
$\mathcal{U}^{ht}$	$\mathcal{U}$ in high T limit: quasi-
	harmonic theory
U_P	particle velocity
U_S	shock velocity
v	velocity of sound
v_b	bulk sound velocity
v_i	i_{th} sound velocity; $(i =$
	$1, 2, 3)$
v_{ij}	v arising from lattice
	motion in association
	with C_{ij}
v_m	mean sound velocity
v_{nnn}	next nearest neighbor re-
	pulsion, a term of ϕ_v
v_p	v arising from longitudi-
	nal motion
v_s	shear sound velocity
V	volt
V	specific volume V
$\mathcal{V}$	internal potential energy
V_a	V at ambient conditions
V_c	volume of unit cell
V_i	volume of single atom in V_c
V_M	molecular volume
V_r	$\sum V_i/V_c$
V_0	specific volume at $P = 0$ or $\eta = 1$
$\mathbf{w}$	$-(1/\alpha K_T)(\partial K_T/\partial T)_P = \delta_T - K'$
w	$V_0/V = \rho/\rho_0 = 1/\eta$
W	$= -\mathbf{w}/\alpha K_T$
W	Watt
x	arbitrary variable
X	arbitrary variable
y	time dependent amplitude of wave
y	dimensionless $h\omega/kT$
y_i	y for mode ω_i
y_j	y for mode ω_j
Y	relative volume change $(V - V_0)/V_0$
Y_i	relative length change $(L_i - L_0)/L_0$
z	depth in earth
z	integration variable
$\mathbf{z}$	atomic number
$\mathcal{Z}$	partition function (sum of all quantum energy levels)
Z_1	ionic charge: atom 1
Z_2	ionic charge: atom 2
$\mathcal{Z}_i$	i^{th} component of $\mathcal{Z}$
$Z(w)$	frequency distribution: Debye theory
$3kT$	thermal energy per atom
$3pN$	modes in crystal

AUTHOR INDEX

There are...so many kinds of voices in the world, and none of them is without signification.

—Paul...I Corinthians 14:10

SUBJECT INDEX

In the business of scholarship,
evidence is far more flexible than opinion.

—Hugh Nibley

www.ingramcontent.com/pod-product-compliance
Lightning Source LLC
Chambersburg PA
CBHW071306180926
51613CB00001B/136
* 9 7 8 0 1 9 5 0 5 6 0 6 8 *